# NEUROMETHODS

**Program Editors: Alan A. Boulton and Glen B. Baker**

## Series I: Neurochemistry

1. **General Neurochemical Techniques**
   Edited by *Alan A. Boulton* and *Glen B. Baker*, 1985
2. **Amines and Their Metabolites**
   Edited by *Alan A. Boulton, Glen B. Baker*, and *Judith M. Baker*, 1985
3. **Amino Acids**
   Edited by *Alan A. Boulton, Glen B. Baker*, and *James D. Wood*, 1985
4. **Receptor Binding**
   Edited by *Alan A. Boulton, Glen B. Baker*, and *Pavel Hrdina*, 1986
5. **Enzymes**
   Edited by *Alan A. Boulton, Glen B. Baker*, and *Peter H. Yu*, 1986

# NEUROMETHODS □ 1

## General Neurochemical Techniques

# NEUROMETHODS

## Series I: Neurochemistry
### Program Editors: Alan A. Boulton and Glen B. Baker

# NEUROMETHODS □ 1

## General Neurochemical Techniques

*Edited by*

*Alan A. Boulton and Glen B. Baker*

# Humana Press • Clifton, New Jersey

Library of Congress Cataloging-in-Publication Data
Main entry under title:

General neurochemical techniques.

(Neuromethods ; 1. Series I, Neurochemistry)
Includes bibliographies and index.
1. Neurochemistry--Technique.  2. Neurophysiology--
Technique.  I. Boulton, A. A. (Alan A.)  II. Baker,
Glen B. , 1947-   .  III. Series: Neuromethods ; 1.
IV. Series: Neuromethods. Series I, Neurochemistry.
[DNLM: 1. Neurochemistry--methods. W1 NE337G v. 1 /
WL 104 G326]
QP356.3.G46  1986            591.1'88         85-30503
ISBN 0-89603-075-X


© 1985 The Humana Press Inc.
Crescent Manor
PO Box 2148
Clifton, NJ 07015

Printed in the United States of America

# Contributors

L. BAUCE • *Department of Medical Physiology, The University of Calgary, Calgary, Alberta, Canada*

A. J. BLUME • *Roche Institute of Molecular Biology, Nutley, New Jersey*

SCOTT T. BRADY • *Department of Cell Biology, University of Texas Health Science Center at Dallas, Dallas, Texas*

J. DISTURNAL • *Department of Medical Physiology, The University of Calgary, Calgary, Alberta, Canada*

A. J. GREENSHAW • *Psychiatric Research Division, University of Saskatchewan, Saskatoon, Saskatchewan, Canada*

L. HERTZ • *Department of Pharmacology, University of Saskatchewan, Saskatoon, Saskatchewan, Canada*

B. H. J. JUURLINK • *Department of Anatomy, University of Saskatchewan, Saskatoon, Saskatchewan, Canada*

L.-W. S. LEUNG • *Department of Psychology, University of Western Ontario, London, Ontario, Canada*

PETER LIPTON • *Department of Physiology, University of Wissonsin, Madison, Wisconsin*

J. F. MACDONALD • *Department of Pharmacology, University of Toronto, Toronto, Ontario*

R. K. MISHRA • *Departments of Psychiatry and Neurosciences, McMaster University, Hamilton, Ontario, Canada*

K. MULLIKIN-KILPATRICK • *Roche Institute of Molecular Biology, Nutley, New Jersey*

MIKLÓS PALKOVITS • *First Department of Anatomy, Semmelweis University Medical School, Budapest, Hungary*

A. PELLEGRINO DE IRALDI • *Instituto de Biologia Celular, Buenos Aires, Argentina*

Q. J. PITTMAN • *Department of Medical Physiology, The University of Calgary, Calgary, Alberta, Canada*

GAVIN P. REYNOLDS • *Department of Pathology, Queen's Medical Centre, University of Nottingham, Nottingham, UK*

C. RIPHAGEN • *Department of Medical Physiology, The University of Calgary, Calgary, Alberta, Canada*

GEORGINA RODRÍGUEZ DE LORES ARNAIZ • *Facultad de Medicina, Instituto de Biologia Celular and Facultad de Farmacia y Bioquitnica, Universidad de Buenos Aires, Buenos Aires, Argentina*

HARVEY SARNAT • *Depatments of Pediatrics, Pathology, and Clinical Neurosciences, University of Calgary, Calgary, Alberta, Canada*

TIMOTHY SCHALLERT • *Department of Psychology and Pharmacology, University of Texas at Austin, Austin, Texas*

QUENTIN R. SMITH • *Laboratory of Neurosciences, National Institute on Aging, National Institutes of Health, Bethesda, Maryland*

K. SONNENFIELD • *Roche Institute of Molecular Biology, Nutley, New Jersey*

S. SZUCHET • *Department of Neurology, University of Chicago, Chicago, Illinois*

C. H. VANDERWOLF • *Department of Psychology, University of Western Ontario, London, Ontario, Canada*

W. L. VEALE • *Department of Medical Physiology, The University of Calgary, Calgary, Alberta, Canada*

W. WALZ • *Department of Physiology, University of Saskatchewan, Saskatoon, Saskatchewan, Canada*

RICHARD E. WILCOX • *Department of Pharmacology, University of Texas at Austin, Austin, Texas*

# Contents

CHAPTER 1
MICRODISSECTION OF INDIVIDUAL BRAIN NUCLEI AND AREAS
Miklós Palkovits

CHAPTER 2
SUBCELLULAR FRACTIONATION
Georgina Rodríguez de Lores Arnaiz and
   Amanda Pellegrino de Iraldi

CHAPTER 3
BRAIN SLICES: Uses and Abuses
Peter Lipton

CHAPTER 4
CELL AND TISSUE CULTURES
L. Hertz, B. H. J. Juurlink, S. Szuchet, and W. Walz

CHAPTER 5
MONOCLONAL ANTIBODIES
Ram K. Mishra, Debra Mullikin-Kilpatrick, Kenneth H. Sonnenfeld,
Chandra P. Mishra, and Arthur J. Blume

CHAPTER 6
IDENTIFICATION OF CENTRAL TRANSMITTERS:
Microiontophoresis and Micropressure Techniques
J. F. MacDonald

CHAPTER 7
ELECTRICAL AND CHEMICAL STIMULATION
OF BRAIN TISSUE IN VIVO
A. J. Greenshaw

CHAPTER 8
PERFUSION TECHNIQUES FOR
NEURAL TISSUE
Q. J. Pittman, J. Disturnal, C. Riphagen, W. L. Veale, and L. Bauce

CHAPTER 9
BRAIN ELECTRICAL ACTIVITY IN RELATION
TO BEHAVIOR
C. H. Vanderwolf and L.-W. S. Leung

CHAPTER 10
NEUROTRANSMITTER-SELECTIVE BRAIN LESIONS
Timothy Schallert and Richard E. Wilcox

CHAPTER 11
METHODS TO DETERMINE BLOOD–BRAIN
BARRIER PERMEABILITY AND TRANSPORT
Quentin R. Smith

CHAPTER 12
AXONAL TRANSPORT METHODS AND APPLICATIONS
Scott T. Brady

CHAPTER 13
NEUROCHEMICAL STUDIES IN HUMAN
POSTMORTEM BRAIN TISSUE
Gavin P. Reynolds

CHAPTER 14
HISTOCHEMICAL MAPPING OF VERTEBRATE BRAINS
FOR STUDY OF EVOLUTION
Harvey B. Sarnat

# Chapter 1

# Microdissection of Individual Brain Nuclei and Areas

## MIKLÓS PALKOVITS

## 1. Introduction

Several macro- and microdissection methods for sampling brain regions have been reported. Large regions, such as cortex, striatum, and hypothalamus, can be separated either in fresh whole brains (*in situ* preparation) or dissected out from brain slices with or without a microscope. When more than 2 mg tissue weight is dissected from the brain, the procedure is called *macrodissection*. Rat brain can be macrodissected *in situ* into seven (Glowinski and Iversen, 1966) or seventeen (Gispen et al., 1972) major regions.

Parallel to the development of highly sensitive biochemical assays able to detect neurotransmitters, enzymes, and receptors in picomole quantities came the development of simple but reproducible microdissection techniques. A new approach in neurochemistry and neuropharmacology has been achieved by the use of individually microdissected and anatomically characterized nuclei of the central nervous system. The emergence of sophisticated neurochemical techniques, together with the discovery of several new neuroactive substances in the brain, coincided with the introduction of a new microdissection technique—"micropunch" technique (Palkovits, 1973)—to obtain over 200 discrete nuclei or areas of the central nervous system.

Micropunch technique is microdissection of nuclei from brain sections with a hollow needle. Since its introduction in 1973

1

(Palkovits, 1973), the technique has been adopted by several laboratories. It offers an order of magnitude better structural resolution than was available previously, so neurobiologists have begun to focus on brain nuclei instead of more heterogenous brain regions. The technique itself is rather simple, and this is the most important reason for its popularity. It fulfills the following requirements:

(1) The method is fine enough to allow sampling of discrete brain nuclei separately. The lowest limit is about 10 μg wet weight, which corresponds to a tissue pellet of 100 μm radius from a 300-μm thick section.

(2) The procedure is rapid. Hundreds of brain samples may be dissected in an experiment within a limited period of time.

(3) The microdissection can be performed either from frozen or irradiated brains in order to minimize postmortem changes, but it is also possible from fresh brain if samples from "living tissue" are required (for tissue cultures).

(4) The technique can be employed without any complicated instrumentation.

(5) The technique is reproducible. The standard error caused by the microdissection may be kept at a level commensurate with general neurobiological standards (SEM = ±10% of the mean).

(6) The microdissection may be easily validated by microscopic controls.

Microdissected brain nuclei can be used for many purposes. The major types of studies are the following:

(1) Biochemical mapping for the distribution of various substances (neurotransmitters, neurohormones, enzymes, and receptors) in the central nervous system.

(2) Measurement of concentrations of the above substances and changes in these following various types of treatments or experimentally induced alterations or in pathological conditions.

(3) Biochemical analysis of substances in relation to their precursors and metabolic products in discrete brain nuclei or areas in which they are synthesized or stored.

(4) Biosynthesis of neuroactive substances can be measured in individual brain nuclei using in vitro systems.

(5) Subcellular distribution and release of substances can be investigated in separated brain nuclei.

(6) Cells from anatomically characterized brain nuclei can be cultured and applied to in vitro developmental studies using light or electron microscopic immunohistochemistry, to electrophysiological studies, and to studies for measuring contents, biosynthesis, or release of substances in or from the cultured neurons.

## 2. Tools and Devices for Microdissection

The microdissection technique is composed of two procedures: brain sectioning and the microdissection itself. The instruments needed are related to these two procedures.

### 2.1. Instruments for Brain Sectioning

(a) The brains can be sectioned free hand with stainless steel razor blade on an elastic stage (on a rubber sheet or a large rubber stopper).

(b) Templates, tissue choppers, or vibratomes may also be used for slicing fresh brains. The template may be fashioned from a block of Plexiglas® (Palkovits and Brownstein, 1983). A hole the shape of the brain is milled into the block, and parallel slots are cut perpendicular to the long axis of the brain at 1-mm intervals. The brain is placed into the depression in the template and sectioned by razor blades pressed into the slots. Various types of tissue choppers are commercially available: Sorvall TC2 Tissue Sectioner; McIlwain Tissue Chopper (Mickle Laboratory Engineering Co.). Vibratomes (Oxford Instruments) are also generally used for fresh brain slicing.

(c) Frozen brains may be sliced by cryostats. Most commercially available cryostats are suitable. The only requirement is that well-oriented (side-symmetric) 300 µm thick serial sections can be routinely cut. Therefore, the specimen holder is fixed, or can be fixed, to a ball joint and there should be ample space for manipulation within the cryostat with both hands. The Cryo-cut (American Optic) or Minotome (IEC) cryostats are the most comfortable for this purpose.

### 2.2. Tools for Microdissection

(a) Microscopes, magnifying glasses, and loupes. Any kind of stereo- or dissecting microscopes can be used. There are the following requirements: little (6–20-fold) magnification, upper illumination with variable focused cool light (fiber optic system), and free space under the microscope tube high enough (10–15 cm minimum) for the stage and for the manipulation with the dissecting needles (Fig. 1). Instead of the microscope, a magnifying lens may be used. At least five-times magnification, upper illumination, and free space under the lens are needed. A loupe is convenient to use if the fivefold magnification is provided by a single magnifying lens.

(b) Cold or elastic stages. Microdissection from frozen brains is performed on a cold stage under the microscope (Fig. 1). A box

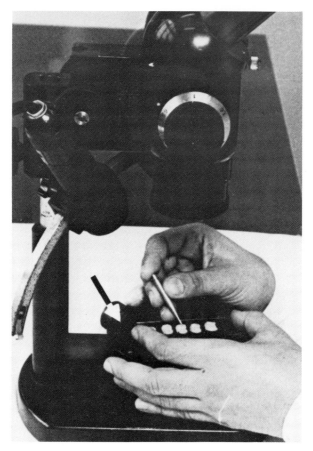

Fig. 1.   Microdissection under a stereomicroscope. Dry ice container is indicated by the arrow.

or dish (Petri dish) filled with dry ice powder or circulating cooling solution is suitable. The top of the stage should be metal (or a metal sheet should be placed on it) to provide good heat conductivity. This cover should be black or painted a dark color to provide a contrasting background for the sections placed on glass slides (Figs. 1 and 4). Microdissection from fresh brain is most successful on elastic (rubber) stages. A large black stopper is the most convenient and provides an excellent contrasting background.

(c) Microknives. A variety of microknives can be used for microdissection of brain areas. The Castroviejo scalpel (Moria-Dugast, Paris), Graefe-knife, or iridectomy scalpel are the best for that purpose (Cuello and Carson, 1983).

(d) Micropunch needles. These hollow needles are constructed of hard stainless steel tubing mounted in a thicker handle (Fig. 2). Their inner diameters vary from 0.2 to 2.0 mm. For comfortable handling, needles should be 4.5–5.0 cm long with a thinner end; at least 5 mm long, so that the tip of the needle is visible under the microscope. The end of the needle should be sharpened. Needles can be equipped with a stylet (Palkovits and Brownstein, 1983). It should be fitted well into the lumen of the needle and reach 2–3 mm beyond the tip of the needle.

## 3. Microdissections From Fresh Brains

Fresh microdissected brain tissue can be used for many purposes, mainly for in vitro studies. Microdissection or micropunch procedures are performed from brain slices. Fresh brain sections should be as thin as possible, around 0.5–1.0 mm. Brains can be sliced immediately after removal from the skull without prior treatment, or after a perfusion of the brain with ice-cold Krebs solution and ethylenediamine tetraacetate–DAB (Cuello and Carson, 1983).

Fresh brain slices (coronal serial sections are generally used for microdissection) may be made manually or with the aid of mechanical devices, such as templates, tissue choppers, or vibratomes. The manual brain slicing may be carried out with a wet razor blade on a black elastic plate under a magnifier or a stereomicroscope. The procedure for fresh brain sectioning with instruments has been described in detail by Cuello and Carson (1983). Freshly cut brain slices for in vitro studies cannot be stored—microdissection must be performed within minutes.

Nuclei from fresh brain slices can be removed by microdissection (Zigmond and Ben-Ari, 1976; Cuello and Carson, 1983) or micropunch (Jacobowitz, 1974; Palkovits and Brownstein, 1983) techniques. In fresh brain sections, the white matter and the ven-

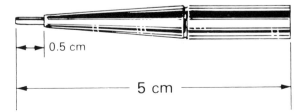

0.5 cm

5 cm

Fig. 2. Microdissection needle.

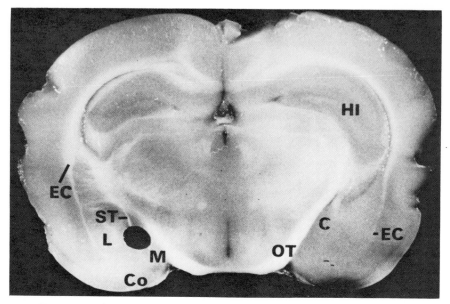

Fig. 3.   The microdissected central amygdaloid nucleus (left side). Unstained melting frozen section. Key: **C**, central amygdaloid nucleus; **Co**, cortical amygdaloid nucleus; **EC**, external capsule; **HI**, hippocampus; **L**, lateral amygdaloid nucleus; **M**, medial amygdaloid nucleus; **OT**, optic tract; and **ST**, stria terminalis.

tricles serve as landmarks. By using upper illumination and black plates under the sections for contrasting background, the white myelinated structures can be clearly recognized (Fig. 3). By using transillumination of tissue slides (Cuello and Carson, 1983) the white matter appears dark and the nuclei transparent. Vital staining of the sections with methylene blue before microdissection improves the orientation (Zigmond and Ben Ari, 1976).

Microdissections of particular brain areas are performed with special microknives. The major steps of that procedure, serial micrographs, and maps of fresh rat brain slices have been published recently by Cuello and Carson (1983). This approach can be modified depending upon the facilities of individual laboratories. For identifying all relevant structures, several other maps are also suitable either for rats or other species (Table 1).

Fresh brain nuclei can be removed from sections that have been placed on a black rubber plate. The tip of the micropunch needle is fixed on the section exactly over the nucleus to be microdissected and the needle is pressed strongly into the rubber plate. The elasticity of the plate pushes the tissue pellet back into the lumen of the needle. The tissue pellet can be blown out of the

TABLE 1
Stereotaxic Maps of Avian and Mammalian Brains

| Species | Authors | Year | Illustrated brain regions[a] |
|---|---|---|---|
| Pigeon | Karten and Hodos | 1967 | B |
| Chicken | Yoshikawa | 1968 | B |
| | Tienhoven and Juhász | 1972 | FM |
| Mouse | Sidman et al. | 1971 | B |
| | Slotnick and Leonard | 1975 | F |
| | Broadwell and Bleier | 1976 | H |
| Rat | de Groot | 1959 | H |
| | de Groot | 1963 | B |
| | König and Klippel | 1963 | F |
| | Wünscher et al. | 1965 | S |
| | Albe-Fessard et al. | 1966 | D |
| | Fifkova and Marsala | 1967 | B |
| | Szentágothai et al. | 1968 | H |
| | Hurt et al. | 1971 | M |
| | Palkovits | 1975 | H |
| | Pellegrino and Cushman | 1967 | B |
| | Pellegrino et al. | 1979 | B |
| | Palkovits | 1980 | B |
| | Simson et al. | 1981 | B |
| | Paxinos and Watson | 1982 | B |
| Guinea pig | Luparello | 1967 | F |
| Opossum | Oswaldo-Cruz and Rocha-Miranda | 1967 | D |
| | Oswaldo-Cruz and Rocha-Miranda | 1968 | B |
| Rabbit | Monnier and Gangloff | 1961 | B |
| | Fifkova and Marsala | 1967 | B |
| Goat | Yoshikawa | 1968 | B |
| Sheep | Yoshikawa | 1968 | B |
| Cow | Yoshikawa | 1968 | B |
| Horse | Yoshikawa | 1968 | B |
| Pig | Yoshikawa | 1968 | B |
| Cat | Jasper and Ajmone-Marsan | 1960 | D |
| | Bleier | 1961 | H |
| | Snider and Niemer | 1961 | B |
| | Verhaart | 1964 | B |
| | Fifkova and Marsala | 1967 | B |
| | Berman | 1968 | S |

TABLE 1 (*Continued*)
Stereotaxic Maps of Avian and Mammalian Brains

| Species | Authors | Year | Illustrated brain regions[a] |
|---------|---------|------|------------------------------|
| Dog | Lim et al. | 1960 | B |
|  | Dua-Sharma et al. | 1970 | B |
| Tupaia | Tigges and Shantha | 1969 | B |
| Collithrix | Stephan et al. | 1980 | B |
| Squirrel | Gergen and MacLean | 1962 | B |
|  | Emmers and Akert | 1963 | B |
| Macaca | Snider and Lee | 1961 | B |
|  | Shantha et al. | 1968 | B |
| Cebus | Eidelberg and Saldias | 1960 | B |
|  | Manocha et al. | 1968 | B |
| Chimpanzee | de Lucchi et al. | 1965 | B |
| Baboon | Davis and Huffman | 1968 | B |
| Human | Schaltenbrand and Wahren | 1977 | B |

[a]B = atlas for the whole brain; F = forebrain only; D = diencephalon only; H = hypothalamus only; S = brainstem only; M = mesencephalon only.

needle. For micropushing from fresh brain slices, micropunch needles with stylets (Palkovits and Brownstein, 1983) are recommended. In that case, the tissue pellet is just pushed out of the needle with the stylet. The micropunch procedure should be as short as possible (within minutes after sectioning); otherwise, the sections are desiccated and further studies are influenced by the postmortem changes.

## 4. Micropunch Technique on Frozen Brain Slices

Punching individual brain nuclei with a hollow needle from frozen sections cut in a cryostat has become the most routinely used microdissection technique in many laboratories. Technical guides and maps for rat brains have been previously reported in detail (Palkovits, 1973, 1975, 1980, 1983a, 1983b; Palkovits and Brownstein, 1983).

Serial sections are cut from frozen brain. After removal, the brain is frozen on a microtome specimen holder with dry ice. The vertical orientation of the brain on the specimen holder is crucial for sectioning of the brain at correct planes, for keeping the coordinates, the side-symmetry, i.e., for the correct topographical identification of brain nuclei. Coronal sections of 300 μm thickness are cut in a cryostat at −10°C. (Brains from larger species can be cut with a thickness of up to 500 μm.) Sections are placed on histological slides (3–6 sections per slide). The glass slides are removed from the cryostat for a few seconds to be thawed, then placed on dry ice until the micropunch sampling. Sections can be stored in a container (histological slide box) filled with dry ice.

Special hollow micropunch needles of varying inner diameter (between 0.2 and 2.0 mm) are used (Fig. 2). The inner diameter of needle used for the dissection depends on the size of the brain nuclei to be punched out. It is essential that the needle be smaller than the smallest diameter of the nucleus. To keep them cold during the microdissection procedure, sections on slides are placed on a cold stage under the microscope (Fig. 1). Micropunching should be accomplished quickly since the visibility will gradually diminish owing to the accumulation of frost on the section surface.

At the first step of the micropunch procedure, the tip of the needle is fixed onto the section in an oblique position (Figs. 1 and 4A), exactly over the nucleus to be removed. (During the manipulation, the tip of the needle should be visible under the micro-

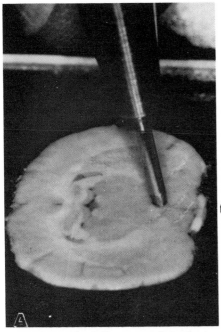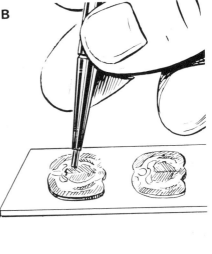

Fig. 4. Micropunch technique, 300 μm thick brain section. **A:** Fixation of the needle tip with an angle on the section over the nucleus to be removed; **B:** Micropunch by pressing of the needle in vertical position.

scope.) Then, the needle should be brought into a vertical position (Fig. 4B), pressed into the tissue, rotated slightly, and then quickly withdrawn. The temperature of the sections is of major importance. If they are too cold, they are prone to break and if they are too warm, they are melted beneath the needle.

The dissected tissue pellet can be blown out of the needle, or pushed out with a stylet into a tube or dish, or onto the tip of a microhomogenizer. If several samples are punched from the same brain nucleus, the pellets need not be blown out separately—they may be collected in the lumen of the needle and removed all at once. After micropunch, sharp-edged holes remain in the section (Figs. 3 and 5).

Microdissected brain nuclei can be used for many purposes. For biochemical measurements, tissue pellets are homogenized. Small glass homogenizers for 20–200 µL volumes of solvent are commercially available. Homogenizations may be performed either manually, with a high-speed homogenizer, or by sonication. In the latter case, a multitube sonicator is very useful.

Pellets are usually too small for direct weighing, so their protein contents are measured. An aliquot (5–10 µL) of the well-suspended tissue homogenate is removed for protein determination, which may be carried out with the Folin phenol reagent (Lowry et at., 1951).

# 5. Micropunch Technique on Fixed Brain Sections

The micropunch technique may also be used for removal of brain nuclei from fixed brains. Microdissections from chemically fixed brains are performed mainly for histological and electron microscopic investigations, whereas microwave-irradiated brains are prepared for biochemical measurements when postmortem changes must be avoided.

## 5.1. Micropunch for Electron Microscopical Studies

Dissection of nuclei from fixed brain provides for better topographical orientation and recognition of small cell groups under an electron microscope. Brains are fixed by perfusion techniques suitable for electron microscopic procedures. Slicing can be performed free-hand as for fresh brains, but more easily. Nuclei may be punched out on a black elastic plate with common or special hollow needles. Special needles are equipped with a marker in the lumen—the needle wall is pressed sharply into the lumen on

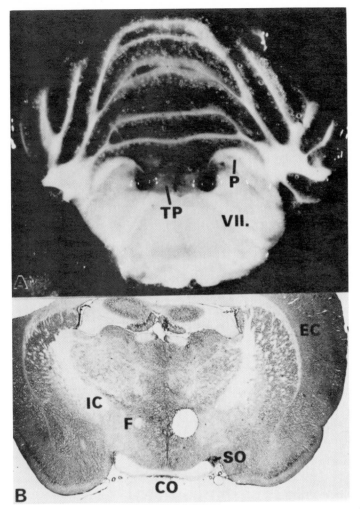

Fig. 5. Validating the microdissection. **A:** Bilateral microdissection of the locus ceruleus. Melting frozen section. **B:** Unilaterally microdissected paraventricular hypothalamic nucleus. Section stained by hematoxylin-eosin. Key: **CO,** optic chiasm; **EC,** external capsule; **F,** fornix; **IC,** internal capsule; **P,** pedunculus cerebellaris superior; **SO,** supraoptic nucleus; **TP,** dorsal tegmental pontine nucleus; **VII,** facial nerve.

one side (Palkovits and Brownstein, 1983). At the time of dissection, the needle is oriented over the section, indicating a direction with the marker that can be recognized on the pellet during embedding and after sectioning.

## 5.2. Micropunch From Microwave-Irradiated Brains

To avoid postmortem changes, the heads of the animals are irradiated by focused microwaves. These brains need special care. They are difficult to remove from the skull because they break into pieces easily. They need to be sectioned in a cryostat but at a lower temperature ($-15°C$) than that used for fresh brains. It is helpful to cool the knife blade intermittently with dry ice. Topographical orientation is quite difficult since the "cooked" brain is almost equally grayish. For micropunch, the smallest diameter punches (0.2–0.3 mm) are rarely used, and 0.5 mm or larger needles with stylet are recommended. It is best to cool the needle tip on dry ice for several seconds before punching. Sometimes, the needle tip should be cooled also after punching because the brain tissue easily smears onto the internal wall of the needle.

# 6. Validating the Microdissection Technique

There are several ways to validate the microdissections after micropunch. This can be controlled microscopically either on unstained (Fig. 5A) or stained (Fig. 5B) sections.

## 6.1. Validation of Micropunch on Unstained Sections

Immediately after microdissection, sections are allowed to thaw and can be transilluminated or illuminated from above on a black background. An excellent identification of brain areas and bundles can be achieved even with low-power magnification based on the differential optical density of myelinated and unmyelinated areas (Fig. 5A). These sections, however, desiccate rapidly, and, therefore, cannot be stored.

## 6.2. Validation of Micropunch on Stained Sections

The 300 μm-thick sections used for microdissection are too thick for histological control, and they have to be cut into thinner sections. After microdissection, sections can be fixed with 4% formalin, embedded, and sliced into thin (10–20 μm thick) sections and stained. The sections need careful handling because they are prone to shrinkage and curling during the histochemical procedure. Therefore, it is difficult to cut thin sections in the same plane as the plane of the original thick section. It is much more simple to prepare frozen sections. After micropunch, the 300 μm thick sections are refrozen on a gelatin block with a flat upper surface and thin sections are cut, fixed, stained, and validated under the microscope (Fig. 5B). Microdissection techniques from tissue blocks have been published (Eik-Nes and Brizzee, 1965; Schumpf et al., 1974). The punch is performed on a block fixed in a microtome, and controlled by subsequent sectioning under a microscope, punch by punch. The technique is rather complicated and has never been accepted for general practice.

# APPENDIX

## List of Stereotaxic Maps of Avian and Mammalian Brains Suitable for Microdissection Guiding

Microdissection of brain nuclei may be performed in any species. Since the micropunch technique has been developed for individual rat brain nuclei, detailed microdissected guides and maps are available only for rat brains. A complete guide and map for dissecting over 200 nuclei from 72 coronal (300 μm thick) serial sections has been published (Palkovits, 1980). Detailed descriptions of micropunching of nuclei from larger brain areas have been reported separately: areas from the cerebral cortex (Palkovits et al., 1979), hypothalamus (Palkovits, 1975), diencephalon (Brownstein et al., 1975), limbic system (Palkovits et al., 1974a), and the lower brainstem (Palkovits et al., 1974b) can be removed with the aid of these guides.

In the last 20 yr several stereotaxic maps have been published (Table 1). Detailed atlases are available for the whole brain or larger brain regions, such as the hypothalamus or lower brainstem of the most frequently used laboratory animals and domestic animals. For microdissection, only stereotaxic maps based on serial sections with exact coordinates and magnifications are recommended. These are summarized in Table 1. If such maps are not available, any other atlas may be used, but with caution. In these cases, the validation of microdissection is important after each micropunch. Detailed maps without stereotaxic coordinates for vertebrates' brains are the following: Neary and Northcutt (1983) for frog; Herrick (1910) for amphibia and reptilia; Jacobowitz and Palkovits (1974), Palkovits and Jacobowitz (1974), and Bleier et al. (1979) for rat; aus der Mühlen (1966) for guinea pig; Loo (1931) for opossum; and Briggs (1946) for sheep. Micro-dissections of nuclear areas from human brain have been recently reported by Aquilonius et al. (1983) and Stevens and Bird (1983).

# References

Albe-Fessard D., Stutinsky S., and Libouban S. (1966) *Atlas Stéréotaxique du Diéncéphale du Rat Blanc,* CNRS, Paris.

Aquilonius S.-M., Eckernäs S-Å., and Gillberg P. (1983) Large Section Cryomicrotomy in Human Neuroanatomy and Neurochemistry, in *Brain Microdissection Techniques* (Cuello A.C., Ed.), pp. 155–170. Wiley, Chichester.

Berman A. L. (1968) *The Brainstem of the Cat: A Cytoarchitectonic Atlas with Stereotaxic Coordinates,* Univ. of Wisconsin Press, Madison.

Bleier R. (1961) *The Hypothalamus of the Cat. A Cytoarchitectonic Atlas with Horsley-Clark Co-ordinates,* Johns Hopkins Press, Baltimore.

Bleier R., Cohn P., and Siggelkow I. R. (1979) A Cytoarchitectonic Atlas of the Hypothalamus and Hypothalamic Third Ventricle of the Rat, in *Handbook of the Hypothalamus,* Vol. 1, *Anatomy of the Hypothalamus* (Morgane P.J. and Panksepp J., Eds.), pp. 137–220. Dekker, New York.

Briggs E. A. (1946) *Anatomy of the Sheep's Brain,* Angus & Robertson, London.

Broadwell R. D., and Bleier R. (1976) A cytoarchitectonic atlas of the mouse hypothalamus. *J. Comp. Neurol.* **167,** 315–340.

Brownstein M., Kobayashi R., Palkovits M., and Saavedra J. M. (1975) Choline acetyltransferase levels in diencephalic nuclei of the rat. *J. Neurochem.* **24,** 35–38.

Cuello A. C., and Carson S. (1983) Microdissection of Fresh Rat Brain Tissue Slices, in *Brain Microdissection Techniques* (Cuello A. C., Ed.), pp. 37–125. Wiley, Chichester.

Davis R., and Huffman R. D. (1968) *A Stereotaxic Atlas of the Brain of the Baboon* (Papio papio), Univ. Texas Press, Austin.

Dua-Sharma S., Sharma S., and Jacobs H. L. (1970) *The Canine Brain in Stereotaxic Coordinates,* MIT Press, Cambridge.

Eidelberg E. and Saldias C. A. (1960) A stereotaxic atlas for cebus monkeys. *J. Comp. Neurol.* **115,** 103–123.

Eik-Nes K. B. and Brizzee K. R. (1965) Concentration of tritium in brain tissue of dogs given $(1,2-{}^{3}H_2)$-cortisol intravenously. *Biochim. Biophys. Acta* **37,** 320–333.

Emmers R. and Akert K. (1963) *A Stereotaxic Atlas of the Brain of the Squirrel Monkey* (Saimiri sciureus), Univ. of Wisconsin Press, Madison.

Fifkova E. and Marsala J. (1967) Stereotaxic Atlases for the Cat, Rabbit, and Rat, in *Electrophysiological Methods in Biological Research* (Bures J., Petrán M., and Zachar J., Eds.), pp. 653–731. Academic Press, New York and London.

Gergen J. A. and MacLean P. D. (1962) *A Stereotaxic Atlas of the Squirrel Monkey's Brain (Saimiri sciureus)*, US Publ. No. 933, Washington.

Gispen W. H., Schotman P., and De Kloet E. R. (1972) Brain RNA and hypophysectomy: A topographical study. *Neuroendocrinology* **9**, 285–296.

Glowinski J. and Iversen L. L. (1966) Regional studies of catecholamines in the rat brain. *J. Neurochem.* **13**, 655–669.

de Groot J. (1959) The rat hypothalamus in stereotaxic coordinates. *J. Comp. Neurol.* **113**, 389–400.

de Groot J. (1963) *The Rat Brain in Stereotaxic Coordinates*, North Holland, Amsterdam.

Herrick C. J. (1910) The morphology of the forebrain in Amphibia and Reptilia. *J. Comp. Neurol.* **20**, 413–547.

Hurt G. A., Hanaway J., and Netsky M. G. (1971) Stereotaxic atlas of the mesencephalon in the albino rat. *Confin. Neurol.* **33**, 93–115.

Jacobowitz D. M. (1974) Removal of discrete fresh regions of the rat brain. *Brain Res.* **80**, 111–115.

Jacobowitz D. M. and Palkovits M. (1974) Topographic atlas of catecholamine- and acetylcholinesterase-containing neurons in the rat brain. I. Forebrain (telencephalon, diencephalon). *J. Comp. Neurol.* **157**, 13–28.

Jasper H. H. and Ajmone-Marsan C. (1960) *Stereotaxic Atlas of the Diencephalon of the Cat.* NRC Canada, Ottawa.

Karten H. J. and Hodos W. (1967) *A Stereotaxic Atlas of the Brain of the Pigeon* (Columba livia), Johns Hopkins Press, Baltimore.

König J. F. and Klippel R. A. (1963) *The Rat Brain: A Stereotaxic Atlas of the Forebrain and Lower Parts of the Brain Stem*, Williams and Wilkins, Baltimore.

Lim R. K. S., Liu C., and Moffitt R. (1960) *A Stereotaxic Atlas of the Dog's Brain*, CC Thomas, Springfield, Ill.

Loo Y. T. (1931) The forebrain of the opossum, *Didelphis virginiana*. II. Histology. *J. Comp. Neurol.* **52**, 1–148.

Lowry O. H., Rosebrough N.Y., Farr A. L., and Randall R. J. (1951) Protein measurement with the Folin phenol reagent. *J. Biol. Chem.* **193**, 265–275.

de Lucchi M. R., Dennis B. J., and Adey W. R. (1965) *A Stereotaxic Atlas of the Chimpanzee Brain* (Pan satyrus), Univ. California Press, Berkeley.

Luparello T. J. (1967) *Stereotaxic Atlas of the Forebrain of the Guinea Pig*, Williams and Wilkins, Baltimore.

Manocha S. L., Shantha T. R., and Bourne G. H. (1968) *A Stereotaxic Atlas of the Brain of the Cebus* (Cebus apella) *Monkey*, Oxford Univ. Press, Oxford.

Monnier M. and Gangloff H. (1961) *Atlas for Stereotaxic Brain Research on the Conscious Rabbit*, Elsevier, Amsterdam.

aus der Mühlen K. (1966) *The Hypothalamus of the Guinea pig; A Topographic Survey of Its Nuclear Regions*, S. Karger, New York.

Neary T. J. and Northcutt R. G. (1983) Nuclear organization of the bull-frog diencephalon. *J. Comp. Neurol.* **213,** 262–278.

Oswaldo-Cruz E. and Rocha-Miranda C. E. (1967) The diencephalon of the opossum in stereotaxic coordinates. II. The ventral thalamus and hypothalamus. *J. Comp. Neurol.* **129,** 39–48.

Oswaldo-Cruz E. and Rocha-Miranda C. E. (1968) *The Brain of the Opossum (Didelphis marsupialis). A Cytoarchitectonic Atlas in Stereotaxic Coordinates,* Inst. Biofisica Univ. Fed. do Rio de Janeiro, Rio de Janeiro.

Palkovits M. (1973) Isolated removal of hypothalamic or other brain nuclei of the rat. *Brain Res.* **59,** 449–450.

Palkovits M. (1975) Isolated Removal of Hypothalamic Nuclei for Neuroendocrinological and Neurochemical Studies, in *Anatomical Neuroendocrinology* (Stumpf W. E. and Grant L. D., Eds.), pp. 72–80. S. Karger, Basel.

Palkovits M. (1980) *Guide and Map for the Isolated Removal of Individual Cell Groups from the Rat Brain* (Hungarian text), Akadémiai Kiadó, Budapest.

Palkovits M. (1983a) Neuroanatomical Techniques, in *Brain Peptides* (Krieger D. T., Brownstein M. J. and Martin J., eds), pp. 495–545. Wiley, Chichester.

Palkovits M. (1983b) Stereotaxic Map, Cytoarchitectonic and Neurochemical Summary of the Hypothalamic Nuclei, Rat, in *Endocrine System. Monographs on Pathology of Laboratory Animals* (Jones T. C., Mohr U., and Hunt R. D., eds), pp. 316–330. Springer-Verlag, Berlin.

Palkovits M. and Brownstein M. J. (1983) Microdissection of Brain Areas by the Punch Technique, in *Brain Microdissection Techniques* (Cuello A. C., ed), pp. 1–36. Wiley, Chichester.

Palkovits M., Brownstein M., and Saavedra J. M. (1974) Serotonin content of the brain stem nuclei in the rat. *Brain Res.* **80,** 237–249.

Palkovits M. and Jacobowitz D. M. (1974) Topographic atlas of catecholamine- and acetylcholinesterase-containing neurons in the rat brain. II. Hindbrain (mesencephalon, rhombencephalon). *J. Comp. Neurol.* **157,** 29–42.

Palkovits M., Saavedra J. M., Kobayashi R. M., and Brownstein M. (1974b) Choline acetyltransferase content of limbic nuclei of the rat. *Brain Res.* **79,** 443–450.

Palkovits M., Záborszky L., Brownstein M. J., Fekete M. I. K., Herman J. P., and Kanyicska B. (1979) Distribution of norepinephrine and dopamine in cerebral cortical areas of the rat. *Brain Res. Bull.* **4,** 593–601.

Paxinos G. and Watson C. (1982) *The Rat Brain in Stereotaxic Coordinates,* Academic Press, Sydney.

Pellegrino L. J. and Cushman A. J. (1967) *A Stereotaxic Atlas of the Rat Brain,* Appleton-Century-Crofts, New York.

Pellegrino L. J., Pellegrino A. S., and Cushman A. J. (1979) *A Stereotaxic Atlas of the Rat Brain,* 2nd ed. Plenum Press, New York.

Schaltenbrand G. and Wahren W. (1977) *Atlas for Stereotaxy of the Human Brain*, Georg Thieme, Stuttgart.

Schumpf M., Waser P., Lichtensteiger W., Langmann H., and Schlup D. (1974) Standardized excision of small areas of rat and mouse brain with topographical control. *Biochem. Pharmacol.* **23,** 2447–2449.

Shantha T. R., Manocha S. L., and Bourne G. H. (1968) *A Stereotaxic Atlas of the Jawa Monkey Brain* (Macaca Irus), Williams and Wilkins, Baltimore.

Sidman R. L., Angevine J. B. Jr., and Taber Pierce E. (1971) *Atlas of the Mouse Brain and Spinal Cord*, Harvard Univ. Press, Cambridge.

Simson E. L., Jones A. P., and Gold R. M. (1981) Horizontal stereotaxic atlas of the albino rat brain. *Brain Res. Bull.* **6,** 297–326.

Slotnick B. M. and Leonard C. M. (1975) *A Stereotaxic Atlas of the Albino Mouse Forebrain.* DHEW Publ., Rockville.

Snider R. S., and Lee J. C. (1961) *A Stereotaxic Atlas of the Monkey Brain* (Macaca mulatta), Univ. Chicago Press, Chicago.

Snider R. S., and Niemer W. T. (1961) *A Stereotaxic Atlas of the Cat Brain*, Univ. Chicago Press, Chicago.

Stephan H., Baron G., and Schwerdtfeger W. K. (1980) *The Brain of the Common Marmoset* (Collithrix jacchus). *A Stereotaxic Atlas*, Springer-Verlag, Berlin.

Stevens T. J. and Bird E. D. (1983) Microdissections of Nuclear Areas From Human Brain Slices, in *Brain Microdissection Techniques* (Cuello A. C., Ed.), pp. 171–183. Wiley, Chichester.

Szentágothai J., Flerkó B., Mess B., and Halász B. (1968) *Hypothalamic Control of the Anterior Pituitary*, Akadémiai Kiadó, Budapest.

van Tienhoven A. and Juhász L. P. (1972) The chicken telencephalon, diencephalon and mesencephalon in stereotaxic coordinates. *J. Comp. Neurol.* **118,** 185–197.

Tigges J. and Shantha T. R. (1969) *A Stereotaxic Brain Atlas of the Tree Shrew (Tupaia glis)*, Williams and Wilkins, Baltimore.

Verhaart W. J. C. (1964) *A Stereotaxic Atlas of the Brain of the Cat*, Van Gorcum, Assen.

Wünscher W., Schober W., and Werner L. (1965) *Architektonischer Atlas vom Hirnstamm der Ratte*, S. Hirzel, Leipzig.

Zigmond R. E. and Ben-Ari Y. J. (1976) A simple method for the serial sectioning of fresh brain slices and removal of identifiable nuclei from stained sections for biochemical analysis. *J. Neurochem.* **26,** 1285–1287.

Yoshikawa T. (1968) *Atlas of the Brains of Domestic Animals*, Univ. Press Tokyo and Pennsylvania State, University Park-London.

# Chapter 2

# Subcellular Fractionation

GEORGINA RODRÍGUEZ DE LORES ARNAIZ
AND
AMANDA PELLEGRINO DE IRALDI

## 1. Introduction

Subcellular fractionation allows for the study of each organelle in isolation from other cellular constituents. Its introduction to the study of brain has provided a powerful tool to contribute to the knowledge of the structure and function of the nervous system.

The present chapter refers to the methods developed for the isolation and purification of different subcellular organelles. Emphasis has been given to those methods related to the preparation of synaptic region structures. Classic methods are described and techniques incorporating significant modifications are included. As a general rule, procedures currently used by workers in the fields of neurochemistry, neuropharmacology, and neurocytology have been selected. Methods for the preparation of synaptosomes, subsynaptic components, myelin, Golgi structures, lysosomes, nuclei, mitochondria, microtubules, and intermediate filaments are described. They include analytical, preparative, and assembly techniques. Specialized reviews on individual structures are mentioned in the corresponding sections. In those articles, more detailed methodological aspects for the isolation and chemical, enzymatic, biochemical, and biophysical characterizations may be found.

## 2. Preparation of Subcellular Particles

The approaches for brain fractionation may differ depending on whether the tissue fractionation will be used for an analytical study or as a preparative tool.

### 2.1. Analytical Procedures

The aim of an analytical approach is to analyze chemical or enzymatic composition of subcellular fractions prepared by differential centrifugation with respect to the original whole homogenate. The same criteria are applied with subfractions obtained by gradient centrifugation with respect to the primary fraction layered on the gradient. In this case, it is possible to establish the distribution profiles of the different intracellular organelles. Marker enzymes or other cellular constituents must be assayed in all fractions obtained. It should be emphasized that no fraction or part is discarded.

Analytical procedures may lead to the establishment of the relative properties of each subcellular fraction with respect to marker enzymes, neurotransmitter uptake, or receptor binding sites. This approach results in the determination of the intracellular localization of the various elements assayed. The percentage content of marker activity is a measure of the distribution in the density gradient of the material initially present in the original fraction. The assay of specific markers of the subcellular particle in question provides knowledge about the relative distribution of the organelles in the various gradient fractions; that is, their relative concentration or enrichment in the fractions.

### 2.2. Preparative Procedures

Preparative procedures may lead to the study of a given organelle in a purified state. In this case, it is important to examine the possible contaminants of the so-called purified fraction.

The isolation procedure of the subcellular particle in question (for a preparative purpose) has to be accompanied by a study of the contamination of the fraction in order to assess its purity. The types of criteria used to assess purity are the observation of typical ultrastructure, the lack of gross contamination, the absence or minimization of marker characteristics of other subcellular particles, and the maximization of marker characteristics of the particle under study. The assays of subcellular markers such as a chemical constituent, an enzyme activity, or a receptor binding capacity are necessary. The concentration of contaminants should become diminished during the course of fractionation procedures.

Furthermore, the homogeneity of a given fraction, especially for small particles such as synaptic vesicles, may be evaluated by means of physical criteria such as sedimentation velocity, equilibrium centrifugation in a density gradient, or mobility in an electric field. The ultrastructural analysis of the fractions must be done with electron microscopy.

## 2.3. Assembly Techniques

Structures like microtubules and intermediate filaments are difficult to isolate with a high yield by current subcellular fractionation procedures. This is caused by their lability and the difficulties in separating them from other structures having similar morphological characteristics.

In assembly techniques, the specific chemical components are isolated in controlled conditions and their assembly is performed. In this way, microtubules and intermediate filaments presenting morphological characteristics similar to those in the intact cell may be prepared.

## 2.4. Technical Considerations

The homogenization of the tissue produces the rupture of the cell plasma membrane. Thus the subcellular organelles will be free in suspension; the integrity of endoplasmic reticulum, Golgi apparatus, cytoskeleton, and microtubular network is lost. As a consequence, intracellular compartmental barriers of the intact tissue disappear. Thus, the subcellular particles will be exposed to a medium that differs from the cell sap in chemical composition. In order to diminish the artifactual conditions created, the homogenization and all subsequent steps during cell fractionation are performed at 0–4°C.

Brain tissue is formed by different types of cells. In order to obtain organelles from cells of a single type, those cells must be separated previous to their disruption.

In the brain there are well-defined regions from an anatomical point of view; in those regions are concentrated neurotransmitters and their enzymes and receptors. For that reason a brain region is selected for specific studies—for instance, the striatum for studying dopaminergic neurotransmission or hypothalamus for noradrenergic neurotransmission. To study cholinergic transmission, the electric organ of electric fishes is the ideal source.

Homogenization of the brain tissue is usually performed in a glass Potter-Elvehjem-type homogenizer; this is equipped with a motor-driven Teflon pestle that rotates at moderately high speeds. A pestle driven by a drilling machine, which moves up and down automatically, and is set to perform one full stroke in 1 min, has also been used. In some techniques, the rupture of the tissue is performed with a Dounce homogenizer. Tissues such as nerves, with a tough sheath, are previously desheathed or homogenized by means of Ultraturrax equipment; for the electric organ, the tissue suspension is forced through steel grids. In order to obtain reproducible results, homogenization conditions must be standardized.

The various cell organelles have different physical properties, such as size, shape, density, and electric charge, that depend on their chemical composition. These differences permit their further separation.

## 3. Synaptosomes

During the homogenization of brain in isotonic sucrose solutions, nerve terminals are torn off the axon and their membranes become resealed. The separate unit formed is currently called a synaptosome. Synaptosomes are synonymous with isolated nerve endings, pinched-off nerve endings, or synaptic bodies. The word synaptosome is the one most commonly used by workers in this field, so it will be used throughout this chapter. However, its use must be reserved for the isolated particle, not for the nerve ending in the intact cell.

Synaptosomes consist of a surrounding plasma membrane, synaptic vesicles, mitochondria, and soluble constituents; in some cases, lysosomes may be observed. Besides the presynaptic membrane, a part of the postsynaptic membrane with the postsynaptic density is often attached to the nerve ending.

Fractionation procedures for isolating synaptosomes imply differential centrifugation, followed by further purification on a sucrose or Ficoll-sucrose density gradient.

Independently, De Robertis et al. (1961) and Gray and Whittaker (1962) were able to demonstrate the presence of synaptosomes in the brain crude mitochondrial fraction. "Crude mitochondrial fraction" is often referred to as "crude synaptosomal fraction." From this fraction, synaptosomes were purified in sucrose density gradients. Those studies were performed using an analytical approach, analyzing the ultrastructure and some biochemical properties of all the separated fractions (De Robertis et al., 1962a; Gray and Whittaker, 1962).

For the preparation of the crude mitochondrial fraction, the brain tissue is homogenized in 0.32*M* sucrose, pH 7.0–7.2, in a manual glass Teflon homogenizer (Potter-Elvehjem type). The homogenate is submitted to differential centrifugation; this implies the running of successive centrifugation steps at increasing speed and time periods. From the crude mitochondrial fraction, the synaptosomes are separated in sucrose or Ficoll gradients.

Although isolated synaptosomes are not completely devoid of contaminating structures, the fractions are pure enough to permit the study of various properties. For instance, synaptosomes are able to carry out glycolysis and oxidative phosphorylation, synthesize different types of molecules, and transport ions and neurotransmitters (*see* Rodríguez de Lores Arnaiz and De Robertis, 1972; Bradford, 1975).

## 3.1. Preparation of the Crude Mitochondrial Fraction

In the procedure of De Robertis et al. (1962a) a rat cerebral cortex homogenate is prepared at 10% in 0.32*M* sucrose and brought to pH 7 with Tris base. The homogenization is performed in a manual glass Teflon Potter-Elvehjem-type homogenizer (clearance not less than 0.25 mm) for two periods of 1 min each, with an interval period between them.

The homogenate is centrifuged at 900*g* for 10 min and the supernatant is removed without disturbing the pellet. Fresh homogenizing medium is added and the suspension is briefly homogenized; this suspension is centrifuged as above, and the supernatant removed; the washing is repeated once more. The sediment is the nuclear fraction (N), which contains nuclei, myelin fragments, capillaries, and cell debris. The resulting supernatants from the three centrifugations are mixed and spun down at 11,500*g* for 20 min; the clear supernatant is removed and fresh homogenizing medium is added to the pellet. This is resuspended by brief homogenization and recentrifuged (11,500*g*, 20 min). The resulting pellet (Mit) contains free mitochondria, synaptosomes, and myelin fragments.

When nuclear and mitochondrial fractions are sedimented, a fluffy layer appears, floating on the surface of the pellet. This is removed carefully by vacuum aspiration with the aid of a Pasteur pipet, in order to avoid the removal of part of the sediment.

In the procedure of Gray and Whittaker (1962), the brain homogenate is centrifuged at 1000*g* for 10 min, with two washings; the sediment ($P_1$) contains nuclei, large myelin fragments, and cell debris. Pooled supernatants are centrifuged at 10,000*g* for

20 min, with one washing; the sediment ($P_2$) is the crude mitochondrial fraction and contains mitochondria, synaptosomes, and small myelin fragments.

Washing of pellets $P_1$ and $P_2$ is often omitted. This results in shortening the time of the fractionation procedure. It should be mentioned that part of the synaptosomes and mitochondria will remain entrapped in the nuclear pellet when washing steps are omitted. This aspect has been critically discussed previously (McCaman et al., 1965).

## 3.2. Purification of Synaptosomes

### 3.2.1. Sucrose Gradients

In the procedure of De Robertis et al. (1962a), the crude mitochondrial pellet (Mit) is resuspended by brief homogenization in 0.32$M$ sucrose (1 g original fresh tissue/3.3 mL) and layered carefully on top of a discontinuous sucrose gradient prepared with 1.4, 1.2, 1.0, and 0.8$M$ sucrose solutions (all of these solutions have been previously neutralized). The gradient is centrifuged at 50,000$g$ for 120 min in a swinging rotor.

This procedure results in four layers (A, B, C, D) situated at the interfaces in the gradient, and one pellet (E). The layers are separated by aspiration under visual control; in each case, half of the clear zone in between the layers is removed carefully to avoid mixing. The fractions are diluted with 0.16$M$ sucrose (to achieve 0.3–0.4$M$ sucrose concentration) and centrifuged at 100,000$g$ for 30 min.

Fraction A situated at the 0.32–0.8$M$ sucrose interface contains myelin fragments of different sizes. Fraction B situated at the 0.8–1.0$M$ sucrose interface contains myelin fragments, small components of microsomal size (vesicles, curved membranes), small fragments, or nerve endings containing synaptic vesicles. Fraction C extending from below the boundary between 1.0 and 1.2$M$ sucrose to the densest region of the gradient is separated from D by a narrow and diffuse interband. This fraction is made up of synaptosomes and their fragments, which contain synaptic vesicles and a few mitochondria.

Fraction D occupies a large volume and extends into the middle of the 1.4$M$ sucrose. It is composed of synaptosomes similar to those of fraction C, but is more abundant in mitochondria.

Fraction E is a rather grayish compact pellet made up almost exclusively of typical, free mitochondria showing different degrees of swelling. A few synaptosomes and darker bodies that may represent lysosomes are occasionally found in this pellet.

In the procedure of Gray and Whittaker (1962), the crude

mitochondrial pellet $P_2$ is resuspended in 0.32$M$ sucrose and centrifuged into a density gradient consisting of equal volumes of 0.8 and 1.2$M$ sucrose at 53,000$g$ for 120 min. The fraction floating between 0.32 and 0.8$M$ sucrose ($P_2A$) contains myelin fragments, and that floating between 0.8 and 1.2$M$ sucrose contains synaptosomes ($P_2B$). The pellet below 1.2$M$ sucrose is formed by free mitochondria ($P_2C$). Fractions $P_2A$ and $P_2B$ are separated, diluted, and centrifuged at 100,000$g$ for 60 min to obtain the corresponding pellets.

A combination of discontinuous sucrose gradients and continuous cesium chloride gradients have also been used to isolate synaptosomes (Kornguth et al., 1969, 1971).

In addition, continuous sucrose gradients have been employed to separate synaptosomes from the CNS (Bretz et al., 1974; Laduron et al., 1983).

Synaptosomes have also been purified by using a vertical rotor; this rotor reduces that column of liquid through which the subcellular particles must migrate during the gradient separation stage (Wood and Wyllie, 1981). The cerebral cortex $P_2$ fraction is resuspended in 0.32$M$ sucrose and layered on a sucrose density gradient comprising 0.8, 1.2, and 3.0$M$ sucrose. This gradient is centrifuged by using a vertical rotor (Sorval TV 850, $8 \times$ 36 mL) at 201,000$g$ for approximately 25 min (2.5 $\times$ $10^{10}$ rad$^2$/s). The synaptosomal fraction ($P_2B$) is removed from the 0.8–1.2$M$ sucrose interface, diluted with 0.32$M$ sucrose, and centrifuged at 97,000$g$ for 30 min. The resulting pellet contains purified synaptosomes (Wood and Wyllie, 1981).

### 3.2.2. Ficoll Gradients

The use of Ficoll-sucrose discontinuous gradients has proven to be very suitable for the preparation of metabolically active synaptosomes; with these gradients, isoosmolarity is maintained and centrifugation time is shortened.

Kurokawa et al. (1965) have described a method for the rapid isolation of synaptosomes from guinea pig cerebral cortex. In this method, the tissue is homogenized in 0.32$M$ sucrose and centrifuged at 1000$g$ for 10 min; the resulting supernatant is centrifuged at 14,500$g$ for 15 min. This sediment—the crude mitochondrial fraction—is resuspended in 0.32$M$ sucrose and layered on the top of a gradient consisting of two steps with 3 and 13% Ficoll dissolved in 0.32$M$ sucrose. The gradient is centrifuged at 20,900$g$ for 15 min in a swinging rotor, and two fractions and a pellet are separated. The fraction layering at 3% Ficoll is composed of myelin and some microsomes; that layering at 13% Ficoll contains synaptosomes; the pellet is composed of mitochondria and some synaptosomes. The layers are separated, diluted with 0.32M sucrose, and centrifuged at 105,400g for 20 min (Kurokawa et al., 1965). Similar Ficoll-sucrose gradients have also been used for the isolation of brain synaptosomes (Abdel-Latif, 1966; Cotman and Matthews, 1971).

Another rapid method for the preparation of metabolically active synaptosomes has been described by Booth and Clark (1978). In this method, rat forebrains are homogenized in 0.32M sucrose containing 1 mM potassium EDTA and 10 mM Tris-HCl, pH 7.4, in a Dounce-type glass homogenizer (0.1 mm clearance). The homogenate is spun down at 1300g for 3 min, and the resulting supernatant is centrifuged at 17,000g for 10 min to separate the crude mitochondrial pellet. This pellet is gently homogenized in a Potter-type homogenizer (clearance 0.375 mm) and introduced into a centrifuge tube; above this, a solution containing 7.5% Ficoll, 0.32M sucrose, and 50 μM potassium EDTA, pH 7.4, is layered. On top of this, isolation medium is layered. The tubes are centrifuged at 99,000g for 30 min in a swing-out rotor. Myelin and synaptosomes lay at the first and second interfaces, respectively, and the free mitochondria appear in the pellet. The synaptosomal fraction is separated, diluted, and pelleted at 5500g for 10 min (Booth and Clark, 1978).

### 3.2.3. Different Types of Synaptosomes

The use of discontinuous sucrose gradients permits the separation of different types of synaptosomes. Rat brain nerve endings rich in acetylcholine and cholineacetyltransferase equilibrate at 0.8–1.0 and 1.0–1.2M sucrose interfaces in discontinuous sucrose gradients, whereas those rich in glutamate decarboxylase (involved in GABA synthesis) equilibrate at the 1.2–1.4M sucrose interface (*see* Rodriguez de Lores Arnaiz and De Robertis, 1972 for ref.).

Different populations of synaptosomes that store labeled NA and GABA may be separated from rat striatum. For this purpose, a 0.32M sucrose homogenate is layered onto a continuous sucrose gradient (0.32–1.5M sucrose) and centrifuged for 2 h at 130,000g. The peak of GABA-storing synaptosomes appears at 0.95M sucrose and the peak of NA-storing synaptosomes will appear at a denser level of sucrose (1.1M). If the hypothalamus is the starting tissue, the distribution of GABA- and NA-storing synaptosomes is similar (Iversen and Snyder, 1968).

With the use of linear, continuous density gradients of sucrose (1.5–0.32M), hypothalamus and midbrain synaptosomes that store 5-hydroxytryptamine (5-HT) are separable from those storing GABA; synaptosomes storing 5-HT are slightly separable from those storing NA (Kuhar et al., 1971).

Three different populations of nerve endings have been resolved by zonal isopycnic centrifugation. The postnuclear supernates of forebrain homogenates in 0.32$M$ sucrose were sampled on a shallow sucrose gradient extending linearly between the densities 1.11 and 1.20, and overlaid by 0.25$M$ sucrose; the gradient rested on a 60% sucrose cushion and centrifugation was carried out at 30,000 rpm for 25 h. With this approach, cholinergic nerve endings (containing acetylcholine) were identified at an equilibrium density of 1.137; nerve endings with high glutamate decarboxylase activity and the ability to accumulate exogenous GABA equilibrate at an average density of 1.149, and adrenergic endings that accumulate exogenous catecholamines equilibrate at an average density of 1.165 (Bretz et al., 1974).

The use of other continuous sucrose gradients also permits a separation of different types of synaptosomes; rat cortex synaptosomes rich in choline acetyltransferase equilibrate at 0.9–1.0$M$ sucrose; synaptosomes presenting 5-HT-uptake capacity equilibrate at 1.1$M$ sucrose, and GABAergic nerve endings have a higher density (1.2$M$ sucrose) (Laduron et al., 1983).

Two populations of synaptosomes obtained from the brains of rats have been separated from myelin and mitochondria using Percoll to generate continuous density gradients (Shank and Campbell, 1984). The pellet $P_2$ was resuspended in 1 mL 0.3$M$ sucrose buffered to pH 7.7 with 8 m$M$ Tris-HCl (or 5 m$M$ NaHCO$_3$), then mixed into a solution containing 28 mL of the buffered sucrose solution and 4 mL of Percoll (Pharmacia, Uppsala). Centrifugation at 32,000$g$ for 40 min (in an angle rotor) separated three fractions. The uppermost fraction, containing essentially myelin, was removed and discarded.

The middle band was also removed, and added to buffered sucrose solution; it was rich in low-density synaptosomes and contained small amounts of myelin fragments and a few free mitochondria. To the remainder of the gradient, buffered sucrose solution was added to bring the volume to 28 mL; 6 mL of Percoll was then gently mixed into the solution. Centrifugation as described above resulted in a discrete band of material near the top and nondiscrete bands of material in other portions of the gradient. The top band contained small, high-density synaptosomes (diameter 1 μm) (Shank and Campbell, 1984).

The centrifugation of the crude mitochondrial fraction supernate at 20,000$g$ for 30 min permits the separation of a sediment (Mic 20) that contains small synaptosomes and is rich in 5-HT (Pellegrino de Iraldi et al., 1968) and histamine (Kataoka and De Robertis, 1967).

Mossy fiber endings of the cerebellum may be isolated in considerable yield if a Dounce-type homogenizer is employed to homogenize the cerebellum (Balacz et al., 1974; Tapia et al., 1974).

3.2.3.1. CHOLINERGIC NERVE ENDINGS　The electric organ of *Torpedo marmorata* may be used for the preparation of pure cholinergic subcellular fractions and is a valuable material for the study of nicotinic cholinergic transmission.

Unlike synaptosomal preparations obtained from brain, synaptosomes prepared from Torpedo electric organ do not remain attached to fragments of postsynaptic membranes, as evidenced by their appearance in electron micrographs.

A method for the preparation of pure cholinergic synaptosomes from Torpedo electric organ has been described (Isräel et al., 1976; Morel et al., 1977). Slices of electric organ (15–20 g) are finely chopped; the mince is suspended in Torpedo physiological solution consisting of 280 m$M$ NaCl, 3 m$M$ KCl, 1.8 m$M$ Mg Cl$_2$, 3.4 m$M$ CaCl$_2$, 1.2 m$M$ sodium phosphate buffer (pH 6.8), 5.5 m$M$ glucose, 300 m$M$ urea, and 100 m$M$ sucrose. After equilibration with O$_2$, 5 m$M$ NaHCO$_3$ is added to adjust the pH to 7.0–7.1. The suspension is stirred for 30 min at 4°C and successively forced through stainless steel grids of 1000–500 and 200 μm squares opening sideways. The suspension is then filtered through a nylon gauze (50 μm square openings). The filtrate is centrifuged at 6000$g$ for 20 min; the resulting pellet is resuspended in the physiological medium, vigorously shaken, and layered onto a discontinuous sucrose gradient. A pure synaptosomal fraction is separated at the 0.3–0.5$M$ sucrose interface (Isräel et al., 1976; Morel et al., 1977). When it is necessary to start with larger amounts of tissue, the modification described by Li and Bon (1983) may be used. This technique allows the use of 150 g dissected electric organs and gives a good yield of nerve endings of equivalent purity. Modifications introduced are the use of a Ca$^{2+}$-free physiological solution to reduce acetylcholine release, a greater number of passages through steel grids and nylon sieves, the inclusion of a low-speed differential centrifugation, and the use of continuous sucrose gradients (instead of discontinuous gradients) (Li and Bon, 1983).

### 3.3. Fractionation of Frozen Brain Tissue

In order to perform studies on the neurochemical pathology of human brain disorders it may be necessary to employ stored frozen preparations.

Synaptosomes isolated from fresh rat or fresh human brain are more active than those prepared from frozen tissue. Several combinations of freezing and thawing of the brain tissue produce inactive synaptosomal preparations. It was observed that rapid freezing (by immersion in liquid nitrogen) and/or slow thawing (in a 4°C room) severely impairs the further performance of incubated synaptosomes. In contrast, synaptosomes from tissue frozen slowly (in a −70°C freezer) and thawed rapidly (in a 37°C bath) show relatively good retention of ultrastructural and metabolic characteristics (Hardy et al., 1983).

### *3.4. Synaptosomal Characterization*

The presence of synaptosomes may be checked by assaying the uptake or the release of labeled neurotransmitters and the different markers suggested for the subsynaptic structures (see below). Synaptosomal integrity may be revealed by the presence of osmotically labile lactate dehydrogenase (E.C. 1.1.1.27) activity. From a morphological point of view, synaptosomes are identified with the electron microscope.

## 4. Synaptic Vesicles

The hypothesis about the participation of synaptic vesicles in the storage and liberation of neurotransmitters has been supported by studies performed after the development of subcellular fractionation techniques. Those studies permitted the isolation of synaptic vesicles and their further biochemical and pharmacological characterization (De Robertis and Rodríguez de Lores Arnaiz, 1969; Israël et al., 1979; Whittaker and Roed, 1982).

In order to isolate synaptic vesicles, it has been crucial to use hypotonic treatment of the isolated nerve endings (De Robertis et al., 1962b). When a subcellular fraction containing synaptosomes is homogenized in hypotonic conditions, synaptosomes undergo breakage. Synaptic vesicles, intraterminal mitochondria, and the terminal axoplasm are liberated; thus, the surrounding synaptosomal membrane remains empty. The membrane that constitutes the synaptic vesicles resists the osmotic shock, probably owing to the small size of the vesicles. The treatment was first applied to a crude mitochondrial fraction and synaptic vesicles were separated by differential centrifugation (De Robertis et al., 1962b, 1963). Thereafter, Whittaker et al. (1964), also using a hypoosmotic treatment of the crude mitochondrial fraction and gradient centrifugation, described a method to isolate synaptic vesicles. Different technical modifications were introduced to improve the yield of vesicles or to better preserve vesicle characteristics. In most methods the first step is the osmotic rupture of the nerve endings.

The osmotic shock may be applied to the crude mitochondrial fraction or to the synaptosomal fraction isolated in sucrose gradient or Ficoll–sucrose gradient.

Brain synaptic vesicles may be separated from other synaptosomal constituents on the basis of their buoyant density in sucrose gradients. Synaptic vesicles equilibrate at $0.4M$ sucrose; this is also observed in synaptic vesicles from peripheral ganglia or electric tissue.

## 4.1. Hypoosmotic Treatment of the Crude Mitochondrial Fraction and Separation of Synaptic Vesicles

In the technique described by De Robertis et al. (1962b, 1963), the crude mitochondrial fraction (Mit) is suspended in water (1 g original tissue/10 mL) and homogenized in a Potter-Elvehjem homogenizer for two periods of 1 min each. This suspension is centrifuged at 20,000$g$ for 30 min. The pellet ($M_1$) contains mitochondria, myelin fragments, and synaptosomal membranes. The supernatant is centrifuged at 100,000$g$ for 30 min; the pellet ($M_2$) contains essentially synaptic vesicles and membrane fragments.

In the procedure described by Whittaker et al. (1964), the crude mitochondrial fraction ($P_2$) is homogenized in water (1 g original tissue/2 mL) and centrifuged at 10,000$g$ for 20 min. The pellet ($W_p$) contains myelin, mitochondria, and damaged synaptosomes. The cloudy supernatant ($W_s$) is layered onto a density gradient consisting of 0.4, 0.6, 0.8, 1.0, and $1.2M$ sucrose solutions; the gradient is centrifuged at 53,000$g$ for 120 min. The hazy layer formed in $0.4M$ sucrose consists of synaptic vesicles (Whittaker et al., 1964).

The centrifugation of the $M_1$ supernatant on $0.5M$ sucrose also permits the separation of a purified synaptic vesicle fraction that bands at the $0.5M$ sucrose interface (Lapetina et al. 1967).

The effect of experimental conditions during hypoosmotic treatment on the further separation of subsynaptic structures has been studied by several authors (see below).

Modifications to the original methods for the isolation of synaptic vesicles from brain include a change in fractionation schedule (Zimmermann, 1982), and the addition of salts (De Lorenzo and Freedman, 1978) or $D_2O$ (Nagy et al., 1977) to the isolation medium. Media other than sucrose, such as sodium diatrizoate (Tamir et al., 1976), may be used for preparation of the density gradient in order to achieve dense solutions without increasing viscosity.

Column chromatography in DEAE-Sephadex, Sepharose 6 B, or porous glass beads may be employed for the further purification of isolated synaptic vesicles. Because of the differences in particle charge, sucrose gradient electrophoresis and continuous free-flow electrophoresis may be applied for the separation of vesicles from other subcellular particles (*see* Zimmermann, 1982).

Brain tissue contains nerve terminals that belong to many different types of neurotransmitter systems. For this reason, the synaptic vesicle fraction from brain is heterogenous. In the following sections, several methods that permit the isolation of homogeneous populations of synaptic vesicles are reviewed.

## *4.2. Cholinergic Synaptic Vesicles*

The proportion of cholinergic nerve endings in brain has been estimated to be about 15%. In order to achieve a pure cholinergic synaptic vesicle fraction from brain, a possible approach is to isolate first an enriched cholinergic nerve ending fraction. Cerebral cortex is better than total brain tissue because it is rich in cholinergic synaptic endings.

Cholinergic synaptic vesicles may be isolated from bovine superior cervical ganglion. Prior to the homogenization, the ganglia are submitted to digestion with collagenase. Other interesting starting tissues that have been used are the submandibular gland of the cat and the guinea pig small intestine, with a high acetylcholine content in the myenteric plexus.

Fractions enriched in coated vesicles, plain synaptic vesicles, and flocculent material may be prepared from a crude synaptosome fraction isolated from brain (Kadota and Kadota, 1973). The method requires the preparation of the subcellular fractions as described by Gray and Whittaker (1962) and De Robertis et al. (1962a). The crude synaptosome fraction is shocked and submitted to differential centrifugation and salting out with ammonium sulfate. After the osmotic shock to the crude synaptosome fraction, KCl is included (to achieve 10 m$M$ final concentration) to facilitate the removal of mitochondria, partially broken synaptosomes, and membrane fragments, which sediment in the following centrifugation. The material is centrifuged at 20,000$g$ for 30 min; the resulting supernatant is submitted to differential centrifugation to separate three fractions: two fractions containing synaptic vesicles, one of which is a mixture of coated vesicles and plain synaptic vesicles (55,000$g$, 60 min), and the other composed of plain synaptic vesicles and flocculent material (80,000$g$, 75 min). A nonvesicular fraction containing presynaptic cytoplasm may be precipitated with ammonium sulfate (Kadota and Kadota, 1973). Choline is associated with the former, whereas acetylcholine is concentrated in the latter fraction (Kamiya et al., 1974).

Sephadex columns can be employed for the partial purification of synaptic vesicles and flocculent material isolated from crude synaptosomes. The passage of the 55,000$g$ fraction pellet through a DEAE-Sephadex column permits the removal of contaminating membrane fragments (Kadota and Kadota, 1973; Kamiya et al., 1974).

Synaptic vesicles can be isolated from guinea pig cerebral cortex by means of a continuous $D_2O:H_2O$ (1:1)–sucrose gradient. Two populations of synaptic vesicles are separated between the layers of 0.2–0.3$M$ and 0.3–0.5$M$ sucrose. The less dense vesicles have a higher acetylcholine content than the more dense vesicles, which are rich in catecholamines. The vesicle fractions may be purified in the presence of 1 m$M$ EGTA by chromatography on columns of glass beads of controlled pore size. With this step, the vesicles are separated from the membranes and soluble protein contaminants (Nagy et al., 1977).

To isolate a pure fraction of cholinergic vesicles, the electric organ of the electric fishes seems to be the best selection. The electric organ receives a very rich and purely cholinergic innervation. The suspending medium has to be isoosmotic, with an osmolarity of about 800 mosmol. Such osmolarity is achieved with additions of NaCl, KCl, glycine, and urea to sucrose or dextrane solutions. Other additions may be chelators (EDTA or EGTA), protease inhibitors, and antioxidant substances. The addition of the inhibitor of acetylcholinesterase, eserine, to the suspending medium, to protect against hydrolysis of free acetylcholine, may be desirable.

In the technique devised by Tashiro and Stadler (1978), cholinergic synaptic vesicles are purified from electric organ of *Torpedo marmorata* using a two-step purification on discontinuous and continuous sucrose density gradients. The frozen tissue is crushed in a mortar; a solution containing 0.4$M$ NaCl, 0.3 m$M$ phenylmethylsulfonyl fluoride, and 20 m$M$ Tris, pH 7.4, is added and the suspension is squeezed through four layers of cheesecloth. The extract is centrifuged at 10,000$g$ for 30 min; the resulting supernatant is loaded onto a discontinuous sucrose-NaCl density gradient consisting of 0.6$M$ sucrose-0.1$M$ NaCl and 0.2$M$ sucrose-0.3$M$ NaCl. The gradient is centrifuged at 67,700$g$ for 2.5

h in a swinging rotor. The fraction layering at the interface between the two sucrose layers is collected and diluted with 0.4$M$ NaCl. This suspension is loaded onto a continuous isoosmotic sucrose-NaCl gradient formed in a zonal rotor (Tashiro and Stadler, 1978), overlaid with 0.4$M$ NaCl, and centrifuged at 171,800$g$ for 3 h. Aliquots of the vesicle peak preparation are chromatographed on a Sepharose column (Tashiro and Stadler, 1978).

In the procedure developed by Carlsson et al. (1978), synaptic vesicles from elasmobranch electric organ are separated by differential centrifugation, flotation equilibrium in sucrose density gradients, and permeation chromatography on glass bead columns. Frozen tissue is smashed to a fine powder under liquid nitrogen; the powder is added to a solution containing 0.4$M$ NaCl, 10 m$M$ Hepes, and 10 m$M$ EGTA (pH 7.0), with constant stirring. The suspension is homogenized in a Waring blender for 2 min. The homogenate is centrifuged at 12,000$g$ for 30 min, and the resulting supernatant at 100,000$g$ for 8 h. The 100,000$g$ pellet is resuspended in buffered 0.8$M$ sucrose and pipeted into the bottom of a centrifuge tube; a solution of 0.5$M$ sucrose and 0.15$M$ NaCl is layered on top, followed by another of 0.45$M$ sucrose and 0.175$M$ NaCl. The tube is then filled with a solution of 0.2$M$ sucrose and 0.3$M$ NaCl. All solutions are buffered with 10 m$M$ Hepes and 10 m$M$ EGTA, pH 7. The gradients are centrifuged at 100,000$g$ for 10 h in a fixed-angle rotor. Samples are collected and layered on top of pore glass bead columns (Carlsson et al., 1978).

## 4.3. Monoaminergic Synaptic Vesicles

Two types of monoaminergic synaptic vesicles have been identified in terminal axons from the peripheral and central nervous systems: a predominant population of vesicles 40–60 nm in size, formed by electron lucent and dense-cored vesicles, and a minor population of granulated vesicles of about 80–90 nm. Ultrastructural and histochemical studies support the concept that dense-cored and electron lucent small vesicles are transient states of a unique organelle (Pellegrino de Iraldi, 1983). Small and large granulated vesicles have been found in the perikaryon and the large ones in the nerve trunks. After osmium tetroxide or glutaraldehyde osmium tetroxide fixation, both kinds of granulated vesicles may be identified in the peripheral nervous system, whereas in the CNS only the large vesicles present a dense core. In the CNS, the small granulated vesicles appear electron lucent unless they are fixed in potassium permanganate (Hökfelt, 1971). Thus, small granulated vesicles may be absent from the electron micrographs of fractions from the CNS enriched in monoaminergic neurotransmitters.

For the isolation of monoaminergic synaptic vesicles, differential and gradient centrifugation have been applied to several tissues (for review, *see* Smith, 1972; Lagercrantz and Klein, 1982). Large dense-cored vesicles were the first NA storage particles isolated from the nerves by differential centrifugation (von Euler and Hillarp, 1956). In the brain stem, a vesicle fraction containing NA and 5-HT was obtained by differential centrifugation from the mitochondrial fraction isolated according to Gray and Whittaker (1962) and exposed to controlled conditions of hypotonicity and sonication (Maynert et al., 1964).

In the rat hypothalamus, a vesicle fraction was isolated by differential centrifugation of a crude mitochondrial fraction exposed to osmotic shock, as described by De Robertis et al. (1962a), except that the homogenizing medium contained $3 \times 10^{-4}M$ tranylcypromine to inhibit monoamine oxidase (De Robertis et al., 1965). Electron microscopy of the fractions fixed in osmium tetroxide showed vesicles of different sizes and shapes; small round electron lucent, larger elongated electron lucent, and large granulated synaptic vesicles, similar to those observed in nerve endings of intact tissue and mitochondrial fraction. Noradrenalin was 6.3 times higher in the hypothalumus than in the total brain. In the synaptic vesicle fraction the concentration of NA by mg protein was 5.3 times higher in the hypothalamus than in the total brain (De Robertis et al., 1965).

By means of differential and continuous sucrose gradient centrifugation techniques previously described (De Potter et al., 1970; Chubb et al., 1970), it is possible to separate two populations of NA-storage particles in the hypothalamus (Belmar et al., 1974). For this purpose, the hypothalami are quickly dissected out and chilled in cold 0.32*M* sucrose, pH 7.4, in 10 m*M* potassium phosphate buffer. The tissue is homogenized in the same medium with a Teflon glass homogenizer (0.05 mm clearance). The homogenate is then centrifuged according to Whittaker and Barker (1972). The $P_2$ fraction submitted to hyposmotic shock is put on a linear density gradient from 0.32 to 2.0*M* sucrose and centrifuged at 105,000*g*, 150 min (Chubb et al., 1970). After perforation of the bottom of the tube, 13 fractions (1 mL each) are collected. When fraction $P_2$ is applied over the linear gradient, a peak of NA is formed at 1.0–1.3*M* sucrose, corresponding to synaptosomes. After hyposmotic shock, the particle-bound peak of NA is resolved in two peaks of lower density—one located at 0.99*M* sucrose and the other at 0.84*M* sucrose. These results suggest that two types of NA-storage particles contained in the synaptosomes of the hypothalamus are separated (Belmar et al., 1974).

Subcellular fractionation of the guinea pig cerebral cortex (Thomas, 1979) has been performed with the modification of the method of Gray and Whittaker (1962). All the solutions contain 1 mM pargyline, 1 mM pyrogallol, and 1 mM EDTA-Na$_2$, pH 6.5. The nuclear fraction (800$g$, 10 min) is washed only once. Supernatant fluid is then centrifuged at 20,000$g$ for 20 min to yield the mitochondrial (P$_2$) fraction. The P$_2$ fraction is resuspended in water (2 mL/g tissue) and subjected to density gradient centrifugation, as described by Whittaker et al. (1964), but with an additional layer of 1.4$M$ sucrose containing the inhibitors placed at the bottom of the density gradient, which is then centrifuged at 93,000$g$ for 2 h. It was found that bound NA formed a peak in subfraction E (0.4–0.6$M$ sucrose), with smaller amounts in subfractions D and F (0.4 and 0.6–0.8$M$ sucrose). Electron microscopy of subfractions prepared by density gradient centrifugation of shocked P$_2$ fractions appeared essentially as detailed by Whittaker (1969), except for subfractions D and E. The latter, from the 0.4–0.6$M$ sucrose interface, contained large numbers of vesicles about 50 nm. Fraction D (0.3–0.4$M$ sucrose) contained negligible particulate material (Thomas, 1979).

The isolation by differential and gradient centrifugation of heavy NA-storage particles from bovine splenic nerve and stellate ganglia, and heavy and light NA-storage particles from dog spleen (De Potter et al., 1970; Chubb et al., 1970) have been described. Tissues were homogenized in a Potter-Elvehjem homogenizer, and the homogenate filtered through surgical gauze and centrifuged successively at increasing velocities of up to 30,000 rpm. Selected particulate fractions were resuspended in 0.25$M$ sucrose and layered over continuous or discontinuous density gradients ranging from 0.3 to 1.7$M$ sucrose. The gradients were centrifuged at 39,000 rpm for 60 or 150 min. In bovine splenic nerve and stellate ganglia, only one type of NA-containing particle with an equilibrium density of 1.16, which corresponds to a sucrose molarity of 1.2, was found. In the dog spleen, two types of NA-storage particles were present, with densities of 1.25 and 1.178 (De Potter et al., 1970; Chubb et al., 1970).

Small and large granulated vesicles have been isolated by differential and gradient centrifugation from rat vas deferens and cat spleen (Bisby and Fillenz, 1971). Vasa deferentia were desheathed and the inner epithelium peeled off. The minced tissues were homogenized in 10 vol of 0.32$M$ sucrose containing $10^{-3}M$ EDTA and 0.1$M$ phosphate buffer, pH 7.4. Tissue homogenates were successively centrifuged at 600$g$, 10 min, 10,000$g$, 30 min, and 100,000$g$, 120 min, resulting in the formation of three pellets. The 100,000$g$ pellet was resuspended in 0.3$M$ sucrose and layered on top of continuous (0.4–1.4$M$) or discontinuous (0.4–1.5$M$) sucrose gradients and centrifuged at 53,000$g$ for 3 h. Density gradients of rat vas deferens had a single low-density peak of NA at 0.6$M$ sucrose (small dense-cored vesicles), whereas those of cat spleen had an additional peak of NA at 1.1$M$ sucrose (large dense-cored vesicles) (Bisby and Fillenz, 1971).

A method to isolate small noradrenergic vesicles from rat seminal ducts following castration has been reported by Fried et al. (1978). The seminal ducts are carefully dissected, desheathed, and put in 0.25$M$ sucrose in 0.05$M$ potassium phosphate buffer, pH 7.2–7.3, minced, and homogenized by 30 passes of a Teflon–glass homogenizer. The homogenate is centrifuged at 10,000$g$ for 30 min. The supernatant is layered on continuous sucrose density gradients (0.25–0.8$M$) and centrifuged at 280,000$g$ for 90 min in a swing-out bucket. Usually no bands will be seen in the gradients after centrifugation. The fraction with a density of 0.4–0.5$M$ sucrose has the highest concentration of NA and is enriched 12-fold from the original homogenate to the final fraction (Fried et al., 1978).

Large dense-cored vesicles have been purified from bovine splenic nerve by Lagercrantz and Klein (1982). Bovine splenic nerves are desheathed, minced, and homogenized by Ultra Turrax at a relatively low setting for 2.5 min. The vesicle isolation medium contains 0.3$M$ sucrose, 0.02$M$ potassium phosphate, pH 7.2–7.4, 1 mM MgCl$_2$. The homogenate is centrifuged at 10,000–14,000$g$ for 15 min. The supernatant is layered on sucrose D$_2$O density gradients ranging from 0.3 to 1.2 with a step of 1.5$M$ sucrose in the bottom of the gradient. Above the gradient, a buffer zone with 2 mL of buffered isosmotic sucrose is layered. The gradients are centrifuged at 280,000$g$ for 90 min in swing-out buckets. The particulate NA is found at the interface between the gradient and the bottom step (between 1.2 and 1.5$M$ sucrose) (Lagercrantz and Klein, 1982).

## 4.4. Peptidergic Synaptic Vesicles

Purinergic nerves present large opaque synaptic vesicles with an electron lucent halo between the granular core and the vesicle membrane. This halo is less prominent than that present in large dense-cored vesicles of adrenergic and cholinergic nerves (Burnstock and Iwayama, 1978). It was subsequently found by inmunocytochemistry that neuropeptides assumed to function as neurotransmitters or modulators are stored in large opaque vesicles. The presence of these peptides was also demonstrated in large dense-cored vesicles of some adrenergic and cholinergic nerves (Hökfelt et al., 1980). In some cases, neuropeptide reactivity was also found in small synaptic vesicles, predominantly at the periphery of these vesicles. By differential and gradient centrifugation, enkephalins, vasoactive intestinal peptide, somatostatin, neurotensin, and thyrotrophin-releasing factor have been localized in large granulated vesicles (Fried, 1982).

A fraction of small synaptic vesicles (36–58 nm) enriched in substance P has been isolated from rat brain (Cuello et al., 1977) by osmotic shock of a mitochondrial fraction according to the procedure of De Robertis et al. (1962a).

By means of the subcellular fractionation technique described by Lagercrantz and Klein (1982), it has been possible to observe that enkephalin and NA in bovine splenic nerve and vasointestinal peptide and acetylcholine in cat submandibular gland are contained in organelles with similar sedimentation properties; thus they may coexist in large vesicles (Fried et al., 1981).

Subsynaptosomal particles containing substance P, detected by radioimmunoassay, have been prepared by osmotic lysis of rat brain stem synaptosomes and chromatography on a calibrated column of controlled pore glass beads of nominal pore size, 300 nm. Immunoreactive substance P migrates on this column with particles of an apparent mean diameter of 177 nm (Floor et al., 1982).

### 4.5 Characterization of Synaptic Vesicles

The presence of synaptic vesicles in subcellular fractions can be assessed by the enrichment in various chemical neurotransmitters. The different populations of synaptic vesicles may be identified with the electron microscope. The various types of monoaminergic vesicles may be demonstrated with histochemical techniques at the level of electron microscopy (Hökfelt, 1971). Dopamine β-hydroxylase (E.C. 1.14.2.1) is a good marker for catecholamine storage vesicles, especially the larger ones.

## 5. Synaptosomal Membranes

### 5.1. Brain Synaptosomal Membranes

The hyposmotic treatment of the brain crude mitochondrial fraction that had originally been described for the separation of synaptic vesicles (*see* section 4) also permits the preparation of purified synaptosomal membranes (synaptosomal membranes, nerve ending membranes, synaptic ghosts, and synaptic plasma membranes are synonymous terms.) These membranes are complex structures consisting of a presynaptic membrane, a postsynaptic portion, and a synaptic thickening connecting the two plasma membrane components.

The use of the same sucrose gradient centrifugation schedule developed for the purification of synaptic vesicles led Whittaker et al. (1964) to isolate a purified synaptosomal membrane fraction. In their discontinuous sucrose gradient, synaptosomal membranes banded at 0.6–0.8 and 0.8–1.0M sucrose interface.

The approach followed in this laboratory (Rodríguez de Lores Arnaiz et al., 1967) consisted of the purification of a synaptosomal membrane fraction from the $M_1$ pellet prepared from brain (De Robertis et al., 1962b, 1963). The $M_1$ pellet is composed of mitochondria, myelin, and synaptosomal membranes, many of the latter containing the attachment of part of the subsynaptic membranes. In this technique a freshly prepared $M_1$ pellet is resuspended in 0.32M sucrose neutralized to pH 7 with Tris base (1 g fresh original tissue/3.3 mL) by brief homogenization. This suspension is layered on a discontinuous sucrose gradient prepared with 1.2, 1.0, 0.9, and 0.8M sucrose (all pH 7). The gradient is centrifuged at 50,000g for 120 min in a swinging rotor. Four layers, $M_1$ 0.8, $M_1$ 0.9, $M_1$ 1.0, and $M_1$ 1.2, and a pellet $M_1$p are observed. The layers produced at the various interfaces are separated by aspiration under visual control. They are then diluted with 0.16M sucrose (to achieve 0.3–0.4M sucrose) and spun down at 100,000g for 30 min.

Fraction $M_1$ 0.8, situated at the 0.8M sucrose interface, contains myelin fragments of different sizes; fraction $M_1$ 0.9, situated at the boundary between 0.8 and 0.9M sucrose, is composed of myelin fragments and predominantly synaptosomal membranes, similar to those found in the $M_1$ 1.0 layer. Fraction $M_1$ 1.0 is situated in the 0.9–1.0M sucrose boundary and consists of round or oval membrane profiles, most of which correspond in size to nerve endings. The synaptic nature of many of these membranes is evidenced by the presence of thickened portions and the attachment of the subsynaptic structures. Fraction $M_1$ 1.2 is more abundant and compact than $M_1$ 1.0, being situated at the boundary between 1.0 and 1.2M sucrose, and contains essentially synaptosomal membranes. Fraction $M_1$p is a pellet that consists mainly of mitochondria swollen to varying degrees by the osmotic shock. Some synaptic membranes appear trapped in between (Rodríguez de Lores Arnaiz et al., 1967).

In order to prepare synaptosomal membranes, Cotman and Matthews (1971) have performed the hypotonic treatment on a synaptosomal fraction isolated in Ficoll-sucrose gradients. In this technique, rat forebrain is homogenized in 0.32$M$ sucrose and submitted to differential centrifuged. The crude nuclear fraction is removed by centrifugation at 1100$g$ for 5 min; the supernatant is centrifuged at 17,000$g$ for 10 min. The pellet (crude mito-chondrial fraction) is resuspended in 10% sucrose by hand homogenization. This suspension is applied to a two-step discon-tinuous Ficoll-sucrose gradient consisting of layers of 13 and 7.5% Ficoll in 0.32$M$ sucrose. The gradient is centrifuged at 63,000$g$ for 45 min and the synaptosomal fraction is obtained at the interface of the 7.5 and 13% Ficoll-sucrose layers. The synaptosomal band is removed, diluted with 10% sucrose, and pelleted. The synaptosomal pellet is resuspended in a small volume of 10% su-crose and osmotically shocked in 6 m$M$ Tris, pH 8.1, for 1.5 h. After osmotic shock, the fraction is concentrated by centrifuga-tion, resuspended, and applied to a discontinuous sucrose gradi-ent prepared with 25, 32.5, 35, and 38% sucrose. The gradient is centrifuged at 50,000$g$ for 90 min and the synaptosomal mem-branes are collected at the 25 and 32.5% sucrose interface (Cotman and Matthews, 1971).

The pH of the medium during osmotic shock is a critical vari-able for the further gradient purification of synaptosomal mem-brane. If synaptosomes are osmotically shocked in unbuffered media, a slightly acidic pH will result; if the osmotic shock is carried out at neutral, or even better, at alkaline pH (pH 8.1 with Tris), for 1.5 h at 4°C, a higher resolution of synaptosomal mem-branes from mitochondria has been observed (Cotman and Mat-thews, 1971).

In order to achieve a better separation between synaptosomal membranes and mitochondria, a histochemical reaction to form a dense formazan in the mitochondrial compartment can be used (Cotman and Taylor, 1972; Davis and Bloom, 1973). In this method, mitochondrial succinate dehydrogenase in the presence of succinate catalyzes the reduction of iodonitrotetrazolium to a formazan pigment within the mitochondria. The isopycnic den-sity of the mitochondria is increased, and they are better sepa-rated from synaptosomal membranes in the sucrose gradients (Cotman and Taylor, 1972). A similar schedule includes an NADH-iodonitrotetrazolium incubation step between the hypos-motic disruption of synaptosomes and the further separation of synaptosomal membranes from mitochondria in a discontinuous sucrose gradient (*see* Cotman, 1974).

A combined flotation–sedimentation density gradient centrifugation has also been described for the isolation of synaptosomal membranes (Jones and Matus, 1974). In this procedure, the rat brain homogenate is centrifuged at 800$g$ for 20 min and the resulting supernatant is centrifuged at 9000$g$ for 20 min, with one washing. The pellet is lysed in 5 m$M$ Tris-HCl buffer, pH 8.1, and incubated at 0°C for 30 min. The lysate is made up to 34% sucrose and delivered into the ultracentrifuge tube; a 28.5% sucrose solution is overlaid above the sample phase and a small volume of 10% sucrose solution overlaid about that. The gradient is centrifuged at 60,000$g$ for 110 min. Three fractions are recovered, one from each interface and one from the pellet. Synaptosomal membranes appear at the interface of 28.5 and 34% sucrose layers (Jones and Matus, 1974).

Synaptosomal membranes can also be prepared by means of a technique that involves the use of sodium diatrizoate (Tamir et al., 1976). For this purpose, a synaptosomal fraction prepared in a sodium diatrizoate gradient is suspended in 0.32$M$ sucrose containing 0.1 m$M$ EDTA and 1 m$M$ phosphate, pH 7.5, for 30 min. A 2$M$ sucrose solution is added to bring the sucrose concentration to 0.32$M$. Membrane fragments are collected by centrifugation at 20,200$g$ for 30 min. This pellet is suspended in sucrose–phosphate and then fractionated in a discontinuous gradient of sodium diatrizoate (22, 18, 16, 14, and 10%) in half-strength Krebs-Ringer's solution without calcium. The gradient is centrifuged at 29,000$g$ for 20 min. Five fractions at interfaces and a pellet are collected, diluted with 0.32$M$ sucrose, and centrifuged at 21,000$g$ for 20 min. Pellets are washed three times by resuspension in 0.32$M$ sucrose and recentrifuged (Tamir et al., 1976).

A technique involving cavitation lysis of synaptosomes for the preparation of synaptosomal membranes has been recently described (Meflah et al., 1984). Frozen bovine caudate nuclei were used for the preparation of synaptosomal membranes according to Jones and Matus (1974) except that the 9000$g$ pellet was lysed under a high-nitrogen pressure (50 atm, 20 min), followed by decompression, before the usual discontinuous sucrose gradient centrifugation. The membranes were washed twice. These authors have both right-side-out and inside-out plasma membrane vesicles; the former are considered to be predominant with the single cavitation lysis method (Meflah et al., 1984).

### 5.1.1. Presynaptic Membranes

When subcellular fractionation techniques are performed on brain tissue, a portion of the postsynaptic membrane remains attached to the presynaptic membrane. In contrast, if fractionation is done on electric organ tissue, the postsynaptic membrane will not accompany the presynaptic membrane. Thus, if it is necessary to run studies exclusively in the presynaptic membrane, the electric organ of elasmobranchs is a good selection. This tissue is purely cholinergic in nature.

Presynaptic membranes have been isolated from torpedo synaptosomes, and disrupted by osmotic shock at alkaline pH, and freeze-thawing (Morel et al., 1982). Synaptosomes of electric organ of *Torpedo marmorata* are prepared as previously described (Israël et al., 1976; Morel et al., 1977). The synaptosomal fraction is diluted with Torpedo physiological medium consisting of 280 mM NaCl, 3 mM KCl, 1.8 mM $MgCl_2$, 3.4 mM $CaCl_2$, 1.2 mM Na phosphate buffer (pH 6.8), 5.5 mM glucose, 300 mM urea, and 100 mM sucrose. After equilibration with $O_2$, $NaHCO_3$ (4 mM) is added to adjust the pH to 7.0–7.2. Synaptosomes are pelleted (10,000$g$, 20 min) and the pellet is resuspended in the lysis buffer consisting of 0.1 mM EDTA and 5 mM Tris, pH 8.0 (Stadler and Tashiro, 1979), by several passages through a hypodermic needle, and frozen overnight ($-70°C$). After thawing at 4°C, lysis of synaptosomes is completed by addition of lysis buffer. This suspension is spun down (360,000$g$ for 30 min); the resulting pellet is resuspended and layered onto a discontinuous gradient of 1.2, 1.0, 0.8, and 0.6M sucrose in the lysis buffer (0.1 mM EDTA, 5 mM Tris, pH 8.0). The gradients are centrifuged at 40,000 rpm for 240 min in a swinging rotor. The bands at 0.6–0.8 and 0.8–1.0M sucrose interfaces contain the presynaptic membranes, and that at 1.0–1.2M sucrose interface contains the postsynaptic membranes (Morel et al., 1982).

Presynaptic membranes may be purified by immunoadsorption (Miljanich et al., 1982). Synaptosomes from the electric organ of electric fishes are prepared, and antisynaptic vesicle antiserum is added; the synaptosomes are washed to remove unbound antiserum. Anti-rabbit immunoglobulin antibody-coated polyacrylamide beads are added and the beads washed to remove unbound material. The bead-bound material is then lysed, sonicated, and washed free of unbound material (Miljanich et al., 1982).

## 5.2. Junctional Complexes

Junctional complexes comprise the two synaptic membranes, the synaptic cleft with the intersynaptic filaments, and the subsynaptic web. Their isolation has been achieved by means of Triton X-100 treatment of membrane fractions. This detergent leaves the junctional complexes intact while solubilizing or detaching the adjacent plasma membranes. Junctional complexes, synaptic complexes, and synaptic junctions are synonymous terms.

The Triton procedure has been applied to several fractions isolated from whole hemispheres or cerebral cortex of rat brain. The tissue is homogenized in 0.32$M$ sucrose and submitted to subcellular fractionation to isolate a crude mitochondrial fraction, Mit, a crude synaptic membrane $M_1$ fraction after osmotic shock and differential centrifugation, and a purified synaptosomal membrane fraction $M_1$ 1.0 (according to De Robertis et al., 1962a, 1963; Rodríguez de Lores Arnaiz et al., 1967, respectively). The fractions are rehomogenized in 0.32$M$ sucrose containing 0.1–0.5% (v/v) Triton X-100. Five minutes thereafter, they are centrifuged at 100,000$g$ for 60 min, and the sediments contain the isolated junctional complexes. Best results are obtained when Triton X-100 is used at 0.1% (v/v) concentration and the treatment is applied to fraction $M_1$ 1.0 of synaptosomal membranes (Fiszer and De Robertis, 1967; De Robertis et al., 1967).

The integrity of isolated junctional complexes is dependent on the amount of Triton used, the pH, and the presence of $Ca^{2+}$. A critical variable in Triton treatment is the Triton-to-protein ratio in the presence of 3 m$M$ $CaCl_2$. Large increases in the amount of Triton X-100, or decreases in $Ca^{2+}$, result in a decrease of the amount of synaptic complexes isolated. The most favorable condition for preservation of synaptic complexes is relatively low Triton concentration in the presence of $Ca^{2+}$ (Cotman et al., 1971).

### 5.3. Postsynaptic Densities

In synapses of the CNS there is a prominent density, the postsynaptic density (PSD), situated on the inner surface of the postsynaptic membrane. This is a disk-shaped structure (some have a large perforation or hole) (Cohen and Siekevitz, 1978; Peters and Kaisermann-Abramot, 1969). Isolation of postsynaptic densities is based on the solubilization of most of the synaptic membrane with detergents, and the further purification of insoluble detached PSDs in sucrose density gradients. The biochemical and morphological properties of the isolated PSDs will be dependent on the particular isolation conditions employed.

A method for the isolation of PSDs has been based on the treatment of synaptosomal membranes with $n$-lauroylsarcosinate, with which most synaptosomal membranes are solubilized; $n$-lauroylsarcosinate in 10 m$M$ Bicine, pH 7.5, is added to the membrane preparation to achieve a final detergent concentration of 3% (w/v). The suspension is incubated at 4°C for 10 min and layered on a sucrose gradient of 1.0, 1.4, and 2.2$M$ sucrose, all 0.05 m$M$ in $CaCl_2$, pH 7.0. The gradient is centrifuged at 63,600$g$ for 75 min and PSDs are collected at the 1.4 to 2.2$M$ sucrose interface, diluted with 0.1 m$M$ EDTA, and pelleted at 78,500$g$ for 20 min (Cotman et al., 1974).

Cohen et al. (1977) have described two techniques for the purification of PSDs—a short and a long procedure. These procedures are based on fractionation methods previously described in Whittaker's and De Robertis' laboratories, and require the use of Triton X-100. A homogenate of rat or dog cerebral cortex is prepared in 0.32$M$ sucrose containing 1 m$M$ $NaHCO_3$, 1 m$M$ $MgCl_2$, and 0.5 m$M$ $CaCl_2$, and is centrifuged at 1475$g$, 10 min; the pellet is resuspended and centrifuged at 755$g$, 10 min. The resulting supernatants are successively centrifuged at 755$g$, 10 min, and 17,300$g$, 10 min, with one washing. The sediment is resuspended in 0.32$M$ sucrose containing 1 m$M$ $NaHCO_3$ and layered over a sucrose gradient formed with 0.85, 1.0, and 1.2$M$ sucrose. The gradient is run at 100,000$g$ for 120 min. From the band layering at 1.0–1.2$M$ sucrose, synaptosomes are separated by centrifugation for 20 min at 48,200$g$. The pellet is resuspended in 0.5% Triton X-100 in 0.16$M$ sucrose, containing 6 m$M$ Tris-HCl, pH 8.1, stirred for 15 min in the cold, and centrifuged at 48,200$g$. The pellet is layered over a density gradient of 1.0, 1.5, and 2.0$M$ sucrose and centrifuged at 275,000$g$ for 120 min; PSDs are separated at the 1.5–2.0$M$ sucrose interface (Cohen et al., 1977).

The general schedule in the long procedure is similar, except that the Triton X-100 treatment is applied to a synaptosomal membrane fraction instead of a synaptosomal fraction. As determined by electron microscopy, chemistry, and gel electrophoresis of the proteins, short and long procedures give uniform preparations of PSD fractions (Cohen et al., 1977).

The use of the Triton X-100 method for the isolation of PSDs from various brain regions has permitted the separation of two fractions enriched in different types of PSDs. PSDs isolated from cerebrum and midbrain would originate from the asymmetric or Gray type I synapses, and those isolated from cerebellum would originate from symmetric or Gray type II synapses (Carlin et al., 1980).

A procedure based on phase partitioning for the preparation of PSDs has been described (Gurd et al., 1982). A synaptosomal fraction is prepared according to the technique of Cotman and Taylor (1972), employing an incubation step with succinate and iodonitrotetrazolium to diminish mitochondrial contamination (Davis and Bloom, 1973). Synaptosomal membranes from the sucrose gradient are pelleted, resuspended in water, and homogenized in the aqueous two-phase polymer system with a Dounce homogenizer. The two-phase system contains polyethylene glycol 6000 (5% w/w) and Dextrane T 500 (6% w/w) containing the mild neutral detergent 1-$O$-$n$-octyl-β-$D$-glucoside (1%). The phases are separated by brief centrifugation. Purified PSDs that banded at the interface between the phases are collected and pelleted (Gurd et al., 1982).

### 5.4. Characterization of Synaptosomal Membranes

Synaptosomal membranes may be identified by their size (which is similar to that of the nerve ending), the presence of the accompanying postsynaptic density, and their receptor binding properties.

As marker enzymes, $Na^+$, $K^+$-activated ouabain-sensitive ATPase (E.C. 3.6.1.4), acetylcholinesterase (E.C. 3.1.1.7), and 5'-nucleotidase (E.C. 3.1.3.5) may be used. However, these enzymes are not exclusively localized in synaptic membranes, but are also present in other membranous structures.

## 6. Myelin

The myelin sheath consists of a compact multilayered structure that may be separated from other cell structures because of its unique properties. Myelin is characterized by its high content of lipids (70%) and a protein composition simpler than other membranes; furthermore, relatively few enzymes appear associated.

For the isolation of myelin, whole brain of adult rats or fresh mammalian brain white matter may be used. The areas containing an important proportion of fiber tracts, such as white matter, will give the highest yields of isolated myelin.

Myelin fragments from the sucrose homogenates can be isolated by their large size and low density; they will sediment with nuclei and mitochondria during differential centrifugation. The amount of myelin present in the nuclear and mitochondrial fractions will vary depending on the area of the CNS selected for the fractionation procedure. Among the various membranous structures present in a brain homogenate, myelin fragments have the lowest intrinsic density. In density gradient centrifugation, myelin will layer above 0.8$M$ sucrose; nuclei, mitochondria, and microsomes will migrate through 0.8$M$ sucrose.

Myelin may be isolated by centrifugation of a total brain homogenate on a sucrose gradient. The homogenate prepared in

0.32*M* sucrose may be overlayered on a more dense sucrose solution and the myelin allowed to migrate down to the interface, or the homogenate may be used to form the denser layer, with myelin allowed to rise to the interface; zonal centrifugation and Ficoll-sucrose gradients have also been used (*see* Norton, 1974a, for references).

Another schedule employs a differential centrifugation scheme; myelin will be found preferentially in the crude mitochondrial fraction, and variable proportions will be present in the nuclear fraction. The relative proportion of myelin between these fractions may be influenced by the rupture procedure and the degree of myelinization of the animal. From the crude mitochondrial pellet, also containing synaptosomes, myelin may be separated by means of a discontinuous sucrose gradient. These methods are described in this chapter in the section on synaptosomes (De Robertis et al., 1962a; Gray and Whittaker, 1962). Purification steps include water shocks and low-speed centrifugation steps to remove microsomes (*see* Norton, 1974).

The procedure currently used for the isolation of myelin is that described by Norton and Poduslo (1973). The brains of adult rats are homogenized in 0.32*M* sucrose with a Dounce homogenizer by using first a loose pestle and then a tight pestle. The homogenate is layered over 0.85*M* sucrose in ultracentrifuge tubes and centrifuged at 75,000*g* for 30 min. The layers of crude myelin at the interface of the two sucrose solutions are collected with a Pasteur pipet. The myelin layers are suspended in water by homogenization, diluted with water, and centrifuged at 75,000*g* for 15 min. The myelin pellets are again dispersed in water and centrifuged at 12,000*g* for 10 min; the loosely packed pellets are again dispersed in water and centrifuged at 12,000*g* for 10 min. These myelin pellets are suspended in 0.32*M* sucrose, layered over 0.85*M* sucrose, and centrifuged at 75,000*g* for 30 min. The purified myelin is removed from the interface with a Pasteur pipet (Norton and Poduslo, 1973).

In order to prepare myelin from 15- or 20-d-old rats, modifications regarding the relationship between tissue and homogenizing medium and the dilution volume during osmotic shock steps are recommended (Norton and Poduslo, 1973).

To eliminate contaminating microsomes, CsCl or sucrose gradients have been used. By means of these gradients it is possible to separate myelin from smaller but denser material (Norton, 1974). Methods described for brain are suitable for spinal cord and the peripheral nervous system.

## 6.1. Myelin Subfractions

For the preparation of myelin subfractions, myelin is first isolated from fresh brain according to Norton and Poduslo (1973). The purified myelin fraction is homogenized in 0.32M sucrose and centrifuged on a new discontinuous sucrose gradient at 75,000g for 30 min. This new gradient is made up of 0.62M sucrose, layered over 0.70M sucrose, and the resulting two layers and the pellet are collected. Each is carefully removed, homogenized in 0.32M sucrose, and centrifuged on separate discontinuous gradients prepared as above. With this procedure, most of the material in each fraction is collected from the same place as with the preceding gradient, the two other positions being practically free of material. The material that floats on 0.62M sucrose is called light myelin, the material on 0.70M sucrose is called medium myelin, and that which floats on 0.85M sucrose in the original myelin purification is called heavy myelin (Matthieu et al., 1973).

## 6.2. Myelin Characterization

Myelin enzyme markers are 2´,3´-cyclic nucleotide 3-phosphohydrolase (E.C. 3.1.4.16) and cholesterol ester hydrolase. Myelin presence in subcellular fractions may be checked with the electron microscope.

# 7. Golgi Structures

## 7.1. Isolation of Golgi Structures From Brain

The Golgi complex is an important component of the intact neuron. It is composed of stalks of six of more closely apposed flattened disk-shaped cisternae, and clusters of tubules and vesicles; the ends of the disk cisternae are frequently dilated. These complexes, or dictyosomes, are involved in glycosylation, selective proteolysis, sulfation, phosphorylation, and addition of fatty acids (*see* Rothman, 1981). A compartmentation of the dictyosome in two (Rothman, 1981) or three compartments (Griffiths et al., 1983; Quinn et al., 1983) with specific functions and characteristic markers has been postulated.

The isolation of dictyosomes is based on gentle homogenization and differential and gradient centrifugation. The Golgi complex is equilibrated in a band having a density of 1.16. The isolated components have been recovered at the 0.8/1.0M sucrose interface.

The presence of $MgCl_2$ in the homogenizing medium and the gradient solutions is considered by some authors to be an important element to maintain aggregated dictyosome components. However, Morré et al. (1970) consider that gentle homogenization is the critical parameter to avoid the separation of stacked tubules and associated vesicles.

Although the Golgi complex was first described in neurons, most of the techniques for its isolation were elaborated using other tissues, such as epididymus and liver (Schneider and Kuff, 1954; Fleischer, 1974).

In the CNS, Golgi membranes were isolated for the first time in a fraction obtained by differential and density gradient centrifugation (Seijo and Rodríguez de Lores Arnaiz, 1970). Rat cerebral cortex is homogenized in 0.32$M$ sucrose at pH 7.0 (with Tris base). Primary fractions are obtained by modification of a method previously described (De Robertis et al., 1962a). The homogenate is successively centrifuged at 900$g$, 10 min (2 washings); 7500$g$, 20 min (1 washing) (mitochondria, some synaptosomes); 20,000$g$, 30 min (1 washing) (synaptosomes, membranes); and 100,000$g$, 60 min (microsomes). After osmotic shock (De Robertis et al., 1963) of the 20,000$g$ pellet, and recentrifugation at 20,000$g$, 30 min, a pellet containing synaptosomal membranes and mitochondria is obtained. The centrifugation of synaptosomal and synaptosomal membrane pellets on discontinuous gradients as previously described (De Robertis et al., 1962a; Rodríguez de Lores Arnaiz, et al. 1967) permits the separation of fractions at the 0.8–1.0$M$ sucrose interface enriched in thiamine pyrophosphatase. In the same fraction, tubular structures and typical curved auricular-shaped membranes are concentrated (Seijo and Rodríguez de Lores Arnaiz, 1970). Similar curved membranes have been observed in Golgi material isolated from liver and other cell types (Fleischer, 1974; Tennekoon et al., 1983).

An enriched Golgi fraction from total rat brain has been obtained by Deshmukh et al. (1978), with a modification of the methods described for liver (Fleischer, 1974; Morré et al., 1970). The brain is homogenized in sucrose 0.32$M$ in 0.05$M$ Hepes (pH 6.5), containing 5 m$M$ $MgSO_4$, 25 m$M$ KCl, 25 m$M$ EDTA, or 2.5 m$M$ EGTA, 14 m$M$ mercaptoethanol, and 1% dextran (mol wt, 170,000), with 10 gentle up and down strokes of a motor-driven Aldridge homogenizer (800 rpm) with a 0.01 in. clearance. The molarity of the sucrose is adjusted to 0.5$M$. The homogenate is centrifuged at 9000$g$ for 10 min. The resulting pellet is rehomogenized in 0.5$M$ sucrose with a Teflon glass homogenizer of 0.026 in. clearance and centrifuged as before. A part of the combined supernatant fractions is removed, and its sucrose concentration adjusted to 0.32$M$ and centrifuged at 17,000 rpm for 30 min. The rest of the supernatant is used for further fractionation as follows. The molarity of the sucrose is adjusted to 0.72$M$. A 15-mL portion of the supernatant is layered over a discontinuous gradient consisting of 1.17, 1.13, and 0.88$M$ sucrose solutions. On top of the supernatant, a cushion of 0.32$M$ sucrose solution is placed. All sucrose solutions contain all the ingredients of the homogenizing medium in the same concentrations. The gradient is centrifuged for 120 min at 25,000 rpm in a swinging rotor. At the 0.88–1.13$M$ interface, which is within the range of the characteristic Golgi density (1.11–1.15), various types of Golgi structures and the highest specific activity of Golgi enzymes are recovered (Deshmukh et al., 1978).

More recently (Tennekoon et al., 1983), the concentration of thiamine pyrophosphatase and cerebroside sulfotransferase, which is also a marker of Golgi membranes, in a fraction obtained from rat forebrains, has been reported. Forebrains from 5–10 rats 21 d of age were homogenized in 4 vol of 0.32$M$ sucrose. The supernatants obtained after an 800$g$, 10 min centrifugation were centrifuged at 6000$g$, 20 min. The pellet was discarded and the supernatant was put over a cushion of 6 mL sucrose (1.6$M$) and centrifuged for 60 min at 100,000$g$. The fraction rich in Golgi markers was recovered from the interface. The electron microscopic study of this fraction showed the presence of smooth vesicles measuring 0.1–0.5 μm in diameter, with some profiles having barbell- or auricular-shaped structures, which are considered to be characteristic of the Golgi apparatus in isolated fractions (Fleischer, 1974; Morré et al., 1970).

## 7.2. Isolation of Golgi-Like Structures From Sciatic Nerve

In the peripheral nervous system, Golgi-like tubular structures enriched 5–6-fold in thiamine pyrophosphatase were recovered from compressed nerves (Pellegrino de Iraldi and Rodríguez de Lores Arnaiz, 1970) using a procedure similar to that employed by Schneider and Kuff (1954) for the epididymus.

Segments of cat sciatic nerve 5 mm in length near the site of compression were taken. Segments 10 mm long, from the homologous noncompressed contralateral side and distant sides of the crush, were used as controls. The epineurium and the perineurium were removed. The dissected segments were homogenized in a Potter-Elvehjem homogenizer in 0.25$M$ sucrose containing 0.34$M$ NaCl. After centrifuging at 500$g$ for 10 min, 300 μL of the supernatant was placed over the gradients prepared with 750, 750, 1000, and 1000 μL of 1.10, 0.95, 0.636, and 0.335$M$ sucrose, respectively; all the solutions contained 0.34$M$ NaCl. The gradients were centrifuged in the swinging rotor for 60 min at 39,000 rpm. The fractions $S_2$ (0.25–0.636$M$ sucrose) and G (0.636–0.957$M$ sucrose) were obtained. Fraction G obtained from control nerves contained vesicular structures of different sizes (30–180 nm), some microtubules, and a few larger tubules. In the crushed nerve, vesicular structures of about 80–90nm in diameter intermingled with some larger vesicles, and many tubular structures (60 nm wide) stacked in parallel were observed. Fraction S appeared more heterogeneous than G.

In compressed nerves, tubular and vesicular elements are disposed in a manner similar to the morphological organization of the Golgi complex of the perikaryon, and the same characteristics are present in fraction G. Although the Golgi apparatus is a structure classically belonging to the perikaryon, these results suggested its development in compressed nerves (Pellegrino de Iraldi and Rodríguez de Lores Arnaiz, 1970).

### 7.3. Characterization of Golgi Structures

The isolated dictyosomes and their components are characterized by their morphology, staining, and histochemical properties, and by their biochemical constituents, such as thiamine pyrophosphatase, glycosyltransferases, and ceramide sulfotransferase.

## 8. Lysosomes

Lysosomes have been defined as the most important subcellular organelle in the dynamics of recycling cellular constituents (De Duve and Wattiaux, 1966); these organelles contain hydrolases with optimal pHs in the acid range.

Lysosomal fractions can be prepared from brain homogenates by differential and density gradient centrifugation. In differential centrifugation procedures, most of the lysosomes appear in mitochondrial and microsomal fractions. There is also a small number of lysosomes in axons and nerve endings that are more difficult to isolate. Lysosomal marker enzymes are recovered in all the primary subcellular fractions.

Discontinuous and continuous sucrose gradients are effective in purifying lysosomes from a crude mitochondrial fraction. Brain lysosomes are generally denser than mitochondria. Homogenization of brain tissue must be performed in sucrose solution (0.3$M$).

The presence of $Mg^{2+}$ (0.01$M$) in the suspending medium produces clumping of the cerebral lysosomes, thus causing them to sediment with a heavy particulate fraction (Koenig, 1974).

If a 20% sucrose homogenate of brain is centrifuged at 800$g$ for 10 min and the supernatant at 3300$g$ for 10 min, a heavy mitochondrial pellet is separated. The resulting supernatent centrifuged at 16,000$g$ for 20 min leads to the separation of a light mitochondrial pellet. This fraction is enriched 2–3-fold over the original homogenate in activities of lysosomal hydrolases (Koenig, 1974).

Lysosomal hydrolases have been found associated with endoplasmic reticulum and structures of the Golgi apparatus (Koenig, 1974), such enzymes probably representing newly formed molecules on the way to their storage form, the lysosome itself (*see* Koenig, 1974, for references).

A crude mitochondrial (or microsomal) fraction may be used to purify lysosomes. The crude mitochondrial fraction resuspended in 0.32$M$ sucrose is introduced onto the top of a gradient prepared with 1.4, 1.2, 1.0, and 0.8$M$ sucrose. The gradient is centrifuged at 63,500$g$, 120 min, and four layers and a pellet are obtained. This pellet is enriched in lysosomal marker enzymes (Koenig et al., 1964).

Subcellular fractions prepared from rat brain according to De Robertis et al. (1962a) have been used to study the distribution of lysosomal enzymes. It was observed that acid phosphatase and β-glucuronidase present different distribution patterns. This observation supports the hypothesis of the existence of different lysosomal populations in brain tissue (Mordoh, 1965).

Primary fractions containing lysosomes may also be subfractionated by centrifugation on a continuous sucrose density gradient (Sellinger and Hiatt, 1968). The original fraction is resuspended in 0.25$M$ sucrose and placed on top of a continuous 32–52% sucrose gradient and centrifuged for 2.5 h at 25,000 rpm. Lysosomal enzymes show a tendency toward a bimodal distribution. No migration of the main lysosomal peak toward regions of higher density occurs upon prolonged centrifugation (20 h). A bimodal distribution of lysosomal enzymes is observed when a crude lysosomal fraction derived from the cerebral cortex, the thalamus, or the hypothalamus is subjected to this continuous gradient (Sellinger and Hiatt, 1968).

A similar procedure employs a discontinuous gradient of 0.9 and 1.4$M$ sucrose to separate myelin (at 0.9$M$ sucrose), a nerve-ending plus mitochondria fraction (at 1.4$M$ sucrose), and a lysosomal pellet (Koenig, 1974).

Lysosomes become more buoyant after intrathecal administration of Triton WR-1339 into immature rats or less buoyant after the administration of basic dyes. In these cases, lysosomes cosediment with myelin and nerve endings, respectively (*see* Koenig, 1974, for references).

## 8.1. Lysosomal Characterization

Lysosomal particles may be identified in subcellular fractions by their natural fluorescence, cytochemical staining for acid hydrolase activities, and metachromatic staining with acridine orange (Koenig, 1974). As lysosomal markers, acid phosphatase (E.C. 3.1.3.2), β-glucuronidase (E.C. 3.2.1.31), β-glucosidase (E.C.3.2.1.21), β-galactosidase (E.C. 3.2.1.23), N-arylamidase, N-acetyl β-D-glucosaminidase (E.C. 3.2.1.30), arylsulfatase (E.C. 2.7.7.16), acid DNAase, and acid RNAase (E.C. 2.7.7.16) may be assayed. Lysosomal ultrastructure in subcellular fractions may be studied by electron microscopy.

# 9. Nuclei

Nuclear fractions prepared from brain after differential centrifugation contain nuclei or large fragments of nuclei from neurons and glial cells. The crude nuclear fractions also contain endothelial cells and some contaminating myelin fragments, mitochondria, and nerve endings.

Rat brain nuclei have been isolated by means of differential centrifugation and discontinuous gradients of hypertonic sucrose (Løvtrup-Rein and McEwen, 1966). The tissue is homogenized by hand in a Potter homogenizer provided with a loosely fitting Teflon pestle (clearance 0.2 mm) in $0.32M$ sucrose containing 1 mM $MgCl_2$, 1 mM potassium phosphate, and 0.25% (w/v) Triton (ph 6.5). The homogenate is filtered through cheesecloth and centrifuged at $850g$ for 10 min; the pellet is resuspended in the same medium without Triton and centrifuged at $850g$ for 10 min. The sediment is washed once more ($600g$, 8 min). The crude nuclear pellet is resuspended and mixed with a medium composed of $2.39M$ sucrose, 1 mM $MgCl_2$, and 1 mM potassium phosphate (pH 6.5), and centrifuged at $53,500g$ for 45 min. The nuclei are obtained at the bottom. The pellet is resuspended in $2.0M$ sucrose and layered over a discontinuous gradient of 2.8, 2.6, 2.4, and $2.2M$ sucrose. The gradient is centrifuged at $75,000g$ for 30 min and the bands are collected. The neuronal nuclei stay above $2.8M$ sucrose, whereas glial nuclei are heavier than $2.8M$ sucrose (Løvtrup-Rein and McEwen, 1966).

The purity and yield of nuclear fractions is improved by the use of detergents. Detergents enhance the lysis of erythrocytes and the rupture of mitochondria and other cytoplasmic constituents (see Løvtrup-Rein and McEwen, 1966).

In the procedure described by Siakotos (1974), the brain is cut into sections and placed into cold 0.4$M$ sucrose; the sections are chopped, drained free of blood, and homogenized in a loose-fitting (0.01 in. clearance) Potter-Elvehjem homogenizer. The homogenate is filtered and centrifuged at 1000$g$ for 30 min; the supernatant is recentrifuged at 1000$g$ for 30 min. Both supernatants are discarded. The pellets are pooled, resuspended, and centrifuged at 1,000$g$ for 30 min. The pellet is resuspended in a solution containing 0.4$M$ sucrose, 0.1 m$M$ CaCl$_2$, 1 m$M$ ATP, and 0.5 m$M$ 2-$N$-morpholinoethanesulfonic acid buffer (pH 6.1), mixed with 1 vol 2.0$M$ sucrose, and centrifuged at 7000$g$ for 45 min. This pellet is passed through a Sephadex G-25 column; the filtrate is centrifuged at 23,500$g$ for 10 min, resuspended, and recentrifuged at 23,500$g$ for 10 min. This pellet is passed through a Sephadex G-25 column and recentrifuged at 23,500$g$ for 20 min; this pellet is resuspended in 1.2$M$ sucrose and cycled through a linear gradient of 1.2–2.25$M$ sucrose at 40,000$g$ for 20 min. Purified nuclei are collected in the pellets (Siakotos, 1974).

Postmortem changes are evident with brain, the stability of glial nuclei being different from that of neuronal nuclei. In this differential stability, species differences are seen, as revealed by phase contrast studies. In bovine brain nuclear preparations, the predominance of large nuclei with one prominent nucleolus (probably derived from neurons) has been shown. In a homologous nuclear fraction prepared from human brain, smaller binucleolar nuclei (probably derived from glial cells) were observed (*see* Siakotos, 1974).

Another approach for the further study of neuronal nuclei separated from glial cell nuclei is to first perform the cell separation of glia from neurons before starting the subcellular fractionation.

## 10. Mitochondria

Mitochondria are the organelles responsible for oxidative metabolism and ATP formation. The content of mitochondrial protein is high in brain, a fact consistent with the high values observed in oxygen and glucose consumption by brain tissue.

In isolation procedures currently employed for brain tissue, glial mitochondria will not be separated from neuronal mitochondria. Mitochondria are more abundant in neurons than in glial cells. Therefore, if gray matter is used as starting tissue for a fractionating procedure, neuronal mitochondria will predominate with respect to glial mitochondria.

Isolation of brain mitochondria involves first the homogenization of brain tissue in isotonic sucrose solution. This is possible since the outer mitochondrial membrane is not destroyed during homogenization in isotonic media; this membrane also resists homogenization in hyposmotic conditions. By differential centrifugation in isotonic sucrose, most mitochondria will cosediment with synaptosomes and myelin fragments (crude mitochondrial fraction). A significant proportion of mitochondria sediments in the nuclear fraction; such mitochondria may be separated from the nucleous by washing the nuclear pellet.

From the crude mitochondrial pellet, mitochondria may be separated by the use of discontinuous sucrose gradients (De Robertis et al., 1962a; Gray and Whittaker, 1962), continuous Ficoll-sucrose gradients (Tanaka and Abood, 1964), or discontinuous Ficoll-sucrose gradients (Abdel-Latif, 1966; Autilio et al., 1968).

In the technique developed by Clark and Nicklas (1970), brain tissue is homogenized in 0.25$M$ sucrose containing 0.5 m$M$ K$^+$-EDTA and 10 m$M$ Tris (pH 7.4) in a Dounce homogenizer. The homogenate is centrifuged at 2000$g$, 3 min, and the resulting supernatant at 12,000$g$, 8 min. This pellet is resuspended in a 3% Ficoll medium (3% Ficoll, 0.12$M$ mannitol, 0.03$M$ sucrose, 25 μ$M$ K$^+$-EDTA, pH 7.4), and layered onto a 6% Ficoll medium (6% Ficoll, 0.24$M$ mannitol, 0.06$M$ sucrose, 50 μ$M$ K$^+$-EDTA, pH 7.4). This gradient is centrifuged at 11,500$g$, 30 min; mitochondria are recovered in the pellet (Clark and Nicklas, 1970).

In the method used by Bernard and Cockrell (1979), brain cortex and cerebellum are homogenized in medium A (0.3$M$ mannitol, 5 m$M$ potassium glycylglycine, and 0.1 m$M$ potassium EDTA, pH 7.4). Homogenization is performed with a loosely fitting Teflon glass homogenizer (two strokes) and thereafter with a tightly fitting Teflon pestle (two strokes) with fresh medium A. The homogenate is centrifuged successively at 1500$g$ for 10 min and at 10,000$g$ for 10 min. The supernatant and white layer of the pellet are decanted and the pellet is resuspended in the salt medium (0.2 m$M$ mannitol, 8 m$M$ potassium glycylglycine, 40 m$M$ KCl, 40 m$M$ NaCl, and 70 μ$M$ potassium EDTA, pH 7.4). This suspension is homogenized by hand with a loosely fitting pestle and centrifuged at 1500$g$ for 2 min; the supernatant is recentrifuged (1500$g$, 2 min). The resulting supernatant is centrifuged at 8000$g$ for 10 min to sediment mitochondria, and washed once

(8000g, 10 min). Light layer mitochondria are carefully decanted and the pellet is used (Bernard and Cockrell, 1979).

### 10.1. Synaptosomal Mitochondria

In order to prepare intraterminal (or synaptosomal) mitochondria, it is necessary to separate synaptosomes from free mitochondria by density gradient centrifugation. Synaptosomes are then submitted to hyposmotic treatment and their components may be separated in a second density gradient.

A procedure for the isolation of synaptosomal mitochondria has been described by Nicholls (1978). A synaptosomal fraction is isolated by a modification of the technique of Cotman and Matthews (1971); a guinea pig cerebral cortex homogenate is prepared in a medium containing 0.32M sucrose, 5 mM Tes 2-[(2-hydroxy-1,1-bis(hydroxymethyl)ethyl) amino]-1-propanesulfonic acid, and 0.5 mM EDTA, pH 7.4, with 10 strokes of a Teflon pestle rotating at 500 rpm (clearance 0.25 mm). The homogenate is centrifuged at 900g, 5 min (washed once), and the combined supernatants are centrifuged at 17,000g, 10 min. The pellet is resuspended and layered on top of a discontinuous Ficoll gradient (4–9 and 12% Ficoll); all solutions are in the isolation medium. The gradient is centrifuged at 75,000g, 30 min. Synaptosomes appear in a double band with the 9% Ficoll; this fraction is resuspended and a suspension of digitonin (obtained by sonication; 0.15 mg/mg synaptosomal protein) in isolation medium is added. The mixture is vortex-mixed vigorously, left at 0°C for 5 min, and sonicated (10 s at 6 μm amplitude). The excess of digitonin micelles are sedimented (5 min at 900g) and the supernatant is layered on 12% (w/v) Ficoll in isolation medium and centrifuged at 75,000g, 60 min. The pellet is washed once (800g, 10 min); it contains intraterminal mitochondria (Nicholls, 1978).

## 11. Microtubules

Microtubules, first described in neurons, are found in all eukaryotic cells either free in the cytoplasm or as part of centrioles, cilia, and flagella. The tubules are 25 nm in diameter, several micrometers long, with a wall 6 nm thick. The heterodimer tubulin, their main component, is present in large amounts in the nervous tissue, neurons, and glial cells, and is not confined to microtubules. Its presence has been reported in plasma membranes and synaptic vesicles (Zisapel et al., 1980). Furthermore it represents a significant percentage of soluble proteins in brain. Thus the isola-

tion of microtubules in subcellular fractions cannot be characterized by the presence of tubulin, but requires the visualization of
microtubules with the electron microscope.

## 11.1. Isolation and Reconstitution of Microtubules

Free cytoplasmic microtubules are labile structures. Their stability
is highly dependent on temperature, pH, and the presence of
monovalent and divalent cations, especially $Ca^{2+}$, and of
nucleotides, especially GTP (Olmstead and Borizy, 1973;
Kirkpatrick et al., 1970). For this reason, an essential requirement
for their purification is to maintain the labile structure long
enough to purify the samples by differential and gradient
centrifugation. EGTA can be used to avoid the effect of calcium on
neurotubule assembly. Organic solvents, particularly hexylene
glycol at slightly acidic pH, stabilize microtubules at low temperature in vitro (Kane, 1965). This solvent has been used by
Kirkpatrick et al. (1970) to isolate microtubules from brain. These
authors use a 16.7% homogenate in 1M hexylene glycol in 20 mM
potassium phosphate, pH 6.4, at 1°C. This homogenate is centrifuged at 48,000g for 30 min and the supernatant is overlayered on
a discontinuous sucrose gradient made by two steps of sucrose
dissolved in hexylene glycol at the concentrations of 1.16 and 1.19
g/mL. The gradient is centrifuged in a swinging rotor at 39,000
rpm for 1 or 4 h. The most highly purified microtubules are obtained from the interface 1.16/1.19 after 1 h of centrifugation. The
purified fraction shows a major band that migrates like purified
tubulin in the SDS gel electrophoresis system (Kirkpatrick et al.,
1970). Another approach is based on the purification of microtubules through repeated cycles of assembly and disassembly of the
extracted cytosolic tubulin (Borizy, et al., 1975; Vallee et al., 1981).

In the last years, taxol, an antimitotic agent derived from the
western yew plant, has been employed in the isolation of
microtubules and microtubule-associated proteins (Vallee, 1982).
This agent dramatically stimulates the polymerization of cytoplasmic microtubules in vitro and in vivo. It promotes microtubule assembly at close 1,1 stoichiometry to tubulin. It is effective in the
presence of elevated concentrations of calcium ion and at low
temperatures, conditions that are normally unfavorable for
microtubule assembly. Taxol in part mimics the effect of
microtubule-associated proteins that promote the assembly of
tubulin in vitro and possibly in vivo (Vallee et al., 1981) in a stoi-
chiometric, rather than a catalytic, manner; it reconstitutes
microtubules directly from brain extracts without previous
purification by repeated cycles of assembly–disassembly (Vallee,

1982). Preparation of extracts is performed at 0–2°C. The buffer used is 0.1$M$ piperazine $N,N$-bis(2-ethanesulfonic acid), pH 6.6, containing 10 m$M$ EGTA, 10 mM MgSO$_4$, and 1 m$M$ GTP. One to three grams of calf cerebral cortex (gray matter) or corpus callosum (white matter) is homogenized in 1.5 vol of microtubule assembly buffer minus GTP in a Potter-Elvehjem Teflon glass homogenizer at 2000 rpm. The homogenate is centrifuged at 30,000 rpm for 15 min and the pellet is discarded. The supernatant is then centrifuged at 180,000$g$ for 90 min and the pellet again discarded. Taxol is added to 20 $\mu M$ and GTP to 1 m$M$. The solution is warmed at 37°C for 10–15 min and the microtubules that are formed are centrifuged at the same temperature for 25 min through a cushion of 5% sucrose in microtubule assembly buffer-containing 20 $\mu M$ taxol in either a swinging bucket or a fixed-angle rotor. At this stage the pellet contains microtubules as pure as those obtained by several cycles of assembly–disassembly purification without taxol (Vallee, 1982).

## 12. Intermediate Filaments

Neurofilaments and glial filaments are intermediate (8–9 nm) filaments present in neurons and glial cells, respectively, where they represent an important component of the cytoskeleton (Schlaepfer and Zimmerman, 1981). Both kinds of filaments are difficult to distinguish after the isolation procedures since they are very similar in diameter (Wuerker, 1970); however, they are immunologically and biochemically distinct (Liem et al., 1978). The neurofilaments are composed of a triplet of polypeptides with mol wts of 200,000, 150,000, and 70,000; glial filaments are composed of a single subunit with a mol wt of 50,000 daltons (Shelanski and Liem, 1979; Liem et al., 1978; Vallee, 1982).

### 12.1. Isolation and Assembly

Intermediate filaments have been isolated from nervous tissue using the axonal flotation technique (Shelanski et al., 1971; De Vries et al., 1972; Yen et al. 1976). This technique uses the presence of myelin around the axons to float them away from capillaries, cells, and other components of the nervous tissue. According to Yen et al. (1976), white matter is homogenized in 0.85$M$ sucrose containing 0.03$M$ KCl in 0.02$M$ phosphate buffer, pH 6.5, in a Dounce homogenizer. This homogenate is centrifuged at 10,000 rpm for 15 min. The myelinated axons floating at the top are collected and homogenized in the homogenizing medium.

The flotation is repeated several times until the pellet is free of red or pink color. Demyelinated axons are homogenized and purified in a discontinuous sucrose density gradient (Yen et al., 1976).

Neurofilaments can be solubilized in low-strength ionic buffer, leaving intact the glial filaments (Liem et al., 1978; Schlaepfer, 1978).

To avoid the disruption of neurofilaments by long exposure to hypotonic solutions, two methods have been reported by Liem et al. (1978). In the first, brain white matter is suspended in 0.85$M$ sucrose in solution A (10 m$M$ phosphate buffer, pH 6.5, containing 0.1$M$ NaCl and 5 m$M$ EDTA). The material is homogenized in a Dounce homogenizer and the axons are floated as described by Yen et al. (1976). Demyelination is done by brief exposure to low-ionic strength solution (0.1$M$ phosphate buffer, pH 6.5) for 1 h; then the pH of the solution is raised to pH 8.8.

The material is mixed with an equal volume of 0.85$M$ sucrose in 0.01$M$ Tris, pH 8.8, and centrifuged at 10,000 rpm for 15 min; the pellet is collected and homogenized in 0.01$M$ Tris buffer, pH 8.8. This homogenate is layered on 0.85$M$ sucrose and centrifuged at 270,000 rpm for 30 min. The pellet is homogenized in solution A and applied to a sucrose gradient of 1.0, 1.5, and 2.0$M$ sucrose; all these sucrose solutions are made in solution A. The gradient is centrifuged at 270,000 rpm for 60 min. Filaments are recovered at the 1.5–2.0$M$ sucrose interface.

In the second method, the exposure to low ionic strength is avoided. After flotation in 0.85$M$ sucrose in solution A, the myelinated axons are homogenized in 1% Triton X-100 in solution A—a treatment that does not solubilize neurofilaments. The suspension is layered in 0.85$M$ sucrose in solution A and centrifuged at 270,000 rpm for 30 min. The material floating at the top is rehomogenized in Triton solution and layered again in 0.85$M$ sucrose in solution A to maximize the yield. The pellet is resuspended in 1% Triton and applied to a discontinuous density gradient of 1.0, 1.5, and 2.0$M$ sucrose solutions, all in solution A, and centrifuged at 270,000$g$ for 60 min. The interface between 1.5 and 2.0$M$ sucrose portions contains the filament-rich fraction that is a mixture of neural and glial filaments (Liem et al., 1978).

To avoid contamination of neurofilaments with glial filaments, their isolation has been performed using intradural spinal nerve roots as starting material, since astroglia are absent from these nerve roots (Liem et al., 1978). Intermediate filaments of definite glial origin have been isolated from human CNS astrocytes (Goldman et al., 1978) from the gliosed white matter of cases of adrenoleuckodystrophy.

The simultaneous separation and purification of neurofilaments and glial filaments and their subsequent reassembly has been performed (Liem, 1982). Brain white matter is homogenized in 0.85$M$ sucrose in a 10 m$M$ phosphate buffer, pH 6.8, containing 0.1$M$ NaCl (solution A). The myelinated axons are floated as described by Yen et al. (1976) and the flotation is repeated three times. Myelin is removed by homogenization in 1% Triton X-100 as described by Liem et al. (1978). The solution is stirred overnight at 4°C and subsequently centrifuged through a layer of 0.85$M$ sucrose in solution A at 270,000$g$ for 30 min. The overnight treatment with Triton X-100 eliminates the need for repeated homogenization of the floated pad to maximize the yield. The pellet is resuspended in solution A and centrifuged at 10,000$g$ and the pellet is dissolved in 10 m$M$ phosphate buffer, pH 7.4, made up in 8$M$ urea containing 1% mercaptoethanol. The filament suspension is homogenized with a Dounce homogenizer and centrifuged at 10,000$g$ for 30 min to clarify the solution. Some 85–90% of the protein is soluble in 8$M$ urea. The filament solution is layered on a hydroxylapatite column in which the glial proteins and the neurofilament triplets are completely separated. Various other impurities in the preparation, including tubulin, can be removed by adjusting the elution buffers.

The proteins reassemble into intermediate filaments with the removal of the denaturant, and reassembly is used as the final step in the purification of filament proteins. The reassembly is dependent on ionic strength and pH. This dependence is greater for neurofilaments than for the glial filaments (Liem, 1982).

## 13. Final Statements

Given the high complexity of nerve cells, it is useful to have simplified systems to study neuronal functioning. Subcellular fractionation techniques provide such simplified systems by separating parts of the neurons and organelles, or parts of them, from the rest of the cellular constituents. The preparation of subcellular fractions is the first step in a complex process that includes biochemical and morphological studies. A finer study of the intrinsic properties of the isolated subcellular structures may then be afforded.

In this chapter, subcellular fractionation techniques developed for the study of the nervous system have been reviewed. Emphasis was given to techniques related to the separation of structural components directly involved in neurotransmission. Procedures requiring the purification of one or more molecular constituents, and the further assembly of the organelle they come from, were also reviewed.

When an organelle is in the intact cell, it is in a dynamic equilibrium with respect to other organelles and/or the other constituents of the cytoplasmic matrix. After the subcellular fractionation of the tissue, those interrelationships will disappear. This situation may provide experimental models that facilitate the knowledge of functional relationships between organelles, or between an organelle and some special molecular component.

## Acknowledgment

The financial support of the Consejo Nacional de Investigaciones Científicas y Técnicas, Argentina, is gratefully acknowledged.

## References

Abdel-Latif A. A. (1966) A simple method for isolation of nerve-ending particles from rat brain. *Biochim. Biophys. Acta* **121,** 403–406.

Autilio L. A., Appel S. H., Pettis P., and Gambetti P. L. (1968) Biochemical studies of synapses in vitro. I. Protein synthesis. *Biochemistry* **7,** 2615–2622.

Balácz R., Hajós F., Johnson A., Tapia R., and Wilkin G. (1974) Subcellular fractionation of rat cerebellum: An electron microscopic and biochemical investigation. III. Isolation of large fragments of cerebellar glomeruli. *Brain Res.* **70,** 285–299.

Belmar J., De Potter W. P., and De Schaepdryver A. F. (1974) Subcellular distribution of noradrenaline and dopamine-β-hydroxylase in the hypothalamus of the rat. Evidence for the presence of two populations of noradrenaline storage particles. *J. Neurochem.* **23,** 607–609.

Bernard P. A. and Cockrell R. S. (1979) The respiration of brain mitochondria and its regulation by monovalent cation transport. *Biochim. Biophys. Acta* **548,** 173–186.

Bisby M. A. and Fillenz M. (1971) The storage of endogenous noradrenaline in sympathetic nerve terminals. *J. Physiol.* (Lond.) **215,** 163–179.

Booth R. F. G. and Clark J. B. (1978) A rapid method for the preparation of relatively pure metabolically competent synaptosomes from brain. *Biochem. J.* **176,** 365–370.

Borizy G. G., Marcum J. M., Olmsted J. B., Murphy D. B., and Johnson K. N. (1975) Purification of tubulin associated with high molecular weight proteins from porcine brain and characterization of microtubule assembly in vitro. *Ann. N. Y. Acad. Sci.* **253,** 107–132.

Bradford H. F. (1975) Isolated Nerve Terminals as an In Vitro Preparation for the Study of Dynamic Aspects of Transmitters Metabolism and Release, in *Handbook of Psychopharmacology*, Vol. 1 (Iversen L. L., Iversen S. D., and Snyder S. H., eds.) pp 191–252, Plenum, New York.

Bretz U., Baggiolini M., Hauser R., and Hodel C. (1974) Resolution of three distinct populations of nerve endings from rat brain homogenates by zonal isopycnic centrifugation. *J. Cell Biol.* **61,** 466–480.

Burnstock G. and Iwayama T. (1971) Fine Identification of Autonomic Nerves and Their Relation to Smooth Muscle, in *Histochemistry of Nervous Transmission* (Eränkö O., ed.) *Prog. Brain Res.* **34,** pp. 389–404, Elsevier, Amsterdam.

Carlin R. K., Grab D. J., Cohen R. S., and Siekevitz P. (1980) Isolation and characterization of postsynaptic densities from various brain regions: enrichment of different types of postsynaptic densities. *J. Cell Biol.* **86,** 831–843.

Carlsson S. S., Wagner J. A., and Kelly R. B. (1978) Purification of synaptic vesicles from Elasmobranch electric organ and the use of biophysical criteria to demonstrate purity. *Biochem.* **17,** 1188–1199.

Chubb I. W., De Potter W. P., and De Schaepdryver A. F. (1970) Evidence for two types of noradrenergic storage particles in dog spleen. *Nature* (Lond.) **228,** 1203–1204.

Clark J. B. and Nicklas W. J. (1970) The metabolism of rat brain mitochondria. *J. Biol. Chem.* **245,** 4724–4731.

Cohen R. S., Blomberg F., Berzins K., and Siekevitz P. (1977) The structure of postsynaptic densities isolated from dog cerebral cortex. *J. Cell Biol.* **74,** 181–203.

Cohen R. S. and Siekevitz P. (1978) Form of the postsynaptic density. A serial study. *J. Cell Biol.* **78,** 36–46.

Cotman C. W. (1974) Isolation of Synaptosomal and Synaptic Plasma Membrane Fractions, in *Methods in Enzymology,* Vol. 31, Part A (Fleisher S. and Packer L., eds.) pp. 445–452 Academic, New York.

Cotman C. W., Levy W., Banker G., and Taylor D. (1971) An ultrastructural and chemical analysis of the effect of Triton X-100 on synaptic plasma membranes. *Biochim. Biophys. Acta* **249,** 406–418.

Cotman C. W. and Matthews D. A. (1971) Synaptic plasma membranes from rat brain synaptosomes. Isolation and partial characterization. *Biochim. Biophys. Acta* **249,** 380–394.

Cotman C. W. and Taylor D. (1972) Isolation and structural studies on synaptic complexes from rat brain. *J. Cell Biol.* **55,** 696–711.

Cotman C. W., Banker G., Churchill L., and Taylor D. (1974) Isolation of postsynaptic densities from rat brain. *J. Cell Biol.* **63,** 441–455.

Cuello A. C., Kanazawa T. M., and Iversen L. L. (1977) Substance P: Localization in synaptic vesicles in rat central nervous system. *J. Neurochem.* **29,** 747–751.

Davis G. A. and Bloom F. (1973) Isolation of synaptic junctional complexes from rat brain. *Brain Res.* **62,** 135–153.

De Duve C. and Wattiaux R. (1966) Functions of lysosomes. *Ann. Rev. Physiol.* **28,** 435–492.

De Lorenzo R. J. and Freedman S. D. (1978) Calcium dependent neurotransmitter release and protein phosphorylation in synaptic vesicles. *Biochim. Biophys. Res. Commun.* **80,** 183–192.

De Potter W. P., Smith A. D., and De Schaepdryver A. F. (1970) Subcellular fractions of splenic nerve: ATP, chromogranin A, and dopamine-β-hydroxylase in noradrenergic vesicles. *Tissue Cell* **2,** 529–546.

De Robertis E., Azcurra J. M., and Fiszer S. (1967) Ultrastructure and cholinergic binding capacity of junctional complexes isolated from rat brain. *Brain Res.* **5,** 45–56.

De Robertis E., Pellegrino de Iraldi A., Rodríguez de Lores Arnaiz G., and Gómez C. J. (1961) On the isolation of nerve endings and synaptic vesicles. *J. Biophys. Biochem. Cytol.* **9,** 229–235.

De Robertis E., Pellegrino de Iraldi A., Rodríguez de Lores Arnaiz G., and Salganicoff L. (1962a) Cholinergic and noncholinergic nerve endings in rat brain. I. Isolation and subcellular distribution of acetylcholine and acetylcholinesterase. *J. Neurochem.* **9,** 23–35.

De Robertis E., Rodríguez de Lores Arnaiz G., and Pellegrino de Iraldi A. (1962b) Isolation of synaptic vesicles from nerve endings of the rat brain. *Nature* (Lond.) **194,** 794–795.

De Robertis E., Pellegrino de Iraldi A., Rodríguez de Lores Arnaiz G., and Zieher L. M. (1965) Synaptic vesicles from the hypothalamus. Isolation and norepinephrine content. *Life Sci.* **4,** 193–201.

De Robertis E. and Rodríguez de Lores Arnaiz G. (1969) Structural Components of the Synaptic Region, in *Handbook of Neurochemistry* Vol. 2 (Lajtha A., ed.), pp. 365–392, Plenum, New York.

De Robertis E., Rodríguez de Lores Arnaiz G., Salganicoff L., Pellegrino de Iraldi A., and Zieher L. M. (1963) Isolation of synaptic vesicles and structural organization of the acetylcholine system within brain nerve endings. *J. Neurochem.* **10,** 225–235.

Deshmukh D. S., Bear W. D., and Soifer D. (1978) Isolation and characterization of an enriched Golgi fraction from rat brain. *Biochim. Biophys. Acta* **542,** 284–295.

De Vries G. H., Norton W. T., and Rainic C. S. (1972) Axon isolation from mammalian nervous system. *Science* **175,** 1370–1372.

Fiszer S. and De Robertis E. (1967) Action of Triton X-100 on ultrastructure and membrane-bound enzymes of isolated nerve endings from rat brain. *Brain Res.* **5,** 31–44.

Fleischer B. (1974) Isolation and Characterization of Golgi Apparatus and Membranes from Rat Liver, in *Methods in Enzymology,* Vol. 31, Part A (Fleischer S. and Packer L., eds.) pp. 180–191, Academic, New York.

Floor R., Grad O., and Leeman S. E. (1982) Synaptic vesicles containing substance P purified by chromatography on controlled pore glass. *Neurosci.* **7,** 1655–1662.

Fried G. (1982) Neuropeptide Storage Vesicles, in *Neurotransmitter Vesicles* (Klein R. L., Lagercrantz H., and Zimmermann H., eds.) pp. 361–374, Academic, New York.

Fried G., Lagercrantz H.,and Hökfelt T. (1978) Improved isolation of small noradrenergic vesicles from rat seminal ducts following castration. A density gradient and morphological study. *Neuroscience* **3**, 1271–1291.

Fried G., Lundberg J. M., Hökfelt T., Lagercrantz H., Fahrenkrug J., Lundgren G., Holmstedt B., Brodin E., Efendic S., and Terenius L. (1981) Do Peptides Coexist with Classical Transmitters in the Same Neuronal Storage Vesicles?, in *Chemical Neurotransmission: 75 Years* (Stjärne L., Hedqvist P., Lagercrantz H., and Wennmalm Å., eds.) pp. 105–111, Academic, New York.

Goldman J. E., Schaumburg H. H., and Norton W. T. (1978) Isolation and characterization of glial filaments from human brain. *J. Cell Biol.* **78**, 426–440.

Gray E. G. and Whittaker V. P. (1962) The isolation of nerve endings from brain; an electron microscopic study of the cell fragments derived by homogenization and centrifugation. *J. Anat.* (Lond.), **96**, 79–88.

Griffiths G., Quinn P., and Warren G. (1983) Dissection of the Golgi complex. I. Monensin inhibits the transport of viral membrane proteins from medial to trans Golgi cisternae in baby hamster kidney cells infected with Semliki Forest virus. *J. Cell Biol.* **96**, 835–849.

Gurd J. W., Gordon-Weeks P., and Howard Evans W. (1982) Biochemical and morphological comparison of postsynaptic densities prepared from rat, hamster, and monkey brains by phase partitioning. *J. Neurochem.* **39**, 1117–1124.

Hardy J. A., Dodd P. R., Oakley A. E., Perry R. H., Edwardson J. A., and Kidd A. M. (1983) Metabolically active synaptosomes can be prepared from frozen rat and human brain. *J. Neurochem.* **40**, 608–614.

Hökfelt T. (1971) Ultrastructural Localization of Intraneuronal Monoamines: Some Aspects on Methodology, in *Histochemistry of Nervous Transmission* (Eränkö O., ed.) *Progr. Brain Res.* **34**, pp. 213–222, Elsevier, Amsterdam.

Hökfelt T., Johansson O., Ljungdahl Å., Lundberg J. M., and Schultzberg M. (1980) Peptidergic neurons: Review Article *Nature* (Lond.) **284**, 515–521.

Israël M., Dunant Y., and Manaranche R., (1979) The present status of the vesicular hypothesis. *Progr. Neurobiol.*, **13**, 237–275.

Israël M., Manaranche R., Mastour-Frachon P., and Morel N. (1976) Isolation of pure cholinergic nerve endings from the electric organ of *Torpedo marmorata. Biochem J.* **160**, 113–115.

Iversen L. L., and Snyder S. H. (1968) Synaptosomes: Different populations storing catecholamines and gamma-aminobutyric acid in homogenates of rat brain. *Nature* (Lond.) **220**, 796–798.

Jones D. H. and Matus A. I. (1974) Isolation of synaptic plasma membranes from brain by combined flotation-sedimentation density gradient centrifugation. *Biochim. Biophys. Acta* **356**, 276–287.

Kadota K. and Kadota T. (1973) Isolation of coated vesicles, plain synaptic vesicles, and flocculent material from a crude synaptosome fraction of guinea pig whole brain. *J. Cell Biol.* **58,** 135–151.

Kamiya H., Kadota K., and Kadota T. (1974) Distribution of choline and acetylcholine in coated vesicles and plain synaptic vesicles. *Brain Res.* **76,** 367–370.

Kane R. E. (1965) The mitotic apparatus. Physical–chemical factors controlling stability. *J. Cell Biol.* **25,** 137–144.

Kataoka K. and De Robertis E. (1967) Histamine in isolated nerve endings and synaptic vesicles of rat brain cortex. *J. Pharmacol. Exp. Therap.* **156, 114–125.**

Kirkpatrick J. B., Haynes L., Thomas V. L., and Howley P. M. (1970) Purification of intact microtubules from brain. *J. Cell Biol.* **47,** 384–394.

Koenig H. (1974) The Isolation of Lysosomes from Brain, in *Methods in Enzymology* Vol. 31, Part A (Fleisher S. and Packer L., eds.) pp. 457–477, Academic, New York.

Koenig H., Gaines D., McDonald T., Gray R., and Scott J. (1964) Studies of brain lysosomes. I. Subcellular distribution of five acid hydrolases, succinate dehydrogenase, and gangliosides in rat brain. *J. Neurochem.* **11,** 729–743.

Kornguth S. E., Flangas A. L., Siegel F. L., Geison R. L., O'Brien J. F., Lamar C. Jr., and Scott G. (1971) Chemical and metabolic characteristics of synaptic complexes from brain isolated by zonal centrifugation in a cesium chloride gradient. *J. Biol. Chem.* **246,** 1177–1184.

Kornguth S. E., Anderson J. W, and Scott G. (1969) Isolation of synaptic complexes in a cesium chloride gradient. Electron microscopic and immunohistochemical studies. *J. Neurochem.* **16,** 1017–1022.

Kuhar M. J., Shaskan E. G., and Snyder S. H. (1971) The subcellular distribution of endogenous and exogenous serotonin in brain tissue: comparison of synaptosomes storing serotonin, norepinephrine, and $\gamma$-aminobutyric acid. *J. Neurochem.* **18,** 333–343.

Kurokawa M., Sakamoto T. and Kato M. (1965) A rapid isolation of nerve-ending particles from brain. *Biochim. Biophys. Acta* **94,** 307–309.

Laduron P. M., Janssen P. F. M., and Ilien B. (1983) Analytical subcellular fractionation of rat cortex: resolution of serotonergic nerve endings and receptors. *J. Neurochem.* **41,** 84–93.

Lagercrantz H. and Klein R. L. (1982) Isolation of Noradrenergic Vesicles, in *Neurotransmitter Vesicles* (Klein R. L., Lagercrantz H., and Zimmermann H., eds.) pp. 89–118, Academic, New York.

Lapetina E. G., Soto E. F., and De Robertis E. (1967) Gangliosides and acetylcholinesterase in isolated membranes of the rat brain cortex. *Biochim. Biophys. Acta* **135,** 33–43.

Li Z.-Y., and Bon C. (1983) Presence of a membrane-bound acetylcholinesterase form in a preparation of nerve endings from *Torpedo marmorata* electric organ. *J. Neurochem.* **40,** 338–349.

Liem R. (1982) Simultaneous separation and purification of

neurofilament and glial filament proteins from brain. *J. Neurochem.* **38**, 142–150.

Liem R. K. H., Yen S. H., Salomon G. D., and Shelanski M. L. (1978) Intermediate filaments in nervous tissue. *J. Cell Biol.* **79**, 637–645.

Løvtrup-Rein, H. and McEwen, B. S. (1966) Isolation and fractionation of rat brain nuclei. *J. Cell Biol.* **30**, 405–415.

Matthieu J. -M., Quarles R. H., Brady R. O., and Webster H. de F. (1973) Variation of proteins, enzyme markers, and gangliosides in myelin subfractions. *Biochim. Biophys. Acta* **329**, 305–317.

Maynert E. W., Levi R., and De Lorenzo A. J. D. (1964) The presence of norepinephrine and 5-hydroxytryptamine in vesicles from disrupted nerve-ending particles. *J. Pharmacol. Exp. Therap.* **144**, 385–392.

McCaman R. E., Rodríguez de Lores Arnaiz G., and De Robertis E. (1965) Species differences in subcellular distribution of choline acetylase on the CNS. A study of choline acetylase, acetylcholinesterase, 5-hydroxytryptophan decarboxylase, and monoamine oxidase in four species. *J. Neurochem.* **12**, 927–935.

Meflah K., Harb J., Duflos Y., and Bernard S. (1984) 5'-Nucleotidase from bovine caudate nucleus synaptic plasma membranes: specificity for substrates and cations; study of the carbohydrate moiety by glycosidases. *J. Neurochem.* **42**, 1107–1115.

Miljanich G. P., Brasier A. R., and Kelly R. B. (1982) Partial purification of presynaptic plasma membranes by immunoadsorption. *J. Cell Biol.* **94**, 88–96.

Mordoh J. (1965) Subcellular distribution of acid phosphatase and β-glucuronidase activities in rat brain. *J. Neurochem.* **12**, 505–514.

Morel N., Israël M., Manaranche R., and Mastour-Frachon P. (1977) Isolation of pure cholinergic nerve endings from *Torpedo* electric organ. Evaluation of their metabolic properties. *J. Cell Biol.* **75**, 43–55.

Morel N., Manaranche R., Israël M., and Gulik-Krzywicki T. (1982) Isolation of a presynaptic plasma membrane fraction from *Torpedo* cholinergic synaptosomes: evidence for a specific protein. *J. Cell Biol.* **93**, 349–356.

Morré D. J., Hamilton R. L., Mollenhauer H. H., Mahley R. W., Cunningham W. O., Cheetham R. D., and Lequire V.S. (1970) Isolation of a Golgi apparatus rich fraction from rat liver. I. Method and morphology. *J. Cell Biol.* **44**, 484–491.

Nagy A., Várady G., Joó F., Rakoneczay Z., and Pilc A. (1977) Separation of acetylcholine- and catecholamine-containing synaptic vesicles from brain cortex. *J. Neurochem.* **29**, 449–459.

Nicholls D. G. (1978) Calcium transport and proton electrochemical potential gradient in mitochondria from guinea pig cerebral cortex and rat heart. *Biochem. J.* **170**, 511–522.

Norton W. T. (1974) Isolation of Myelin from Nerve Tissue, in *Methods in Enzymology* Vol. 31, Part A (Fleisher S. and Packer L., eds.) pp. 435–444, Academic, New York.

Norton W. T. and Poduslo S. E. (1973) Myelination in rat brain: Method of myelin isolation. *J. Neurochem.* **21**, 749–757.

Olmsted J. B. and Borizy G. G. (1973) Microtubules. *Ann. Rev. Biochem.* **42,** 507–540.

Pellegrino de Iraldi A. (1983) Compartmentation of Monoaminergic Synaptic Vesicles. Physiological Implications, in *Neural Transmission, Learning, and Memory* (Caputto R. and Ajmone Marsan C., eds.) pp. 65–79, Raven, New York.

Pellegrino de Iraldi A. and Rodríguez de Lores Arnaiz G. (1970) Thiamine pyrophosphatase activity in a fraction rich in Golgi-like structures from crushed sciatic nerve of the cat. *J. Neurochem.* **17,** 1601–1606.

Pellegrino de Iraldi A., Zieher L. M, and Jaim Etcheverry G. (1968) Neuronal Compartmentation of 5-Hydroxytryptamine Stores, in *Advances in Pharmacology* Vol. 6, Part A. (Costa E. and Sandler M., eds.) pp. 257–270. Academic, New York.

Peters A. and Kaisermann-Abramof I. R. (1969) The small pyramidal neuron of the rat cerebral cortex. The synapses upon dendritic spines. *Z. Zellforsch. mikrosk. Anat.* **100,** 487–506.

Quinn P., Griffiths G., and Warren G. (1983) Dissection of the Golgi complex. II. Density separation of specific functions in virally infected cells treated with monensin. *J. Cell Biol.* **96,** 851–856.

Rodríguez de Lores Arnaiz G., Alberici M., and De Robertis E. (1967) Ultrastructural and enzymic studies of cholinergic and noncholinergic synaptic membranes isolated from brain cortex. *J. Neurochem.* **14,** 215–225.

Rodríguez de Lores Arnaiz G. and De Robertis E. (1972) Properties of the Isolated Nerve Endings, in *Current Topics in Membranes and Transport* (Bronner F. and Kleinzeller A., eds.) pp. 237–272, Academic, New York.

Rothman J. E. (1981) The Golgi apparatus: two organelles in tandem. *Science* **213,** 1212–1219.

Schlaepfer W. W. (1978) Observations on the disassembly of isolated mammalian neurofilaments. *J. Cell Biol.* **76,** 50–56.

Schlaepfer W. W., and Zimmerman U-J. P. (1981) Calcium mediated breakdown of glial filaments and neurofilaments in rat optic nerve and spinal cord. *Neurochem. Res.* **6,** 243–245.

Schneider W. C. and Kuff E. L. (1954) On the isolation and some biochemical properties of the Golgi substance. *Am. J. Anat.* **94,** 209–224.

Sellinger O. Z. and Hiatt R. A. (1968) Cerebral lysosomes. IV. The regional and intracellular distribution of arylsulfatase and evidence for two populations of lysosomes in rat brain. *Brain Res.* **7,** 191–200.

Seijo L. and Rodríguez de Lores Arnaiz G. (1970) Subcellular distribution of thiamine pyrophosphatase in rat cerebral cortex. *Biochim. Biophys. Acta* **211,** 595–598.

Shank R. P., and Campbell G. LeM (1984) $\alpha$-Ketoglutarate and malate uptake and metabolism by synaptosomes: further evidence for an astrocyte-to-neuron metabolic shuttle. *J. Neurochem.* **42,** 1153–1161.

Shelanski M. L., Albert S., DeVries G. H., and Norton W. T. (1971) Isolation of filaments from brain. *Science* **174,** 1242–1245.

Shelanski M. L. and Liem R. K. H. (1979) Neurofilaments. *J. Neurochem.* **33,** 5–13.

Siakotos A. N. (1974) The Isolation of Nuclei from Normal Human and Bovine Brain, in *Methods in Enzymology* Vol. 31, Part A (Fleisher S. and Packer L., eds.) pp. 452–457, Academic, New York.

Smith A. D. (1972) Subcellular localization of noradrenaline in sympathetic neurons. *Pharmacol. Rev.* **24,** 435–457.

Stadler H. and Tashiro T. (1979) Isolation of synaptosomal plasma membranes from cholinergic nerve terminals and a comparison of their proteins with those of synaptic vesicles. *Eur. J. Biochem.* **101,** 171–178.

Tamir H., Rapport M. M., and Roizin L. (1974) Preparation of synaptosomes and vesicles with sodium diatrizoate. *J. Neurochem.* **23,** 943–949.

Tamir H., Mahadik S. P., and Rapport M. M. (1976) Fractionation of synaptic membranes with sodium diatrizoate. *Anal. Biochim.* **76,** 634–647.

Tanaka R. and Abood L. G. (1964) Studies on adenosine triphosphatase of relatively pure mitochondria and other cytoplasmic constituents of rat brain. *Arch. Biochem. Biophys.* **105,** 554–562.

Tapia R., Hajós F., Wilkin G., Johnson A. L., and Balázs R. (1974) Subcellular fractionation of rat cerebellum: an electron microscopic and biochemical investigation. II. Resolution of morphologically characterized fractions. *Brain Res.* **70,** 285–289.

Tashiro T. and Stadler H. (1978) Chemical composition of cholinergic synaptic vesicles from *Torpedo marmorata* based on improved purification. *Eur. J. Biochem.* **90,** 479–487.

Tennekoon G., Zaruba M., and Wolinsky J. (1983) Topography of cerebroside sulfotransferase in Golgi-enriched vesicles from rat brain. *J. Cell Biol.* **97,** 1107–1112.

Thomas D. V. (1979) Subcellular distribution of noradrenaline in guinea pig cerebral cortex and spleen. *J. Neurochem.* **32,** 1259–1267.

Vallee R. B. (1982) A taxol-dependent procedure for the isolation of microtubules and microtubules associated proteins (MAPs). *J. Cell Biol.* **92,** 435–442.

Vallee R. B., Di Batolomeis M. J., and Theurkauf W. (1981) A protein kinase bound to the projection portion of MAPs (microtubule associated proteins). *J. Cell Biol.* **90,** 568–576.

von Euler U. S. and Hillarp N. A. (1956) Evidence for the presence of noradrenaline in submicroscopic structures of adrenergic axons. *Nature* (Lond.) **177,** 44–45.

Whittaker V. P. (1969) The Synaptosomes, in *Handbook of Neurochemistry* Vol. 2 (Lajtha A., ed.) pp. 327–364, Plenum, New York.

Whittaker V. P. and Barker L. A. (1972) The Subcellular Fractionation of Brain Tissue with Special References to the Preparation of Synaptosomes and Their Component Organelles, in *Methods of Neurochemistry*, Vol. 2, (Fried R. L., ed.) pp. 1–52, Dekker, New York.

Whittaker V. P., Michaelson I. A., and Kirkland R. J. A. (1964) The separation of synaptic vesicles from nerve ending particles (synaptosomes). *Biochem. J.* **90,** 293–303.

Whittaker V. P. and Roed I. S. (1982) New Insights into Vesicle Recycling in a Model Cholinergic System, in *Neurotransmitter Interaction and Compartmentation* (Bradford H. F., ed.) pp. 151–173, Plenum, New York.

Wood M. D. and Wyllie M. G. (1981) The rapid preparation of synaptosomes, using a vertical rotor. *J. Neurochem.* **37,** 795–797.

Wuerker R. (1970) Neurofilaments and glial filaments. *Tissue Cell* **2,** 1–9.

Yen S. D., Dahl D., Schachner M., and Shelanski M. L. (1976) Biochemistry of the filaments of brain. *Proc. Natl. Acad. Sci. USA* **73,** 529–523.

Zimmermann H. (1982) Isolation of Cholinergic Nerve Vesicles, in *Neurotransmitter Vesicles* (Klein R. L., Lagercrantz H., and Zimmermann H., eds.) pp. 241–268, Academic, New York.

Zisapel N., Levi M., and Gozes I. (1980) Tubulin: an integral protein of mammalian vesicle membranes. *J. Neurochem.* **34,** 26–34.

# Chapter 3

# Brain Slices

## Uses and Abuses

### Peter Lipton

## 1. Introduction

This chapter is designed to provide a scientist who is not familiar with the brain slice preparation with some of the information needed to evaluate its suitability for a particular task. It is obviously not possible in a book chapter to describe all the uses to which brain slices have been put, and so one is faced with the options of describing one's own studies, embroidered perhaps by those of others when necessary, of lighting on specific topics and hoping that these will satisfy some of the readers, or, as I have chosen to do, of trying to evaluate the suitability of the slice preparation for as wide a variety of uses as possible. None of these approaches can be completely satisfactory. I chose the latter because, ultimately, we must determine the applicability of information we get from the slice to events in the intact brain. This, of course, is true for all model systems, but seems particularly urgent for brain studies in which the difficulty of studying the intact organ has led to an abundance of model systems and in which the tissue is so fragile. It thus seems important not only to describe, but also evaluate different ways of preparing and experimenting on tissue, even if such an evaluation must be subjective.

    With this purpose, then, I shall discuss the overall integrity of the slice preparation as it compares to both the intact brain and to another often-used preparation–synaptosomes. I shall then consider different methods of preparation, incubation, and electrical stimulation of slices. Finally I shall discuss two major questions that have been approached via the brain slice. Here I shall rely quite strongly, but not exclusively, on studies we have carried out in this laboratory.

## 1.1. Why the Brain Slice?

Finding a suitable tissue preparation has been the key to progress in many areas of biology: In general one looks for the simplest preparation that retains the properties of the physiological system. The problem, of course, is to retain these properties. It is of course ideal to experiment on the intact physiological system. Although there have been major advances in techniques for measuring transmitter release *in situ* (e.g., Soubrie et al., 1984), in non-invasive ways of measuring aspects of energy metabolism (Jobsis et al., 1977; Sokoloff, 1980), and in the development of the isolated perfused brain, a vast number of workers still study brain function by working on what may be termed *reduced preparations*. These include cultured neurons and glia, brain slices, and synaptosomes. In some cases these systems are used simply because they are easier to use than the *in situ* organ. However, there appear to be at least two scientifically valid reasons for using the reduced systems. First, culture systems and synaptosomes can be very important when the investigator wants to assign functions to either neurons or to glia, or to localize a function to synaptic endings. Here, though, technological developments may well weaken the justification for using the reduced systems. As an example, several workers have used isolated glial cells or glial cultures to analyze the compartmentation of glutamate metabolism (e.g., Hertz et al., 1979). However, recent studies on the intact brain do much the same job, histochemically (Norenberg and Martinez-Hernandez, 1979). The second justification for using reduced systems is the ability to alter the extracellular *milieu* of the tissue. This is sorely lacking in brain, where the quite formidable blood–brain barrier severely limits the ability to alter ion and metabolite concentrations surrounding brain cells, and makes quantitating any such changes almost impossible. Thus, the reduced preparation becomes invaluable when the study involves altering the composition of the extracellular environment, in particular with species such as ions and metabolites that do not cross the blood–brain barrier. Furthermore, although push–pull cannulation is becoming more widespread, collection and measurement of cell products is still very difficult in the intact preparation.

It must be stressed that any decision to use a reduced preparation is made at a cost. (This cost may be directly to the investigator: It is not infrequent for peer reviewers to look askance at proposals using brain slices or synaptosomes!) Thus, basal metabolism and, in many cases, the morphology of the reduced preparations are less intact than the *in situ* brain. In addition, spontaneous neural activity is generally lacking. These factors and their consequences will be discussed in the following section. Fortunately, the impaired basal function does not seem to alter many important phenomena in brain slices, including evoked electrical activities (Lynch and Schubert, 1980), neurotransmitter metabolism and release (Auerbach and Lipton, 1985), and the dependence of evoked activities on metabolism (Lipton and Whittingham, 1979).

## 2. The Slice Preparation

### 2.1. Comparison of Basal Energy Metabolism in Synaptosomes, Slices, and In Situ Brain

The brain slice preparation was first used by Warburg and coworkers in the 1920's and, soon after, there followed a spate of studies on respiration of the slices. These were nicely summarized by Quastel (1939) who noted that slice respiration rates were always about half those of *in situ* rates. Quastel expressed the hope that incubation conditions would eventually be found that would elevate these rates. However, this optimism has not been realized, and in the ensuing near-half century, basal respiration of slices has remained between one-third and one-half that of the intact brain, despite many heroic efforts at improvement. Respiratory rates of synaptosomes, on a fresh weight basis, are similar to those of slices. In general, a variety of metabolic indicators show that energy metabolism in these reduced preparations is noticeably compromised with respect to the intact brain. Synaptosomal energy metabolism is further compromised with respect to brain slices, especially when studies are done at physiological temperatures. Table 1 compares some key metabolic parameters in synaptosomes, brain slices, and the intact brain. It is apparent that slice values of ATP and phosphocreatine (PCr) are approximately half those of intact brain, for both rat and guinea pig; furthermore, cell $K^+$ is quite reduced in the slice. It is interesting, though, that the ratios of high-energy phosphates:PCr/ATP and the "energy charge" are quite similar in slices and *in situ*. It is apparent that energy-related parameters in the synaptosomes are, for the most part, less physiological than in the slice; this is espe-

TABLE 1
Comparison of Energy Metabolism-Related Parameters in
Synaptosomes, Brain Slices, and the Intact Brain[a]

| Parameter | Preparation | | |
|---|---|---|---|
| | Synaptosomes | Brain slice | Intact brain |
| $O_2$ consumption ($\mu$mol/g tissue/min) | 1.2[b,d,e] | 1.3[f] | 4.2[h] |
| | | 1.6[g] | |
| ATP ($\mu$mol/g tissue) | 1.3[j] | 1.5[c] | 2.91[c,i] |
| | 0.4[c] | 1.4[f] | 2.60[r] |
| PCr ($\mu$mol/g tissue) | 1.3[j] | 3.6[c] | 5.6[c,i] |
| | | 3.0[i] | 5.0[p] |
| PCr/ATP | 1.0[j] | 2.5[i] | 1.9[c] |
| ATP/ADP | 1.5[c] | 5.0[f] | 9.0[q] |
| | 2.8[j] | | |
| ($K^+$) (mmol/L cell $H_2O$) | 72[k] | 101[h] | 145[o] |
| | | 113[m] | 127[n] |

[a]All slice data is at 37°C. Temperatures for synaptosomal data are noted in footnotes d–r.
[b]Assumes a synaptosomal ($H_2O$:protein) of 3.2 $\mu$L/mg (from Scott and Nicholls, 1980) and a protein:dry wt. ratio of 1:2.
[c]Assumes a protein:wet wt. ratio of 1:10 (our observations).
[d]Scott and Nicholls (1980), guinea pig, 30°C.
[e]Booth et al. (1983), rat, 37°C.
[f]Rolleston and Newsholme (1967), guinea pig.
[g]Benjamin and Verjee (1980), rat.
[h]Ghajar et al. (1981), rat.
[i]Whittingham et al. (1984), guinea pig.
[j]Rafalowska et al. (1980), rat, 22°C.
[k]Pastusko et al. (1981), rat, 22°C.
[l]Lipton and Robacker (1983), guinea pig.
[m]Kass and Lipton (1982), rat.
[n]Baethmann and Sohler (1975), rat.
[o]Lipton and Robacker, unpublished, guinea pig.
[p]Pulsinelli and Duffy (1983), rat.
[q]Siesjo et al. 1983), rat.
[r]Whittingham (1980), guinea pig.

cially true at 37°C, where ATP and ATP/ADP values are far lower than in the slice; even at 25°C, synaptosomal $K^+$ is far lower than slice $K^+$. There is a need for further measurements of synaptosomal energy-related parameters at physiological temperatures.

We have concluded elsewhere (Lipton and Whittingham, 1984) that appoximately half the lowering of energy metabolites in the slice is caused by destruction of cells at the periphery of the tissue, and the other half results from alteration of the remaining, intact cells. The nature of this alteration is not known. It does not appear, however, to prevent quite normal electrophysiological and metabolic responses to stimuli and altered conditions. On the other hand, it presents a disquieting problem that pervades all studies using brain slices.

## 2.2. Comparison of Morphology of Synaptosomes, Brain Slices, and Intact Brain

Slices always appear to have a zone of about 50–100 μm at the cut edge in which cells are badly damaged (Garthwaite et al., 1979; Froetscher et al., 1981; Misgeld and Froetscher, 1982). Thus, the outer regions of slices should be considered dead tissue; extracellular space measurements indicate a region of this size that is permeable to sucrose and insulin and may well correspond to the Futhermore, it is not possible to obtain good electrophysiological responses from the top 50 or so μm of slices (Bak et al., 1980; our observations). The more important issue is the ultrastructure of the slice interior.

First, it is important to note that there are quite marked differences in the preservation of ultrastructure in slices from different brain regions and different animals. These differences seem to follow a pattern; it appears that the more sensitive a brain region is to anoxic damage, the less its slice ultrastructure is preserved. As examples, rat cerebellar slices are extremely well preserved (Garthwaite et al., 1979), olfactory cortex slices are fairly well preserved (Fig. 1), whereas striatal (Bak et al., 1980) and hippocampal (our observations) slices from rat are quite poorly preserved. This is in line with the relative sensitivities of these regions to ischemia in the intact brain (Pulsinelli et al., 1982). Guinea pig hippocampal slices are, in contrast to rat, quite nicely preserved (Figs. 2 and 3), and guinea pig hippocampal slices are less susceptible to anoxia than rat hippocampal slices (our observations).

Unfortunately, no studies have resolved the issue of whether the ultrastructural changes in the slice occur during incubation of the slice or during fixation; both are reasonable possibilities. In particular, the same mechanisms that render a tissue sensitive to anoxia may also render it sensitive to the fixative. The long diffusion pathways for fixative in the slice (150–250 μm) compared with those *in situ* make it reasonable that the slice would be less rapidly fixed than the intact brain. On the other hand, the large extracellular spaces seen in slices (Fig. 2) are probably partially real, since slice extracellular space measured with tracers is some-

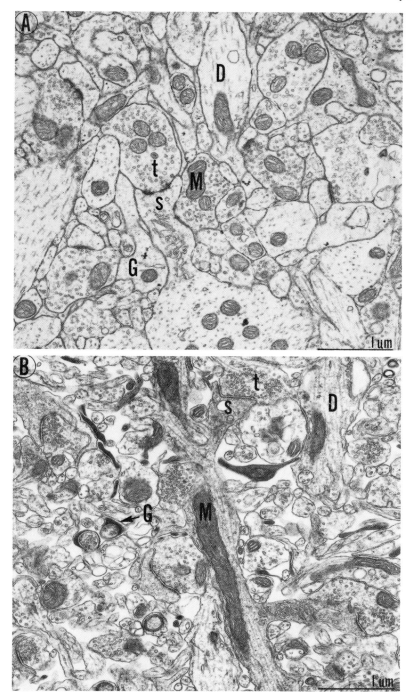

what larger than *in situ* extracellular space (Katzmann and Pappius, 1973; Lipton and Whittingham, 1984).

Figures 1–3 are electron micrographs of intact brain and slices from opossum, rat, and guinea pig. These were taken by Ms. Sherry Feig, along with L. Haberly (opossum and rat) and J. Keating (guinea pig). They demonstrate fairly general features of the *in situ* to in vitro ultrastructural transition. Figure 1 compares olfactory or pyriform cortex neuropil from the intact (opossum) brain with that from the (rat) olfactory cortex slice. Figure 2 shows neuropil from the stratum radiatum of the CA1 region of the guinea pig hippocampal slice. The ultrastructures of the two slice preparations are quite similar. The dendrites are well preserved, with intact tubules and spines, and the major difference from dendrites of the intact brain is the darker appearance of the cytosolic ground substance.

Synaptic boutons and synapses are also well preserved, for the most part, but in the hippocampal slice there are a large num-

Fig. 1.   Electron micrographs of layers 1a of olfactory cortex from opossum fixed *in situ* (A), and from the rat slice (B). (A) The anesthetized opossum was perfused, via the ascending aorta, with a 1/1.25% mixture of paraformaldehyde and glutaraldehyde and subsequently with a mixture of double this strength. Tissue blocks were then cut and fixed overnight in the dilute fixative. They were then further fixed and stained with 2% $OsO_4$, in a 5% sucrose-containing phosphate buffer, and stained *en bloc* with uranyl acetate and lead citrate; they were then dehydrated in increasing concentrations of ethyl alcohol (Haberly and Feig, 1983). (B) Olfactory cortex slices from the rat were prepared using a vibratome and were cut to 300 μm. The slices were incubated in glucose-fortified Krebs buffer at 25°C for 2 h. They were then fixed overnight in 2/2.5% paraformaldehyde/glutaraledehyde at 0–2°C. They were rinsed, fixed, stained, and dehydrated as described for panel A. Although this slice was incubated at 25°C, there are no dramatic differences noted in the ultrastructure of slices incubated at 37°C.

Both micrographs are from the 1a neuropil layer of the olfactory cortex. The dendrites (D) are somewhat darker in the slice than in the neuropil. Mitochondria (M) are very well preserved in the slice, but glial processes (G) are poorly preserved. Thus, there are few distinct glial profiles and there are clear signs of degenerating glial processes in the regions of synaptic boutons (G). There are many quite distended and structureless profiles in the slice preparation. These may well represent swollen glial processes. Terminals (t) and dendritic spines (s) are generally well preserved in the slice. Very noticeable is the far larger extracellular space in the slice preparation (courtesy of L. Haberly and S. Feig, 1983, unpublished).

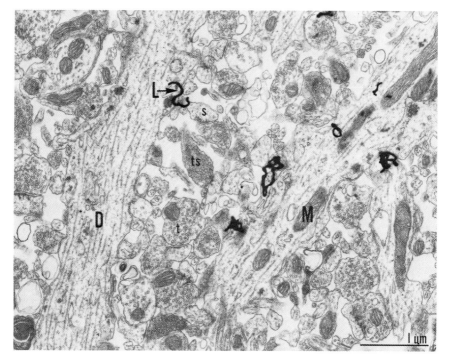

Fig. 2.   Electron micrograph of the stratum radiatum in the CA1
region of the guinea pig hippocampal slice. The slice was prepared by
slicing the hippocampus free-hand with a Gillette Blue-Blade. It was incubated for 4 h at 36°C and during the last 3 min it was exposed to 20
μCi/mL of ($^3$H) leucine. It was then rinsed for 30 s and fixed in 2/2.5%
paraformaldehyde/glutaraldehyde, gently bubbled with air at 1–2°C.
Further fixation, staining, and dehydration were as described for 1a.
The thin section was coated with Ilford L-4 emulsion and developed for
6.5 wk. The slice was about 400 μm thick and the section was taken 200
μm from the surface. The region shown is approximately 200 μm from
the pyramidal cell layer, into the stratum radiatum of CA1. Symbols are
the same as for Fig. 1. In these sections there are clear signs of
degenerating terminals (ts). Otherwise the structure is very similar to
that of the olfactory cortex slice from the rat. (L) are developed silver
grains and presumably represent newly synthesized proteins or proteins in the process of being synthesized. It is apparent that they are localized predominantly around the outer edges of dendrites (micrographs courtesy of J. Keating and S. Feig).

ber (perhaps 30–40%) of degenerating terminals with darkened
cytoplasm and a very high vesicle density. These structures appear also in the CA3 region (Froetscher et al., 1981). However, a
large number of synaptic terminals are clearly intact. The most
striking difference from the intact brain is the increased area of

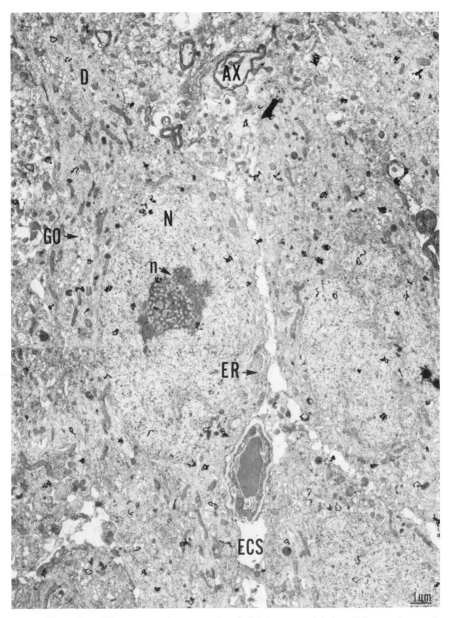

Fig. 3. Electron micrograph of CA1 pyramidal cell layer from the guinea pig hippocampal slice. This micrograph is taken from the same tissue as that used in Fig. 2. It represents the cellular layer of the CA1 region. The magnification is approximately one fourth the magnification in the preceding micrographs. The nucleus (N) and the nucleolus (n) are very similar to the structures in *in situ* fixed preparation (Haberly and Feig, 1983). Myelinated axons (AX) appear normal, as do the proximal

the intercellular space and, particularly in olfactory cortex, the association of this expanded space with degenerating glial structures in the proximity of synapses. This, coupled with the lack of well-identified glial processes and the very swollen glial elements, suggest that much of the increased intercellular space may result from ruptured glial processes. It is undetermined whether the spaces or the synaptic damage in the neuropil are related to synapses containing one particular neurotransmitter type. For example, glutamate leads to pronounced glial swelling and it is possible that the aldehyde fixatives release glutamate from endings, causing local accumulation and, hence, glial process swelling.

Figure 2 demonstrates another interesting feature of the neuropil: the association of radiolabeled leucine, very probably incorporated into protein, with the periphery of dendritic shafts and, in at least two cases, with dendritic spines. In these studies (Keating, Feig, and Lipton, unpublished), slices were exposed to $^3$H-leucine for only 3 min, so that the label should represent locally synthesized protein. Thus, the results are consistent with studies of Steward and Levy (1982) showing localization of polysomes near spines in dendrites of intact hippocampus, and suggest that these polysomes may well be quite active. Kiss (1977) demonstrated synthesis of proteins in dendrites of intact hippocampus. This synthesis emphasizes that the neuronal elements of the slice neuropil appear fairly normal, whereas the glial processes appear badly damaged.

Figure 3 shows the soma and proximal dendrites of CA1 pyramidal cells from the guinea pig hippocampal slice. As in the neuropil, there is abnormally abundant extracellular space (ECS); it lies between the normally tightly opposed soma. The cell nuclei are normal, in that chromatin is not clumped; the nucleolus appears vacuolated, but this is also the case *in situ* (Haberly and Feig, 1983). Mitochondria throughout the cells appear quite normal. Both endoplasmic reticulum and golgi vacuoles are swollen compared with their appearance *in situ* (Peters et al., 1975). There are no obvious membrane discontinuities. This montage is quite respresentative of the general appearance of the pyramidal cell

Fig. 3 (*Cont.*).   dendrites (D). The golgi bodies (GO) and the endoplasmic reticulum (ER) are both swollen compared with their appearance in tissue fixed *in situ* (Peters et al., 1975; Haberly and Feig, 1983); so, too, is the extracellular space (ECS), which normally is not observed at all between pyramidal cells. Degenerating structures, which may well be glial elements, are apparent in these spaces. Developed silver grains are visible throughout the cells, including the nuclei.

layer; however, there are regions in which the cells are condensed and look far less healthy than these. Neither we nor others have produced a quantitative study of cell "abnormalties" in hippocampal or other brain slices.

Although not shown here, the ultrastructure of synaptic boutons in synaptosomal preparations is not noticeably different from that of the synaptic endings in the slice preparaton (Arnaiz and DeRobertis, 1971), although the preparations generally contain a large number of synaptosomes with disrupted membranes (Deutsch et al., 1981). They also contain a significant number of glial profiles.

Thus, it is apparent that the ultrastructure of the interior of brain slices is, in general, somewhat compromised with respect to the intact tissue. It is not clear, however, how much of this alteration is caused by fixation artifacts and how much is caused by true structural alterations in the slice, though such changes as glial process degeneration are too rapid to be caused by fixation. In our studies we tried several different fixative concentrations and modes of applying the fixative before settling on the one described in the legend. However, the impossibility of perfusing the slice means that the diffusion path for the fixative must be a long one, and this has traditionally led to poor fixation of brain tissue. Indeed, only with the advent of perfusion techniques has brain ultrastructure been able to be well preserved (Karlsson and Schulz, 1965).

A final note: If this author were faced with the choice of using hippocampal slices from the rat or the guinea pig, he would choose the latter in almost all cases. Thus, all of our data on ultrastructure (this chapter and unpublished micrographs), levels of high-energy phosphates (Lipton and Whittingham, 1982; Kass and Lipton, 1982), rates of protein synthesis (Lipton and Heimbach, 1978; Raley and Lipton, unpublished), and glycogen levels (Hurtenbach and Lipton, unpublished) show that slices from the guinea pig appear healthier than those from the rat. The only caveat is that there are far more extant data on intact rat brain than intact guinea pig brain. This is sometimes an advantage and must be weighed against the better preservation of the guinea pig slice.

## 2.3. Uses of the Brain Slice Preparation

In spite of the suboptimal preservation of the cells of the brain slice preparation, it has become extensively used over the last decade and a half. Thus, many important responses of brain tissue can be observed in the slice preparation and, in addition, membrane potentials and cell electrophysiology are quite normal in a large number of the cells. These factors have been exploited in a wide variety of studies and some of these are listed in Table 2, which demonstrates the many kinds of problems for which the slice preparation has been used (in spite of the nagging "how physiological is it?" question). The table is one in which topics that are most informatively studied with brain slices have been selected. It is not meant to be comprehensive, but it does highlight major areas. It is designed to give the reader an idea of the variety of topics that can be successfully studied with this preparation.

It is apparent that the slice has become very useful for detailed electrophysiological studies, largely because of the accessibility and ease of identification of cells. Nearly all this work has followed the landmark studies on olfactory cortex and hippocampus slices in the late 1960's and early 1970's, in which normal electrophysiological pathways were first activated in slices (Yamamoto and McIlwain, 1966; Skrede and Westgaard, 1971). It is also apparent from the table that synaptic neurochemistry and energy metabolism are widely studied using the slice. The large amount of space in the table that is devoted to the hippocampus is indicative of the relative amount of effort devoted to that preparation.

## 3. General Slice Technique

### 3.1. Preparing the Slice

There are many variations on the theme of preparing the brain slice to obtain a healthy preparation. The reader interested in a specific slice should consult the appropriate reference in Table 2; there are also very useful reviews (McIlwain, 1975; Dingledine et al., 1980; Alger et al., 1984). The guiding principle is to do the task as rapidly and with as little physical trauma as possible; in general, workers avoid anesthetics since one of the virtues of the slice is that studies can be done in the absence of drugs. I shall confine the experimental descriptions to rodents, in particular the rat and the guinea pig.

#### 3.1.1. Removal of Brain

The first step is decapitation, generally with a guillotine, and this is followed as rapidly as possible by removing the skull and exposing the brain. The brain is then removed, as hemi-sections, and put into ice-cold buffer that has been preoxygenated and is glucose-fortified. There has been some discussion as to whether ice-cold buffer is appropriate for this and the subsequent steps of preparation; some workers have argued that better preservation is attained when temperatures of 10–15°C or room temperatures are used. We have not found this for rat hippocampal slices, but, interestingly, there is no firm evidence that maintaining a low temperature helps. Indeed, guinea pig hippocampal slices perform identically in terms of energy metabolism and electrophysiology whether they are prepared at 0–2 or 37°C (Whittingham et al., 1984). An investigator beginning a study would do well to test different preparation temperatures.

The next step is to separate the brain region that is to be sliced, and the method will clearly depend on the region. For the *hippocampus*, the half-brain is turned upside down, the brain stem is retracted from the caudal region of the brain, and the now-exposed hippocampus (a pale, bean-like structure) is dissected out by gently lifting the rostral edge with a spatula and cutting along the caudal contour. Care must be taken not to damage the ventral surface, since this destroys conduction through the dentate gyrus. Isolating the *cortex* and *cerebellum* is straightforward; for the *striatum* a coring device that punches out a cylinder containing the basal ganglia has been developed (Lighthall et al., 1981), but direct dissection from the cerebrum is also effective. For the *brain stem,* the particular structure to be studied is generally in an appropriate transverse slice.

### 3.1.2. Slicing the Brain

There are three basic methods used to slice the brain region. These are the hand-held razor blade or bow cutter, the automatic tissue chopper, and the vibratome. Although there are no definitive studies on which of these is most appropriate, the tissue chopper should generally be avoided. Thus both striatal (Bennett et al., 1983) and cerebellar (Garthwaite et al., 1979) slices suffer clear morphological damage when prepared by the chopper as opposed to a hand-held slicer. The vibratome or the adapted "Vibroslice" (Jeffreys, 1981) has been successfully used for preparations from the brain stem, and, although most cumbersome, appears to be as gentle as the hand-guided blades. It has the virtue of being able to make very thin slices. The slicer is safe if he or she uses the hand-held (or guided) Gillete Blue-Blade (McIlwain, 1975; Lipton and Whittingham, 1979) or the adapted bow cutter (Garthwaite et al., 1979) (*see* Alger et al., 1984 for an extended discussion of this question).

TABLE 2
Uses of Brain Slices From Different Regions: Selected Phenomena
Studied Using the Brain Slice

| Region of slice origin | Phenomenon | Reference[a] |
|---|---|---|
| Brain stem | Monoamine metabolism and release | Elks et al. (1979a,b) |
| | Membrane electrophysiology | Hendersen et al. (1982) |
| | Effects of pH on membrane electrophysiology | Fukada and Loeschke (1977) |
| Cerebellum | Membrane electrophysiology and dendrite properties | Llinas and Sugimori (1980) |
| Cochlear nucleus | Membrane electrophysiology and transmitters | Oertel (1983); Hirsch and Oertel (1984) |
| Hippocampus | Pathway stimulation and circuitry | Skrede and Westgaard (1971) |
| | Neurotransmitter identification and action | Malthe-Sorensen et al. (1979) |
| | | Corradetti et al. (1983) |
| | | Langmoen et al (1981) |
| | Membrane electrophysiology | Dingledine and Langmoen (1980) |
| | Synaptic transmission electrophysiology | Brown and Johnston (1983) |
| | Dendritic electrophysiology | Turner and Schwartzkroin (1983) |
| | Actions of anoxia/ hypoglycemia | Lipton and Whittingham (1982) |
| | | Kass and Lipton (1982) |
| | | Cox and Bachelard (1982) |
| | Regulation of protein synthesis | Lipton and Heimbach (1978) |
| | Epilepsy | Traub and Wong (1983) |
| | Long-term potentiation | Lynch et al. (1983) |

*(Continued)*

TABLE 2
(*Continued*)

| Region of slice origin | Phenomenon | Reference[a] |
|---|---|---|
| Hypothalamus | Monoamine metabolism and release | Hyatt and Tyce (1984) |
| | Membrane electrophysiology and hormone secretion | Hatton (1984) |
| Neocortex | Metabolic response to intense stimulation | McIlwain (1975) |
| | Neurotransmitter release | Potashner (1978) |
| | Neuronal electrophysiology | Connors et al. (1982) |
| | Protein synthesis | Dunlop et al. (1975) |
| Olfactory cortex | Pathway stimulation | Richards and Sercombe (1970) |
| | Neurotransmitter identification | Bradford and Richards (1976) |
| | Membrane electrophysiology | Scholfield (1978) |
| | Circuitry | Bower and Haberly (1984) |
| Striatum | Membrane electrophysiology | Kitai and Kita (1984) |
| | Neurotransmitter release and metabolism | Milner and Wurtman (1984) |

[a]References are selected from, in most cases, tens of possibilities. Those selected are chosen because they are either recent or important. The aim is to provide the potential user with a useful starting point for the study of a particular phenomenon.

The orientation of the slice again depends on the region from which it is cut. For the hippocampus, slices should be cut at an angle of 70–90° to the longitudinal axis in order to best preserve the electrophysiological pathways; inhibitory pathways are better preserved with the blade slightly off the perpendicular (Rawlins and Green, 1977). Much-improved electrophysiological recordings have been obtained from the neocortical slice by cutting slices as coronal sections (Connors et al., 1982; Stafstrom et al., 1982), rather than tangentially, as they had been cut until the end of the 1970's.

Slice thickness must be limited at the low end by the quite large (75–100 μm) region that is damaged on the edge of the slice and, at the high end, by the need to maintain adequate oxygenation and substrate supply. The latter imposes a practical limit of about 500 μm (Harvey et al., 1974; Fujii et al., 1982; Lipton and Whittingham, 1984) when experiments are done at 37°C. This is especially true for metabolic studies in which the core anoxia is not tolerable. Thicker slices (600–800 μm) may be used for electrophysiological studies and will yield larger evoked responses in the hippocampus. Apparently the slight anoxia is more than compensated for by the retention of more components of the afferent pathway.

### 3.1.3. Incubation of the Slice

If metabolic studies are to be done, it is essential to maintain the slice at its normal *in situ* temperature. For example, the susceptibility of the slice to alterations in energy metabolism is extremely sensitive to reductions in temperature; evoked responses in rat hippocampal slices will be irreversibly lost after 10 min of anoxia at 36°C (Kass and Lipton, 1982), but they will survive at least 45 min of anoxia at 30°C (Jahn and Lipton, unpublished results). Electrophysiological responses are larger and sustained for longer incubation periods when experiments are run between 30 and 35°C, and so studies are often done at this temperature (e.g., White et al., 1979; Collingridge et al., 1983; Mueller et al., 1984). This should, however, be avoided unless absolutely necessary because there is little point in introducing an additional distinction between the slice and the *in situ* preparation. In some cases, though, it is very useful. For example, Corradetti et al. (1983) incubated hippocampal slices at 32°C to measure evoked glutamate release because spontaneous glutamate release should be greatly attenuated at this temperature (Langmoen and Anderson, 1981).

Almost all workers incubate slices in a bicarbonate-buffered modified Krebs buffer. The buffer must be well oxygenated and is usually saturated with 95% $O_2$/5% $CO_2$. Our usual buffer composition is: NaCl, 124 m$M$; NaHCO$_3$, 26 m$M$; KCl, 3 m$M$; KH$_2$PO$_4$, 1.4 m$M$; CaCl$_2$, 2.4 m$M$; MgSO$_4$, 1.2 m$M$; and glucose, 4 m$M$. Care must be taken to ensure that the system geometry does not lead to a significant desaturation at the point where the slice is located. This may occur in more cases than is realized and whenever a set-up is initiated, $O_2$ at the slice position should be measured using an oxygen electrode.

Although phosphate and Hepes buffers have been used when manometry is being performed, they should be avoided at all other times. Glucose concentration is almost always maintained at 10 mM, but it is worth noting that blood glucose is about 4 mM. We usually use the latter concentration and this should be done whenever regulation of function by metabolism is being studied. For example, lowering glucose from 10 to 5 mM inhibits hippocampal slice transmission during perforant-path stimulation at 2 Hz (Cox and Bachelard, 1982). Using 10 mM glucose will give quite different results from using 4 mM glucose.

Even more important, and perhaps surprisingly so, is the need to ensure adequate anerobiosis for studies in which effects of oxygen deprivation are being studied (Kass and Lipton, 1982). Thus, though $O_2$ saturations of 60–80% are quite consistent with maximal oxygenation for most slice thicknesses, $O_2$ at the slice surface must be lowered to less than 2% if full effects of anoxia are to be observed (our accidental observations). In order to maintain such an $O_2$ tension, special precautions must be taken, even if nitrogenated buffer is provided to the slice; we perfuse the slices very rapidly (75–90 mL/min). Another possibility, if slower flow is necessary, is to create a low $O_2$ tension in the air above the slice chamber.

### 3.1.3. "Souping Up"

Buffers in which brain slices are incubated are often somewhat "souped up." As discussed, temperatures are lowered and glucose is elevated. In addition, $K^+$ and $Ca^{2+}$ are elevated above normal values (5 rather than 3 mM; 2.0 rather than 1.2 mM) (Alger et al., 1984). These maneuvers are designed to provide enhanced electrical responses, and they do. The "souping up" practice, however, seems questionable. Although they make electrophysiology easier to carry out, the buffer changes may be producing responses that do not normally occur. It seems advisable to avoid such changes whenever possible, and, in general, it is possible. The slice literature is also replete with descriptions of efforts to enhance slice metabolic performance with organic substrates or colloids. These include effects of adenosine (Thomas, 1957), inert macromolecules such as dextrans (Keesey et al., 1965), and culture media such as Eagle's M.E.M. None of these have provided noticeable enhancement, although a recent preliminary communication from Schwartzkroin suggests that 10% fetal calf serum may improve slice longevity (Alger et al., 1984).

## 3.2. Incubation Chambers for Slices

For metabolic and ion flux studies, slices may be suspended freely in buffer in which the aeration keeps them "gently dancing." Several "transfer holders" have been described (e.g., Lund-Andersen, 1974; Teichberg et al., 1981) for experiments in which the slices need to be moved from solution to solution; these are particularly valuable for studies in which efflux of an ion or molecule from the slice is being measured. The holder is usually a plastic cylinder with a bottom made of nylon net or bolting cloth (Auerbach and Lipton, 1985); the aeration tube is attached to it.

Several different "chamber" designs have been developed for electrophysiological studies, and the best approach would be for the reader to consult original papers (Table 2). Figure 4 shows chambers we use for (a) electrophysiological recordings (Lipton and Whittingham, 1979) and (b) combined slice-stimulation/metabolic/optical studies of brain slices (Lipton, 1973b). The slice is totally immersed in the perfusing buffer and, for the chamber in which electrophysiological recordings are being made, the buffer flows past the slice at a very rapid rate (between 25 and 90 mL/min). This rapid-flow chamber is ideal for studies in which rapid changes of the slice milieu (20–30 s) are necessary. It also provides good oxygenation and anerobiosis (see above). It is not useful for intracellular recordings because the slice is not stable enough; it is, however, stable enough for extracellular recordings that remain constant for several hours.

Slow-flow chambers must be used for intracellular electrophysiological studies of slices; they are also used in other studies. There are two alternative approaches. One is to again have the slice totally immersed in buffer, but to have the buffer moving very slowly across the slice (Nicoll and Alger, 1981) or for the buffer to be aerated and not move at all (Teyler, 1980). The other approach is the so-called interface incubation in which the bottom of the slice is exposed to the buffer but the top of the slice is exposed to warmed, humidified 95% $O_2$/5% $CO_2$ (Schwartzkroin, 1975; Haas et al., 1979). Here again, the flow of the buffer is slow. Such chambers are now available commercially (Frederick Haer, Brunswick, Maine; Stoelting, Chicago, IL) with feedback temperature control. The slow-flow chambers are necessary for intracellular recordings and it is crucial that the chamber be designed well enough to ensure good temperature control and oxygenation; this is quite difficult at the low flow rates. The virtue of the "interface" chamber is that phenomena such as long-term potentiation are best seen using it; evoked responses tend to be larger also. The relatively high resistance of the slice extracellular fluid probably accounts for both of these observations. Although the slow-flow chambers are widely used, it is worth noting that metabolite measurements have not been made on slices so incubated. Metabolite measurements have been made on slices that are rapidly perfused (Lipton and Heimbach, 1978; Lipton and Whittingham, 1982).

The chamber in Fig. 4B allows the simultaneous measurement of slice respiration and slice fluorescence, reflectance, or light absorption. In addition, the slice can be field-stimulated (*see* section 4). Thus, effects of different treatments—including electrical stimulation—on important parameters of energy metabolism and on cell volume can be measured. The very slow flow (about 0.5 mL/min) combined with the small chamber volume (about 1 mL) allows for the respiration of one slice to be measured. The fact that the chamber is covered allows for good oxygenation and temperature control, in spite of the slow flow. The capabilities of this chamber design are shown in Fig. 5, which demonstrates the effects of field-stimulation on respiration, pyridine nucleotide fluorescence, and reflectance of a guinea pig neocortical slice (Lipton, 1973b).

# 4. Electrical and Chemical Stimulation of the Slice

## 4.1. Introduction

Table 2 shows many of the studies in which slices have been used. It does not include the large number of studies in which metabolism is measured in the absence of imposed electrical activity of any kind. In describing the specific uses of slices, I shall omit those studies as well as those in which electrophysiological properties alone were studied. I shall restrict myself to work in which interactions between electrical activity and metabolism have been explicitly studied. All these studies involve some form of stimulation of the slice.

The history of this field falls into two phases. The first began in the 1930's when brain slice respiration was shown to be dramatically stimulated by high KCl concentrations (Ashford and Dixon, 1935), and it received a great boost when McIlwain and coworkers were able to produce a similar effect by electrically stimulating the slice with high-frequency, high-voltage pulses (McIlwain, 1951). This led to a vast number of studies in which effects of whole slice electrical acitivity on metabolism were investigated, and these continue unabated to the present. The second phase began in the late 1960's and early 1970's when McIlwain, Yamamoto, Skrede, and Westgaard developed the olfactory cortex slice, and, most importantly, the hippocampal slice preparations. These preparations were notable because one could stimulate intact neural pathways and record postsynaptic responses.

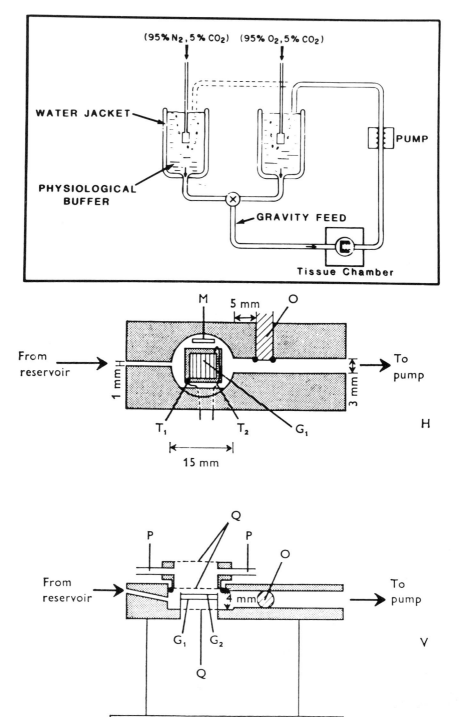

(95% $N_2$, 5% $CO_2$)  (95% $O_2$, 5% $CO_2$)

WATER JACKET

PUMP

PHYSIOLOGICAL BUFFER

GRAVITY FEED

Tissue Chamber

M  5 mm  O

From reservoir

To pump

1 mm

3 mm

$T_1$  $T_2$  $G_1$  H

15 mm

Q

P  P  O

From reservoir

To pump

4 mm

$G_1$  $G_2$  V

Q

Stand

Fig. 4. Perfusion chambers for brain slices. A. Chamber and perfusion set-up for studying hippocampal slice electrophysiology. Slices are maintained in the plastic chamber. They are seated on removable grids of nylon bolting cloth (Nitex) and held at one end under a plastic grid bar. The chamber volume is 5 mL and its diameter is 2 cm. Buffer is maintained in large reservoirs that are heated so that reservoir buffer temperature during circulation is about 41°C. Chamber temperature is between 36 and 37°C. The composition of the two reservoirs may be different; the figure shows the situation when studies of anoxia are being carried out. Buffer flows by gravity through small bore (1 mm) capillary tubing into the chamber and is pumped back up to the reservoirs by a peristaltic pump. The reservoir height and tube resistance are adjusted to give a flow of between 25 and 90 mL/min. The latter is only needed when irreversible effects of anoxia are being studied. The grid and slice are submerged beneath about 4 mm of buffer; at least this depth is necessary for the study of irreversible effects of anoxia (see text). Electrodes are inserted through the open top of the chamber and the preparation is viewed through a dissecting microscope via reflected light.

B. Chamber for studying the effects of whole slice stimulation on slice oxygen consumption and optical properties. The chamber is designed so that the slice is held between two grids ($G_1$ and $G_2$) that are the opposite poles of a stimulator. The grids are made from Teflon-coated silver wire that is stripped by heat on the side adjacent to the slice. Grid terminals ($T_1$ and $T_2$) are connected to wires that are inserted through the plastic chamber wall. A magnetic stirring bar, M, is placed against one wall of the chamber and is stirred throughout the experiment. This is absolutely essential for steady readings of respiration. The slice is viewed, optically, through quartz windows, Q, that are in the bottom of the chamber and form the surfaces of the chamber top. This latter screws on to an "O-ring" seal and is water-warmed through inflow and outflow pipes (P) connected to a Haake circulator. The oxygen electrode, O, is screwed in down-stream from the slice and buffer flows through the closed system very slowly and steadily (0.5 mL/min), driven by an LKB peristaltic pump. The temperature in the chamber is maintained at 37°C by having the reservoir outlet very close to the chamber and by the thermal insulation provided by the thick plastic and the water-jacketed top. H is a horizontal section; V is a vertical midsection. The chamber was constructed at the Johnson Research Foundation, University of Pennsylvania, Philadelphia, by Alan Bonner and, although difficult to construct, was extremely effective (from, Lipton, 1973a).

The slice is illuminated by a Hg arc and viewed from above using a Leitz ultrapak. The Hg light is filtered to produce the desired input wavelength. Using two photomultiplier tubes, the output from the incident light can be measured at two different wavelengths. This allows reflectance (366–366 nm) and pyridine nucleotide fluorescence (366–450 nm) to be recorded simultaneously. Respiration is measured by recording the oxygen tension at the electrode. Providing the flow is constant, the fall in oxygen tension as the buffer moves past the slice will be proportional to the rate of slice respiration. The lag time for these measurements is several minutes because of the the slow flow. However, this slow flow is absolutely necessary to allow a measurable fraction of buffer oxygen to be consumed by the slice.

They opened a new neurochemical era by allowing measurements of effects of physiological stimulations on metabolism and of metabolic changes on transmission in normal neural pathways. However, because metabolic effects of physiological stimuli are often very small, and so not possible to measure, and because many neural pathways still cannot be stimulated directly in slices (in particular monoaminergic and peptidergic pathways), the "whole slice" stimulation approach has survived and does, indeed, still thrive. In particular, it is widely used to study the release and metabolism of neurotransmitters.

## 4.2. Stimulation of the Whole Slice

### 4.2.1. Electrical Stimulation

McIlwain and coworkers first showed that subjecting slices to alternating voltage pulses of several tens of volts and frequencies of 5–50 Hz produced metabolic changes, in particular an increase in respiration, which were inhibitable by tetrodotoxin and, hence, appeared to represent $Na^+$-channel-dependent responses to the pulses (McIlwain and Bachelard, 1971). This mode of exciting the slice has been widely applied to the study of metabolic changes and transmitter release.

There are two experimental approaches. The first is to put one or two slices between the two poles of a grid electrode. Designing the grids is extensively described by McIlwain (1975). Figure 4B shows a chamber we developed for this purpose. The mode of stimulation and the responses in Fig. 5 illustrate several features of this form of stimulation. The electrodes are best designed so that the wire is insulated except where it is in contact with the tissue; Teflon-coated Ag wire is very suitable (Medwire, Valhalla, NY). Pulses of alternating polarity, or whose polarity is changed every few seconds, are absolutely necessary to prevent electrode polarization; alternating square wave pulses are most effective (*see* McIlwain and Joanny, 1963 for an extensive discussion of the effectiveness of different pulse forms). Pulses may be sustained for many minutes; they produce quite steady effects, which are generally reversible.

An alternative design, which is widely used when large masses of tissue are required, is to suspend many slices in a vessel in which two wires are placed several millimeters apart. These wires form the two electrodes, and the slices in solution respond to the pulses as well as they do in the grid structures. Again, details of chamber design are nicely described by McIlwain (1975). These chambers can also be designed to allow superfusion of the slice

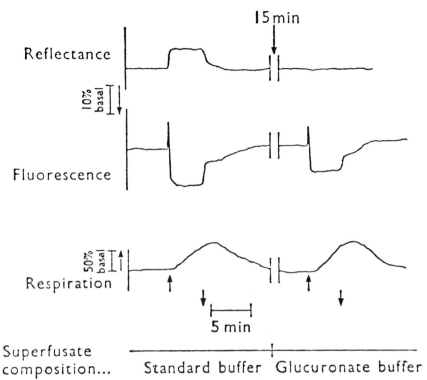

Fig. 5. Effects of electrical stimulation on respiration and optical properties of the guinea pig neocortical slice. These studies were carried out in the chamber described in Figure 4B. The top trace represents the reflectance of the slice. This is inversely related to cell volume (Lipton, 1973a). The middle trace is the fluorescence of reduced pyridine nucleotides in the slice, and the bottom trace is the slice respiration. The kinetics of the latter do not reflect the kinetics of the change in respiration; the steady-state value does represent the steady change in respiration. The slice was electrically stimulated at 50 Hz and 10 V during the time periods between the arrows. In normal buffer this stimulation causes a decrease in reflectance, corresponding to cell swelling; it causes a rapid oxidation of pyridine nucleotide (mitochondrial) followed by a long-lasting reduction (cytosolic). This latter appears to result from activation of glycogenolysis. It also causes about a 75% increase in respiration. Glucuronate buffer is buffer in which the NaCl has been replaced by Na glucuronate. Glucuronate is an impermeable anion and it is apparent that it prevents the reflectance change with little effect on the metabolic changes. This is consistent with the conclusion that the reflectance change results from water flow following an increased uptake of NaCl (Lipton, 1973a). The optical technique allows very rapid changes to be measured non-invasively and, so, is very powerful. It should be readily adaptable to measurements of pH, $Ca^{2+}$, and membrane-potential indicating dyes (from Lipton, 1973b).

with subsequent collection of the effluent, and this is very convenient for studies in which efflux of neurotransmitters or ions are measured. A well-designed chamber of this kind is described for the measurement of release of endogenous dopamine from the striatum (Milner and Wurtman, 1984). It allows effluxes from a large mass of tissue to be measured.

### 4.2.2. Chemical Stimulation

There are two widely used methods of stimulating slices chemically; that is, without applying electric fields. These are (1) by increasing the concentration of extracellular $K^+$ to between 15 and 60 mM or sometimes higher (this should be done by isoosmotic replacement of NaCl with KCl), or (2) by using veratridine or a mixture of veratrum alkaloids. Concentrations of between 1 and 50 $\mu M$ are used. As with electrical stimulation, both of these additions produce large increases in respiration and release of transmitters. The techniques here are trivial. They simply involve a change of buffer composition, and chamber design should allow rapid changes to be made. The "stimulating" conditions can be maintained for many minutes.

### 4.2.3. Effects of Electrical Stimulation and Chemical Stimulations on Tissue

Because both of these techniques are used to try to mimic effects of electric activity on cell function, it is important to note and compare the effects each has on the properties of the slice. Both methods produce a profound membrane depolarization of neurons. Although spiking during electrical stimulation has not been measured, a few minutes of 10 V, 100 Hz stimulation depolarizes the membrane by about 40 mV (Hillman et al., 1963). This is similar to the depolarization produced by 60 mM $K^+$ in neurons of olfactory slices (Scholfield, 1978). The depolarization by electrical stimulation is almost certainly a result of the large decrease in intracellular $K^+$ that accompanies the electrical stimulation (Keesey et al., 1965). Both of these techniques produce isoosmotic tissue swelling, although the effect of 60 mM $K^+$ is far greater than that of electrical stimulation. 15 mM $K^+$ produces the same swelling as that caused by maximal electrical stimulation, and 60 mM $K^+$ produces 4–5 times this amount (Okamoto and Quastel, 1970a,b; Lipton, 1973a).

There is good evidence that a large portion of the high $K^+$-induced swelling is in the glia (Hertz, 1973; Bourke et al., 1983), so the neuronal volume may not change much in this paradigm. Electrical stimulation causes a large, approximately equal increase in $Na^+$ and a decrease in $K^+$ content (Keesey et al., 1965).

Stimulation by high $K^+$ increases $Na^+$ and $K^+$, along with the large volume increase (Hertz, 1973). Much of this increase in $K^+$ appears to be glial (Walz and Hertz, 1983), as does the $Na^+$ increase (Bourke et al., 1983), so that high $K^+$ probably does not alter neuronal ion balance significantly. Increasing $K^+$ above 25 m$M$ causes a decrease in ATP and an even larger decrease in phosphocreatine; ATP values drop to about two thirds of their basal value when $K^+$ is increased to 50 m$M$ (Rolleston and Newsholme, 1967). Electrical stimulation at 100 Hz produces a similar decline in ATP in cortical slices (Okamoto and Quastel, 1970a; Jones and Banks, 1970).

Like the other depolarizing treatments, veratridine activates brain slice respiration by up to 50%. Low (5 $\mu M$) concentrations of veratridine produce increases of sodium and decreases of potassium of about equal magnitude. Above 5 $\mu M$ there is significant swelling and a loss of ATP (Okamoto and Quastel, 1970b; Lipton, 1973a; Lipton and Heimbach, 1978). Thus, 50 $\mu M$ veratridine, a drug concentration that is used quite extensively, reduces ATP by 55% (Lipton and Heimbach, 1978, unpublished data) and increases cell volume by almost 100% (Lipton, 1973a). All of these effects are blocked by tetrodotoxin at 1 $\mu g/mL$ and are thought to be neuronal, although it has recently been shown that glial cells do have veratridine-sensitive $Na^+$ channels (Reiser et al., 1983; Bowman et al., 1984).

## 4.3. Stimulation of Intact Pathways

Up until the last two years, the hippocampus and the olfactory cortex were the only preparations in which intact neural pathways could be stimulated to give postsynaptic responses. As indicated in Table 2, cochlear nucleus and neocortex are now being developed very nicely in this direction and there are preliminary reports that postsynaptic responses to stimuli can be obtained in striatum (Kitai and Kita, 1984). In our hippocampal studies we record extracellular potentials and these are adequate for most experiments aimed at answering neurochemical questions. Intracellular recordings in this preparation have been extensively described elsewhere (e.g., Dingledine et al., 1980). We place two tungsten microelectrodes (Frederick Haer, bipolar electrodes) on the pathway being stimulated and activate them via bipolar pulses of 200 $\mu s$ duration and 20–80 V amplitudes, generated by a WPI anapulse stimulator. The electrodes should be inserted about 150–200 $\mu m$ into the tissue (about a third to halfway into the slice) and, once inserted, should not be moved to another nearby position. Such a move usually destroys the pathway. The response is recorded by placing a 1–3 MΩ tungsten electrode (many workers use glass electrodes), also purchased from Haer, in the portion of the slice containing either the dendrites or the somata of the postsynaptic cells. The electrode should be lowered about 100 μm into the slice. The response may be optimized by small changes in depth and by moving the electrode within a given region. The electrode output is fed through a differential amplifier (WPI or Grass P15), and the output with respect to a reference electrode is monitored on a storage oscilloscope. A typical response is shown in Fig. 6.

There are now A–D converters that will allow good signal analysis of the evoked responses when coupled to a computer (e.g., RC Electronics, Santa Barbara, California). The magnitude of the response can be measured in two ways. The field excitatory postsynaptic potential (epsp) is best measured as the maximal slope of the first phase of the response (*see* Fig. 6A). The population spike is best measured as the average of the two peak-to-trough voltages of the spike (*see* Fig. 6B). It is important to be explicit about the measurement method being used. Various factors can produce artifactual changes in the measured response. Chief among these are temperature changes (relatedly), flow changes, electrode movements relative to the slice, and finally, changes in buffer oxygenation. If these factors are taken into consideration, stable recordings can be made from the hippocampal (and olfactory cortex) slice for many hours. The hippocampal slice can withstand stimulus frequencies of up to 0.5 Hz with no noticeable attentuation. Above this frequency, evoked responses in both the dentate gyrus (DG) and CA1 begin to fall to lower, but steady, values. DG is less able to conduct rapid impulses, at least in the guinea pig (Keating, unpublished). Frequencies above 3–4 Hz can rarely be sustained for longer than a minute and high frequency stimulation produces strange response patterns for short durations (Keating, unpublished).

There are, unfortunatey, no reported measurements of effects of pathway stimulation on metabolic parameters and ion concentrations (but, *see* section 6). It is generally assumed that these stimuli mimic in vivo pathway activity, but it should be noted that a very large number of axons are usually activated by these artifactual stimuli. This is probably rarely, if ever, true in vivo.

## 4.4. Comparison of Mechanisms for Evoking Transmitter Release

A major use of slices is to study stimulation-induced transmitter metabolism and release. This should ideally be done by stimulating direct pathways. However it has only been possible, to date, to use direct-pathway stimulation to measure release of *acidic amino acid transmitters* from hippocampal (Malthe-Sorensen et al., 1979; Wieraszko and Lynch, 1979; Corradetti et al., 1983) and olfactory cortex (Bradford and Richards, 1976) slices. Even in these preparations, as will be discussed later, measurements of effects of *normal* stimuli on release of *endogenous* transmitters have only been possible with sophisticated and expensive technology (Corradetti et al., 1983). It is thus necessary to use whole-slice stimulation for most problems. The preceding discussion indicated that all the usual ways of producing whole slice stimulation cause alterations in energy metabolism and, perhaps, ion balance. In that sense none of them are satisfactory. The question for the potential user is, then, whether one method is more physiological than the others.

Chemical methods are simpler, and because veratridine acts to open sodium channels, it would appear to be more physiological than high $K^+$ (Minchin, 1980). However, there is an important aspect of veratridine's action that makes it inadvisable to use, at least for amino acid neurotransmitters. Although veratridine-evoked release of catecholamines and acetylcholine is $Ca^{2+}$-dependent, as expected, the release of amino acid transmitters is actually greater in the absence of $Ca^{2+}$ than it is when $Ca^{2+}$ is present (Levi et al., 1980; Minchin, 1980). Stimulated release of adenosine is independent of extracellular $Ca^{2+}$ (Jonzon and Fredholm, 1984). Although the explanation might well be intriguing (Cunningham and Neal, 1981) (see below), the result strongly militates against the use of veratridine for release studies of amino acid and purine transmitters.

*A priori,* electrical field stimulation appears more physiological than stimulation by high $K^+$. However, this may not be the case. There is a good correlation between $Ca^{2+}$ uptake and adrenaline release caused by high $K^+$ (Kilpatrick et al., 1982) and, indeed, the $Ca^{2+}$ dependence of release of adenosine and GABA is stronger when high $K^+$ is used than when the slices are electrically stimulated (Szerb, 1979, 1982; Jonzon and Fredholm, 1984). High $K^+$ and electrical field stimulation produce qualitatively similar increases in release, synthesis, and accumulation of GABA in cortical slices (Szerb, 1979, 1982). For 5-hydroxytryptamine (5-HT; serotonin) also, field stimulation (Elks et al., 1979a) and high $K^+$ (Elks et al., 1979a; Auerbach and Lipton, 1985) activate release, synthesis, and accumulation of the transmitter to approximately the same extent. Dependence of release on tryptophan in the medium is far stronger during $K^+$ stimulation than during electrical field stimulation, and this is the only apparent difference between the two stimuli. Although few studies have assiduously compared the two modes of stimulation, the following

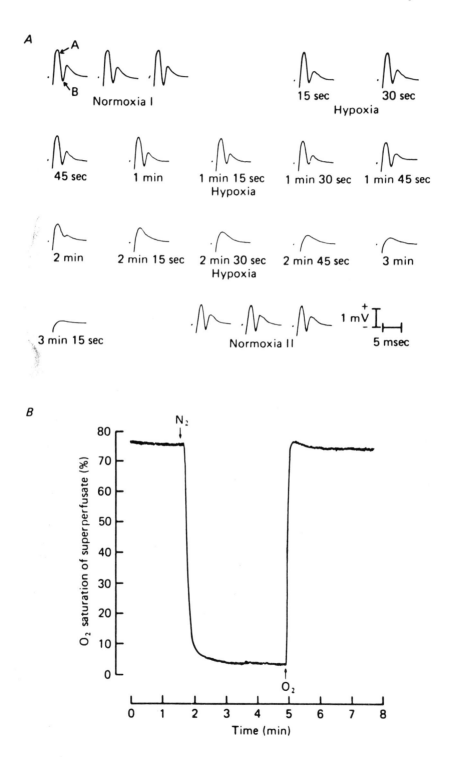

suggests that high $K^+$ may be closer to the physiological situation: It is apparent that electrical stimulation increases tissue $Na^+$ and decreases tissue $K^+$ concentrations (*see* section 4.2.3), and this is, in fact, thought to account for the $Ca^{2+}$-independent release of amino acid transmitters (Cunningham and Neal, 1981; Szerb, 1982). Pathway-evoked glutamate release is completely dependent upon extracellular $Ca^{2+}$ (Malthe-Sorensen et al., 1979), so that this altered ion balance probably does not occur during normal pathway stimulation. Thus, electrical field stimulation appears to functionally alter cell $Ca^{2+}$ in a way in which normal stimulation does not. This probably does not occur with high $K^+$ stimulation. This is suggested both by the apparent lack of an increased $Na^+$ and decreased $K^+$ in neurons, as discussed above (section 4.2.3), and by the complete extracellular $Ca^{2+}$ dependence of amino acid transmitter release by levels of high $K^+$. Thus, insofar as transmitter regulation is concerned, the $K^+$-depolarization might better mimic action potentials than does the electric field depolarization. High levels of $K^+$ will have an effect on the glia that electrical stimulation does not (Hertz, 1982) and this may confound metabolic measurements. The paucity of glial volume in, at least, hippocampal neuropil (Matthews et al., 1976) reduces the potential severity of this problem. Although the issue is by no means settled (there is a dearth of critical studies), the above considerations should be borne in mind by the worker who is about to begin a study of neurotransmitter regulation.

## 5. Neurotransmitter Studies

The slice is being widely used to determine transmitters in different pathways and also to determine modes of regulation of trans-

Fig. 6.   Field potential and chamber $O_2$ tension during anoxia in the dentate gyrus of the guinea pig hippocampal slice. (Top) Each trace denotes a typical field potential. The perforant path (*see* Fig. 9) is stimulated with bipolar tungsten electrodes and a recording electrode (tungsten) is placed in the dentate granule cell layer. (A) denotes the field epsp and (B) is the population spike. Recordings were made in the chamber described in Fig. 4. The bottom trace shows the time-course of oxygen tension in the chamber when the tap is suddenly turned to allow flow from the nitrogenated reservoir. Note the very rapid transition time, even at the flow rate of 25 mL/min. The field potential decays with time, with the population spike decreasing before the epsp. At 3 min, 15 s, only the stimulus artifact remains. The reponse recovers fully after 15 min of normoxia (from Lipton and Whittingham, 1979).

mitter release. I shall provide examples of approaches to these problems.

## 5.1. Application of Drugs and Putative Transmitters

There are four ways of applying neuroactive agents to the slice. The simplest is to add the substance to the bath. Other methods are the "nano-drop" method, in which a small drop of the agent is added to the top of a slice in an interface chamber, iontophoresis (Dingledine et al., 1980), and pressure ejection through microelectrodes. The pros and cons of these methods are well reviewed by Alger et al. (1984). The applications of the drugs or transmitters may be made while electrically stimulating afferent pathways (White et al., 1979), or while recording in the absence of pathway stimulation (Segal, 1980; Hablitz and Langmoen, 1982; Collingridge et al., 1983). A virtue of the iontophoresis or pressure application method is that drugs or transmitters may be applied to specific cell regions, for example dendrites or cell somata, which can provide important information (Segal, 1980). Bath application is most convenient for studying drug effects on evoked responses, but care must be taken to allow for effects of desensitization. It is therefore essential to monitor effects of the added substance over time.

## 5.2. Identifying Putative Transmitters by Release Studies

Malthe-Sorensen et al. (1979) incubated 600-$\mu$m thick guinea pig hippocampal slices at 37°C and loaded them with (2,3-$^3$-H)-D-aspartate. This D-isomer is nonmetabolizable and is used as a glutamate analog. The slice was then put into a perfusion cell and slowly perfused (0.25 mL/min). The effluent was collected. Release of $^3$H label into the superfusate was evoked by stimulating the Schaeffer collaterals in the stratum radiatum of CA1. The key in these types of experiments is to evoke the release of a measurable amount of radioactive label above background levels via *direct pathway stimulation* (this is a trivial problem with whole slice stimulation) (Nadler et al., 1977). Weiraszko and Lynch (1979) did this by withdrawing superfusate from directly above the stimulated region via a push–pull cannula. They stimulated at 3 Hz. Malthe-Sorensen et al. (1979) did this by using a slow perfusion rate, collecting the total effluent, and stimulating very rapidly (25 Hz). By stimulating for 15 s, they were able to maintain a short-lived but large evoked ($^3$H) efflux that was (still only) 50% above background levels, but was readily measurable. They showed that this release was completely $Ca^{2+}$-dependent and that neither leucine nor GABA were released by the stimulation.

Using a mass spectrometric analysis, it has recently been possible to measure release of *endogenous* glutamate and aspartate from the same CA1 slice region with more "physiological" stimulus frequencies of 2 Hz (Corradetti et al., 1983). At 32°C, at which basal release is suppressed, release of both glutamate and aspartate was increased about twofold over resting values and the effects were completely $Ca^{2+}$-dependent. Mass spectrometry was also used to demonstrate that stimulation increased the synthesis of glutamate and aspartate from glucose. This measurement of release of endogenous transmitter at low frequencies is far more satisfactory than the studies of Mathe-Sorenson et al. (1979), but does require a sophisticated and expensive analysis. However, this approach seems very promising and analysis of the glutamate and aspartate released (200 pmol/5 min/slice) should be possible on the more accessible high-pressure liquid chromatograph.

## 5.3. Regulation of Monoamine Release and Metabolism

Measurable monoamine release has not yet been evoked by direct pathway stimulation in a slice preparation. Relatively detailed studies of regulation of serotonin (5-hydroxytryptamine; 5-HT) release (Elks et al., 1979), dopamine release (Milner and Wurtman, 1984), and GABA release (Szerb, 1982) via electrical field stimulation have been reported. We studied the release of serotonin via high-$K^+$ stimulation (Auerbach and Lipton, 1985). The aim of the study was to test a model describing the dependence of serotonin release on its metabolism; that is, its synthesis and breakdown rates. The model is shown and described in Fig. 7. The most important predictions of this model are that release of transmitter during a sustained depolarization is biphasic and that the early (first 5 min or so) phase is largely dependent on Sro (the amount of serotonin in the releasable pool at the onset of depolarization) and the net entry of $Ca^{2+}$ into the cytosol and the second, constant-release-rate phase (10–40 min after depolarization begins) depends only on the rate at which serotonin is resupplied to the releasable pool. This in turn is postulated to depend on the net rate of serotonin synthesis during depolarization. Thus, the model predicts that the steady-state release of serotonin is a strong function of the rate of metabolism of serotonin. This is important because serotoninergic neurons generally discharge at a steady rate for prolonged periods. The model thus suggests that regulation of serotonin release might best be achieved by regulating its metabolism.

To test the model, we measured the kinetics of the release of serotonin from rat hippocampal slices during high $K^+$-induced depolarization. The major experimental problem is to measure the small amount of endogenous serotonin that is released during 1- and 2-min intervals. The slices from two complete hippocampi are distributed into two transfer holders, as described in section 3.2. In order to compare results from the two holders, the slices must be distributed so that alternating slices are put in each holder. This is because of the very steep gradient of serotonin content along the hippocampal axis (Auerbach and Lipton, 1982). Slices are preincubated for 60 min. This allows serotonin levels in the slice to become steady at values that are, in fact, very close to *in situ* values (Auerbach and Lipton, 1982). At this point slices are transferred into either normal buffer or high-$K^+$ buffer. They are transferred every minute and then at longer intervals, as shown in Fig. 8. Tissue (at the end of the experiment) and buffers are analyzed for serotonin and 5-hydroxyindoleacetic acid (5-HIAA) by HPLC (Auerbach and Lipton, 1985). Buffer is kept on ice in the dark until analysis. This should not be for more than 1 d. Release during a 1-min interval from a whole hippocampus in 18 m$M$ $K^+$ buffer is about 0.5 ng. Using a buffer volume of 2 mL, 50 $\mu$L of this, containing about 12 pg of serotonin, can be analyzed on the HPLC. With *careful treatment*, the noise level of this method of analysis corresponds to about 2 pg of monoamine. In order to obtain reliable results, 50 $\mu M$ EDTA must be included in all buffers to prevent oxidation of the amines; ascorbate, which is often used, is not suitable because it inhibits serotonin synthesis and release. In addition, care must be taken to see that all slices are continually "dancing"; this is important. With these precautions, very reliable measurements of tissue and released serotonin and 5-HIAA can be made so that synthesis and release of the monoamine can be measured accurately.

Figure 8 shows two key results that substantiate the predictions made by the model. Figure 8a shows that increasing $K^+$ greatly enhances the early (3 min) rate of serotonin release, but has little effect on the final, steady-state rate of release. This is consistent with the fact that the primary effect of elevating $K^+$ from 30 to 60 m$M$ is to increase the influx of $Ca^{2+}$. Figure 8b shows that when the level of the serotonin precursor trytophan is increased in the medium , there is both an increase in the rate of serotonin synthesis and, importantly, in the steady-state rate of release. The early release phase is also augmented (but not nearly as much as the later phase) because of increased serotonin in the releasable pool. Thus, predictions of the model are fulfilled and

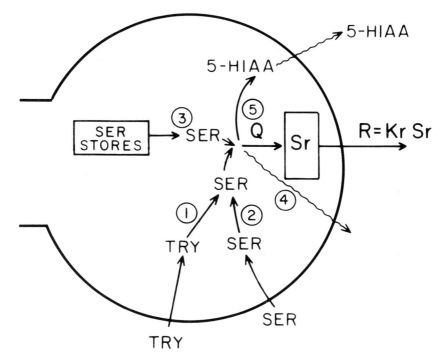

Fig. 7.  A model for the serotoninergic nerve terminal. Abbreviations: SER, serotonin; TRY, tryptophan; 5-HIAA, 5-hydroxyindolacetic
acid; Kr, release rate constant (/min) Sr, releasable pool of serotonin
(ng/mg prot); Q, rate of supply of serotonin to releasable pool (ng/mg
prot min); R, rate of release of serotonin from the releasable pool (ng/mg
prot min); SER STORES, serotonin in nerve terminal that is not in releasable pool and not directly available for transfer into the releasable pool.
Serotonin may be transferred into the releasable pool either from
(1) newly synthesized serotonin; (2) serotonin that has newly been taken
up by a high-affinity uptake system; or (3) from serotonin that has been
maintained in a storage compartment. Serotonin is released by depolarization at a rate equal to KrSr. Alternatively, serotonin may (4) move out
of the nerve ending in a depolarization-independent fashion; or (5) be
broken down into 5-HIAA. Kr is potential-dependent and serotonin
synthesis, breakdown, and transfer into Sr may, also, be potential-
dependent (from Auerbach and Lipton, 1985).

the slice is able to provide important information about the regulation of release of a neurotransmitter. The model and mode of
analysis should be quite applicable to other transmitters. The
question of the physiological applicability of high $K^+$ depolarization has been discussed above and, of course, remains an enigma.
However, it is encouraging that the calculated rates of serotonin

turnover, synthesis, and levels in the presence of 30 m$M$ K$^+$ are very similar to those observed during normal activity in the intact mouse brain (Tracqui et al., 1983). The ability to alter the environment, and its usefulness, are apparent in the slice studies.

## 6. Neural Activity and Energy Metabolism

### *6.1. Introduction*

One of the first, and still major, uses of the slice preparation has been to study the effects of electrical activity on energy metabolism and ion balance. This has been discussed in earlier sections and a particular experimental design was illustrated in Fig. 4. All

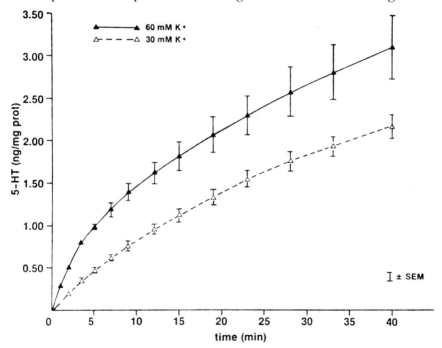

Fig. 8. Serotonin (5-HT) release from rat hippocampal slices by high-K$^+$ stimulation. A. Effects of different K$^+$ concentrations on release of serotonin. Rat hippocampal slices were preincubated for 60 min in normal buffer. They were then transferred to buffers containing 30 or 60 m$M$ K$^+$. The K$^+$ was substituted isoosmotically for Na$^+$ by replacing NaCl with KCl. Incubations were continued for 40 min; samples were moved between collection vials at the times indicated by the data points. All buffers contained 10 μ$m$ Try and 10 μm imipramine to prevent reuptake of serotonin. Values are means ±SEM for six experimental samples. △, K$^+$ = 30 m$M$; (▲), K$^+$ = 60 m$M$.

of these studies have involved whole slice stimulations (e.g., Fig. 5); so far there have been no good analyses of the metabolic consequences of discrete pathway stimulation in the slice. This will no doubt be remedied quite soon. Perhaps a more important problem, and one that *has* been addressed in the slice, is the effect of altered energy metabolism on electrical activity in discrete pathways (Lipton and Whittingham, 1979, 1982; Cox and Bachelard, 1982; Kass and Lipton, 1982; Fredholm et al., 1984). These more recent studies have taken over from earlier studies in which whole slice stimulation was used. In those early studies, the respiratory response to such stimulation was used as an index of the electrical activity of the slice and effects of different metabolic conditions on activity were studied in this way (McIlwain, 1962).

## 6.2. Effects of Anoxia on Synaptic Transmission

The two major issues being addressed in these kinds of studies are (1) the substrate dependence of synaptic transmission, and (2) the effects of anoxia (or ischemia) on synaptic transmission. The power of the hippocampal slice preparation is illustrated by studies of the effects of anoxia on synaptic transmission. The slice not only allows the extracellular environment of the tissue to be altered, but also, very importantly, allows the measurement of levels of high-energy metabolites in the same tissue region as that in

Fig. 8 *(Cont.)*

|  | 30 mM $K^+$ | 60 mM $K^+$ |
|---|---|---|
| Rates of serotonin synthesis (ng/mg protein/40 min) | 4.29 | 4.62 |
| Rates of serotonin breakdown | 3.40 | 5.41 |

B. Effects of altering extracellular tryptophan concentration on serotonin release in 60 mM $K^+$ buffer. Conditions and protocol were as described in the legend to Fig. 8. In these studies tissues were exposed to different concentrations of tryptophan for the final 30 min of the preincubation period and were then transferred to 60 mM $K^+$ buffer containing that same tryptophan concentration. All buffers contained 10 $\mu M$ imipramine and 10 $\mu M$ methiothepin. Values are means $\pm$ SEM for six experimental observations. (▲), O $\mu M$ tryptophan; (△), 2 $\mu M$ tryptophan; (●), 10 $\mu M$ tryptophan.

|  | 0 $\mu M$ Try | 2 $\mu M$ Try | 10 $\mu M$ Try |
|---|---|---|---|
| Rates of serotonin synthesis (ng/mg protein/40 min) | 2.76 | 3.49 | 5.52 |
| Rates of serotonin breakdown | 5.25 | 4.92 | 6.50 |

which synaptic transmission is being measured. This latter property was exploited in studies of the role played by ATP levels in the inhibition of synaptic transmission by anoxia (Lipton and Whittingham, 1982). In studies of whole brain, it had long been observed, unexpectedly, that the decay of electrical activity during anoxia preceded any measured reduction in brain ATP levels (Albaum et al., 1953; Siesjo et al., 1975). We monitored synaptic transmission in the dentate gyrus of the guinea pig hippocampal slice; as shown in Fig. 5, this transmission decays reversibly during anoxia (Lipton and Whittingham, 1979). We measured slice ATP levels at different times during anoxia. To do this we froze the tissue in liquid nitrogen and then extracted and analyzed ATP, as described by Lowry and Passonneau (1972). As shown in Table 3, when we analyzed ATP in the whole slice we found that it had not fallen at the time that synaptic transmission had begun to decay. This was consistent with the results on whole brain. However, when we went on to assay the ATP in the microdissected neuropil, where transmission was occurring (Fig. 9), we found that it had indeed fallen at the time that transmission had begun to decay. Thus, by exploiting the ability to microdissect the hippocampal slice, one can obtain quite important information that otherwise remains hidden. The ramification of this finding is

TABLE 3
ATP and Phosphocreatine (PCr) in Trimmed Slice and
Molecular Layer During Normoxia and Hypoxia[a]

|  | Normoxia | Hypoxia |
|---|---|---|
| Trimmed slice values, nmol/mg protein | | |
| ATP | $18.9 \pm 0.8$ | $17.0 \pm 1.2$ |
| PCr | $37.6 \pm 1.5$[b] | $31.1 \pm 2.7$[b] |
| Molecular layer values, nmol/mg protein | | |
| ATP | $23.5 \pm 1.1$[c] | $19.6 \pm 0.8$ |
| PCr | $55.6 \pm 3.1$[c] | $36.0 \pm 1.7$[c] |

[a]Tissue was frozen either immediately before exposure to hypoxia (normoxia) or during hypoxia at the time the evoked potential showed its first significant decrease (mean = 26 s of hypoxia perfusion). Twelve experiments for each value; values are means ±SEM. Differences between normoxic and hypoxic means are significant at:
[b]$P < 0.025$.
[c]$P < 0.005$.

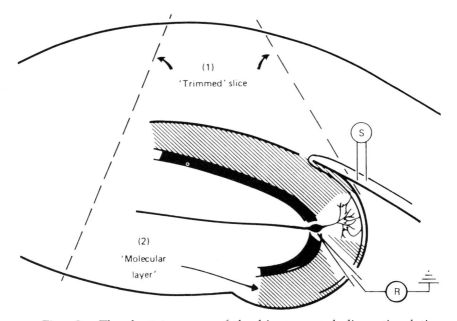

Fig. 9. The dentate gyrus of the hippocampal slice; stimulation and microdissection. The shaded area denotes the molecular layer of the dentate gyrus and the dark band is the layer of dentate granule cells. If the slice is microdissected at −20°C, the shaded region can be freed from the rest of the slice (Table 3 data are obtained in this way). If the lyophilized slice is microdissected, then most of the dark band stays with the shaded region, since this is the natural cleavage line of the tissue. The "trimmed" slice is the region taken for analysis of whole slice metabolites. It includes approximately two thirds of the slice, excluding only CA3 and CA2. The stimulating electrodes (S) are on the perforant path and the recording electrode (R) is in the dentate granule cell layer. The "cell" that is shown is a dentate granule cell and the dendritic tree shows the region where synapses are made with the incoming perforant path axon branches (from Lipton and Whittingham, 1982).

that metabolic correlates of neuronal activity can only be meaningfully studied by analyzing the regions in which the activity is occurring. This is not, however, always done (Fredholm et al., 1984), since it is somewhat (but not very) time-consuming. However, it is *essential* for these kinds of studies. Slice microdissection is most simply done by freezing and then lyophilizing the slice. Care must be taken during this process; if the slice becomes wet at any point, metabolite levels will decay profoundly. We precool the lyophilizing flasks and surround them with dry ice for the first hour of the lypohilizing process. After the slice is dried, it is very easy to remove the molecular layer of the dentate gyrus; it falls away when cut gently with a dissecting knife. The CA1 region can also be microdissected; the stratum radiatum, with the pyramidal cells, falls away along a natural cleavage. It is notable, though, that the microdissected pieces include much of the cell body layers of the respective regions. These cannot be separated from the neuropil in lyophilized tissue. If such a separation is necessary, then the dissection should be carried out at $-20°C$ (Lipton and Whittingham, 1982). This preserves metabolites and the tissue retains a consistency that is excellent for fine dissection. The dissector, however, should be well-clothed for such a procedure!

The accessibility of the slice milieu to ion changes has been exploited to deal with another important problem related to cerebral anoxia. Irreversible anoxic brain damage follows 10–15 min of exposure to anoxia (Siesjo, 1981). It has been postulated that the damage may result from $Ca^{2+}$ entry into the neurons during anoxia (Hass, 1981; Siesjo, 1981). However, this has been impossible to address by direct experiments *in situ*. We have developed the rat hippocampal slice as a model for irreversible anoxic damage and have been able to test this hypothesis (Kass and Lipton, 1982). Ten minutes of anoxia irreversibly abolishes synaptic transmission in the dentate gyrus and in the CA1 region. This is shown for the former in Fig. 10A. We directly investigated the role of $Ca^{2+}$ entry in this phenomenon by removing $Ca^{2+}$ from the bathing medium during the anoxia. As shown in Fig. 10B, this manipulation completely protects the transmission from irreversible damage. Thus, the accessibility of the slice allows significant progress to be made toward elucidating an important mechanism.

We have recently combined the microdissection approach with measurements of $Ca^{2+}$-uptake during anoxia and have demonstrated differential uptake into dentate gyrus (DG) and CA1 regions of the hippocampal slice. The CA1 region, which is more sensitive to anoxia than DG, shows a larger increase in $Ca^{2+}$-uptake during anoxia than does DG (Kass and Lipton, 1985, in preparation).

## 7. Conclusion

The brain slice preparation has been used to address many extremely important neurochemical questions. This has been particularly true for the hippocampal slice—surely one of the best model systems to have been discovered by neurobiologists. Although able to provide many results, quite readily, the preparation has distinct drawbacks and these have to be borne in mind by scientists who work with it. In this chapter I have tried to point

ADULT ∟

PREANOXIA
(NORMAL BUFFER)

PREANOXIA
(0 Ca⁺⁺, 10 mM Mg⁺⁺)

PREANOXIA
(NORMAL BUFFER)

10 MINUTES OF ANOXIA
(NORMAL BUFFER)

1 HOUR RECOVERY
(NORMAL BUFFER)

## OmM Ca⁺⁺ TREATMENT

**PREANOXIA**

**10 MINUTES OF ANOXIA**
(0 mM Ca⁺⁺)

**1 HOUR AFTER ANOXIA**

Fig. 10. Effects of 10 min anoxia on long-term recovery of field potentials in the dentate gyrus of the rat hippocampal slice. Recordings were made as for the guinea pig hippocampal slice, described in Fig. 6.

A. The anoxia was induced while the slice was superfused with normal buffers. Note that although the postsynaptic field potential does not recover at all, the presynaptic potential recovers fully from 10 min of anoxia (from Kass and Lipton, 1982).

B. $Ca^{2+}$ is removed from the buffer during exposure to anoxia. It is apparent that the field potential recovers almost completely 1 h after the anoxia (there is no presynaptic potential in this recording). In recent studies, in which $Ca^{2+}$ is removed for 5 min before and after anoxia as well as during the anoxia, we find consistent *complete* recovery of the response even after 20 min of anoxia (from Kass and Lipton, 1985, submitted for publication).

out some of these drawbacks, the caution with which one should use a particular method, and, also, the great power of the preparation when not abused.

# References

Alger B. E., Dhanjal S. S., Dingledine R., Garthwaite J., Henderson G., King G. L., Lipton P., North A., Schwartzkroin P. A., Sears T. A., Segal M., Whittingham T. S., and Williams J. (1984) Brain Slice Methods, in *Brain Slices* (Dingledine R., ed.), Plenum, New York.

Arnaiz G. R. de L. and DeRobertis E. (1971) Properties of isolated nerve endings. *Curr. Top. Membr. Trans.* **3**, 237–272.

Ashford C. A. and Dixon K. C. (1935) The effect of potassium on the glycolysis of brain tissue with reference to the Pasteur effect. *Biochem. J.* **29**, 157–168.

Auerbach S. and Lipton P. (1985) Regulation of serotonin release from the *in vitro* rat hippocampus; effects of alterations in levels of depolarization and in rates of serotonin metabolism. *J. Neurochem.* **44**, 1116–1130.

Auerbach S. and Lipton P. (1982). Vasopressin augments depolarization -induced release and synthesis of serotonin in hippocampal slices. *J. Neurosci.* **2**, 477–482.

Baethmann A. and Sohler K. (1975) Electrolyte and fluid spaces of rat brain *in situ* after infusion with dinitrophenol. *J. Neurobiol.* **6**, 73–84.

Bak I. J., Misgeld U., Weiler M., and Morgan E. (1980) The preservation of nerve cells in rat neostriatal slices maintained *in vitro;* a morphological study. *Brain Res.* **197**, 341–353.

Benjamin A. M. and Verjee Z. H. (1980) Control of aerobic glycolysis in the brain *in vitro. Neurochem. Res.* **5**, 921–934.

Bennett G. W., Sharp T., Marsden C. A., and Parker T. L. (1983) A manually operated brain tissue slicer suitable for neurotransmitter release studies. *J. Neurosci. Methods* **7**, 107–115.

Booth R. F. G., Harvey S. A. K., and Clark J. B. (1983) Effects of *in vitro* hypoxia on acetylcholine synthesis by rat brain synaptosomes. *J. Neurochem.* **40**, 106–110.

Bourke R. S., Kimelberg H. K., Daze M., and Church G. (1983) Swelling and ion uptake in cat cerebrocortical slices: control by neurotransmitters and ion transport mechanisms. *Neurochem. Res.* **8**, 5–24.

Bower J. M. and Haberly L. B. (1984) Single pyramidal neurons in piriform cortex receive excitatory inputs from facilitating and nonfacilitating synapses associated with different fiber systems. *Soc. Neuroscience* (Abs.) **10**, 546.

Bowman C. L., Kimelberg H. K., Frangakis M. V., Netter Y., and Edwards C. (1984) Astrocytes in primary culture have chemically activated sodium channels. *J. Neurosci.* **4**, 1527–1534.

Bradford H. F. and Richards C. D. (1976) Specific release of endogenous glutamate from piriform cortex stimulated *in vitro. Brain Res.* **105,** 168–172.

Brown T. H. and Johnston D. (1983) Voltage clamp analysis of mossy fiber synaptic input to hippocampal neurons. *J. Neurophysiol.* **50,** 487–507.

Collingridge G. L., Kehl S. J., and McLennan H. (1983) The antagonism of amino acid-induced excitation of rat hippocampal CA1 neurones *in vitro. J. Physiol.* (Lond.) **334,** 19–31.

Connors B. W., Gutnick M. J., and Prince D. A. (1982) Electrophysiological properties of neocortical neurons *in vitro. J. Neurophysiol.* **48,** 1302–1320.

Corradetti R., Moneti G., Moroni F., Pepeu G., and Wieraszko A. (1983) Electrical stimulation of the stratum radiatum increases the release and neosynthesis of aspartate, glutamate and γ-aminobutyric acid in hippocampal slices. *J. Neurochem.* **41,** 1518–1525.

Cox D. W. G. and Bachelard H. S. (1982) Attentuation of evoked field potentials from dentate granule cells by low glucose, pyruvate + malate, and sodium fluoride. *Brain Res.* **239,** 527–534.

Cunningham J. and Neal M. J. (1981) On the mechanism by which veratridine causes a calcium-dependent release of γ-aminobutyric acid from brain slices. *Br. J. Pharmacol.* **73,** 655–667.

Deutsch C., Drown C., Rafalowska U., and Silver I. A. (1981) Synaptosomes from rat brain; morphology, compartmentation and transmembrane pH and electrical gradients. *J. Neurochem.* **36,** 2063–2072.

Dingledine R., Dodd J., and Kelly J. S. (1980) The *in vitro* brain slice as a useful neurophysiological preparation for intracellular recording. *J. Neurosci. Methods* **2,** 323–362.

Dingledine R. and Langmoen L. A. (1980) Conductance changes and inhibitory actions of hippocampal recurrent IPSP's. *Brain Res.* **185,** 277–287.

Dunlop D. S., Van Elden W., and Lajtha A. (1975) Optimal conditions for protein synthesis in incubated slices of rat brain. *Brain Res.* **99,** 303–318.

Elks M. L., Youngblood W. W., and Kizer J. S. (1979a) Synthesis and release of serotonin by brain slices; effect of ionic manipulations and cationic ionophores. *Brain Res.* **172,** 461–469.

Elks M. L., Youngblood W. W., and Kizer J. S. (1979b) Serotonin synthesis and release in brain slices; independence of tryptophan. *Brain Res.* **172,** 471–486.

Fredholm B. B., Dunwiddie T. V., Bergman B., and Lindstrom K. (1984) Levels of adenosine and adenine nucleotides in slices of rat hippocampus. *Brain. Res.* **295,** 127–136.

Frotscher M., Misgeld U., and Nitsch C. (1981) Ultrastructure of mossy fiber endings in *in vitro* hippocampal slices. *Exp. Brain Res.* **41,** 247–255.

Fujii T., Baumgartl H., and Lubbers D. W. (1982) Limiting section thickness of guinea pig olfactory cortical slices studied from tissue $pO_2$ values and electrical activities. *Pflugers Arch.* **393**, 83–87.

Fukada Y. and Loeschke. H. H. (1977) Effect of $H^+$ on spontaneous neuronal activity in the rat medulla oblongata *in vitro. Pflugers Arch.* **371**, 125–134.

Garthwaite J., Woodhams P. L., Collins M. J., and Balazs R. (1979) On the preparation of brain slices; Morphology and cyclic nucleotides. *Brain Res.* **173**, 4373–377.

Ghajar J. B. G., Plum F., and Duffy T. E. (1982) Cerebral oxidative metabolism and blood flow during acute hypoglycemia and recovery in anesthetized rats. *J. Neurochem.* **38**, 397–408.

Haas H. L., Schaerer B., and Vosmansky M. (1979) A simple perfusion chamber for the study of nervous tissue slices *in vitro. J. Neurosci. Methods.* **1**, 323–325.

Haberly L. and Feig S. (1983) Structure of the piriform cortex of the opossum. II. Fine structure of cell bodies and neuropil. *J. Comp. Neurol.* **216**, 69–88.

Hablitz J. J. and Langmoen I. A. (1982) Excitation of hippocampal pyramidal cells by glutamate in the guinea-pig and rat. *J. Physiol.* (Lond.) **325**, 317–331.

Harvey J. A., Scholfield C. N., and Brown, D. A. (1974) Evoked surface positive potentials in isolated mammalian olfactory cortex. *Brain Res.* **76**, 235–245.

Hass W. K. (1981) Beyond Cerebral Blood Flow, Metabolism and Ischemic Thresholds; Examination of the Role of Calcium in the Initiation of Cerebral Infarction, in *Cerebral Vascular Disease,* vol. 3. (Meyer J. S., Reivich M., Ott E., and Arabinar A., eds.) pp. 3–17. Excerpta Medica, Amsterdam.

Hatton G. I. (1984) Hypothalamic Neurobiology, in *Brain Slices* (Dingledine R., ed.), pp. 341–374. Plenum, New York.

Henderson G., Pepper C. M., and Shefner S. A. (1982) Electrophysiological properties of neurons contained in the locus coeruleus and mesenphalic nucleus of the trigeminal nerve *in vitro. Exp. Brain Res.* **45**, 29–37.

Hertz L. (1973) $K^+$ *Effects on Metabolism in the Adult Mammalian Brain* In Vitro. Arhus, Copenhagen.

Hertz L. (1982) Astrocytes, in *Handbook of Neurochemsitry,* Vol. 1, 2nd Ed. (Lajhta A., ed.) pp. 319–355. Plenum, New York.

Hertz L., Bock E., and Schousboe A. (1979) GFA content, glutamate uptake and activity of glutamate metabolizing enzymes in differentiating mouse astrocytes in primary cultures. *Dev. Neurosci.* **1**, 226–238.

Hillman H. H., Campbell W. J., and McIlwain H. (1963) Membrane potentials in isolated and electrically stimulated mammalian cerebral cortex. *J. Neurochem.* **10**, 325–339.

Hirsch J. A. and Oertel D. (1984) Intracellular recordings from brain slices of the dorsal cochlear nucleus of the mouse. *Soc. Neurosci.* (Abs.) **10**, 842.

Hyatt M. C. and Tyce G. M. (1984) Effects of estradiol on the basal and evoked efflux of norepinephrine and 5-hydroxytryptamine from slices of rat hypothalamus. *Life Sci.* **35,** 2269–2274.

Jeffreys J. G. R. (1981) The Vibroslice, a new vibrating blade tissue slicer. *J. Physiol.* (Lond.) **324,** 2P.

Jobsis F. F., Keizer J., LaManna J. C., and Rosenthal M. (1977) Reflectance spectrophotometry of cytochrome aa₃ *in vivo J. Appl. Physiol.* **43,** 858–872.

Jones C. T. and Banks P. (1970) The effect of electrical stimulation and ouabain on the uptake and efflux of L-($^{14}$C) valine in chopped tissue from guinea pig cerebral cortex. *Biochem. J.* **118,** 801–812.

Jonzon B. and Fredholm B. B. (1984) Release of purines, noradrenaline and GABA from rat hippocampal slices by field stimulation. *J. Neurochem.* **44,** 217–224.

Karlsson U. and Schulz R. L. (1965) Fixation of the central nervous system for electron microscopy by aldehyde perfusion. *J. Ultrastruct. Res.* **12,** 160–186.

Kass I. S. and Lipton P. (1982) Mechanisms involved in irreversible damage to the *in vitro* hippocampal slice. *J. Physiol.* (Lond.) **332,** 459–472.

Katzmann R. and Pappius H. M. (1973) *Brain Electrolytes and Fluid Metabolism.* Williams & Wilkins, Baltimore.

Keesey J. C., Wallgren H., and McIlwain H. (1965) The sodium, potassium and chloride of cerebral tissues; maintenance, change on stimulation and subsequent recovery. *Biochem. J.* **85,** 289–300.

Kilpatrick D. L., Slepetis R. J., Corcoran J. J., and Kirshner N. (1982) Calcium uptake and catecholamine secretion by cultured bovine adrenal medullary cells. *J. Neurochem.* **38,** 427–435.

Kiss J. (1977) Synthesis and transport of newly formed proteins in dendrites of rat hippocampal pyramidal cells. An electron microscope autoradiographic study. *Brain Res.* **124,** 237–250.

Kitai S. T. and Kita H. (1984) Electrophysiological Study of the Neostriatum in Brain Slice Preparations, in *Brain Slices* (Dingledine R., ed.) Plenum, New York, pp. 285–296.

Langmoen I. A. and Anderson P. (1981) The Hippocampal Slice *In Vitro.* A Description of the Technique and Some Examples of the Opportunities It Offers, in *Electrophysiology of Isolated Mammalian CNS Preparations.* (Kerkut G. A. and Wheal H; eds.) pp. 51–105 Academic, New York.

Langmoen I. A., Segal M., and Anderson P. (1981) Mechanisms of norephinephrine actions on pyramidal cells *in vitro. Brain Res.* **208,** 349–363.

Levi G., Gallo V., and Raiteri M. (1980) A reevaluation of veratridine as a tool for studying the depolarization-induced release of neurotransmitters from nerve endings. *Neurochem. Res.* **5,** 281–295.

Lighthall J. W., Park M. R., and Kitai S. T. (1981) Inhibition in slices of rat neostriatum. *Brain Res.* **212,** 182–187.

Lipton P. (1973a) Effects of membrane depolarization on light scattering by cerebal cortical slices. *J. Physiol.* (Lond.) **231,** 365–383.

Lipton P. (1973b) Effects of membrane depolarization on nicotinamide-nucleotide fluorescence in brain slices. *Biochem. J.* **136**, 999–1009.

Lipton P. and Heimbach C. J. (1978) Mechanisms of extracellular potassium stimulation of protein synthesis in guinea pig hippocampal slices *J. Neurochem.* **31**, 1299–1307.

Lipton P. and Robacker K. M. (1983) Glycolysis and brain function; Ko$^+$-stimulation of protein synthesis and K$^+$ uptake require glycolysis. *Fed Proc.* **42**, 873–880.

Lipton P. and Whittingham T. S. (1982) Reduced ATP concentration as a basis for synaptic transmission failure during hypoxia in the *in vitro* guinea pig hippocampus. *J. Physiol.* (Lond.) **235**, 52–65.

Lipton P. and Whittingham T. S. (1979) The effect of hypoxia on evoked potentials in the in vitro hippocampus. *J. Physiol.* (Lond.) **287**, 427–438.

Lipton P. and Whittingham T. S. (1984) Energy Metabolism and Brain Slice Function, in *Brain Slices,* (Dingledine R., ed.) pp. 113–153. Plenum, New York.

Llinas R. and Sugimori M. (1980) Electrophysiological properties of *in vitro* Purkinje cell dendrites in mammalian cerebellar slices. *J. Physiol.* (Lond.) **305**, 197–213.

Lowry O. H. and Passonneau J. V. (1972) *A Flexible System of Enzyme Analysis.* Academic, New York.

Lund-Andersen H. (1974) Extracellular and intracellular distribution of inulin in rat brain cortex slices. *Brain Res.* **65**, 239–254.

Lynch G. and Schubert P. (1980) The use of *in vitro* brain slices for multidisciplinary studies of synaptic function. *Ann. Rev. Neurosci.* **3**, 1–22.

Lynch G., Kessler M., Halpin S., and Baudry M. (1983) Biochemical effects of high-frequency synaptic activity studied with *in vitro* slices. *Fed. Proc.* **42**, 2886–2890.

Malthe-Sorenson D., Skrede K. K., and Fonnum F. (1979) Calcium-dependent release of D-$^3$H aspartate evoked by selective electrical stimulation of excitatory afferent fibers to hippocampal pyramidal cells *in vitro. Neurosci.* **4**, 1255–1263.

Matthews D. A., Cotman C. A., and Lynch G. A. (1976) An electron microscope study of lesion-induced synaptogenesis in the dentate gyrus of the adult rat. *Brain Res.* **115**, 1–15.

McIlwain H. (1951) Metabolic response in vitro to electrical stimulation of sections of mammalian brain. *Biochem. J.* **49**, 382–393.

McIlwain H. (1962) Electrical Pulses and the *In Vitro* Metabolism of Cerebral Tissues, in *Neurochemistry* (Eliot K. A. C., Page I. H., and Quastel J. H., eds.) Thomas, Springfield.

McIlwain H. (1975) Electrical Stimulation of the Metabolism of Isolated Neural Tissues, in *Practical Neurochemistry,* (McIlwain H., ed.) pp. 159–190. Churchill Livingstone, New York.

McIlwain H. and Joanny P. (1963) Characteristics required in electrical pulses of rectangular time-voltage relationships for metabolic change and ion movements in mammalian cerebral tissues. *J. Neurochem.* **10**, 313–323.

McIlwain H. and Bachelard H. S. (1971) *Biochemistry and the Central Nervous System.* Churchill Livingstone, Edinburgh.

Milner J. D. and Wurtman R. J. (1984) Release of endogenous dopamine from electrically stimulated slices of rat striatum. *Brain Res.* **301,** 139–142.

Minchin M. C. W. (1980) Veratrum alkaloids as transmitter-releasing agents *J. Neurosci. Methods* **2,** 111–124.

Misgeld U. and Froetscher M. (1982) Dependence of the viability of neurons in hippocampal slices on oxygen supply. *Brain Res. Bull.* **8,** 95–100.

Mueller A. L., Taube J. S. and Schwartzkroin P. A. (1984) Development of hyperpolarizing inhibitory potentials and hyperpolarizing response to γ-aminobutyric acid in rabbit hippocampus studied *in vitro. J. Neurosci.* **4,** 860–867.

Nadler J. V., White W. F., Vaca K. W., Redburn D. A., and Cotman C. W. (1977) Characterization of putative amino acid transmitter release from slices of rat dentate gyrus. *J. Neurochem.* **29,** 279–290.

Nicoll R. A. and Alger B. E. (1981) A simple chamber for recording from submerged brain slices. *J. Neurosci. Methods* **4,** 153–156.

Norenberg M. D. and Martinez-Hernandez A. (1979) Fine structural localization of glutamine synthetase in astrocytes of rat brain. *Brain Res.* **161,** 303–310.

Oertel D. (1983) Synaptic responses and electrical properties of cells in brain slices of the mouse anteroventral cochlear nucleus. *J. Neurosci.* **3,** 2043–2053.

Okamoto K. and Quastel J. H. (1970a) Water uptake and energy metabolism in brain slices from rat. *Biochem. J.* **120,** 25–36.

Okamoto K. and Quastel J. H. (1970b) Tetrodotoxin-sensitive uptake of ions and water by slices of rat brain *in vitro. Biochem. J.* **120,** 37–47.

Pastuszko A., Wilson D. F., Erecinska M., and Silver I. A. (1981) Effects of *in vitro* hypoxia and lowered pH on potassium fluxes and energy metabolism in rat brain synaptosomes. *J. Neurochem.* **36,** 116–123.

Peters A., Palay S. L., and Webster H. deF (1975) *The Fine Structure of the Nervous System: The Neurons and Supporting Cells.* Saunders, Philadelphia.

Potashner S. J. (1978) The spontaneous and electrically evoked release from slices of guinea pig cerebral cortex of endogenous amino acids labelled via metabolism of D-(U-C14) glucose. *J. Neurochem.* **31,** 177–186.

Pulsinelli W. A., Brierly J. B., and Plum F. (1982) Temporal profile of neuronal damage in a model of transient forebain ischemia. *Ann. Neurol.* **11,** 497–498.

Pulsinelli W. A. and Duffy T. E. (1983) Regional energy balance in rat brain after transient forebrain ischemia. *J. Neurochem.* **40,** 1500–1503.

Quastel J. H. (1939) Respiration in the central nervous system. *Physiol. Rev.* **19,** 135–183.

Rafalowska U., Erecinska M., and Wilson D. F. (1980) Energy metabolism in rat brain synaptosomes from nembutal-anesthetized and non-anesthetized animals *J. Neurochem.* **34,** 1380–1386.

Rawlins J. N. P. and Green K. F. (1977) Lamellar organization in the rat hippocampus. *Exp. Brain Res.* **28,** 335–344.

Reiser G., Loffler F., and Hamprecht B. (1983) Tetrodotoxin-sensitive ion channels characterized in glial and neuronal cells from rat brain. *Brain Res.* **261,** 335–340.

Richards C. D. and Sercombe R. (1970) Calcium, magnesium and the electrical activity of guinea pig olfactory cortex *in vitro J. Physiol.* (Lond.) **211,** 571–584.

Rolleston F. S. and Newsholme E. A. (1967) Control of glycolysis in cerebral cortical slices. *Biochem. J.* **104,** 524–533.

Scholfield C. N. (1978) Electrical properties of neurons in the olfactory cortex slice *in vitro. J. Physiol. (Lond.)* **275,** 535–546.

Schulze E. (1980) Untersuchungen zur Transmitterhypothese von Taurin. PhD Thesis, University of Gottingen, 116–131.

Schwartzkroin P. A. (1975) Characteristics of CA1 neurons recorded intracellularly in the hippocampal *in vitro* slice preparation. *Brain Res.* **85,** 423–436.

Scott I. D. and Nicholls D. G. (1980) Energy transduction in intact synaptosomes. *Biochem. J.* **186,** 21–33.

Segal M. (1980) The action of serotonin in the rat hippocampal slice preparation. *J Physiol.* (Lond.) **303,** 423–439.

Siesjo B. K. (1981) Cell damage in the brain, a speculative synthesis. *J. Cereb. Blood Flow Metab.* **1,** 155–185.

Siesjo B. K., Ingvar M., Folbergova J., and Chapman A. (1983) Local circulation and metabolism in bicuculline-induced status epilepticus; relevance for development of cell damage. *Adv. Neurol.* **34,** 217–230.

Siesjo B. K., Johannsson H., Norberg K. and Salford L. (1975). Brain Function Metabolism and Blood Flow in Moderate and Severe Arterial Hypoxia, in *Brain Work,* (Alfred Benzon Symposium VIII), pp. 101–125, Munksgaard, Copenhagen.

Skrede K. K. and Westgaard R. H. (1971) The transverse hippocampal slice; a well-defined cortical structure maintained *in vitro. Brain Res.* **35,** 589–593.

Sokoloff L. (1980) The ($^{14}$C) Deoxyglucose Method for the Quantitative Determination of Local Cerebral Glucose Utilization. Theoretical and Practical Considerations, in *Cerebral Metabolism and Neural Function* (Passonneau, J. V. and Hawkins, R. A., eds.), pp. 319–330. Williams & Wilkins, Baltimore.

Soubrie P., Reisne T. D., and Glowinski, J. (1984) Functional aspects of serotonin transmission in the basal ganglia; a review and an *in vivo* approach using the push–pull cannula technique. *Neuroscience* **13,** 605–625.

Stafstrom C. E., Schwindt P. C. and Crill W. E. (1982) Negative slope conductance due to a persistent subthreshold sodium current in cat neocortical neurons *in vitro. Brain Res.* **236,** 221–226.

Steward O. and Levy W. B. (1982) Preferential localization of polyribosomes under the base of dendritic spines in granule cells of the dentate gyrus. *J. Neurosci.* **2,** 284–291.

Szerb J. C. (1979) Relationship between $Ca^{2+}$-dependent and independent release of ($^3$H) GABA evoked by high $K^+$, veratridine or electrical stimulation from rat cortical slices. *J. Neurochem.* **32,** 1565–1573.

Szerb J. C. (1982) Effect of nipecotic acid, a γ-aminobutyric acid transport inhibitor, on the turnover and release of γ-aminobutyric acid in rat cortical slices. *J. Neurochem.* **39,** 850–858.

Teichberg V. I., Goldberg O., and Luini A. (1981) The stimulation of ion fluxes in brain slices by glutamate and other excitatory amino acids. *Mol. Cell. Biochem.* **39,** 281–295.

Teyler T. J. (1980) Brain slice preparation: Hippocampus. *Brain Res. Bull.* **5,** 391–403.

Thomas J. (1957) The composition of isolated cerebral tissues: Purines. *Biochem. J.* **66,** 655–658.

Tracqui P., Morot-Gaudry Y., Staub J. F., Brezillon P., Staub S. M., Borgouin S., and Hamon M. (1983) Model of brain serotonin metabolism. II. Physiological interpretation. *Am. J. Physiol.* **244,** R206–215.

Traub R. D. and Wong R. K. S. (1983) Synchronized burst discharge in the disinhibited hippocampal slice. II. Model of the cellular mechanism. *J. Neurophysiol.* **49,** 459–471.

Turner D. A. and Schwartzkroin P. A. (1983) Electrical characteristics of dendrites and dendritic spines in intracellularly-stained CA3 and dentate neurons. *J. Neurosci.* **3,** 2381–2394.

Walz W. and Hertz L. (1983) Intracellular ion changes of astrocytes in response to extracellular potassium. *J. Neurosci. Res.* **10,** 411–423.

White W .F., Nadler J. V., and Cotman C. W. (1979) The effect of acidic amino acid antagonists on synaptic transmission in the hippocampal formation *in vitro. Brain Res.* **164,** 177–194.

Whittingham T. S. (1980) Investigation of events leading to neuronal transmission failure in the hippocampal slice during anoxia. PhD Thesis, Univ. of Wisconsin.

Whittingham T. S., Lust W. D., Christakis D. A., and Passonneau J. V. (1984) Metabolic stability of hippocampal slice preparations during prolonged incubation. *J. Neurochem.* **43,** 689–696.

Wieraszko A. and Lynch G. (1979) Stimulation-dependent release of possible transmitter substances from hippocampal slices studied with localized perfusion. *Brain Res.* **160,** 372–376.

Yamamoto C. and McIlwain H. (1966) Electrical activities on thin sections from the mammalian brain maintained in a chemically-defined medium *in vitro J. Neurochem.* **13,** 1333–1343.

# Chapter 4

# Cell and Tissue Cultures

L. Hertz, B. H. J. Juurlink, S. Szuchet, and W. Walz

## 1. Introduction

The methodology of tissue culture was introduced more than three quarters of a century ago by Harrison (1907) as "a method by which the end of a growing nerve could be brought under direct observation while alive, in order that a correct conception might be had regarding what takes place as the fiber extends during embryonic development from the nerve center out to the periphery" (Fig. 1). Thus, from the very beginning a major purpose of tissue culture studies has been to provide information about events occurring in vivo, and many of the early tissue culture studies were carried out to enable direct observations of living cells.

Cultured cells also offer a number of other advantages over other preparations in investigating nervous system function and development. Two major advantages are the ability to manipulate living cells and rigorously control the extracellular environment in which these cells reside, i.e., the culture medium. The presently used culture media consist mainly of balanced salt solutions containing glucose, amino acids, vitamins, and, in general, antibiotics. Serum is added to supplement the chemically defined medium with lipids, growth factors, and trophic factors, as well as essential trace elements. The identity of these trophic factors, which may be of importance during normal development as well

as during regeneration, are generally unknown. They are, however, currently being identified by a number of laboratories, and
completely chemically defined media are being formulated for a
number of cell types, also from the nervous system (e.g., Barnes
and Sato, 1980; Bottenstein, 1983). Whether such completely
chemically defined media are of major advantage depends upon
the type of question one is interested in answering. For studies of
trophic factors they are often essential, but it should be remembered that such media are often formulated to give maximal cell
proliferation and may not necessarily allow cells to differentiate or
maintain their terminally differentiated state.

Culture methods also offer the advantage of working with
one type of viable cells. The issue of homogeneity must be
stressed, especially when biochemical properties are being exam-

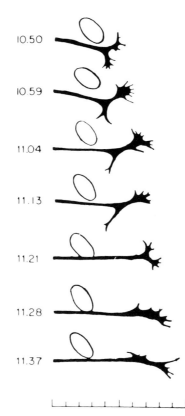

Fig. 1.  Movements and changes in shape of the growing tip of a
frog embryo nerve fiber during 47 min, at 4 d in culture. An erythrocyte
serves as marker (after Harrison, 1907—from Murray 1965).

ined, because these are bulk properties and, therefore, cannot be discriminated on a per cell basis. The different cell types may be selected on the basis of certain cell biological properties, e.g., proliferation or lack of proliferation at specific developmental stages in vivo (primary cultures), or they may be obtained from transformed cells (cell lines). One can thus analyze the properties of live cells of one cell type in isolation from other cell types, often difficult to do with other preparations. This ability is of special importance for studies of the nervous system, which consists of a multitude of intricately interwoven cells of different types. In addition, culture methods make possible selectively recombining particular cell types at will, enabling one to examine cell-to-cell interactions. This has been done elegantly in primary cultures with the examination of the cellular interactions of neurons, Schwann cells, and fibroblasts during peripheral nervous system myelination or with examination of oligodendrocytes, astrocytes, and neurons during central nervous system myelination (section 4.4). To counterbalance these advantages it should, however, be kept in mind that successful establishment of primary cultures from the central nervous system generally requires that one initiates the cultures with immature, relatively undifferentiated cells, thus bringing up questions of whether such immature cells will differentiate in a similar manner in vitro as in vivo. This question is even more crucial in the case of cell lines that are constituted of transformed, dedifferentiated cells.

The final feature we would like to consider is that cultured cells often live in a flat two-dimensional universe, i.e., they constitute a monolayer. This offers a considerable number of advantages when examining membrane phenomena, such as ion fluxes or transmitter binding. On the other hand, this may induce novel properties or exaggerate certain properties of the cells in question as compared to cells living in a three-dimensional universe in vivo or in reaggregating cultures (Morris and Moscona, 1971; Kozak et al., 1978; Seeds, 1983).

On the whole, the advantage of culture methods more than compensates for the disadvantages. By careful comparison of the behavior of cells grown in cultures with other neural preparations, much knowledge has been gained about the function of the nervous system and its constituent cell types. In this chapter we will discuss different forms of cultures (primary cultures, cell lines) as well as the preparation of some of these cultures and examples of studies that can be carried out using cultures from the nervous system.

## 2. Primary Cultures

### 2.1. Definitions and Principles

Cells taken directly from the organism and kept in vitro for more than 24 h are considered to be *primary cultures* (*see* Fedoroff, 1977 for tissue culture terminology). Such cultures may consist of fragments of tissue and are then referred to as explant cultures, or they may be derived from a tissue dissociated into single cells before planting into the culture vessel. These latter are named cell cultures because the tissue organization has been disrupted. Cell cultures may consist of more than one cell type (mixed cultures) that under certain conditions may form reaggregates (e.g., Garber, 1977), or be of principally one cell type (monotypic cultures). Primary neural cell cultures are currently widely used for neurochemical investigations and have advantages over the other cell culture type (cell lines, *see* section 2.4) because they consist of normal diploid cells (established for human fetal neural cells by Icard et al. (1981) and for primary mouse astroglial cultures by B. H. J. Juurlink, L. Hertz, and H. C. Wang (unpublished observations). They also more closely reflect the metabolism and function of cells found in vivo (Schousboe, 1977; Hertz, 1979; Hertz and Chaban, 1982; Walz and Hertz, 1983b).

With a few exceptions (Pontén and Westermark, 1980), to establish primary neural cultures from the central nervous system the starting material must be obtained from immature tissue at a developmental stage when the cells that are to be cultured are still proliferating (astroglial cells), or are in the early postmitotic stages of differentiation (neurons). One explanation for this requirement is that tissue isolation and dissociation procedures used for establishing primary neural cultures allow selective survival during the dissociation procedures of small, undifferentiated cells with few cell processes, and that dividing cells may rapidly populate a culture dish. A consideration, therefore, of the ontogenesis of specific neural cell populations is useful for determining strategies to be used for the successful establishment of primary neural cultures. Some of these considerations have been reviewed by Hertz et al. (1985) and will not be considered here. They are, for example, the reason for the choice of tissue from animals of different ages for the establishment of monotypic cultures of different cell types (neurons, astrocytes, oligodendrocytes). In contrast to the central nervous system, neurons from peripheral nervous tissue can survive in cultures even if obtained from adult animals (Scott, 1977).

## 2.2. Explant Cultures

For a long time the predominant culture type used in the neurosciences was the explant (or organotypic) culture, i.e., cultures that have been obtained by transferring small pieces of tissue from the animal to the tissue culture vessel. A large amount of such work was carried out, mainly in Europe, during the first decades of this century (reviewed by Von Mihalik, 1935). In North America, studies of the nervous system using tissue culture methodologies were pioneered by Margaret Murray who has comprehensively reviewed this area (Murray, 1965, 1971). Under appropriate conditions explant cultures can develop morphologically, physiologically (Crain, 1976), and biochemically (Lehrer et al., 1970) in a similar manner as in vivo and explant cultures have made a large contribution in the area of electrophysiology (Crain, 1976; Nelson and Lieberman, 1981).

## 2.3. Cell Cultures

A number of strategies can be used for obtaining particular types of cell cultures. These have been reviewed in Hertz et al. (1985), and only some of the strategies will be briefly mentioned here. If one selects an appropriate region of the developing nervous system at a time when mainly one cell type is proliferating, then one can choose dissociation and culture conditions that will further select for that one cell type. As an example, the major population of proliferative cells in the mouse or rat newborn cerebrum and in 15-d-old embryonic chick cerebrum is astroglial precursor cells. The use of vigorous dissociation procedures on these tissues results in the destruction of the majority of the differentiating process-bearing neurons and selective survival of small undifferentiated cells (mainly astroglial precursors). Planting of such dissociated cells into culture results in a mainly astrocytic culture (Hertz et al., 1982, 1985). If one chooses tissues at an earlier stage of development, when large numbers of young, relatively undifferentiated postmitotic neurons are formed and only a few glial precursor cells are proliferating, this results in cultures that initially have a mainly neuronal composition. As time goes by the glial precursor cells will, however, proliferate, resulting in a mixed neuronal–glial culture (Booher and Sensenbrenner, 1972). This is illustrated in Fig. 2. By the judicious use of a cytotoxic agent, such as cytosine arabinoside, which inhibits DNA synthesis and hence kills cells in the S phase of the cell cycle with little or no effect on nondividing cells (Sotelo et al., 1980), one can eliminate the majority of proliferating glial cells in such cultures, re-

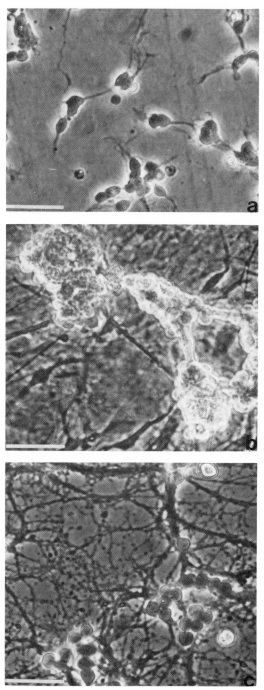

Fig. 2. Phase contrast micrographs of living cultures of mouse cerebellar interneurons prepared as in section 2.3.2 but less densely seeded. The cultures were grown for 30 h (**a**); 14 d without cytosine arabinoside treatment (**b**); or 14 d with cytosine arabinoside treatment between d 3 and 4 (**c**). Bar equals 25 μm (from Yu, 1980).

sulting in a monotypic neuronal culture (e.g., Dichter, 1978; Yu et al., 1984). Another means of preventing astrocytic overgrowth is to employ a defined serum-free medium designed to maintain neuronal survival, but not to maintain astrocytic growth (Skaper et al., 1979; Bottenstein et al., 1980; Bottenstein, 1983). In such cultures there is generally more than one type of neuron present. By choosing regions of the nervous system in which principally one class of neurons is forming at a certain time, such as the rodent postnatal cerebellum (in which the majority of cells being formed are the granule cell interneurons), one can further restrict the neuronal cell types surviving in culture.

Another means of obtaining monotypic neuronal cultures is by taking advantage of the selective affinity of one cell type for the substratum, or of selective vulnerability. These strategies have been most elegantly used for obtaining almost pure peripheral neuronal cultures (Varon et al., 1973; McCarthy and Partlow, 1976; Barde et al., 1982) and they take advantage of the fact that nonneuronal cells attach more rapidly to collagen-coated Petri dishes than do neuronal cells. By the judicious use of periodic vibrations and timing, the cells in suspension will consist principally of neurons and the attached cells principally of Schwann cells and fibroblasts. If one plants the cells in suspension into another culture vessel, one ends up with a mainly monotypic neuronal culture. The fibroblasts in the initial culture dish can be destroyed by complement-dependent lysis using antibodies against the Thy-1 antigen that is present on the fibroblast cells membrane, but not on the Schwann cell membrane (Brockes et al., 1981). In this way one ends up with a monotypic Schwann cell culture. A slight variation of this strategy has been used to obtain oligodendroglial cultures. This method stems from the observation that planting of dissociated cells from perinatal rodents at very high density results in cultures consisting of an astrocytic monolayer on top of which oligodendroglial cells rest (Labourdette et al., 1980). The oligodendroglial cells have a lesser affinity for their astrocytic substratum than do the astrocytes for their plastic substratum. Hence, using appropriate gyratory agitation, the oligodendrocytes can be selectively removed from the underlying astrocytes and replanted into culture (McCarthy and DeVellis, 1980). The cultures of oligodendrocytes obtained in this manner are by definition no longer primary cultures but could appropriately be named secondary cultures. Their degree of functional differentiation remains to be established.

The final strategy used for obtaining monotypic neural cultures that we would like to consider is a separation of cells using physical parameters prior to planting into culture. Thus brain cell suspensions have been separated into constituent cell types by gradient centrifugation techniques. Relatively homogeneous populations of neurons, astrocytes, and oligodendrocytes can be obtained in this way (Henn, 1980). This method has become of special importance for the isolation of oligodendroglial cells (Szuchet and Stefansson, 1980; Gebicke-Härter et al., 1984; Hertz et al., 1985) because at the moment there are no convenient methods for establishing relatively pure, well-differentiated primary cultures of oligodendroglial cells without prior separation of the cell types in the brain.

In the remaining portion of this section we would like to present selected techniques that will give rise to successful central nervous system monotypic neural cultures. For greater detail, *see* Hertz et al. (1985).

### 2.3.1. Procedure to Obtain Cerebral Cortical Cultures (Hertz et al., 1984b)

This procedure is based on the techniques outlined in Yavin and Yavin (1980), Yu et al. (1984), and Hertz et al.(1985). The advantage of this preparation is that the cultures are highly enriched in well-differentiated GABAergic cortical neurons (Fig. 3) and neurons constitute $\geqslant 90\%$ of the cells in culture. The basic strategy used is to isolate the cerebrum at a time in development when large numbers of neurons have just entered their postmitotic stage of differentiation and only few proliferative glial precursors are present. The majority of these glial precursors are eliminated in culture with the use of cytosine arabinoside. The procedure used is as follows:

1. Take out the cerebral hemispheres aseptically from 15-d-old mouse embryos and place in tissue culture medium (see below) with 20% horse serum. The age of the mouse embryos is rather critical because in older embryos and fetuses a large proportion of the neurons are differentiating and show extensive process formation that reduces the likelihood they will survive the dissociation procedure.

2. Remove the olfactory bulbs, hippocampal formations, basal ganglia, and meninges, thus isolating the neopallium, i.e., the cortical tissue above and lateral to the lateral ventricles. This step is not essential but probably leads to a more homogeneous culture of GABAergic neurons.

3. Replace medium with Puck's solution, i.e., a $Ca^{2+} - Mg^{2+}$ free balanced salt solution. Cut the neopallium

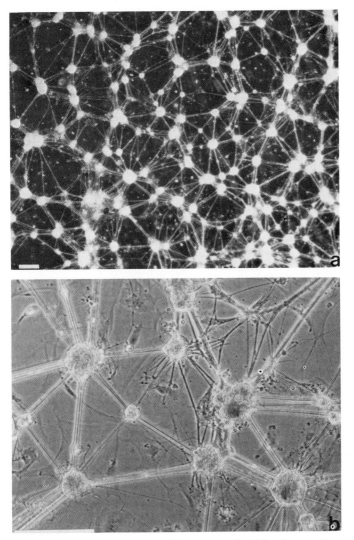

Fig. 3. Photomicrographs of living 3-wk-old culture of mouse cerebral cortical neurons prepared as described in section 2.3.1. Top, **a,** is a dark-field micrograph, and **b,** a phase contrast micrograph. Bar indicates 200 μm.

into 1 mm cubes and add trypsin to a final concentration of 0.2% (Gibco 1:250).
4. Trypsinize for 2 min at room temperature and inactivate trypsin action by adding horse serum to final concentration of 20%.
5. Triturate the tissue with a fire-polished pipet and

centrifuge down the cells for 2 min at 200–900g.

6. Resuspend the pellet in serum-free tissue culture medium where the glucose concentration has been adjusted to 30 m$M$. Filter through a Nitex® mesh of 80 μm pore size. Our medium is a slightly modified Dulbecco's MEM (described in detail in Hertz et al., 1982), aerated with 5% $CO_2$/95% air (v/v). Somewhat different media have been used by other workers.

7. Plant cells in 60-mm Falcon or NUNC tissue culture Petri dishes (3 mL, i.e., cells from one brain per dish) that have been coated by exposure overnight to 12.5 μg/mL of D-polylysine in water.

8. Fifteen minutes later, remove medium with unattached cells (mostly nonneuronal cells) and add 3 mL of fresh medium with 5% horse serum. Incubate at 37°C in a 95/5% (v/v) mixture of atmospheric air and $CO_2$ with a relative humidity of 90%.

9. After 3 d in vitro, expose the cultures to 40 μ$M$ cytosine arabinoside (added as 100 μL of a 1.2 m$M$ solution) to eliminate the majority of proliferating nonneuronal cells.

10. Twenty-four hours later, refeed cultures with 3 mL fresh growth medium but without the cytotoxic agent. The cultures are maintained without further feeding for an additional 1.5–2.5 wk. During this time the cells migrate to form clumps (Fig. 3), but they do not proliferate.

## 2.3.2. Procedure to Obtain Cerebellar Interneuronal Cultures (Meier et al., 1984)

Cerebellar granule cells in culture (Fig. 2) provide a preparation that allows studies of biochemical, physiological, pharmacological, and morphological aspects of glutamatergic neurons.

The procedure is based upon techniques described by Messer (1977), Yu and Hertz (1982), Drejer et al. (1983), and Meier et al. (1983) and takes advantage of the fact that the majority of immature neurons in the early postnatal rodent cerebellum are the cortical interneurons. The preparation proceeds as follows:

1. Take cerebella from 7-d-old mice or rats and place (about 0.3 mL/brain) in a Krebs' buffer (120 m$M$ NaCl; 5 m$M$ KCl; 1.2 m$M$ KH$_2$PO$_4$; 1.2 m$M$ MgSO$_4$; 25 m$M$ NaHCO$_3$; 14 m$M$ glucose) with 0.3% bovine serum albumin and equilibrated with atmospheric air/$CO_2$ (95/5%).

2. Cut the tissue into $0.4 \times 0.4$ mm cubes using a tissue chopper. Triturate the minced tissue with a Pasteur pipet in the Krebs' buffer.

3. Centrifuge briefly at $200g$ and remove supernatant. Resuspend pellet in Krebs' buffer containing 0.025% (w/v) trypsin (2 mL/brain) and incubate for 15 min at 37°C while shaking. Immediately after the incubation add 2 mL/brain of Krebs' buffer containing 0.004% (w/v) DNAse and 0.03% (w/v) soy bean trypsin inhibitor.

4. Centrifuge for 5 min at $200g$. Remove the supernatant and triturate the pellet in Krebs' buffer containing DNAse and trypsin inhibitor (0.3 mL/brain) using a syringe with a long cannula. Leave the suspension on the table for 5 min to allow sedimentation of clumps. Transfer supernatant (dissociated cells) to a tube containing 0.3 mL/brain of the culture medium, a slightly modified Dulbecco's medium (Hertz et al., 1982) containing 25 m$M$ NaHCO$_3$; 1.0 m$M$ glutamine; 24.5 m$M$ KCl; 30 m$M$ glucose; 7 $\mu M$ $p$-aminobenzoic acid; 100 mU/L insulin and 10% (v/v) fetal calf serum or 20% (v/v) horse serum, which is aerated with 5% CO$_2$/95% atmospheric air (v/v). Resuspend remaining pellet in Krebs' buffer plus DNAse and trypsin inhibitor (0.3 mL/brain) and repeat procedure. Centrifuge the combined supernatants at $200g$ for 5 min. Discard supernatant and resuspend pellet in tissue culture medium. The cell density in the suspension is determined by counting and should be $30–35 \times 10^6$ cells/mL. Seed cells in poly-L-lysine (or poly-D-lysine) coated NUNC or Falcon culture dishes (see above) at a density of $1–2 \times 10^6$ cells/cm$^2$. Incubate at 37°C in a 95%/5% (v/v) mixture of atmospheric air and CO$_2$ with a relative humidity of 90%.

5. After 24 hr in vitro, expose cultures to 40 $\mu M$ cytosine arabinoside (*see* section 2.3.1) for 2–3 d to avoid proliferation of glial cells (Fig. 2). When necessary (once a week) medium can be changed by exchanging one half of the culture medium with fresh medium. The cultures can be kept for 2–3 wk and astroglial contamination kept to a minimum (5–10%) by repeating the cytosine arabinoside treatment once a week. If so desired, the cells can be

maintained in a defined serum-free medium. They do not appear to proliferate.

### 2.3.3. Procedure to Obtain Astroglial Cultures

The procedure is based on the techniques outlined in Booher and Sensenbrenner (1972) and Hertz et al. (1982, 1985). The main advantage of this preparation (Fig. 4), which under optimum culturing conditions consists almost exclusively of astroglial cells (95% of all cells) and contains no neurons, is the extensive biochemical and biophysical similarities between these astrocytes grown in vitro and their in vivo counterparts. The basic strategies used in establishing these cultures is: (1) the selection of nervous tissue that consists essentially of maturing neurons with few or no newborn neurons and has astroglial precursor cells as the major proliferative population, (2) the use of vigorous mechanical dissociation procedures that destroy the bulk of the neurons, but allow the survival of small undifferentiated astroglial precursor cells, (3) the employment of a cell filtration step that filters out meningeal remnants and blood vessels, and (4) the use of appropriate culture conditions that lead to preferential survival and proliferation of astroglial precursor cells. The procedure used is as follows:

1. Remove the cerebral hemispheres aseptically from newborn mice and place in 5–10 mL of tissue culture

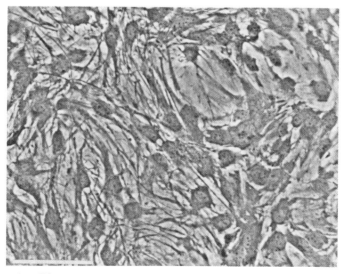

Fig. 4. Phase contrast micrograph of living 4-wk-old primary culture of mouse astrocytes prepared as described in section 2.3.3 and treated for the last 2 wk with 0.25 m$M$ dBcAMP. Bar equals 100 μm.

medium containing 20% horse or calf serum. The medium we use is a slightly modified Dulbecco's MEM (*see* Hertz et al., 1982) although other workers have used somewhat different media. Such differences in medium composition may affect the cells (Moonen et al., 1975).

2. Remove the olfactory bulbs, hippocampal formations, basal ganglia, and meninges, thus retaining the neopallium.

3. Dice tissue into 1 mm cubes and vortex tissue at maximum speed (Scientific Products' Deluxe Mixer) for 1 min and pass cell suspension serially through Nitex® filters of 80 and 10 μm pore sizes.

4. Plant cells into 60-mm Falcon or NUNC tissue culture Petri dishes (cells from one mouse brain into 3–4 Petri dishes).

5. Incubate at 37°C in a 95/5% (v/v) mixture of atmospheric air and $CO_2$ with a relative humidity of 90%.

6. After 3 d, remove medium and feed cultures with fresh medium containing 10% horse serum. From this point on, feed cultures twice weekly with a culture medium containing 5–10% horse serum (or 10% fetal calf serum). After 2 wk of culture when cells are confluent (due to a rapid proliferation), dibutyryl cyclic adenosine monophosphate (dBcAMP) may be added to the culture medium to a final concentration of 0.25 m$M$. This addition of dBcAMP causes pronounced morphological and some biochemical differentiation of the cells. In addition, this tre atment reduces the number of macrophages (the major nonastroglial cell population present in these cultures).

## 2.3.4. Procedure to Obtain Oligodendroglial Cultures

No simple procedures are presently available for the establishment of monotypic oligodendroglial cultures from immature brain as there are for neurons and astrocytes. The most successful oligodendroglial cultures are obtained by isolating young oligodendrocytes by differential centrifugation prior to culturing. Such procedures were pioneered by Fewster and Blackstone (1975) and Poduslo and McKhann (1977). Here we will summarize the procedure developed by Szuchet and coworkers for the isolation and long-term culture of ovine oligodendrocytes (for more details, *see* Szuchet and Stefansson, 1980; Szuchet et al., 1980a,b; Szuchet and Yim, 1984; Hertz et al., 1985). Oligodendro-

130                                                    *Hertz et al.*

cytes isolated and cultured by this procedure are morphologically
(Fig. 5) and biochemically differentiated cells and remain so during their span in vitro. This is evidenced by their morphology and
ultrastructure (Massa et al., 1983, 1984) and their biochemical
(Szuchet et al., 1983) and immunocytochemical properties
(Szuchet et al., 1980; Szuchet and Yim, 1984). Another important
characteristic of these cultures is that they have a high degree of
homogeneity: better than 98% of the cells, as assessed immunocy-
tochemically, are positive for a variety of oligodendrocyte markers (Szuchet et al., 1980b; Szuchet and Yim, 1984). This degree of
homogeneity, essential for biochemical work, is conserved during
the life of the cultures. The cells do not appear to proliferate during the culturing period.

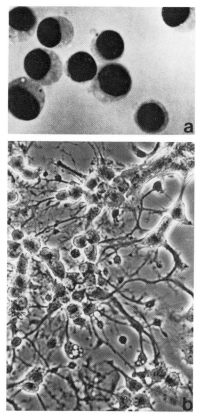

Fig. 5.    Phase contrast micrographs of cultured oligodendrocytes,
prepared as in section 2.3.4. Top, **a,** is stained with orceine and shows
cells grown for 4 d in vitro. Note the lack of processes. Bottom, **b,** shows
attached, living cells after 3 wk in culture. Note extensive network of
processes.

### 2.3.4.1. Procedure for Isolating Oligodendrocytes

Though the method as described uses lamb brains, it can be adapted to other species. The single most important factor for obtaining viable cells is to get the brains to the laboratory quickly, not more than ½ h from the time of sacrificing the animals.

1. Collect 12 lamb brains (3–6 mo old) at an abattoir; place them in a beaker containing Hanks' balanced salt solution (without $Ca^{2+}$ and $Mg^{2+}$; Hanks' solution) plus 4.8 µg/mL garamycin, 0.6 µg/mL fungizone at 4°C and take them to the laboratory. The remaining operation is carried out in a laminar flow hood with the exception of the trypsinization and centrifugation steps. A temperature of 4°C is used throughout unless indicated otherwise.

2. Dissect the white matter (centra ovales and corpora callosa), clean it free of grey matter with a spatula, and place it into four preweighed beakers containing ~80 mL of cold Hanks' each. Determine net weight of white matter.

3. Transfer white matter to a cold stainless steel plate and mince into ~1 mm$^3$ pieces. Divide the minced white matter into 4 × 500 mL Erlenmeyer flasks; add 200 mL of 0.1% trypsin in Hanks' containing 2% (w/v) Ficoll-400 plus 0.2% EDTA, pH 6.0, to each flask; put flasks in a shaking water bath at 37°C, 150 rpm for 2.5 min/g white matter (~60 min).

4. Add 60 mL of 0.60% trypsin inhibitor in Hanks' (cold) to each flask; shake well; leave standing for 2 min and centrifuge at 192$g$ for 3 min (Sorvall GSA rotor).

5. Combine the four pellets into two; wash with 0.0075% trypsin inhibitor in Hanks'; centrifuge at 192$g$ for 3 min; next wash pellets with Hanks'; centrifuge 192$g$ for 3 min; next wash pellets with Hanks'; centrifuge 192$g$ for 3 min; discard supernatants.

6. Resuspend pellets in ~50 mL 0.9$M$ sucrose in Hanks' at half-strength (x/2 Hanks'). Disrupt tissue by gentle passage through a series of screens (nylon mesh except where indicated) with the aid of a stream of the sucrose solution delivered from a wash bottle. Start with 350 µm screen (twice) followed by 210 µm, double 130 µm, double 52 µm (twice), and finally 30-µm stainless steel screen (do not let screens dry; total solvent used ~800 mL); centrifuge this suspension at

850$g$ for 10 min (GSA rotor); during this step myelin floats and the cells pellet ($P_1$).

7. Aspirate supernatant to obtain $P_1$.

8. Resuspend $P_1$ in 6 mL of 0.9$M$ sucrose and apply on top of two preformed linear gradients (1.0–1.15$M$ sucrose + 4% dextran 70 in x/2 Hanks' plus 1 m$M$ EDTA); centrifuge at 431$g$ for 12 min (Sorvall HS-4 rotor) using a rate controller to start and stop the run. Three bands, I, II, and III, from top to bottom of tube, separate on this gradient. Reproducibility in band separation depends critically on maintaining consistent speed and time during the centrifugation step.

9. Fractionate bands carefully. For identification of cells found in each band, *see* Szuchet et al. (1980b). Here we will only refer to cells from band III (interfascicular oligodendrocytes).

10. Dilute contents of band III by slow addition of Hanks' to double the volume; centrifuge at 554$g$ for 10 min (HS-4 rotor); gently aspirate supernatants, leaving a volume of ~5 mL in each tube; measure that volume; remove an aliquot for cell count; add more Hanks' to tubes (~40 mL); centrifuge at 554$g$ for 7 min; discard supernatant.

11. 90–95% of cells in the pellet are oligodendrocytes. The cells are now ready for culture. Yield: 2 × $10^6$ cells/g wet white matter.

### 2.3.4.2. *Procedure for Culturing Oligodendrocytes*

1. Resuspend a pellet of freshly isolated cells in culture medium: 90 mL DMEM (Dulbecco's MEM), 10 mL MEM, 25 mL horse serum supplemented with 2 m$M$ L-glutamine, and 0.3 µg/mL Fungizone plus 2.4 µg/mL Garamycin, to give a concentration of 2 × $10^6$ cells/mL; plate at 10 mL/20 × 100 mm Falcon tissue culture dish and maintain at 37°C in 95% air–5% $CO_2$ (v/v), humidity at saturation.

2. Leave cultures for 4 d; collect supernatants containing floating cells (B3.f oligodendrocytes); centrifuge at 63$g$ for 5 min (International model CL tabletop centrifuge); collect pellet and gently resuspend in fresh culture medium to give a concentration of 2 × $10^6$ cells/mL.

3. Replate on polylysine-coated plastic Petri dishes;

feed twice weekly. If fibroblasts are seen 2 or 3 /d after plating, add $10^{-5}M$ cytosine arabinoside to culture medium and leave for 48 h. Change to normal medium.

## 2.4. Cell Identification in Primary Cultures by the Aid of Macromolecular Markers

It is a relatively simple matter to recognize by morphology neurons, astrocytes, oligodendrocytes, and Schwann cells in intact mature nervous tissue since one has many frames of reference. It is a more difficult (often impossible) task to recognize undifferentiated cells in immature nervous tissue. Similar difficulties are encountered with cultured cells for a number of reasons. Such cells are often immature and do not carry the differentiated features of mature cells found in vivo. In addition, with the culture system many of the frames of reference used for in situ identification are lost. Hence, unequivocal identification of cultured cells is a long-recognized problem in tissue culture. Recently, the recognition of cell-specific antigenic markers (Bock, 1977; Varon, 1978; Raff et al., 1979; Eng and Bigbee, 1978; Schachner et al., 1983) and the availability of ligands directed against them has facilitated the task of identifying cultured cells. Although macromolecular markers are commonly referred to as cell-specific markers, they might better be considered as function-specific markers. The functions of the generally used markers are largely restricted to specific cell types; however, the markers also may be present in other cell types or in small subpopulations of other cell types. For example, the neuron-specific enolase is the enolase isozyme that is relatively chloride-insensitive (Marangos et al., 1982) and was initially considered to be a neuron-specific marker; however, it is also present in other cell types that have excitable membranes and associated chloride fluxes, such as pancreatic islet cells (Schmechel et al., 1978).

Two antigenic markers have proven to be unequivocal markers of astrocytes in the mature central nervous system. These are glutamine synthetase (Norenberg and Martinez-Hernandez, 1979) and glial fibrillary acid protein (GFAP), which is the major intermediate filament protein in astrocytes (Bignami and Dahl, 1977). Several antigenic markers have been found that identify oligodendrocytes. These include galactocerebroside (Raff et al., 1978), myelin basic protein (Sternberger et al., 1978), and Wolfgram protein (Roussel et al., 1978). The presence of these antigenic markers as well as the absence of other antigenic markers (e.g., GFAP) have been of great benefit in identifying oligodendrocytes in vitro.

In general, neurons in culture are easy to recognize because of their unique morphology; however, because other cell types (e.g., oligodendrocytes) may exhibit similar morphology, especially in comparison to the morphology of small interneurons, it is initially important to use other criteria in addition to morphology to characterize the cultures. These criteria may include neurophysiological properties, including electrophysiology, or the presence of neurotransmitter synthesizing enzymes (Sensenbrenner et al., 1973; Mandel et al., 1976; Vernadakis and Arnold, 1980). Antigenic markers restricted to neurons in the mature nervous system include neuron-specific enolase (Block, 1977; Schengrund and Marangos, 1980; Ledig et al., 1982), synaptin, $D_2$ and $D_3$ (Bock, 1977; Garthwaite and Balazs, 1981; Yu, 1980), and the tetanus toxin-binding gangliosides GD1B and GTI (Van Heyningen, 1963; Mirsky et al., 1978; Pettmann et al., 1979; Currie, 1980), as well as the monoclonal antibody A2B5 that binds to the ganglioside GQ and was initially considered to be a specific marker for neurons (Eisenbarth et al., 1979).

It should be recognized that certain antigens are only transiently present (e.g., $M_1$ in astrocytes, Schachner, 1982; Schachner et al., 1983) and that others (e.g., GFAP, neuron-specific enolase) become abundant only at a relatively late developmental stage (Bignami et al., 1980; Schmechel et al., 1980). It should also be kept in mind that certain antigens that mark a particular cell type in the mature nervous system may be present in immature cells of another type. An example of this is nonneuronal enolase that is a marker for nonneuronal cells in the mature nervous system but is also present in immature neurons in which it is ultimately replaced by the neuron-specific enolase (Schmechel et al., 1980). Another example of an apparently anomalous distribution of cell markers in developing cells is a recent report by Choi and Kim (1984) who observed the presence of GFAP within immature myelinating cells in vivo. It may furthermore be possible that the in vitro milieu may cause the expression of an antigenic marker in a cell type in which it would not normally be present. These qualifications apply mainly to relatively undifferentiated cells and not to relatively mature cells found in culture. Despite the above qualifications antigenic markers have proven to be powerful tools in identifying or confirming the identification of cultured cells.

Within the last few years several reports appeared illustrating the use of antigenic markers to differentiate between subclasses of a given cell type or to demonstrate pluripotential cell precursors.

Thus, Schnitzer and Schachner (1982) and Berg and Schachner (1982) reported the binding of A2B5 on the membranes of a small proportion of mature cerebellar astrocytes grown in vitro as well as on some immature oligodendroglial cells and on a large proportion of immature astrocytes of cerebellar cultures. In more detailed studies Raff et al. (1983a,b) reported that optic nerve astrocytes in vitro could be divided into two distinct populations termed type 1 and type 2. Both types are GFAP positive. The type 1 astrocytes have a fibroblast-like morphology and proliferate in culture, especially in response to growth factors, such as epidermal growth factor. They have no detectable surface binding of tetanus toxin or A2B5. In contrast, type 2 astrocytes are smaller cells with processes and they resemble oligodendrocytes at the light microscopic level. Such cells divide infrequently in culture and in addition to being GFAP-positive they also bind tetanus toxin and A2B5 on their surfaces. Type 2 astrocytes develop in culture from GFAP-negative and A2B5-positive cells. Using various culture manipulations, Raff et al. (1983a,b) have presented convincing arguments that GFAP-negative and A2B5-positive cells can give rise to either the A2B5-positive and GFAP-positive type 2 astrocytes or to GFAP-negative and galactocerebroside positive oligodendrocytes, depending upon the signal(s) present within the culture medium. These findings do not appear to be culture artifacts since GFAP-positive and A2B5-positive astrocytes have also been demonstrated *in situ* (Miller and Raff, 1984). It thus appears that the development of the various glial cell populations is a more complicated phenomenon than originally believed. However, the use of cell markers should enable investigators not only to better subdivide the various glial populations, but ultimately through an interplay of in vitro and in vivo experiments to delineate some of the factors that enable a cell to choose to follow a particular lineage of differentiation.

## 3. Cell Lines

Cell lines by definition are cultures that have been serially transplanted (subcultured) from one culture vessel to another for a number of generations (Fedoroff, 1977). Cell lines may consist of normal diploid cells. Cells subcultured from a primary culture fall into this category. Normal diploid cell lines have a limited life span, as is the case for normal human glial cells that start to senesce after 15–30 population doublings (Pontén and McIntyre, 1968; Pontén and Westermark, 1980). For this reason, normal diploid cell lines have only been used to a limited extent in neurobiological work (Sheffield and Kim, 1977; Icard et al., 1981; Walum et al., 1981). The other major category of cell lines is represented by the established cell lines that consist of cells whose genome has been altered such that they become "immortal," i.e., can proliferate indefinitely. Established cell lines are obtained from tumors (either spontaneous, chemically, or virally induced) or from cultures of originally normal cells that have transformed spontaneously or as a result of exposure to a viral or chemical carcinogen. Established cell lines have been widely used. They are generally nondiploid and often, but not always, tumorogenic (Fedoroff, 1977). Neurochemically, they are in general less differentiated than cells in primary culture (e.g., Schousboe, 1977). Clones, i.e., populations derived from one single cell by mitosis (Fedoroff, 1977), using different techniques, can in principle be obtained from primary cultures, diploid cell lines, or established cell lines. Often the term "clonal cell line" is used for an established cell line that has been cloned. Occasional recloning may be required to maintain such a clonal cell line.

Established cell lines offer some advantages to the investigator. Since they can be propagated indefinitely, they cost less to maintain than primary cultures and the investigator has access to presumably identical material from experiment to experiment. There are, however, a number of questions about whether established cell lines after a number of serial propagations are identical to the starting population. For instance, the rat glial cell line $C_6$ has high levels of 2',3'-cyclic nucleotide 3'-phosphodiesterase (CNPase) activity (an oligodendrocytic characteristic) and low glutamine synthetase levels (an astrocytic marker) at the 22nd passage and the inverse characteristics at the 82nd passage (Parker et al., 1980). This also brings up the question of how well-differentiated such cells are. They appear to be less differentiated than normal cells and often exhibit features that are not characteristic for the cell types in question. As an example, neuronal-specific enolase has been demonstrated in a presumably glial cell line (Viñores et al., 1982). Thus, it appears that established cell lines are very useful in defining general cell biological mechanisms and the spectrum of neurobiological mechanisms available to normal neural cells, but that they do not necessarily define the exact functions operating in normal neural cells. This aspect has been discussed in some detail by Hertz et al. (1985).

Another means of establishing cell lines is through the use of temperature-sensitive viruses. This technique has only recently been used for developing established neural cell lines (Giotta et al., 1980; Giotta and Cohn, 1982). It appears to hold great promise because it may combine the benefits of cell immortality with cell normality. Using this procedure, large numbers of transformed cells may be grown at permissive temperatures and may be able to differentiate into normal cells at nonpermissive temperatures. Whether such established cell lines will differentiate into completely normal cells at nonpermissive temperature remains to be established.

A large variety of established cell lines have been described in the literature. Some of these are listed in Table 1. For more de-

TABLE 1
Cell Lines[a]

| Cell line | Species | Source | Cell type | Reference |
|---|---|---|---|---|
| C6 | Rat | MNU[b] -induced CNS tumor | Astroglial | Benda et al., 1968 |
| LRM55 | Rat | ENU[c] -induced spinal glioma | Astroglial | Martin and Shain, 1979 |
| NN | Syrian hamster | Spontaneous transformation of cultured brain cells | Astroglial | Shein et al., 1970 |
| MG138 | Human | Astrocytoma-glioblastoma tumor | Astroglial | Ponten and McIntyre, 1968 |
| 21A | Rat | ENU[c] -induced cerebral tumor | Oligodendroglial | Fields et al., 1975 |
| 33B | Rat | ENU[c] -induced spinal tumor | Oligodendroglial | Fields et al., 1975 |
| G26 | Mouse | Methylcholanthrene-induced CNS tumor | Oligodendroglial | Sundarraj et al., 1975 |
| C1300 | Mouse | Spontaneous spinal tumor | Neuronal | Augusti-Tosco and Sato, 1969 |
| $E_c^t$ | Mouse | Spontaneous transformation of cultured cerebellar cells | Neuronal | Bulloch et al., 1978 |
| B104 | Rat | ENU[c]-induced | Neuronal | Schubert et al., 1974 |
| PC12 | Rat | Spontaneous adrenal pheochromocytoma | Chromaffin cells | Green and Tischler, 1976 |

[a]This table is not comprehensive, but does include several commonly used cell lines.
[b]Methylnitrosourea.
[c]Ethylnitrosourea.

tails, the reader is referred to Schubert et al. (1974); Pfeiffer et al. (1977); Bulloch et al. (1976, 1978); and Greene and Tischler (1982).

# 4. Experimental Use of Cell and Tissue Culture

## 4.1. General Outline

Primary cultures of astrocytes or oligodendrocytes are often prepared in batches containing up to 100 individual cultures whereas neurons are prepared in somewhat smaller batches, i.e., in our laboratory with a maximum of 50–60 cultures. Since no proliferation of neurons or oligodendrocytes occurs during the culturing this means that all the neurons or oligodendrocytes in one batch of cultures represent the same original tissue suspension. We generally regard individual cultures in the same way as individual brain slices. They are not complete replicas of each other and may be regarded as individual experiments. We occasionally prefer to use control and experimental cultures from the same batch and never accept any results before they have been shown in different batches. We have, however, not encountered other than relatively minor quantitative differences between different batches. This is provided cultures of the same age are used, and the development of GAD activity in the cultured cells (Fig. 6) suggests that cerebral cortical neurons should be maintained in culture until they are about 2 wk old. The longevity of these neurons is enhanced when they are not fed subsequent to the feeding with normal medium after the exposure to cytosine arabinoside. This is the major reason for the use of high glucose concentration in the medium. However, glutamine, which initially is present at a concentration of 2 m$M$, becomes virtually exhausted in the medium at the age of 2 wk.

Since astrocytes proliferate vividly during the culturing, virtually all the cells in a given culture will have been born in vitro. This means that these cultures, if anything, are less related to each other than are corresponding cultures of neurons or oligodendrocytes. Again, batch-to-batch variations are small and these cultures can be kept almost indefinitely and are frequently fed. After they have attained an age of 3–4 wk, subsequent biochemical alterations are slight. This applies, e.g., to glutamine synthetase activity (Fig. 7) and to protein formation (White and Hertz, 1981). Exceptions to this are an increase in carbonic anhydrase activity (Schousboe et al., 1980) and of monoamine oxidase B activity (Yu and Hertz, 1982), which continues to occur, especially if the medium has been supplemented with dibutyryl cyclic AMP.

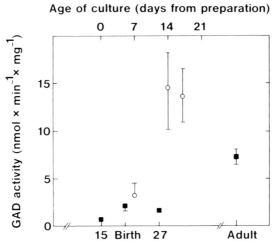

Fig. 6. Development of glutamate decarboxylase (GAD) activity in primary cultures of cerebral cortical neurons, prepared from the brains of 15-d-old mouse embryos, as a function of age in culture (○) compared to the in vivo activities of the enzyme (■) in the brain (neopallium) at comparable ages (from Yu et al., 1984).

Since the cultures of oligodendrocytes we use are prepared from much more mature nervous tissue than are the neuronal and astrocytic cultures, both short- (up to 48 h in culture) and long-term cultures can be utilized for biochemical studies of

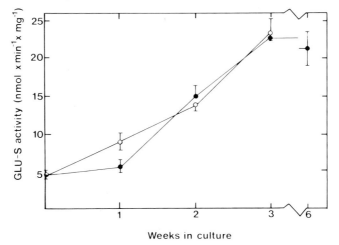

Fig. 7. Development of glutamine synthetase activity in primary cultures of astrocytes, prepared from neonatal mouse brains, as a function of age in culture (○) compared to the in vivo activities of the enzyme in brain (●) at comparable ages (from Hertz et al., 1978).

oligodendrocytic function. The former can be used for probing
the range of metabolic activities expressed by cells shortly after
isolation (Poduslo and McKhann, 1977; Poduslo et al., 1978) and
are also suitable for examining metabolic pathways (Pleasure et
al., 1977; 1981). Incorporation of labeled acetate and glycerol into
different lipid structures in cells of different ages is shown in Fig.
8 as an example of such studies, and the correlation between
these phenomena and events occurring in vivo has been discussed by Hertz et al. (1985).

Most studies reported in the literature have been performed

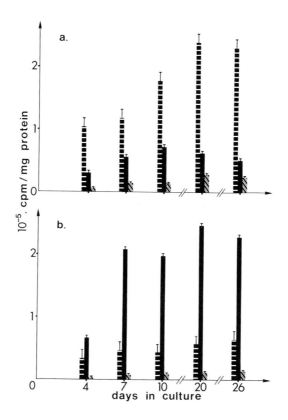

Fig. 8.    Incorporation of [$^{14}$C]-acetate (**a**) and [$^3$H]-glycerol (**b**) into
neutral lipids (▤), phospholipids (▮) and glycolipids (◪) by cultured
oligodendrocytes. The cells were double labeled for a total of 72 h. Excess label was removed, and cells harvested, disrupted, and processed
for thin layer chromatography as described by Szuchet et al. (1983).
Note that acetate is predominantly incorporated into neutral lipids, that
glycerol is mainly incorporated into phospholipids, and that neither of
these compounds is a good precursor of glycolipids.

using intact cells or cell homogenates, but interest is developing in examining specific subcellular fractions. In principle it should be possible to obtain subcellular fractions from any kind of cell culture. The greatest progress toward this goal has been made in the case of oligodendrocytes, where Szuchet et al. (1983, 1984) developed a procedure for isolating plasma membranes. The cells were kept in culture for 4–5 d under conditions of nonattachment; i.e., plated on a substratum that does not favor cell attachment. The rationale for choosing cultured cells rather than freshly isolated cells stems from the fact that the procedure for isolating oligodendrocytes involves a trypsinization step that may lead to hydrolysis of membrane proteins. By maintaining the cells in vitro they are given an opportunity to resynthesize damaged proteins. The resulting plasma membrane fraction is enriched 26-fold in Na-K-ATPase, tenfold in CNPase, and has no mitochondrial or lysosomal membranes. Such preparations should permit a structural, chemical, and biochemical characterization of oligodendrocyte plasma membrane and thus further our understanding of this membrane and its relationship to myelin.

## 4.2. Biochemical and Pharmacological Studies

A confluent culture of astrocytes in a 60-mm dish contains a monolayer of cells, attached to the dish and amounting to 0.3–0.5 mg protein. A corresponding culture of neurons contains about 1 mg protein and that of oligodendrocytes, 0.1–0.2 mg. Such cultures can be used for almost anything that can be done with a brain slice or a brain homogenate. Many enzyme activities, e.g., glutamine synthetase activity, can be determined in individual cultures, whereas others, e.g., pyruvate carboxylase activity, require that a few cultures be pooled.

Neural cells in primary cultures have been used to follow the time course of enzyme expression as illustrated in Figs. 6–7. Since the medium is easily controlled, they are also well suited to study the factors that affect their development. For example, hydrocortisone has been found to induce glutamine synthetase activity in cultured astrocytes (Juurlink et al., 1981; Hallermayer et al., 1981) and glycerol phosphate dehydrogenase (GPDH) activity in cultured oligodendrocytes (McCarthy and DeVellis, 1980). McCarthy and DeVellis (1980) compared the specific activity of 2′,3′-cyclic nucleotide 3′-phosphodiesterase (CNPase), GPDH, and lactic dehydrogenase (LDH) in oligodendrocytic and astrocytic cultures. Not only did they find differences in the expression of these enzymes by the two cell types, but the regulation of these enzymes was also distinct. Thus, hydrocortisone

(HC) induced GPDH in oligodendrocytes but had very limited influence on the same enzyme in astrocytes. This induction of GPDH by HC appears to be species-specific and/or developmentally restricted since Szuchet and Yim (1984) could not detect any significant effect of HC on the expression of GPDH by ovine oligodendrocytes.

Metabolic events in individual cell types can be studied with advantage in cell cultures. Hertz and coworkers thus have studied the metabolic fate of glutamate and glutamine in primary cultures of neurons or astrocytes after exposure to the labeled amino acid (Yu et al., 1982; Hertz et al., 1983b). Szuchet et al. (1983) followed the lipid and glycolipid metabolism of cultured oligodendrocytes by studying the incorporation of [$^{14}$C]-acetate, [$^{3}$H]-glycerol, [$^{3}$H]-galactose, and [$^{35}$S]-sulfate as a function of time in culture for up to 35 d. Cells were double-labeled with either [$^{14}$C]-acetate and [$^{3}$H]-glycerol (Fig. 8) or [$^{3}$H]-galactose and [$^{35}$S]-sulfate for a total of 72 h. After harvesting the cells, lipids and glycolipids were resolved by thin layer chromatography. From these studies it was concluded that after attachment to a substratum, oligodendrocytes started synthesizing membranes that resembled their own plasma membrane more than myelin membrane, and then proceeded to differentiate these membranes to what might become myelin. These observations revealed cell plasticity and suggested that in vitro post-myelination oligodendrocytes behaved more like myelin-forming cells than myelin-maintaining cells, i.e., the cells reenacted early events associated with myelinogenesis. Other precursors can also be used. For example, Szuchet and Yim (1984) and Yim and Szuchet (1984) examined the glycoprotein and protein metabolism of cultured oligodendrocytes as a function of time in culture up to and including 56 d. In this case they used [$^{3}$H]-fucose, [$^{3}$H]-leucine, and [$^{35}$S]-methionine to label the cells. These studies also showed significant changes in the metabolism of oligodendrocytes following cell attachment to a substratum and process extension, i.e., during membrane synthesis.

Cell cultures provide the possibility of studying binding to intact cells. Since the cells adhere strongly to the dish, the binding can be measured simply by exposing the intact culture to the radioactive ligand and subsequently removing the radioactive solution by efficient, but relatively fast, washing. We routinely wash three times with nonradioactive solution, including removal of washing fluid between each wash with a Pasteur pipet. Testing of this procedure on empty dishes has shown an apparent binding corresponding to at most 10% of that to the cells (Walker and Peacock, 1981; L. Hertz and J. S. Richardson, unpublished results). This binding will be included in the nonspecific binding which, however, often is less than 10% of the total binding, suggesting that the plastic dishes are partly protected by the cells from accumulation of radioactive ligands. After the washing, 0.2 mL of Tris solution (or another saline) is added to the dish and the cells are harvested by aid of a thin plastic sheet. The 0.2 mL Tris solution with the cells is added to a test tube containing 0.8 mL 1$M$ NaOH and the solution is left overnight. Protein and radioactivity are determined in the cell digest and radioactivity in the medium, and binding is calculated and expressed per mg protein. Similar procedures have now been quite commonly used (for reference *see* Hertz et al., 1983). Binding to membrane fractions or homogenates can be done either with the aid of a filtration assay (e.g., McCarthy and Harden, 1981; Tardy et al., 1981) or after centrifugation (Hertz and Richardson, 1983). Comparison of, e.g., benzodiazepine binding with and without homogenization is easily feasible in the cultured cells and such a comparison has recently demonstrated that a very large proportion of the bound ligand is lost after homogenization (Hertz and Mukerji, 1980; Hertz et al., 1983; Bender and Hertz, 1984). This is not necessarily because actual binding sites are destroyed, but might reflect partial release of bound radioactive ligand from its binding sites during the centrifugation assay or poor recovery of the binding sites after the filtration assay. Such an impaired recovery of the bound ligand probably explains the conclusion by McCarthy and Harden (1981) that astrocytes in primary cultures have few, if any, binding sites for flunitrazepam. Before the conclusion is accepted that binding sites apparently are lost or poorly recovered after homogenization, the possibility should, however, be excluded that the larger binding in intact cells might be due to an uptake of the ligand. This is in all probability not the case for flunitrazepam binding as (1) affinity is virtually the same before and after homogenization, (2) the Hill plot indicates interaction with one binding site, and (3) an increase in temperature that would be expected to enhance an active uptake, on the contrary, decreases the binding (Bender and Hertz, 1984). With respect to another ligand, i.e., dihydroalprenolol, a β-adrenergic ligand, the situation seems to be different. Again, the binding of the ligand is considerably decreased after homogenization, but *in this case* this difference can be explained by a nonspecific retention of the very lipophilic ligand in the cells (Maderspach et al., 1981). Thus, the binding to the intact cells represents only to a limited extent binding to a high affinity receptor binding site that may explain why

the binding is not potently displaced by such β-adrenergic agonists as isoproterenol (Hertz et al., 1983a). A similar situation
probably exists in the case of many lipophilic drugs, e.g., the
antidepressant doxepin, which also is retained to a very large extent by primary cultures of astrocytes (Hertz et al., 1980). In contrast to the retention of dihydroalprenolol, which may be of little
biological relevance, the corresponding retention of neurotropic
drugs may be of importance, since the phenomenon occurs when
therapeutically relevant doses of the drugs are used. It is, therefore, likely that a corresponding phenomenon will exist also in
vivo.

## 4.3. Physiological Studies

Within the framework of the present chapter, a comprehensive
review of the application of physiological methods to cell cultures
cannot be attempted. We will, therefore, concentrate on the two
methods with which we have personal experience, i.e., ion transport and electrophysiological studies. Ion transport studies with
radiotracers generally are restricted to homogeneous cultures,
i.e., cell lines and monotypic neuronal and astrocytic cultures, but
can with suitable mathematical treatment be applied also to mixed
cultures, e.g., of neurons and astrocytes (Latzkovits et al., 1974).
Electrophysiological methods present single cell techniques and,
therefore, can easily be performed in explant cultures or other
mixed cultures, provided they are accompanied by proper cell
identification.

### 4.3.1. Ion Transport and Water Content

Transport characteristics are easily determined in cultured cells
because of the thinness of the preparation and the lack of problems with extracellular diffusion. Thus, in a recent study,
Larrabee (1984) demonstrated that in intact chains of sympathetic
ganglia of chicken embryos there was a considerable barrier to exchange between the medium and the extracellular space. However, when the corresponding tissue was plated in cultures, and
seeded at relatively low density, this barrier had disappeared. It
was concluded that the barrier probably was a result of restricted
diffusion through narrow and tortuous channels between closely
associated cells in the intact preparation.

Ion contents are determined by the balance between ion
uptake (influx) and ion release (efflux). Contents and fluxes can
be determined by aid of radiotracer or flame photometry analysis.
A key question in this respect is how accurately the intracellular
ion contents can be measured. These methods involve removal of
a large amount of extracellular ions from a small amount of tissue

by washing, usually with isotonic sucrose solution (1° C). Sources of error would be ion loss during the wash or the presence of remaining extracellular ions or externally bound cations. The latter should be sensitive to the $H^+$ concentration of the washing fluid because protons compete very effectively with externally bound cations (Sanui and Rubin, 1979). Control experiments with washing fluids at different pH values suggest that externally bound monovalent cations do not contribute to the measured ion contents. The situation is different for divalent cations such as calcium. For this ion, several internal compartments exist that have to be identified by analysis of influx and efflux kinetics, and by application of metabolic inhibitors and ionophores (Kurzinger et al., 1980). In cases in which the extracellular concentration of an ion is much higher than the intracellular concentration (e.g., sodium), prolonged washing is necessary to remove virtually all extracellular fluid, and this could possibly influence the measured ion content.

Manipulating the experimental conditions during which ion transport is measured is easily possible in cultured cells. We thus analyze potassium uptake into astrocytes with two different techniques, i.e., unidirectional uptake of $^{42}K$ into cells at steady-state or into cells that have been depleted for potassium by preincubation in cold potassium-free medium. The steady-state fluxes are very high (2000 nmol $\times$ mg$^{-1}$ protein $\times$ min$^{-1}$) mainly because of a large diffusional component made possible by the high negative membrane potential and pronounced potassium permeability (Walz et al., 1984). Potassium-depleted cells show virtually no polarization, and the initial phase of the uptake (300 nmol $\times$ mg$^{-1}$ $\times$ min$^{-1}$) represents mainly an active uptake into the cells (Walz and Hertz, 1982). The potassium-depleted cells are not damaged, but they have certain abnormal properties, like an internal sodium content that is increased 5–7 times normal levels. The intracellular potassium content in these cells is very low (Walz and Hertz, 1982), so the unidirectional potassium uptake after subsequent transfer to a potassium-containing medium cannot be caused by homoexchange of potassium. Efflux rates can either be traced by exposing cells that previously have been loaded with radiotracer for different time periods to nonlabeled medium and investigating the remaining radioactivity (Kimelberg, 1981) or by sampling the released activity at constant time intervals. For sodium, which is relatively slowly transported, this procedure allows a separation between extracellular and intracellular sodium based upon efflux kinetics (Walz and Hertz, 1984). For potassium, for which the efflux is extremely rapid, we use a method derived from Boonstra et al. (1981). Control cultures are treated with gramicidin that leads to a breakdown of the potassium gradient, and to a negligible potassium content within the cells. The efflux from these treated cells indicates almost exclusively the time course for the removal of extracellular potassium and can be subtracted from the total efflux curves of normal cultures to obtain the time course of the efflux across the cell membrane (Walz and Hertz, 1983a).

Intracellular water content of astrocytes has been measured in double-labeling experiments by exposing the culture to a rapidly diffusing marker ($^3H_2O$) for total fluid and to an impermeable marker ($^{14}C$-inulin) for extracellular fluid (W. Walz and L. Hertz, unpublished experiments) as described by Chan et al. (1982). The difference between the volume each marker occupies represents the intracellular water space, which in turn can be used as a measure of cellular volume (Macknight and Leaf, 1977). At normal potassium concentrations and during exposure of the cells to the markers for no longer than 2–5 min, permeability of the astrocytic cell membrane to inulin is no problem. However, this method requires draining of the cultures by soaking almost all extracellular fluid away with a fine nozzle connected to an air suction pump in order that an overwhelming amount of extracellular fluid shall not obscure the measurement. In addition, the cell cultures are blotted with facial tissue, a procedure that should be gentle enough that the protein content of the cultures is not affected.

### 4.3.2. Intracellular Electrophysiology

Intracellular electrophysiological measurements in explant cultures were pioneered by Crain (1976) and in cell cultures by Nelson (1975). They offer a unique possibility to correlate electrical behavior, e.g., action potential characteristics, with cell anatomy and function (Ransom and Barker, 1981). During intracellular electrophysiological experiments, the culture should be continuously superfused with medium or saline under as normal conditions as possible, i.e., with a bicarbonate-buffered medium at 37°C. This is not always possible, and some neuronal cultures are damaged by continuous superfusion (Gruol, 1983). Continuous superfusion is preferable in order to reduce problems with evaporation, leakage of KCl from reference electrode (if an agar-bridge is used), and accumulation of $K^+$ outside the cell during impalement. It also prevents the risk of exchange artifacts when the perfusion system is repeatedly turned on and off. It is of advantage to use cell cultures for impalement with microelectrodes without any previous treatment. For practical purposes we plant the cells in a 35-mm culture dish and we use this dish as the experimental chamber. The apparatus developed for this purpose is described in Fig. 9. Whenever possible, saline should be preferred over medium to reduce problems associated with interactions of medium components with the tip of the glass electrode, with the reference electrode, and with ion-exchange resins. We use a saline medium, which is almost identical to that in Eagle's MEM medium, i.e., it has the following composition (in m$M$): 5.4 KCl, 116 NaCl, 26 NaHCO$_3$, 1.0 NaH$_2$PO$_4$, 0.8 MgSO$_4$, 1.8 CaCl$_2$, 10 glucose. The saline solution is well aerated with a 95/5% (v/v) mixture of atmospheric air and CO$_2$.

We presently use either single-channel electrodes (for determination of membrane potential) or dual-channel electrodes, which allow accurate measurement of membrane conductance when used for separate current injection and voltage recording. They can be used also for dye injection to identify the cell type (in heterotypic cultures) after completion of the measurement and for determining intracellular ion activities (Kettenmann et al., 1983a,b). It is possible also to obtain stable recordings from neurons (Nelson et al., 1977) and astrocytes (Bowman et al., 1983)

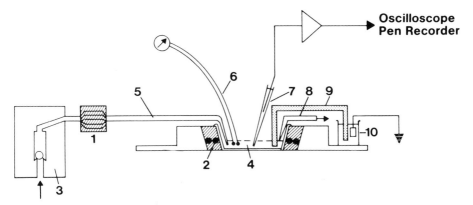

Fig. 9. Experimental technique for intracellular measurements with microelectrodes in tissue culture. The cells remain attached to their culture dish (35 mm diameter) and the dish is continuously superfused with saline. **1**: Heating system in the solution supply system; **2**: Heating ring in an aluminum block surrounding the dish located on the microscope stage; **3**: multiple valve system for exchange of solutions in the supply system; **4**: culture dish (the broken line represents the level of fluid in the dish); **5**: solution supply system; **6**: thermocouple, located in the solution in the dish and connected with a multimeter for temperature readout; **7**: single-channel microelectrode connected with chlorided Ag-AgCl wire with a preamplifier; **8**: suction system for removal of solution (microperspex peristaltic pump provides the necessary vacuum); **9**: agar bridge filled with 3$M$ KCl, and **10**: well filled with 3$M$ KCl, containing a chlorided Ag–AgCl pellet connected with ground.

using two separate microelectrodes. The dual channel electrodes are prepared from Brown-Flaming Pyrex glass (which has a ratio of septum to wall of 1.2:1) of 2.2 mm outer diameter (R & D Scientific Glass Company, Spenverville, MD). It is pulled in a Brown-Flaming microelectrode puller P-77B (Sutter Instrument Company, San Rafael, CA) using a 3-mm square box-filament, 2 mm wide. The heating process takes about 25 s and is terminated by exposure to a pulse of relatively high nitrogen pressure to rapidly cool the filament (Brown and Flaming, 1977). With this method it is possible to get tapers of 6 mm and a channel resistance of 40–60 m$\Omega$. Coupling between two channels is reduced by a layer of plastic rubber around the back-end of the longer channel. The coupling resistance between two channels averages 100 k$\Omega$. With these electrodes it is possible to maintain impalements for up to 1 h from astrocytes, and the membrane potential is indistinguishable from that seen after impalement with single channel electrodes and also similar to that in astrocytes in vivo (*see* Walz et al., 1984). The short taper reduces vibrations of the electrode tip, and the flow rate of the saline can be as high as 6–8 bath vol/min without affecting the recording quality.

A problem with dissociated cell cultures, especially of astrocytes, is the flatness of the cells. For a reasonable success rate, it is necessary to impale the cell with the microelectrode at an almost 90° angle. It is, therefore, necessary to use microscopes that permit a long (10 cm) working distance and a penetration angle of the electrode as close as possible to 90°. Generally, it is useful to tap gently on the baseplate or to turn the capacitance control to its maximum for a fraction of a second once the electrode tip has contact with the cell surface (Purves, 1981). However, the success rate is dramatically increased by using a minimum velocity of 2 $\mu$m/ms during impalement, a velocity that can be accomplished only by using a step motor-driven manipulator (Sonnhof et al., 1982).

An alternative method to determine membrane potential is the use of fluorescent dyes and lipophilic cations to measure transmembrane potential differences (Cohen, 1973; Waggoner, 1979; Freedman and Laris, 1981). This method is free of many of the difficulties associated with cell impalement by electrolyte-filled microelectrodes and it is capable of detecting rapid changes in membrane potential. It is especially useful in small cells and may be, in the future, a serious alternative to electrophysiological methods. It should, however, be kept in mind that application of the dye causes a fall in ATP levels, and an evaluation of its possible problems was recently published (Johnstone et al., 1982). In the future, this method may become very useful for tracking of local variations and spread of activity in tissue culture systems. This might well be useful to analyze spreading depression and gap junction pathways in cultured cells and tissues.

An important advantage of electrophysiological techniques is that they allow measurements of intracellular ion activities by the aid of ion-selective microelectrodes (e.g., Zeuthen, 1981). This method is relatively new for tissue culture work and is best established for the potassium ion (Werrlein, 1981), although measurements of calcium also have been done (Morris and McDonald, 1982). The method has been perfected by Kettenmann et al. (1983b) for cultured oligodendrocytes in mixed cultures with dye injection for subsequent analysis of the cell type.

Voltage clamp techniques have been used in cultured cells only to a limited extent, but a detailed theoretical and practical introduction into voltage clamp techniques in neuronal cultures has been published by Smith et al. (1981). The problem is that the cells have to be isopotential over their total membrane surface, which is not always the case. At the moment voltage clamp techniques seem only possible for spherical neurons with small axons, like dorsal root cells (Ransom et al., 1977). Some investigators (Fukuda et al., 1976) achieve isopotentiality by altering the morphology of the cells by treatment with drugs.

Analysis of membrane noise allows an estimate of the ionic currents carried by individual channels in the cell membrane, and its application to tissue culture was extensively reviewed by Lecar and Sachs (1981). The method involves low-noise instrumentation so that the signal can be recognized above background. Examples of its application are neurons (McBurney and Barker, 1978) and oligodendrocytes (Kettenmann et al., 1982) in explant cultures from mouse spinal cord.

## 4.4. Studies of Cell Interactions

In the previous sections, we have discussed use of monotypic cell cultures for biochemical and physiological work. Even in such cultures interactions may be operating between cells of the same type ("isotypic" cell interactions). In addition, cell culture work has been instrumental for the investigation of interactions between different cell types, brought together in the culture dish ("heterotypic" cell interactions). This procedure allows much better control of the microenvironment than can be obtained in vivo and also a more precise knowledge of the interacting cell populations. A coculture of neurons and oligodendrocytes is shown in Fig. 10.

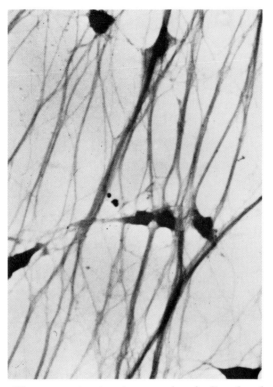

Fig. 10. Phase contrast micrograph of oligodendrocyte–neuron coculture stained with toluidine blue. Ovine oligodendrocytes were maintained in culture for 4–5 d under conditions of nonattachment and then added to a culture of rat dorsal root ganglion neurons as described by Wood et al. (1983) and cultured for 1 wk. Note that oligodendrocytes adhere to the neuronal network and extend long processes.

### 4.4.1. Isotypic Cell Interactions

There are several lines of evidence that suggest that oligodendrocytes, astrocytes, and neurons interact with their own kind both in vivo and in vitro. Interfascicular oligodendrocytes align in rows in close proximity to each other, and both neurons and oligodendrocytes show a strong tendency to form aggregates. Tight junctional complexes between oligodendrocytes were detected in vivo (Massa and Mugnaini, 1982) as well as in vitro (Massa et al., 1983, 1984). Massa et al. (1983, 1984) found that the number of tight junctions increased with time in culture, indicating that synthesis of tight junction particles took place in vitro. These studies show clearly that oligodendrocyte cultures are suited to investigations of the nature of oligodendrocyte–

oligodendrocyte interactions and their effect on oligodendrocyte functions. In particular, these cultures constitute an excellent source with which to inquire into the factors that regulate tight junction formation. Intramembranous assemblies in astrocytes have been demonstrated both in vivo (Brightman et al., 1980) and in cultures (Anders et al., 1981), and astrocytes are electrically coupled (and dye coupled) not only in vivo but also in cell cultures (Moonen and Nelson, 1978; Kettenmann et al., 1983a).

### 4.4.2. Heterotypic Cell Interactions

In vivo astrocytes interact with oligodendrocytes by forming gap junctions. Remnants of these junctional complexes can be found on the plasma membranes of freshly isolated oligodendrocytes (Massa et al., 1984). However, cultured oligodendrocytes in isolation do not express gap junctions, an indication that oligodendrocyte–astrocyte interaction may be essential for the synthesis of these structures. It is in support of this idea that electrical coupling has been demonstrated between oligodendrocytes in explant cultures (Kettenmann et al., 1983a). In spite of this, there was no electrical coupling (or dye coupling) between the two cell types.

Oligodendrocytes may need astrocytes early in their development. This is suggested by the work of Noble (1984), who claims that the in vitro development of premyelination oligodendrocytes differs depending on whether or not astrocytes are present. Nevertheless, as indicated by the studies on monotypic cultures of oligodendrocytes, these cells are well able to maintain several differentiated characteristics in isolation (*see* also Mirsky et al., 1980; Mack and Szuchet, 1980). This behavior of oligodendrocytes is quite distinct from that of Schwann cells that require the continuous input from neurons to remain differentiated (Mirsky et al., 1980).

Interactions involving oligodendrocytes and neurons can be studied at several levels: (1) One may want to address the issue of myelination or remyelination (since in the particular culture system we are employing, most of the oligodendrocytes must have already myelinated once); (2) one may wish to examine whether or not neurons exert trophic or tropic influences on oligodendrocytes; (3) if such influences are found, one may want to proceed a step further by examining whether they require direct physical contact between the two cell types or whether they are mediated by the medium. Alternatively, the reverse question may also be asked: do oligodendrocytes influence neurons?

Wood et al. (1983) used dissociated fetal rat dorsal root ganglion neurons depleted of Schwann cells and fibroblasts (Wood et al., 1980) to assess the capacity of these cells to stimulate ovine oligodendrocytes to proliferate and produce myelin sheath. While their results are still preliminary, they did show that isolated and cultured oligodendrocytes retain the ability to recognize neurons, as evidenced by the fact that oligodendrocytes dispersed and attached throughout the neurite network, and extended long and thin processes (Fig. 10) that stained with antigalactocerebroside serum. Cell proliferation, measured by [$^3$H]-thymidine incorporation, was discrete. Four to six weeks after the addition of oligodendrocytes, myelin formation was detected by Sudan black staining in approximately half the cultures. These initial studies point to the feasibility of using this approach to probe into the nature of oligodendrocyte–neuron interaction.

To the best of our knowledge, cocultures of oligodendrocytes, astrocytes, and neurons have yet to be established. However, in cultures from the peripheral nervous system, it has been demonstrated elegantly that interactions occur among three types of cells, i.e., Schwann cells, neurons, and fibroblasts. From in vivo studies, it has long been suspected, with little direct evidence, that the axon regulates Schwann cell proliferation and myelination (e.g., Friede and Samorajski, 1968). By using recently developed in vitro methods, several laboratories have been able to approach this problem more directly. That the axon does regulate Schwann cell proliferation is strongly supported by the observation that Schwann cells undergo little or no proliferation in vitro unless they are in contact with axons (Wood and Bunge, 1975; McCarthy and Partlow, 1976). In vitro procedures have also enabled investigators to pursue in greater detail this mechanism of regulation. The mitogenic influence of the axon on the Schwann cell requires direct physical contact between the Schwann cell and the axolemma since separation of these two elements by a permeable 6 μm collagen membrane eliminates the mitogenic effect of the axon (Salzer et al., 1980a). Furthermore, this mitogenic stimulus appears to be localized to the axolemma, since axolemma-enriched fractions are mitogenic for Schwann cells (Salzer et al., 1980b; DeVries et al., 1982) and, in addition, trypsinization of intact axons results in the loss of the mitogenic signal (Salzer et al., 1980a), suggesting that the signal is bound to the outer face of the axolemma. This is not to suggest that the axon is the only regulator of Schwann cell numbers, since several other factors have also been demonstrated to affect the proliferative activity of Schwann cells in vitro (Brockes et al., 1981; Varon and Manthorpe, 1982) and may possibly play similar roles in vivo.

Another assumption concerning the roles of the various peripheral nervous system elements that has arisen from in vivo studies is that the cellular interactions necessary for ensheathment and myelination of the axon occur only between the neuron and Schwann cell. In vitro studies by Bunge and Bunge (1978) strongly suggest that a third component is required as well: the extracellular matrix deposited by the fibroblasts. Thus, in vitro when a Schwann cell makes contact with an axon in the absence of extracellular matrix (collagen), ensheathment, or myelination does not occur, whereas if such Schwann cells also are presented with collagen, ensheathment and/or myelination proceeds.

In addition to being used to examine factors involved in the myelination of the peripheral axon, cultures of the peripheral nervous system have been extensively used in delineating some of the trophic factors required for neural cell differentiation (Perez-Polo et al., 1983).

The above examples elegantly demonstrate not only how in vitro approaches may confirm and amplify in vivo observations, but also how they have the ability to demonstrate phenomena not readily evident from in vivo observations alone.

Other types of cell interactions are the possible transfer of metabolites from one cell type to another, and the release of compounds, such as potassium ions, from one cell type and the subsequent action of these compounds upon a second cell type. Thus, potassium ions are released from excited neurons and the resulting increase in its extracellular concentration has profound effects on astrocytes (e.g., Walz and Hertz, 1983b). In a series of elegant experiments Hösli and coworkers (e.g., Hösli and Hösli, 1983) have shown that a depolarizing effect of transmitter amino acids on astrocytes is due to such a neuronal potassium release because it can be observed after application of the transmitter to an area in an explant culture that contains nearby neurons, but not after application to an area containing only astrocytes. Studies of metabolic transfer have apparently not been performed in cultured neural cells but have been successfully carried out in, e.g., fibroblasts. The first demonstration of a direct transfer of metabolites from one cell to the other was done by Subak-Sharpe et al. (1968) in what can be now regarded as classical experiments. Mutant fibroblasts that lack hypoxanthine guanine phosphoribosyl transferase activity and, therefore, cannot incorporate hypoxanthine into nucleic acids when grown alone, can do so when cocultured with wild-type cells. This is thought to be caused by the passage of nucleotides from one cell to the other through intercellular channels. Pitts and Simms (1977) have devised another experimental paradigm to demonstrate metabolic interchange between cells that do not require mutants. The essence of the experiments of Pitts and Simms (1977) is to prelabel cells with a radioactive precursor that is converted to an intermediate metabolite retained by the cells (donor cells). Unused precursor is removed by washing and the cells are cocultured with unlabeled cells (recipient cells). What Pitts and Simms have shown with these experiments is that transfer of metabolites takes place *when* and *only when* cells are in direct physical contact with each other, indicating that this is done through communicating channels. Kasa (1984) has obtained evidence in a similar kind of experiment that choline may be transferred from astrocytes to neurons in which it is used for synthesis of acetylcholine.

## 5. Conclusion

In understanding the biology of a given cell type in the nervous system, the trophic factors that influence it, or the nature of cell-to-cell interactions that modulate it, one must break the system into its parts and then attempt to put it back again, one piece at a time, not unlike the pieces of a puzzle, until a "synthetic whole" is built. Cell cultures seem eminently well suited for such work. The implied assumption in this approach is that knowledge gained from the in vitro system can be extrapolated back to the in vivo situation.

Such an in vitro approach is valid depending upon the extent to which cultured neural cells reflect the characteristics of their in vivo counterparts. For some characteristics, such a comparison is feasible, e.g., a high glutamine synthetase activity in astrocytes, a high potassium conductance in these cells, a high GAD activity in neurons, excitability of these cells, the presence of myelin basic protein, and other proteins (including enzymes) and lipids that are specific for myelination in oligodendrocytes. In these cases, all evidence suggests that the cultured cells, and especially those in primary cultures, do fulfill this requirement. Our knowledge of cell-to-cell interaction in the central nervous system is at its very beginning. But here, too, events should start moving quickly now that procedures for isolating and recombining the individual cell types in culture are available. This will mark the opening of new and exciting avenues in our quest to unravel the cellular and molecular mechanisms that govern events in the CNS.

## Acknowledgments

The authors' research has been supported by grants from the National Multiple Sclerosis Society (to S.S.) and from the Medical

Research Council of Canada (to B.H.J., L. H., and W.W.). Ms. Jackie Bitz is cordially thanked for typing the manuscript.

# References

Anders J. J. and Brightman M. W. (1982) Particle assemblies in astrocytic plasma membranes are rearranged by various agents in vitro and cold injury in vivo. *J. Neurocytol.* **11,** 1009–1029.

Augusti-Tocco G. and Sato G. (1969) Establishment of functioning clonal lines of neurons from mouse neuroblastoma. *Proc. Nat. Acad. Sci. USA* **64,** 311–315.

Barde Y.-A., Edgar D., and Thoenen H. (1982) Culture of Embryonic Chick Dorsal Root and Sympathetic Ganglia, in *Neuroscience Approached Through Cell Culture,* vol. I (Pfeiffer S. E., Ed.), pp. 83–86. CRC Press, Boca Raton, Florida.

Barnes D. and Sato G. (1980) Serum-free cell culture: a unifying approach. *Cell* **2,** 649–655.

Benda P., Lightbody J., Sato G., Levine L. and Sweet W. (1968) Differentiated rat glial cell strain in tissue culture. *Science* **161,** 370–371.

Bender A. S. and Hertz L. (1984) Flunitrazepam binding to intact and homogenized astrocytes and neurons in primary cultures. *J. Neurochem.* **43,** 1319–1327.

Berg G. J. and Schachner M. (1982) Electron-microscopic localization of A2B5 cell surface antigen in monolayer cultures of murine cerebellum and retina. *Cell Tissue Res.* **224,** 637–645.

Bignami A. and Dahl D. (1977). Specificity of the glial fibrillary acidic protein for astroglia. *J. Histochem. Cytochem.* **25,** 466–469.

Bignami A., Dahl D., and Rueger D. C. (1980) Glial Fibrillary Acidic (GFA) Protein in Normal Neural Cells and in Pathological Conditions, in *Advances in Cellular Neurobiology,* vol. 1 (Fedoroff S. and Hertz L., Eds.), pp. 286–310. Academic Press, New York.

Bock E. (1977) Immunochemical Markers in Primary Cultures and in Cell Lines, in *Cell, Tissue, and Organ Cultures in Neurobiology* (Fedoroff S. and Hertz L., Eds.), pp. 407–422. Academic Press, New York.

Booher J. and Sensenbrenner M. (1972) Growth and cultivation of dissociated neurons and glial cells from embryonic chick, rat and human brain in flask cultures. *Neurobiol.* **2,** 97–105.

Boonstra J., Mummery C. L., Tertoolen L. G. J., van der Sang P. T. and de Laat S. (1981) Characterization of $^{42}K^+$ and $^{86}Rb^+$ transport and electrical membrane properties in exponentially growing neuroblastoma cells. *Biochim. Biophys. Acta* **643,** 89–100.

Bottenstein J. E. (1983) Growth Requirements of Neural Cells *In Vitro,* in *Advances in Cellular Neurobiology,* vol. 4 (Fedoroff S. and Hertz L., Eds.), pp. 333–379. Academic Press, New York.

Bottenstein J. E., Skaper S. D., Varon S., and Sato G. H. (1980) Selective survival of neurons from chick embryo sensory ganglionic dissociates utilizing serum-free supplemented medium. *Exp. Cell Res.* **125,** 183–190.

Bowman C. L., Edwards C., and Kimelberg H. K. (1983) Veratridine causes astrocytes in primary culture to become excitable. *Soc. Neurosci. Abstr.* **9**, 448.

Brightman M. W., Anders J. J., and Rosenstein J. M. (1980) Specializations of Non-Neuronal Cell Membranes in the Vertebrate Nervous System, in *Advances in Cellular Neurobiology* vol. 1 (Fedoroff S. and Hertz L., Eds.), pp. 3–29. Academic Press, New York.

Brockes J. P., Fryxell K. J., and Lemke G. E. (1981) Studies on cultured Schwann cells: the induction of myelin synthesis and the control of their proliferation by a new growth factor. *J. Exp. Biol.* **95**, 215–230.

Brown K. T. and Flaming D. G. (1977) New microelectrode techniques for intracellular work in small cells. *Neuroscience* **2**, 813–827.

Bulloch K., Stallcup W. B., and Cohn M. (1976) The derivation and characterization of neuronal cell lines from rat and mouse brain. *Brain Res.* **135**, 25–36.

Bulloch K., Stallcup W. B., and Cohn M. (1978) A new method for the establishment of neuronal cell lines from the mouse brain. *Life Sci.* **22**, 495–504.

Bunge R. P. and Bunge M. B. (1978) Evidence that contact with connective tissue matrix is required for normal interaction between Schwann cells and nerve fibers. *J. Cell Biol.* **78**, 943–950.

Chan P. H., Kerlan R., and Fishman R. A. (1982) Intracellular volume of osmotically regulated C-6 glioma cells. *J. Neurosci. Res.* **8**, 67–72.

Choi B. H. and Kim R. C. (1984) Expression of glial fibrillary acidic protein in immature oligodendroglia. *Science* **223**, 407–409.

Cohen L. B. (1973) Changes in neuron structure during action potential propagation and synaptic transmission. *Physiol. Rev.* **53**, 573–618.

Crain S. M. (1976) *Neurophysiologic Studies in Tissue Culture.* Raven Press, New York.

Currie D. N. (1980) Identification of Cell Type by Immunofluorescence in Defined Cell Cultures of Cerebellum, in *Tissue Culture in Neurobiology* (Giacobini E., Vernadakis A., and Shahar A., Eds.), pp. 75–87. Raven Press, New York.

DeVries G. H., Salzer J. L., and Bunge R. P. (1982) Axolemma-enriched fractions isolated from PNS and CNS are mitogenic for cultured Schwann cells. *Devel. Brain Res.* **3**, 295–299.

Dichter M. A. (1978) Rat cortical neurons in cell culture: culture methods, cell morphology, electrophysiology and synapse formation. *Brain Res.* **149**, 279–293.

Drejer J., Larsson O. M., and Schousboe A. (1983) Characterization of uptake and release processes for D- and L-aspartate in primary cultures of astrocytes and cerebellar granule cells. *Neurochem. Res.* **8**, 231–243.

Eisenbarth G. S., Walsh F. S., and Nirenberg M. (1979) Monoclonal antibody to a plasma membrane antigen of neurons. *Proc. Nat. Acad. Sci. USA* **76**, 4913–4917.

Eng L. F. and Bigbee J. W. (1978) Immunohistochemistry of nervous system specific antigens. *Adv. Neurochem.* **3**, 43–98.

Fedoroff S. (1977) Primary Cultures, Cell Lines, and Cell Strains: Terminology and Characteristics, in *Cell, Tissue and Organ Cultures in Neurobiology* (Fedoroff S. and Hertz L., Eds.), pp. 265–286. Academic Press, New York.

Fewster M. E. and Blackstone S. (1975) In vitro study of bovine oligodendroglia. *Neurobiology* **5**, 316–328.

Fields K. L., Gosling C., Megson M. and Stern P. L. (1975) New cell surface antigens in rat defined by tumors of the nervous system. *Proc. Nat. Acad. Sci. USA.* **72**, 1296–1300.

Freedman J. C. and Laris P. C. (1981) Electrophysiology of cells and organelles: Studies with optical potentiometric indicators. *Int. Rev. Cytol.* Suppl. **12**, 177–246.

Friede R. L. and Samorajski T. (1968) Myelin formation in the sciatic nerve of the rat. A quantiative electron microscopic, histochemical and radioautographic study. *J. Neuropath. Exp. Neurol.* **27**, 546–570.

Fukuda J., Fischbach G. D., and Smith T. G. (1976). A voltage clamp study of the sodium, calcium and chloride spikes of chick skeletal muscle cells grown in tissue culture. *Dev. Biol.* **49**, 412–424.

Garber B. B. (1977) Cell Aggregation and Recognition in the Self-assembly of Brain Tissues, in *Cell, Tissue, and Organ Cultures in Neurobiology* (Fedoroff S. and Hertz L., Eds.), pp. 515–537. Academic Press, New York.

Garthwaite J. and Balas R. (1981) Separation of Cell Types From the Cerebellum and Their Properties, in *Advances in Cellular Neurobiology*, vol. 2 (Fedoroff S. and Hertz L., Eds.), pp. 461–489. Academic Press, New York.

Gebicke-Härter P. J., Althans H. -H., Rittner R., and Neuhoff V. (1984) Bulk separation and long-term culture of oligodendrocytes from adult pig brain. I. Morphological studies. *J. Neurochem.* **42**, 357–368.

Giotta G. J. and Cohen M. (1982) Derivation of Neural Cell Lines With Rous Sarcoma Virus, in *Neuroscience Approached Through Cell Culture* vol. I (Pfeiffer S. E., Ed.), pp. 203–225. CRC Press, Boca Raton, Florida.

Giotta G. J., Heitzmann J., and Cohn M. (1980) Properties of two temperature-sensitive Rous sarcoma virus transformed cerebellar cell lines. *Brain Res.* **202**, 445–458.

Greene L. A. and Tischler A. S. (1976) Establishment of a noradrenergic clonal line of rat adrenal pheochromocytoma cells which respond to nerve growth factor. *Proc. Nat. Acad. Sci. USA* **73**, 2424–2428.

Greene L. A. and Tischler A. S. (1982) PC12 Pheochromocytoma Cultures in Neurobiological Research, in *Advances in Cellular Neurobiology* vol. 3 (Fedoroff S., and Hertz L., Eds.), pp. 374–414. Academic Press, New York.

Gruol D. L. (1983) Cultured cerebellar neurons: Endogenous and exogeneous components of Purkinje cell activity and membrane response to putative transmitters. *Brain Res.* **263,** 223–241.

Hallermayer K., Harmening C., and Hamprecht B. (1981) Cellular localization and regulation of glutamine synthetase in primary cultures of brain cells from newborn mice. *J. Neurochem.* **37,** 43–52.

Harrison R. G. (1907) Observations on the living developing nerve fiber. *Proc. Soc. Exp. Biol. Med.* **4,** 140–143.

Henn F. A. (1980) Separation of Neuronal and Glial Cells and Subcellular Constituents, in *Advances in Cellular Neurobiology* vol. 1 (Fedoroff S. and Hertz L., Eds.), pp. 373–403. Academic Press, New York.

Hertz L. (1979) Functional interactions between neurons and astrocytes. I. Turnover and metabolism of putative amino acid transmitters. *Prog. Neurobiol.* **13,** 277–323.

Hertz L., Bender A. S., and Richardson J. S. (1983a) Benzodiazepine and β-adrenergic binding to primary cultures of astrocytes and neurons. *Prog. Neuropsychopharmacol. Biol. Psychiat.* **7,** 681–686.

Hertz L., Bock E., and Schousboe A. (1978) GFA content, glutamate uptake and activity of glutamate metabolizing enzymes in differentiating mouse astrocytes in primary cultures. *Dev. Neurosci.* **1,** 226–238.

Hertz L. and Chaban G. (1982) Indications for an Active Role of Astrocytes in Potassium Homeostasis at the Cellular Level: Potassium Uptake and Metabolic Effects of Potassium, in *Neuroscience Approached Through Cell Culture,* vol. I (Pfeiffer S. E., Ed.), pp. 157–174. CRC Press, Boca Raton, Florida.

Hertz L., Juurlink B. H. J., Fosmark H., and Schousboe A. (1982) Astsrocytes in Primary Culture, in *Neuroscience Approached Through Cell Culture* vol. 1 (Pfeiffer S. E., Ed.), pp. 175–186. CRC Press, Boca Raton, Florida.

Hertz L., Juurlink B. H. J., Fosmark H., and Schousboe A. (1984c) Preparation of Primary Cultures of Mouse (Rat) Astrocytes, in *A "Cookbook" for Culturing Neuronal and Glial Cells* (Shahar A., Filogamo G., and Vernadakis A., eds.) , p. 81, International Society for Developmental Neuroscience, University of Chieti, Italy.

Hertz L., Juurlink B. H. J., and Szuchet S. (1985) Cell Cultures, in *Handbook of Neurochemistry* vol. 8 (Lajtha A., Ed.), Plenum Press, New York.

Hertz L. and Mukerji S. (1980) Diazepam receptors on mouse astrocytes in primary cultures; displacement by pharmacologically active concentrations of benzodiazepines or barbiturates. *Can. J. Physiol. Pharmacol.* **58,** 217–220.

Hertz L. and Richardson J. S. (1983) Acute and chronic effects of antidepressant drugs on adrenergic function in astrocytes in primary cultures—an indication of glial involvement in affective disorders? *J. Neurosci. Res.* **9,** 173–183.

Hertz L., Richardson J. S., and Mukerji S. (1980) Doxepin, a tricyclic antidepressant, binds to normal, intact astroglial cells in cultures and inhibits the isoproterenol-induced increase in cyclic AMP production. *Can. J. Physiol. Pharmacol.* **58**, 1515–1519.

Hertz E., Yu A. C. H., Hertz L., and Schousboe A. (1984b) Preparation of Primary Cultures of Cortical Mouse Neurons, in *"A Cookbook" for Culturing Neuronal and Glial Cells* (Shahar A., Filogamo G., and Vernadakis A., eds.), pp. 21–22. International Society for Developmental Neuroscience, University of Chieti, Italy.

Hertz L., Yu A. C. H., Potter R. L., Fisher T. E. and Schousboe A. (1983b) Metabolic Fluxes From Glutamate and Towards Glutamate in Neurons and Astrocytes in Primary Cultures, in: *Glutamine, Glutamate and GABA in the Central Nervous System* (Hertz L., Kvamme E., McGeer E. G., and Schousboe A., eds.), pp. 327–342. Alan R. Liss, New York.

Hösli L. and Hösli E. (1983) Electrophysiological and Autoradiographic Studies on GABA and Glutamate Neurotransmission at the Cellular Level, in *Glutamine, Glutamate and GABA in the Central Nervous System* (Hertz L., Kvamme E., McGeer E. G., and Schousboe A., eds.), pp. 441–455. Alan R. Liss, New York.

Icard C., Liepkalns V. A., Yates A. J., Singh N. P., Stephens R. E., and Hart R. W. (1981) Growth characteristics of human glioma-derived and fetal neural cells in culture. *J. Neuropath. Exp. Neurol.* **40**, 512–525.

Johnstone R. M., Laris P. C., and Eddy A. A. (1982) The use of fluorescent dyes to measure membrane potentials: A critique. *J. Cell Physiol.* **112**, 298–301.

Juurlink B. H. J., Schousboe A., Jorgensen O. S., and Hertz L. (1981) Induction by hydrocortisone of glutamine synthetase in mouse primary astrocyte cultures. *J. Neurochem.* **36**, 136–142.

Kasa P. (1984) The Role of Glial Cells in Neuronal Acetylcholine Synthesis, in *Dynamics of Cholinergic Function* (Hanin I., ed.), Plenum Press, New York (in press).

Kettenmann H., Orkand R. K., Lux H. D., and Schachner M. (1982) Single potassium channel currents in cultured mouse oligodendrocytes. *Neurosci. Lett.* **32**, 41–46.

Kettenmann H., Orkand R. K., and Schachner M. (1983a) Coupling among identified cells in mammalian nervous system cultures. *J. Neurosci.* **3**, 506–516.

Kettenmann H., Sonnhof U., and Schachner M. (1983b) Exclusive potassium dependence of the membrane potential in cultured mouse oligodendrocytes. *J. Neurosci.* **3**, 500–505.

Kimelberg H. K. (1981) Active accumulation and exchange transport of chloride in astroglial cells in culture. *Biochim. Biophys. Acta* **646**, 179–184.

Kozak L. P., Dahl D., and Bignami A. (1978) Glial fibrillary acidic protein in reaggregating and monolayer cultures of fetal mouse cerebral hemispheres. *Brain Res.* **150**, 631–637.

Kurzinger K., Stadtkus C., and Hamprecht B. (1980) Uptake and energy-dependent extrusion of calcium in neural cells in culture. *Eur. J. Biochem.* **103**, 597–611.

Labourdette G., Roussel G., and Nussbaum J. L. (1980) Oligodendroglia content of glial cell primary cultures, from newborn rat brain hemispheres, depends on the initial plating density. *Neurosci. Lett.* **18**, 203–209.

Larrabee M. G. (1984) Alanine uptake and release by sympathetic ganglia of chicken embryos. *J. Neurochem.* **43**, 816–829.

Latzkovits L., Sensenbrenner M., and Mandel P. (1974) Tracer kinetic model analysis of potassium uptake by dissociated nerve cell cultures: glial-neuronal interrelationship. *J. Neurochem.* **23**, 193–200.

Lecar H. and Sachs F. (1981) Membrane Noise Analysis, in *Excitable Cells in Tissue Culture* (Nelson P. G. and Liebermann M., eds.), pp. 137–172. Plenum Press, New York.

Ledig M., Tholey G., and Mandel P. (1982) Neuron-specific and non-neuronal enolase in developing chick brain and primary cultures of chick neurons. *Dev. Brain Res.* **4**, 451–454.

Lehrer G. M., Bornstein M. B., Weiss C., Furman M., and Lichtman C. (1970) Enzymes of carbohydrate metabolism in the rat cerebellum developing in situ and in vitro. *Exp. Neurol.* **27**, 410–425.

Mack S., and Szuchet S. (1980) Synthesis of myelin glycosphingolipids by isolated oligodendrocytes in tissue culture. *Brain Res.* **214**, 180–185.

Macknight A. D. C., and Leaf A. (1977) Regulation of cellular volume. *Physiol. Rev.* **57**, 510–573.

Maderspach K., Nemecz G., and Yigiter M. (1981) Binding of fluorescent and radio-labelled alprenolol to intact cultured brain cells and liposomes. *Acta Biol. Acad. Sci. Hung.* **32**, 283–290.

Mandel P., Ciesielski-Treska J., and Sensenbrenner M. (1976) Neurons in vitro, in *Molecular and Functional Neurobiology* (Grispen W. H., Ed.), pp. 111–155. Elsevier, Amsterdam.

Marangos P. J., Polak J. M., and Pearse A. G. E. (1982) Neuron-specific enolase. A probe for neurons and neuroendocrine cells. *Trends Neurosci.* **5**, 193–196.

Martin D. L. and Shain W. (1979) High affinity transport of taurine and β-alanine and low affinity transport of γ-aminobutyric acid by a single transport system in cultured glioma cells. *J. Biol. Chem.* **254**, 7076–7084.

Massa P. T. and Mugnaini E. (1982) Cell junctions and intramembrane particles of astrocytes and oligodendrocytes: a freeze-fracture study. *Neurosci.* **7**, 523–538.

Massa P. T., Szuchet S. and Mugnaini E. (1983) Tight junctions and intramembrane particles of cultured oligodendrocytes. *Biophys. J.* **41**, 69.

Massa P. T., Szuchet S., and Mugnaini E. (1984) Cell-cell interactions of isolated and cultured oligodendrocytes: Tight junction formations and expression of peculiar intramembrane particles. *J. Neurosci.* **4,** 3128–3139.

McBurney R. N. and Barker J. L. (1978) GABA-induced conductance fluctuations in cultured spinal neurons. *Nature* (Lond.) **274,** 596–597.

McCarthy K. D. and De Vellis J. (1980) Preparation of separate astroglial and oligodendroglial cell cultures from rat cerebral tissue. *J. Cell. Biol.* **85,** 890–902.

McCarthy K. D. and Harden T. K. (1981) Indentification of two benzodiazepine binding sites on cells cultured from rat cerebral cortex. *J. Pharmacol. Exp. Ther.* **216,** 183–191.

McCarthy K. D. and Partlow L. M. (1976) Neuronal stimulation of [$^3$H]thymidine incorporation by primary cultures of highly purified non-neuronal cells. *Brain Res.* **114,** 415–426.

Meier E., Drejer J., Hertz L., and Schousboe A. (1984) Culture of Cerebellar Granule Cells, in *A "Cookbook" for Culturing Neuronal and Glial Cells* (Shahar A., Filogamo G., and Vernadakis A., eds.), p. 38. International Society for Developmental Neuroscience, University of Chieti, Italy.

Meier E., Drejer J., and Schousboe A. (1983) Trophic Action of GABA on the Development of Physiologically Active GABA-Receptors, in *CNS-Receptors From Molecular Pharmacology to Behavior* (Mandel P. and DeFeudis, eds.), pp. 47–58. Raven Press, New York.

Messer A. (1977) The maintenance and identification of mouse cerebellar granule cells in monolayer culture. *Brain Res.* **130,** 1–12.

Miller R. H. and Raff M. C. (1984) Fibrous and protoplasmic astrocytes are biochemically and developmentally distinct. *J. Neurosci.* **4,** 585–592.

Mirsky R., Wendon L. M. B., Black P., Stolkin C., and Bray D. (1978) Tetanus toxin: a cell surface marker for neurons in culture. *Brain Res.* **148,** 251–259.

Mirsky R., Winter J., Abney E. R., Pruss R. M., Gaurilovic J., and Raff M. C. (1980) Myelin-specific proteins and glycolipids in rat Schwann cells and oligodendrocytes in culture. *J. Cell Biol.* **84,** 483–494.

Moonen G., Cam Y., Sensenbrenner M., and Mandel P. (1975) Variability of the effects of serum-free medium, dibutyryl-cyclic AMP or theophylline on the morphology of cultured newborn rat astroblasts. *Cell Tiss. Res.* **163,** 365–372.

Moonen G., and Nelson P. G. (1978) Some Physiological Properties of Astrocytes in Primary Cultures, in *Dynamic Properties of Glia Cells* (Schoffeniels E., Franck G., Hertz L. and Tower D. B., eds.), pp. 389–393. Pergamon Press, Oxford.

Morris M. E., and MacDonald J. F. (1982) Measurements of intracellular [$Ca^{2+}$] in cultured mammalian neurons. *Soc. Neurosci. Abstr.* **8,** 909.

Morris J. E., and Moscona A. A. (1971) The induction of glutamine synthetase in cell aggregates of embryonic neural retina: correlations with differentiation and multicellular organization. *Dev. Biol.* **25,** 420–444.

Murray M. R. (1965) Nervous tissues in vitro, in *Cells and Tissues in Culture: Methods, Biology and Physiology* Vol. 2 (Willmer E. N., ed.), pp. 373–455. Academic Press, New York.

Murray M. R. (1971) Nervous Tissues Isolated in Culture, in *Handbook of Neurochemistry* vol. V pt. A (Lajtha, A., ed.), pp. 373–438. Plenum Press, New York.

Nelson P. G. and Lieberman M., Eds. (1981) *Excitable Cells in Tissue Culture.* Plenum Press, New York.

Nelson P. G., Ransom B. R., Henkart M., and Bullock P. N. (1977) Mouse spinal cord in cell culture IV. Modulation of inhibitory synaptic function. *J. Neurophysiol.* **40,** 1178–1187.

Noble M. (1983) Familial and social interrelationships of astrocytes and oligodendrocytes. *Abst. Soc. Neurosci.* **9,** 447.

Norenberg M. D. and Martinez-Hernandez A. (1979) Fine structural localization of glutamine synthetase in astrocytes of rat brain. *Brain Res.* **161,** 303–310.

Parker K. K., Norenberg M. D., and Vernadakis A. (1980) Transdifferentiation of C6 glial cells in culture. *Science* **208,** 179–181.

Perez-Polo J. R., DeVellis J., and Haber B., Eds. (1983) *Growth and Trophic Factors. Progress in Clinical and Biological Research,* vol. 118. Alan R. Liss, New York.

Pettmann B., Louis J. C., and Sensenbrenner M. (1979) Morphological and biochemical maturation of neurons cultured in the absence of glial cells. *Nature* (Lond.) **281,** 378–380.

Pfeiffer S. E., Betschart B., Cook J., Mancini P., and Morris R. (1977) Glial Cell Lines, in *Cell, Tissue, and Organ Cultures in Neurobiology* (Fedoroff S. and Hertz L., eds.), pp. 287–346. Academic Press, New York.

Pitts J. D. and Simms J. W. (1977) Permeability of junctions between animal cells. Intercellular transfer of nucleotides but not of macromolecules. *Exp. Cell Res.* **104,** 153–163.

Pleasure D., Abramsky O., Silberberg D., Quinn B., Parkis J., and Saida T. (1977) Lipid synthesis by an oligodendroglial fraction in suspension culture. *Brain Res.* **134,** 377–382.

Pleasure D., Hardy M., Johnson G., Lisak R., and Silberberg D. (1981) Oligodendroglial glycerophospholipid synthesis: incorporation of radioactive precursors into ethanolamine glycerophospholipids by calf suspension culture. *J. Neurochem.* **37,** 452–460.

Poduslo S. E. and McKhann G. M. (1977) Synthesis of cerebrosides by intact oligodendroglia maintained in culture. *Neurosci. Lett.* **5,** 159–163.

Poduslo S. E., Miller K., and McKhann G. M. (1978) Metabolic properties of maintained oligodendroglia purified from brain. *J. Biol. Chem.* **253,** 1592–1597.

Polak P. E., Szuchet S., and Yim F. H. (1984) Subcellular fractionation of live cells: the oligodendrocyte plasma membrane. *J. Biophys.* **45**, 271a.

Pontén J. and McIntyre E. M. (1968) Long-term culture of normal and neoplastic human glia. *Acta Pathol. Microbiol. Scand.* **74**, 465–486.

Pontén J. and Westermark B. (1980) Cell Generation and Aging of Nontransformed Glial Cells From Adult Humans, in *Advances in Cellular Neurobiology* vol. 1 (Fedoroff S. and Hertz L., eds.), pp. 209–227. Academic Press, New York.

Purves R. D. (1981) *Microelectrode Methods for Intracellular Recording and Iontophoresis.* Academic Press, New York.

Raff M. C., Abney E. R., Cohen J., Lindsay R., and Noble M. (1983b) Two types of astrocytes in cultures of developing rat white matter: differences in morphology, surface gangliosides, and growth characteristics. *J. Neurosci.* **3**, 1289–1300.

Raff M. C., Fields K. L., Hakomori S.-I., Mirsky R. M., Pruss M., and Winter J. (1979) Cell-type-specific markers for distinguishing and studying neurons and the major classes of glial cells in culture. *Brain Res.* **174**, 283–308.

Raff M. C., Miller R. H., and Noble M. (1983a) A glial progenitor cell that develops *in vitro* into an astrocyte or an oligodendrocyte depending on culture medium. *Nature* (Lond.) **303**, 390–396.

Raff M. C., Mirsky R., Fields K. L., Lisak R. P., Dorfman S. H., Silberberg D. H., Gregson N. A., Leibowitz S., and Kennedy M. C. (1978) Galactocerebroside is a specific cell-surface marker for oligodendrocytes in culture. *Nature* (Lond.) **274**, 813–816.

Ransom B. R. and Barker J. L. (1981) Physiology and Pharmacology of Mammalian Central Neurons in Cell Culture, in *Advances in Cellular Neurobiology* vol. 2 (Fedoroff S. and Hertz L., eds.), pp. 83–114. Academic Press, New York.

Ransom B. R., Christian C. N., Bullock P. N., and Nelson P. G. (1977) Mouse spinal cord in cell culture. II. Synaptic activity and circuit behavior. *J. Neurophysiol.* **40**, 1151–1162.

Roussel G., Delaunoy J.-P., Mandel P., and Nussbaum J.-L. (1978) Ultrastructural localization study of two Wolfgram proteins in rat brain tissue. *J. Neurocytol.* **7**, 155–163.

Salzer J. L., Bunge R. P., and Glaser L. (1980a) Studies of Schwann cell proliferation. III. Evidence for the surface localization of the neurite mitogen. *J. Cell. Biol.* **84**, 767–778.

Salzer J. L., Williams A. K., Glaser L., and Bunge R. P. (1980b) Studies of Schwann cell proliferation. II. Characterization of the stimulation and specificity of the response to a neurite membrane fraction. *J. Cell Biol.* **84**, 753–766.

Sanui R. and Rubin H. (1979) Measurement of total, intracellular and surface bound cations in animal cells grown in culture. *J. Cell. Physiol.* **100**, 215–225.

Schachner M. (1982) Cell type-specific surface antigens in mammalian nervous system. *J. Neurochem.* **39,** 1–8.

Schachner M., Sommer J., Lagenaur C., Schnitzer J., and Berg G. (1983) Antigenic Markers of Glia and Glial Subclasses, in *Neuroscience Approached Through Cell Culture* (Pfeiffer S. E., ed.), vol. 2, 115–139. CRC Press, Boca Raton, Florida.

Schengrund C. L., and Marangos P. J. (1980) Neuron-specific enolase levels in primary culture of neurons. *J. Neurosci. Res.* **5,** 305–311.

Schmechel D. E., Brightman M. W., and Marangos P. J. (1980) Neurons switch from non-neuronal enolase to neuron-specific enolase during differentiation. *Brain Res.* **190,** 195–214.

Schmechel D., Marangos P. J., and Brightman M. (1978) Neurone-specific enolase is a molecular marker for peripheral and central neuroendocrine cells. *Nature* (Lond.) **276,** 834–836.

Schnitzer J. and Schachner M. (1982) Cell type specificity of a neural cell surface antigen recognized by the monoclonal antibody A2B5. *Cell Tissue Res.* **224,** 625–636.

Schousboe A. (1977) Differences Between Astrocytes in Primary Cultures and Glial Cell Lines in Uptake and Metabolism of Putative Amino Acid Transmitters, in *Cell, Tissue, and Organ Cultures in Neurobiology* (Fedoroff S. and Hertz L., eds.), pp. 441–446. Academic Press, New York.

Schousboe A., Nissen C., Bock E., Sapirstein V. S., Juurlink B. H. J., and Hertz L. (1980) Biochemical Development of Rodent Astrocytes in Primary Cultures, in *Tissue Culture in Neurobiology* (Giacobini E., Vernadakis A. and Shahar A., eds.), pp. 397–409. Raven Press, New York.

Scott B. S. (1977) Adult mouse dorsal root ganglia neurons in cell culture. *J. Neurobiol.* **8,** 417–427.

Schubert D., Heinemann S., Carlisle W., Tarikas H., Kimes B., Patrick J., Steinbach J. H., Culp W., and Brandt B. L. (1974). Clonal cell lines from the rat central nervous system. *Nature* (Lond.) **249,** 224–227.

Seeds N. W. (1983) Neuronal Differentiation in Reaggregate Cell Cultures, in *Advances in Cellular Neurobiology* vol. 4 (Fedoroff S. and Hertz L., eds.), pp. 57–79. Academic Press, New York.

Sensenbrenner M., Booher J., and Mandel P. (1973) Histochemical study of dissociated nerve cells from embryonic chick cerebral hemispheres in flask cultures. *Experientia* **29,** 699–701.

Sheffield W. O., and Kim S. U. (1977) Myelin basic protein causes proliferation of lymphocytes and astrocytes *in vitro*. *Brain Res.* **132,** 580–584.

Shein S. H., Britva A., Hess H. H., and Selkoe D. J. (1970) Isolation of hamster brain astroglia by *in vitro* cultivation and subcutaneous growth, and content of cerebroside, ganglioside, RNA and DNA. *Brain Res.* **19,** 497–501.

Skaper S. D., Adler R., and Varon S. (1979) A procedure for purifying neuron-like cells in cultures from central nervous tissue with a defined medium. *Dev. Neurosci.* **2**, 233–237.

Smith T. G., Baker J. L., Smith B. M., and Colburn T. R. (1981) Voltage Clamp Techniques Applied to Cultured Skeletal Muscle and Spinal Neurons, in *Excitable Cells in Tissue Culture* (Nelson P. G. and Lieberman M., eds.), pp. 111–136. Plenum Press, New York.

Sonnhof U., Forderer R., Schneider W., and Kettenmann H. (1982) Cell puncturing with a step motor-driven manipulator with simultaneous measurement of displacement. *Pflügers Arch.* **392**, 295–300.

Sotelo J., Gibbs C. J., Gajdusek D. C., Toh B. H., and Wurth M. (1980) Method for preparing cultures of central neurons: cytochemical and immunocytochemical studies. *Proc. Nat. Acad. Sci. USA* **77**, 653–657.

Sternberger N. H., Itoyama Y., Kies M. W., and Webster H. deF. (1978) Immunocytochemical method to identify basic protein in myelin-forming oligodendrocytes in newborn rat CNS. *J. Neurocytol.* **7**, 251–263.

Subak-Sharpe H., Burk R. R., and Pitts J. D. (1968) Metabolic cooperation between genetically marked human fibroblasts in tissue culture. *Nature* (Lond.) **220**, 272–277.

Sundarraj N., Schachner M., and Pfeiffer S. E. (1975) Biochemically differentiated mouse glial cell lines carrying a nervous system-specific cell surface antigen (NS-1). *Proc. Nat. Acad. Sci. USA* **72**, 1927–1931.

Szuchet S., Arnason B. G. W., and Polak P. E. (1980a) Separation of ovine oligodendrocytes into two distinct bands on a linear sucrose gradient. *J. Neurosci. Methods.* **3**, 7–19.

Szuchet S. and Polak P. E. (1983) A simple cell disrupter designed for small cells with relatively large nuclei. *Anal. Biochem.* **128**, 453–458.

Szuchet S., Polak P. E., and Yim F. H. (1984) Subcellular fractionation of live cells: the oligodendrocyte plasma membrane. *J. Biophys.* **45**, 271a.

Szuchet S. and Stefansson K. (1980) *In Vitro* Behavior of Isolated Oligodendrocytes, in *Advances in Cellular Neurobiology* vol. 1 (Fedoroff S. and Hertz L., eds.), pp. 314–346. Academic Press, New York.

Szuchet S., Stefansson K., Wollmann R. L., Dawson G., and Arnason B. G. W. (1980b) Maintenance of isolated oligodendrocytes in long-term culture. *Brain Res.* **200**, 151–164.

Szuchet S. and Yim S. H. (1984) Characterization of a subset of oligodendrocytes separated on the basis of selective adherence properties. *J. Neurosci. Res.* **11**, 131–144.

Szuchet S., Yim S. H., and Monsma S. (1983) Lipid metabolism of isolated oligodendrocytes maintained in long-term culture mimics events associated with myelinogenesis. *Proc. Natl. Acad. Sci. USA,* **80**, 7019–7023.

Tardy M., Costa M. F., Rolland B., Fages C., and Gonnard P. (1981)

Benzodiazepine receptors on primary cultures of mouse astrocytes. *J. Neurochem.* **36,** 1587–1589.

Van Heyningen W. E. (1963) The fixation of tetanus toxin, strychnine, serotonin and other substances by gangliosides. *J. Gen. Microbiol.* **31,** 375–387.

Varon S. (1978) Macromolecular Glial Cell Markers, in *Dynamic Properties of Glia Cells* (Schoffeniels E., Franck G., Hertz L., and Tower D. B., eds.), pp. 93–103. Pergamon Press, Oxford.

Varon S. and Manthorpe M. (1982) Schwann Cells: An *In Vitro* Prospective, in *Advances in Cellular Neurobiology* vol. 3 (Fedoroff S. and Hertz, L., eds.), pp. 35–95. Academic Press, New York.

Varon S., Raiborn C., and Tyszka E. (1973) *In vitro* studies of dissociated cells from newborn mouse dorsal root ganglia. *Brain Res.* **54,** 51–63.

Vernadakis A. and Arnold E. B. (1980) Age-Related Changes in Neuronal and Glial Enzyme Activities, in *Advances in Cellular Neurobiology* vol. 1 (Fedoroff S. and Hertz L., eds.), pp. 230–283. Academic Press, New York.

Viñores S. A., Marangos P. J., and Ko, L. (1982) Butyrate-induced increase in neuron-specific enolase and ornithine decarboxylase in anaplastic glioma cell lines. *Dev. Brain Res.* **5,** 23–28.

Von Mihálik (1935) Über die Nervengewebekulturen mit besonderer Berücksichtigung der Neuronenlehre und der Mikrogliafrage. *Arch. Exp. Zellforsch.* **17,** 119–176.

Waggoner A. S. (1979) Dye indicators of membrane potential. *Ann. Rev. Biophys. Bioeng.* **8,** 47–68.

Walker C. R. and Peacock J. H. (1981) Diazepam binding of dissociated hippocampal cultures from fetal mice. *Dev. Brain Res.* **1:** 565–578.

Walum E., Westermark B., and Ponten J. (1981) Growth-dependent induction of high affinity $\gamma$-aminobutyric acid transport in cultures of normal human brain cell line. *Brain Res.* **212,** 215–218.

Walz W. and Hertz L. (1982) Ouabain-sensitive and ouabain-resistant net uptake of potassium into astrocytes and neurons in primary cultures. *J. Neurochem.* **39,** 70–77.

Walz W. and Hertz L. (1983a) Comparison between fluxes of potassium and of chloride in astrocytes in primary cultures. *Brain Res.* **277,** 321–328.

Walz W. and Hertz L. (1983b) Functional interactions between neurons and astrocytes. II. Potassium homeostasis at the cellular level. *Prog. Neurobiol.* **20,** 133–183.

Walz W. and Hertz L. (1984) Sodium transport in astrocytes. *J. Neurosci. Res.* 11231–11239.

Walz W., Wuttke W., and Hertz L. (1984) Astrocytes in primary cultures: Membrane potential characteristics reveal exclusive potassium conductance and potassium accumulator properties. *Brain Res.* **292,** 367–374.

Werrlein R. J. (1981) Cells and Tissue Culture Systems, in *The Application of Ion-Selective Microelectrodes* (Zeuthen T., ed.), pp. 257–277. Elsevier/North Holland Biomedical Press, Amsterdam.

White F. P. and Hertz L. (1981) Protein synthesis by astrocytes in primary cultures. *Neurochem. Res.* **6**, 353–364.

Wood P. M. and Bunge R. P. (1975) Evidence that sensory axons are mitogenic for Schwann cells. *Nature* (Lond.) **265**, 662–664.

Wood P., Okada E., and Bunge R. P. (1980) The use of networks of dissociated rat dorsal root ganglion neurons to induce myelination by oligodendrocytes in culture. *Brain Res.* **196**, 247–252.

Wood P., Szuchet S., Williams A. K., Bunge R. P., and Arnason B. G. W. (1983) CNS myelin formation in cocultures of rat neurons and lamb oligodendrocytes. *Trans. Am. Soc. Neurochem.* **14**, 212.

Yavin Z. and Yavin E. (1980) Survival and maturation of cerebral neurons on poly(L-lysine) surfaces in the absence of serum. *Dev. Biol.* **75**, 454–459.

Yim F. H. and Szuchet S. (1984) Protein metabolism of cultured oligodendrocytes. A time course study. *Trans. Amer. Soc. Neurochem.* **15**, 231.

Yu A. C. H. (1980) Uptake of glutamine and glutamate by cultured neurons, M.Sc. Thesis, Univ. of Saskatschewan, Canada.

Yu A. C. H., Hertz E., and Hertz L. (1984) Alterations in uptake and release rates for GABA, glutamate and glutamine during biochemical maturation of highly purfied cultures of cerebral cortical neurons, a GABAergic preparation. *J. Neurochem.* **42**, 951–960.

Yu A. C. H. and Hertz L. (1982) Uptake of glutamate, GABA and glutamine into a predominantly GABA-ergic and predominantly glutamatergic nerve cell population in culture. *J. Neurosci. Res.* **7**, 23–35.

Yu A. C. H., Schousboe A., and Hertz L. (1982) Metabolic fate of [$^{14}$C]-labeled glutamate in astrocytes. *J. Neurochem.* **39**, 954–966.

Yu P. H. and Hertz L. (1982) Differential expression of type A and type B monoamine oxidase of mouse astrocytes in primary cultures. *J. Neurochem.* **39**, 1493–1495.

Zeuthen T., Ed. (1981) *The Application of Ion-Selective Microelectrodes.* Elsevier/-North-Holland.

# Chapter 5

# Monoclonal Antibodies

RAM K. MISHRA, DEBRA MULLIKIN-KILPATRICK, KENNETH
H. SONNENFELD, CHANDRA P. MISHRA, AND
ARTHUR J. BLUME

## 1. Introduction

For several years, immunological assays have been used to detect, localize, and quantify extremely small amounts of antigen in complex tissues. The immunoglobulins (antibodies) are proteins of the immune system. Their special structural characteristics allow them to bind specifically to one of many possible foreign molecules (antigens), and thereby render them inactive.

The best characterized and understood immunoglobulin is immunoglobulin G (IgG); other classes of immunoglobulins are IgA, IgM, IgD, and IgE. The immunoglobulin (IgG) molecule contains four polypeptide chains. Two long chains (heavy) are connected by disulfide bonds to form a "Y" configuration (Fig. 1). One light chain is bound by a disulfide bond to each of the arms formed by the two heavy chains. Each arm of the immunoglobulin molecule (Fab fragment) contains an antigen-binding portion at its terminal end. This amino terminal or variable region of each antibody Fab fragment contains hypervariable amino acid sequences that fold to generate a specific site for the antigen. From the data currently available, it can be estimated that an individual can make between one million to one billion different antibody molecules, each with different variable regions. The base of the immunoglobulin is called the Fc fragment because it is easily crystallized when separated from the Fab fragment. Regions of rela-

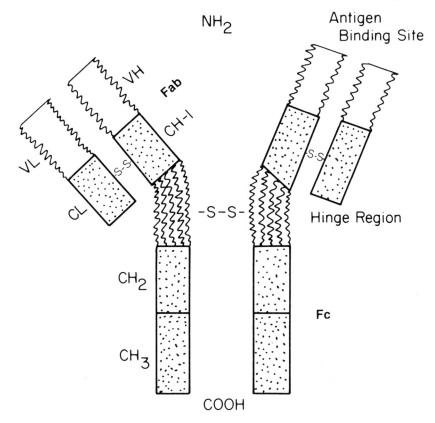

Fig. 1.   The structure of the immunoglobulin molecule. The variable (v) region, also called the Fab region, confers antigen-binding specificity, and the constant region (c), also called Fc Region, is responsible for the overall structure of the molecule and for its recognition by other components of the immune system.

tively constant amino acid sequences of both the Fab and Fc fragments are responsible for the overall structure of the molecule. Recognition of the immunoglobulin by other components of the immune system is mediated through the Fc fragment.

Whether antibodies are used clinically as therapeutic or diagnostic agents, or in basic science as research tools, they possess the ability to bind the specific antigen and produce a biological effect. This unique specificity is a characteristic that determines their usefulness in biology and medicine. Many, if not all, of the uses of antibodies in biology have been hampered by the problem of generating large amounts of highly nonspecific antisera by conventional immunization techniques. Some of the specific problems associated with conventional immunization procedures are as follows:

(a) Specific antisera are unpredictable
(b) Contribution to immunogenicity by minor contaminants
(c) Production of heterogeneous antibodies
(d) Production of low-affinity antibodies
(e) High risk of cross reactivity
(f) Small amounts of the desired antibody for a limited time

In 1975, Kohler and Milstein converted antibody-secreting spleen cells into cloned hybrid cell lines. This technique overcomes many of the problems associated with conventional immunological procedures. It is now possible to use heterogenous immunogens and still derive monospecific antibodies. In theory, monoclonal antibodies could be produced against every immunogenic molecule. In practice, however, the range of the antibodies that can be produced will be limited by the immunogens available, as well as by the sensitivity and discrimination of the assay used to detect them.

In recent years, significant advances in hybridoma technology have so broadened their application that hybridomas have proven useful in almost every aspect of medicine and biology. Monoclonal antibodies have found wide application in basic research areas such as cell and molecular biology, in which their high specificity and binding capability can be exploited for both identifying and purifying antigens, as well as for the alteration of biological processes. Clinically, monoclonal antibodies have important therapeutic and diagnostic value. The applications of monoclonal antibodies range from the very fundamental level of protein purification, to the genetic level, as well as to the treatment of whole organisms. Major applications of monoclonal antibodies have been developed in the fields of cancer chemotherapy, central nervous system disease, heart disease, and infectious disease (e. g., those caused by bacteria, viruses, and parasites). Future research in the monoclonal antibody field will permit a better understanding of the characteristics of antigens involved in the host immune response.

In the following section we describe a general procedure for the production of monoclonal antibodies. No two laboratories use exactly the same technique and many of the variations are based on empirical observations. This implies that there is a great degree of flexibility possible in the production of a hybridoma.

# 2. Immunization

## 2.1. In Vivo Immunization

### 2.1.1. Choice of Mouse Strain

The Balb/c mouse is generally used so that the hybrid myeloma derived from a Balb/c plasmacytoma can be grown in syngeneic animals; ascite tumors grow particularly well in this strain. If there is no immunological response to an antigen in the Balb/c mouse, other strains may be used without any loss of fusion efficiency. Normally, female mice are used because they are more easily handled and are obtained at approximately 6 wk of age.

### 2.1.2. Immunization Schedule

The choice of immunization protocol depends on the type of antigens (e. g., whole cells, crude tissue homogenates or soluble, partially purified, or purified antigens). Most laboratories have their own protocols. The protocol we have chosen to use, shown in Table 1 and Fig. 2, should be used only as a guide.

### 2.1.3. Immunization by Filter Paper Implantation

In vivo immunization has become a routine laboratory procedure for the production of monoclonal antibodies. However, the quantities of immunogen required for injection into host animals are much larger than for in vitro immunization. In addition, each animal should be screened as an antibody producer prior to fusion.

TABLE 1
Immunization Protocols[a]

*Whole cell antigens* (e.g., human neuroblastoma cell SH-SY5Y)
Day 1          $1–5 \times 10^6$ Washed cells in PBS, ip injection
Day 28         $1–5 \times 10^6$ Washed cells in PBS, ip injection
Perform fusion after 3d

*Soluble antigen*
Day 1          10–100 μg sc injection in Freund's complete adjuvant
Day 28         100–200 μg sc injection in Freund's incomplete adjuvant
Three days before fusion, 100 μg, iv

*Purified antigens*
Day 1          1–20 μg sc injection in Freund's complete adjuvant
Day 28         1–20 μg iv injection in incomplete Freund's adjuvant
Fusion follows several days later

[a]If possible, following hyperimmunization the booster injection of the antigen should be given in a few days before the fusion. This improves the yield of antibody-secreting hybrids. The reason for this is not clear, but may be related to a stimulation of memory B cells.

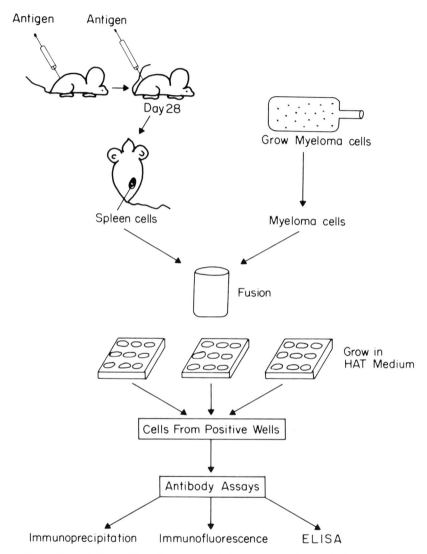

Fig. 2. Schematic diagram of the method used for producing monoclonal antibodies by in vivo immunization.

Sternick and Sturmer (1984) have recently developed a modified in vivo technique to produce monoclonal antibodies against very low amounts of soluble antigens. The antigen is adsorbed onto nitrocellulose filter disks (0.45 μm porosity, 7 mm diameter) and the antigen-coated filter disks are then implanted inside the peritoneal cavity. First, second, and third boosts can be given with antigen-coated nitrocellulose implants at 1 wk intervals after initial immunization. Fusion can then be performed 3–4 d after the last implantation. The quantity used on each filter disk can be as low as 1 ng. Thus, for the rare antigen this technique of immunization is more attractive than the classical method of injection.

## 2.2. In Vitro Immunization

In vitro immunization of spleen cells in culture permits the use of minute quantities of antigen and enables enhanced recovery of specific antigen-activated clones (Fig. 3). Dissociated spleen cells are cultured in the presence of both thymocyte-conditioned medium and antigen. The thymocyte-conditioned medium probably generates lymphokines that stimulate lymphocytes, in the presence of antigen, to promote antibody production.

### 2.2.1. Isolation of Spleen Cells

Mice are sacrificed by $CO_2$ asphyxiation and immersed briefly in 70% ethanol. The spleens are surgically removed under strict sterile conditions in a laminar flow hood and washed in phosphate buffered saline (pH 7.4) (PBS, Gibco Laboratories). Spleen cells are mechanically dissociated by passing them through a 0.05 mm$^2$ mesh stainless steel screen.

### 2.2.2. Immunization of Spleen Cells

Spleen cells are cultured in T-75 flasks containing $1–2 \times 10^8$ spleen cells in 10 mL MEM with 20% fetal calf serum and 10 mL thymocyte-conditioned media and antigen.

The length of time spleen cells are exposed to the antigen and the amount of antigen necessary for best fusion results vary among different antigens. Concentrations of antigen used for in vitro immunization by this method may range from 0.05 to 10 mg/mL. Timing may be more critical than the amount of antigen; therefore, multiple fusions on d 2, 3, and 4 after the initial sensitization may lead to best results (Fig. 3). The spleen cell culture is then continued in 5% $CO_2$ at 37°C. On the appropriate day, the non-adherent cells in the culture are removed, washed with serum-free medium, and used for fusion with the myeloma cells, as described in a later section. (*See* review article on in vitro immunization by Reading, 1982.)

### 2.2.3. Preparation of Thymocyte-Conditioned Medium for Spleen Cell Cultures

Thymocyte-conditioned medium is prepared by removing the thymus gland from 10–15-d-old Balb/c mice. The tissue is then passed through a 0.50 mm$^2$ mesh stainless steel screen. Thymus

## IN VITRO IMMUNIZATION

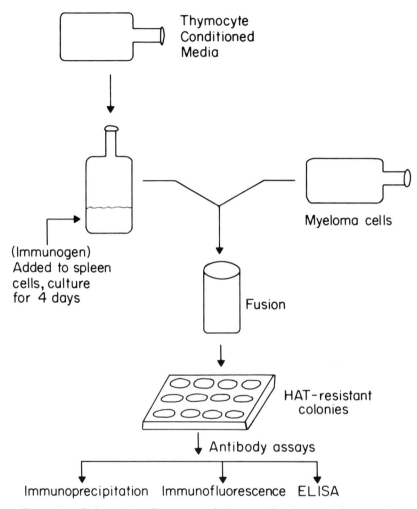

Fig. 3. Schematic diagram of the method used for producing monoclonal antibodies by in vitro immunization.

cells ($4 \times 10^6$ cells/mL) are cultured in Dulbecco's modified Eagle's medium (Gibco Laboratories, Grand Island, NY), containing 5 g glucose/L, 2 m$M$ glutamine, 5 μ$M$ hypoxanthine, and 1 m$M$ pyruvate with 15% fetal calf serum. The thymus cells are cultured for 48 h at 37°C in 5% $CO_2$. The thymocyte-conditioned medium is then harvested by centrifugation and the supernatant is used immediately or stored at $-70$°C.

# 3. Fusion

## 3.1. Myeloma Cell Line

The most commonly used myeloma cell line is P3-NS1/1Ag4 1 (NSI) which can be obtained commercially from Flow Laboratories. Other myeloma cell lines used are shown in Table 2.

Myeloma cells are cultured either in RPMI-1640 or in Iscove's medium supplemented with heat-inactivated calf serum (10%), 2 mM L-glutamine, 10 mM 2-mercaptoethanol, penicillin, and streptomycin. Two weeks before fusion they are grown for 3 d in medium supplemented with 8-azaguanine (2 mg/100 mL) to ensure retention of the HGPRT trait and removal of revertants. Following the 3-d subculture period, the cells are transferred to normal growth media and grown in hypoxanthine, aminopterin, and thymidine (HAT) medium. If cells are either susceptible to 8-azaguanine or resistant to HAT medium, a fresh culture of myeloma cells should be started.

In general, an average fusion requires $5 \times 10^7$ myeloma cells. It is of prime importance to use myeloma cultures in a logarithmic growth phase. Cultures with 95% viability should be maintained and used for fusion. The doubling time for mouse myeloma cells is 12–18 h. This is much shorter than for many mammalian cells. Thus, it is very common to see many attached pairs of recently divided cells in a healthy myeloma culture, and such cells should not be confused with dead or dying myeloma cells. In our experiments for the production of monoclonal antibodies against SH-SY5Y human neuroblastoma cells, we have used PAI-0 myeloma cells for fusion (Stocker et al., 1982). This myeloma cell line has several advantages over the cell lines described in Table 2. The

TABLE 2
Mouse Myeloma-Derived Cell Lines
Available for Fusion

| Cell line[a] | Short term | Immunoglobulins |
|---|---|---|
| P363-Ag8 | X63 | δ,K |
| P3-NSi-Ag4/1 | NS1 | K |
| P3x63-Ag8.653 | 6S3 | None |
| P3-SP2/ Ag14 | SP2 | None |
| FO | FO | None |
| S194/5XXOBU1 | BU1 | None |

[a]Most of these cell lines can be obtained from the American type Culture Collection.

major advantages are: (1) these cells do not require macrophage feeder cells at any stage; and (2) PAI-0 cells not fused with spleen cells do not produce immunoglobulin (Ig) chains.

## 3.2. Preparation of Macrophages

One or 2 d before fusion, $10^4$ peritoneal cells or normal spleen cells are prepared in HAT media and plated in a 96-well microtiter plate (e. g., Falcon catalog #3042) or $5 \times 10^4$ cells in a 24-well cluster plate (e. g., Costar #3524). These cells act as feeders and encourage growth of the hybridoma cells. They should be prepared under strict sterile conditions. Details of the procedure for macrophage preparation are described elsewhere (Fazekas et al., 1980).

There are several protocols available for cell fusion. However, it seems that many of the parameters for fusion are not too critical (e. g., use of serum-free RPMI vs salt buffer). The two most important technical points are the timing of addition of reagents and the fusogen used. Polyethylene glycol (PEG) is deleterious to cells (causes large osmotic changes in the cells); too rapid an addition of PEG can cause cell death. Some of these effects can be overcome by including (2–5%) dimethylsulfoxide (DMSO) in the fusion mixture. We have had good success using PEG 1500, sold by British Drug Houses (BDH), as the fusogen. However, there is still debate regarding the optimum molecular weight of PEG to use.

## 3.3. Fusion Protocol

Our general fusion protocol, in stepwise order, is as follows:

(a) The spleen from the immunized mouse is removed under sterile conditions and washed twice with RPMI-1640 or Buffer A (1 L of Buffer A contains 8 g NaCl, 0.4 g KCl, 1.77 g $Na_2HPO_4$, 0.69 g $NaH_2PO_4$, 2 g glucose, and 0.01 g phenol red, pH 7.2).

(b) The spleen is disaggregated as described above and the cells are pipeted into a sterile 50-mL plastic tube. The tube is left for 5 min to allow clumps to settle. The cells (supernatant) are then transferred to a second tube and centrifuged at 200*g* for 10 min.

(c) The myeloma cells are washed with serum-free RPMI-1640 or Buffer A, and pelleted by centrifugation at 200*g* for 10 min.

(d) The two pellets ($10^8$ spleen cells and $5 \times 10^7$ myeloma cells) are resuspended in serum-free RPMI-1640 or Buffer A, mixed, and recentrifuged at 2500$g$ for 10 min to form a firm pellet. The tube is then transferred to a 37°C water bath or a beaker containing sterile water at 37°C that can be used in a sterile hood.

(e) A 1-mL portion of 50% PEG is added over a 1-min period with gentle stirring. The pipet tip itself could be used for this purpose. After the addition of PEG, the tube is gently stirred by hand for an additional 1 min at 37°C. At this point, one can visualize the formation of a grainy mixture owing to the formation of small clumps of cells.

(f) The fusion is then stopped by addition of serum-free RPMI-1640 or Buffer A. This is done very slowly. Two milliliters of medium are added over a 2-min period and 8 mL over a 4-min period. The diluted fusion is left at room temperature for 10 min. The cells are then gently pelleted and resuspended in 37°C HAT selection medium at $1 \times 10^6$ cells/mL.

(g) The cells are then plated in either 24- or 96-well plates. One milliliter of the cell suspension is added per well if a 24-well plate is used, whereas 0.1 mL/well is added if a 96-well plate is used. The smaller wells (96-well plate) have an advantage in that most wells in which cultures grow will have only one colony of hybrid cells. However, the disadvantage is that the volume of supernatant available for screening the clones is small. Culture plates are placed at 37°C in a 5% $CO_2$ incubator until colonies appear (7–15 d).

## 4. Screening of Clones for Antibody Secretion

Screening clones for antibody secretion is one of the most crucial aspects of hybridoma technology, and it is beyond the scope of this paper to give a comprehensive review of antibody screening procedures. The procedure chosen would obviously depend on the type of antigen used for immunization. The general techniques currently available are described below.

### 4.1. Radioimmunobinding (Radioimmunoprecipitation)

Radioimmunobinding is very useful when a radiolabeled antigen or anti-mouse (e. g., [125]I-rabbit anti-mouse, goat anti-mouse, or protein A) second antibody is readily available. The method can

be adapted for use with whole cells, tissue homogenates, or soluble proteins. We have used this procedure employing various cell lines; i. e., human neuroblastoma SH-SY5Y and MC-1XC cells, rat × mouse neuroblastoma glioma hybrid NG108-15, Wish cells, human endometrial cancer cells (STR), and other cell lines. The cells after harvesting are suspended in RPMI-1640 containing 0.1% BSA and 50 m$M$ HEPES, pH 7.4. Four hundred microliters of cell suspension (2 million cells) are then added to microfuge tubes
(e. g., Sarstedt #72.690) containing 50–100 μL of culture supernatant from wells containing hybridoma clones. After incubation for 1 h at 37°C in a shaking waterbath, the tubes are centrifuged at 15,000$g$ for 5 min. The supernatant is aspirated and the pellets are washed twice with cold buffer by vortexing and recentrifuging. In a volume of 400 mL buffer, $^{125}$I-labeled second antibody is added to each tube. After 2 h of incubation at 0–4°C, the tubes are washed twice without resuspending the pellets. The bound antibody is measured in a gamma counter (*see* Fig. 4). Appropriate controls should be done at the same time using culture fluid from nonhybridoma-producing wells.

For protein antigens that can be obtained radioactively labeled or can be easily iodinated, the procedure involves the incubation of an antibody with radioactive antigen in a buffer (Buffer B) containing 2.5% Triton X-100, 0.5% sodium dodecyl sulfate, 0.1 m$M$ NaCl, 6 m$M$ EDTA, 0.5 mg/mL BSA, and 50 m$M$ Tris, pH 7.4. The antigen–antibody complex is allowed to form by incubating the tubes at 37°C for 1 h (*see* Fig. 4B). After this, a second antibody coupled to sepharose or amylose is added (approximately 10–20 μL), and incubation is carried out at 0–4°C for 2 h. The tubes are then centrifuged and pellets are washed twice with ice-cold Buffer B and the remaining radioactivity is counted.

For the nonradiolabeled protein antigens, 50–100 μL (0.1–0.5 mg/mL antigen) are added to wells of flexible plastic microtiter plates. Protein molecules adhere to the negatively charged plastic and are therefore immobilized. The plates are incubated for 6–8 h at 0.4°C and the solution is removed. The remaining negatively charged empty sites on the plastic wells are then saturated with 0.5% BSA/PBS and incubated for 1 h at 0°C; the plate is then washed with 1% BSA/PBS three times. The media from hybridoma-producing wells (50 μL) is added to each well on the plate. After 1 h of incubation at 37°C, the plates are washed three times with 0.1% BSA/PBS and the radiolabeled anti-mouse antibody is added. The incubation is continued at 0–4°C for 2 h; the plates are then washed twice, the individual wells cut out, and the radioactivity counted.

## RADIOIMMUNOPRECIPITATION

A. For whole cells

2 x 10$^6$ cells
+50-100 µl Hybridoma culture supt.
Incubate for 1 hr at 37°C

Wash 2X with RPM I-BSA-HEPES

Add 400 µl of diluted $^{125}$I-antibody e.g.
(Rabbit antimouse)
Incubate for 2 hrs at 0-4°

Wash 2X

Count the radioactivity
in gamma counter.

B. For $^{125}$I-antigen

$^{125}$I- antigen
+10-20 µl of culture supt.
Incubate for 1 hr at 37°C

Add goat antimouse-amylose
or sepharose coupled antibody

Incubate for 2hrs at 0-4°

Wash 2X

Count the radioactivity

Fig. 4. Schematic diagram of the radioimmunoprecipitation method.

## 4.2. Immunofluorescence Microscopy

The choice of an immunofluorescence-labeling technique used in any individual laboratory is generally based primarily on the availability of the reagents and preference of the investigator. The immunofluorescence technique is relatively simple to perform. Cells are generally grown on sterile glass cover slips that will be mounted on slides. The cells are washed and incubated with PBS containing normal goat serum (10–20%) for 20–30 min. The cover slip is then washed briefly in PBS and incubated with 10–20 µL of hybridoma supernatant for 30–45 min at room temperature.

Following this incubation, the cover slip is washed with PBS. Fluorescein or rhodamine-conjugated goat anti-mouse immuno-globulin (1:20 dilution with PBS) is added, and the cover slip is incubated again. The mounted cover slips are then washed once more in PBS and examined under the fluorescence microscope (*see* Fig. 5). This procedure is described in more detail by Hempstead and Morgan (1983).

## IMMUNOFLUORESCENCE

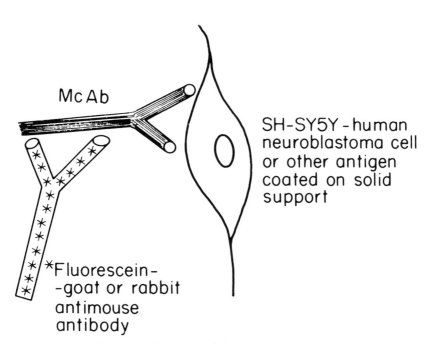

McAb

SH-SY5Y-human neuroblastoma cell or other antigen coated on solid support

*Fluorescein-
-goat or rabbit
antimouse
antibody

Fig. 5.   Schematic diagram of the immunofluorescence technique.

## 4.3. Enzyme-Linked Immunosorbent Assay for Screening Monoclonal Antibodies (ELISA)

The presence of a monoclonal antibody for a specific antigen is determined by incubating the hybridoma supernatant with immobilized antigen. This is followed by washing and the subsequent addition of anti-immunoglobulin antibody conjugated with an enzyme. Further incubation with substrate results in a measurable reaction (e.g., change in optical density, in only those wells in which an antibody specifically binds to an antigen (*see* Fig. 6). The choice of the enzyme and the substrate depends upon the individual characteristics of the assay. In general, soluble antigens (generally proteins) can be attached to a solid phase (e.g., plastic) by adsorption. The soluble antigen (5–10 mg/mL protein conc.) is added to wells of a 96-well microtiter plate (50–200 µL) and incubated at 4°C for at least 5 h. Plates can be stored safely for several weeks at −20°C. In the case of whole cells, they may be used fresh or after fixation with glutaraldehyde. For fresh cells, the microtiter plate should be incubated first for 1–3 h with 0.1% BSA/PBS/NaN$_3$. After washing the plates, 50–100 µL of a cell suspension ($5 \times 10^4$–$2 \times 10^5$) without serum is dispensed into each well and the cells are allowed to settle for 1–3 h (depending upon the cell type).

The antigen-coated plates are then incubated with 200 µL of 0.1% BSA/PBS for 1 h at 4°C. This process allows a blockade of empty electrostatic sites on the polystyrene plates. The buffer is removed after this process and 20–50 µL of hybridoma supernatant are added to triplicate wells. As a positive control, a previously known antiserum should also be added. Negative controls may include supernatant from a nonhybridoma well or PBS alone. The plate is incubated for 1 h at 37°C. Following this, the plate is washed with PBS and the enzyme-labeled antibody (generally a 1:200 dilution) (*see* Table 3 and Fig. 6) is then added to each well and the plate is incubated at room temperature for 1 h.

TABLE 3
Most Commonly Used Enzymes for the ELISA Method

| Enzyme | Substrate | Wavelength, nm |
|---|---|---|
| Horseradish peroxidase | o-Phenyldiamine | 492 |
| E. coli β-galactosidase | o-Nitrophenyl β-D-Galactopyranoside | 414 |
| Alkaline phosphatase | p-Nitrophenyl phosphate | 405 |

## Enzyme-Linked Immunosorbent Assay
## (An ELISA "Sandwich" Method)

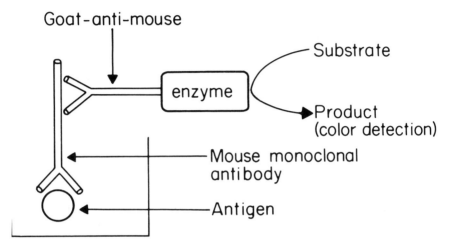

Fig. 6.    Schematic diagram of the enzyme-linked immunosorbent assay and ELISA "sandwich" method.

After incubation with the substrate, the 96-well plate can be read either in an automated vertical-beam reader or by a spectrophotometer at 450 nm. The color development in each well may be stopped by addition of 5*M* sulfuric acid.

The reference well chosen as the blank determines which wells are negative. For better accuracy, several negative controls should be included in the assay.

## 5. Clonal Expansion

### 5.1. Subcloning

Hybidoma cells can be cloned either by limiting dilution or by the semisolid agar technique. Generally, the limiting dilution method is used for cloning and is described briefly here. For efficient cloning, approximately $5 \times 10^5$ hybridoma cells are required. Therefore, the cells cannot be successfully cloned until they are growing well in the initial 24-well plate used after fusion. The hybridoma cells are fed 24–72 h before cloning. The cells are diluted with complete medium to a concentration of 400–600 cells/mL. The diluted cell suspension is added to the first row of eight wells in a 96-well microtiter plate and diluted serially so that the concentration of cells/well ranges from 400 to 1. The microtiter plates should first be layered with $2-3 \times 10^5$ thymocytes (feeder cells) per well as described in the fusion section.

The plates are left at 37°C in 5% $CO_2$ until colonies appear, which is usually 7–21 d. The contents of those wells showing single clones are then transferred to a new 24-well plate in 0.5 mL of media (HAT can now be omitted). The 24-well plates are fed with 0.3–0.5 mL media on the seventh day after transfer. The supernatant is removed on the fifteenth day and reassayed for specific antibody activity. Positive wells are expanded further into 25 cm$^2$ flasks and then into 75 cm$^2$ flasks. In general, 5–10 positive clones are expanded to the 75-cm$^2$ flask stage and the one with the highest antibody titer is chosen for further expansion and injection into animals for the production of ascites fluid. The cloning efficiency of the hybridoma under these conditions is 30–50%.

## 5.2. Cryropreservation of Hybridoma Cells

At the time of subcloning, $1-5 \times 10^6$ hybridoma cells should be frozen to guard against any contamination that might occur in subsequent manipulations of the culture. The freezing mixture contains IMDM or RPMI complete media with 20% fetal calf serum and 10% dimethylsulfoxide (DMSO). The cells should be resuspended ($1-2 \times 10^6$ cells/mL) in this media at 4°C after harvesting and then transferred to a $-$ 20°C freezer. The ampules are left at $-$ 20°C for 40 min and then transferred to dry ice for 10 min. The ampules are finally transferred to a liquid nitrogen freezer for storage. After 1 wk, cells from one or two ampules should be thawed quickly, centrifuged, and resuspended in fresh media for growth to check the efficiency of the freezing procedure.

## 5.3. Ascites Fluid Production

Large amounts of monoclonal antibodies (1–20 mg/mL) can be produced easily by injecting the positive hybridomas into Balb/c mice. Mineral oil or pristane (0.5 mL) is injected intraperitoneally 1 wk prior to the ip injection of $2-5 \times 10^5-1 \times 10^7$ hybridoma cells. After cell injection, the mice are inspected every 24 h for swelling of the abdomen. The ascites fluid is collected using an 18-gage needle in a heparinized tube and centrifuged at 800$g$ for 15 min. The supernatant is collected and filtered through cotton wool (1 cm, loosely packed in the barrel of a 5-mL syringe) to remove fibrin, DNA, and other particles. The filtrate is then centrifuged (100,000$g$, 15 min) to remove aggregates and other particles and, finally, is passed through a 0.45 μm millipore filter. The ster-

# Foreword

Techniques in the neurosciences are evolving rapidly. There are currently very few volumes dedicated to the methodology employed by neuroscientists, and those that are available often seem either out of date or limited in scope. This series is about the methods most widely used by modern-day neuroscientists and is written by their colleagues who are practicing experts.

Volume 1 will be useful to all neuroscientists since it concerns those procedures used routinely across the widest range of subdisciplines. Collecting these general techniques together in a single volume strikes us not only as a service, but will no doubt prove of exceptional utilitarian value as well. Volumes 2 and 3 describe current procedures for the analyses of amines and their metabolites and of amino acids, respectively. These collections will clearly be of value to all neuroscientists working in or contemplating research in these fields. Similar reasons exist for Volume 4 on receptor binding techniques since experimental details are provided for many types of ligand-receptor binding, including chapters on general principles, drug discovery and development, and a most useful appendix on computer programs for Scatchard, nonlinear and competitive displacement analyses. Volume 5 provides procedures for the assessment of enzymes involved in biogenic amine synthesis and catabolism.

Volumes in the NEUROMETHODS series will be useful to neuro-chemists, -pharmacologists, -physiologists, -anatomists, psychopharmacologists, psychiatrists, neurologists, and chemists (organic, analytical, pharmaceutical, medicinal); in fact, everyone involved in the neurosciences, both basic and clinical.

# Preface to the Series

When the President of Humana Press first suggested that a series on methods in the neurosciences might be useful, one of us (AAB) was quite skeptical; only after discussions with GBB and some searching both of memory and library shelves did it seem that perhaps the publisher was right. Although some excellent methods books have recently appeared, notably in neuroanatomy, it is a fact that there is a dearth in this particular field, a fact attested to by the alacrity and enthusiasm with which most of the contributors to this series accepted our invitations and suggested additional topics and areas. After a somewhat hesitant start, essentially in the neurochemistry section, the series has grown and will encompass neurochemistry, neuropsychiatry, neurology, neuropathology, neurogenetics, neuroethology, molecular neurobiology, animal models of nervous disease, and no doubt many more "neuros." Although we have tried to include adequate methodological detail and in many cases detailed protocols, we have also tried to include wherever possible a short introductory review of the methods and/or related substances, comparisons with other methods, and the relationship of the substances being analyzed to neurological and psychiatric disorders. Recognizing our own limitations we have invited a guest editor to join with us on most volumes in order to ensure complete coverage of the field and to add their specialized knowledge and competencies. We anticipate that this series will fill a gap; we can only hope that it will be filled appropriately and with the right amount of expertise with respect to each method, substance or group of substances, and area treated.

Alan A. Boulton
Glen B. Baker

# Preface to Volume 1

When I began neurochemical research in Saskatoon after working for several years in the UK on the chemistry of yeast protoplasts and the levels of amine metabolites in the body fluids of mentally disordered individuals, I remember wondering, which animal should I select, how do I get its brain out, how do I dissect that brain, how do I slice it or obtain subcellular fractions? Later my worries continued, albeit at a rather more specialized level, such as how do I isolate discrete nuclei, perfuse parts of the brain, or lesion or stimulate it? It seemed to me at the time that a book that described general techniques would have been invaluable. For the first volume in this new series on Neuromethods, therefore, we have tried to assemble the techniques most of us use routinely.

In Chapter 1, Palkovits describes the microdissection of fresh, frozen, or fixed (stained) brains yielding as little as 10 μg wet wt. of a specific nucleus from a variety of species. de Lores Arnaiz and Pellegrino de Iraldi in Chapter 2 describe how to break down cellular integrity and isolate synaptosomes, synaptic vesicles, synaptosomal membranes, myelin, Golgi structures, lysosomes, nuclei, mitochondria, microtubules, and filaments. In Chapter 3 Lipton describes the uses and abuses of the brain slice, paying particular attention to its selection, preparation, morphology, integrity, incubation, metabolism, stimulation, and advantages in release studies. Hertz et al. in their contribution describe the advantages of using tissue culture and cultured cells from the cerebral cortex and cerebellum, and describe the isolation of astrocytes and oligodendrocytes and their use in biochemical, pharmacological, and physiological experiments. Mishra et al. in their chapter describe how molecular biology and in particular hybridoma technology can be applied to the preparation of monoclonal antibodies for application in the neurosciences, particularly in the study of neurotransmitter and neuropeptide receptors.

In more physiological terms, MacDonald (Chapter 6) outlines the use of microiontophoresis and micropressure techniques in the identification of central transmitters, describing in some detail the types of release, origins of variability, artifacts, errors, and extra- and intracellular recording; Greenshaw (Chapter 7) describes electrical and chemical ways to stimulate the various parts and pathways of the brain, paying considerable attention to stereotaxic techniques, anesthesia, electrodes, stimulation circuits, wave forms, cannulae, pumps, microinjection, and damage to injection sites, as well, of course, as several examples of working applications.

Pittman et al. in Chapter 8 offer a critical review of perfusion technology; they cover the perfusion medium itself, the extent of tissue damage, and the collection of the perfusate, as well as applications (ventricular, spinal cord, in vitro, cup, and local tissue). The basic concepts and methods used to study brain electrical activity are described in Chapter 9 by Vanderwolf and Leung. In particular they discuss the problems of assessing electrophysiological correlates of behavior, before proceeding to describe recording methods, slow waves, evoked potentials, reticulocortical and neocortical activity, and the effects of drugs on these activities in relation to animal behavior.

After discussing early ablation techniques for creating lesions, Schallert and Wilcox (Chapter 10) move to a detailed discussion of transmitter-selective lesions created by neurotoxins such as 6-hydroxydopamine, 5,7-dihydroxytryptamine, and the aziridinium ion of ethylcholine, as well as mentioning some of the newer toxins such as N-2-chloroethyl-N-ethyl-bromobenzylamine and 1-methyl-4-phenyl-1,2,3,6-tetrahydropyridine, and their likely future uses.

In Chapter 11 Smith provides a critique of earlier single-pass techniques for determining blood–brain barrier permeabilities, and after thorough kinetic analyses concludes that intravenous administration and *in situ* brain perfusion techniques are the most versatile and sensitive for measuring transport of substances into the brain.

In Chapter 12, Brady reviews types of axonal transport, some theories to explain the phenomenon, and the six discrete rate components that exist before describing in detail the three procedures (direct visualization, labeling, and extrinsic markers) for its assessment. Postmortem analyses, collection of human postmortem material in a brain bank, dissection, and safety, as well as some of the variables that affect neurochemical values in postmortem materials, are discussed in the penultimate chapter by Reynolds. The effects of age, cause of death, mental disorder, drug treatment, time of death, and time after death, as well as some methods, are all covered. Finally, in the last chapter Sarnat describes the evolution of the vertebrate brain, the relation of evolution to disease, and the application of histochemistry in comparative neuroanatomy.

I hope that this general techniques volume will prove to be as useful today as its counterpart would have been to me 20 years ago.

**Alan A. Boulton**

ilized fluid is stored at $-20°C$ in 100-$\mu$L aliquots for further use. The cells remaining after the first centrifugation can be passaged up to three times for ascites production; after that, the ascites should be produced from fresh hybridoma cells.

## 6. Purification of the Immunoglobulin

After generating large amounts of ascites fluid or antibody-containing supernatants, the immunoglobulins can be purified by conventional means for further characterization. Starting with 50–100 mL, after thawing, the ascites fluid is precipitated with saturated ammonium sulfate. The precipitate is removed by centrifugation (1000$g$ for 5 min at 4°C). The remaining immunoglobulins in the supernatant are precipitated as above and the two precipitates are combined and solubilized in PBS (0.1 vol of PBS per original volume of ascites fluid). At this stage, the soluble precipitate is analyzed for its antibody properties by passing it over a Sephadex G-200 column equilibrated with PBS. This procedure separates IgM, IgG, and serum albumin. The IgG can be greater than 90% pure at this stage, if the starting ascites fluid contained high concentrations of IgG (*see* Fig. 7).

Further purification of IgM and IgG can be carried out by dialysis and DEAE-cellulose column chromatography. The fractions containing immunoglobulins are pooled and dialyzed against Tris-HCl, pH 7.5. The IgM will be precipitated by this procedure and can be collected by centrifugation (10,000$g$ for 30 min), redissolved in 0.1$M$ Tris-HCl (pH 8.5), and dialyzed against PBS.

The IgG class of antibody can be purified either by a DEAE-cellulose column or by a protein A-Sepharose affinity column. The supernatant collected after dialysis and centrifugation is subjected to a DE52 column equilibrated with 5$M$ Tris-HCl, pH 7.5. The column is eluted with a linear gradient of 0–0.1$M$ NaCl and then with 0.5$M$ NaCl. The fractions containing significant amounts of protein are assayed for antibody properties. Active fractions are pooled, concentrated (1–20 mg/mL), and stored in liquid nitrogen.

Protein A, a constituent of the cell wall of most strains of *Staphylococcus aureus*, has the ability to bind to the Fc regions of IgG and other antibodies. This method of purification is limited, however, to IgG$_2$, since protein A binds very poorly to IgM and IgG$_1$. The ascites fluid or supernatant from a hybridoma is applied to a protein A–Sepharose CL4B column (Pharmacia, Catalog #17-0780-01), and the immunoglobulins are eluted with 5$M$ NaCl or some other suitable buffer.

PURIFICATION OF IMMUNOGLOBULINS

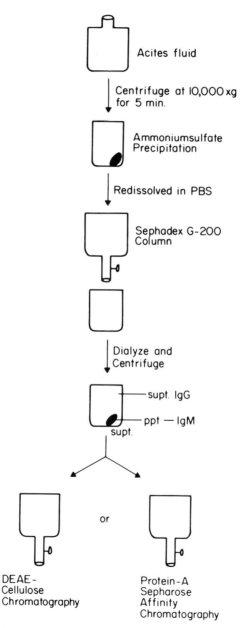

Fig. 7.    Schematic diagram of the method for the purification of immunoglobulins.

## 7. Monoclonal Antibodies to SH-SY5Y Human Neuroblastoma Cells

Using the procedures described earlier in this manuscript, we have been able to produce a series of monoclonal antibodies that can recognize specific surface components of the SH-SY5Y cell. The results in Table 4 show that one group of antibodies specifically binds to SH-SY5Y neuroblastoma cells and not to other neuroblastoma or cancer-related cells. The detailed characterization of this antibody is in progress.

## 8. Application of Monoclonal Antibodies to Neurosciences

The hybridoma technology has been applied to many different biomedical problems, and a complete description is not feasible. Some of the more important uses, however, with reference to the neurosciences are described here.

The hybridoma technique can be used to obtain pure antibodies against cell surface antigens; this applies specifically to neurotransmitter or neuropeptide receptors that are present in minute quantities on cell membranes. To obtain antibodies, spleen cells immunized with crude or partially purified mem-

TABLE 4
Binding of Monoclonal Antibody to Various Cell Lines

| Cell line | | $^{125}I$, counts per minute |
|---|---|---|
| SH-SY5Y (Human neuroblastoma) | Background | 310 |
| | with antibody | 1580[a] |
| J C-2 (Human neurofibroma) | Background | 200 |
| | with antibody | 207 |
| MC-IXC (Human neuroblastoma) | Background | 220 |
| | with antibody | 251 |
| STR (Human stromal) | Background | 185 |
| | with antibody | 181 |
| NG108-15 (neuroblastoma × glioma) hybrid | Background | 194 |
| | with antibody | 216 |

[a]Indicates high binding affinity of the monoconal antibody. The screening procedure utilized the radioimmunoprecipitation method described in the screening section. $5 \times 10^4$ Counts per minute of second antibody anti-mouse were added in the assay.

branes are used for fusing with myeloma cells. The resultant hybridomas are then screened for the production of antireceptor antibody using a procedure that identifies either binding of antibody to receptor or an antibody-mediated biological effect. Monoclonal antibodies have been prepared that are specific for a variety of membrane components, such as acetylcholine receptors, β-adrenergic receptors, histocompatiblity antigens, and antigens to specific particular cell types, including red blood cells, lymphocytes, or different types of tumor cells.

## 9. Potential Uses of Monoclonal Antibodies

(a) Monoclonal antibodies are currently being utilized in an effort to kill tumor cells by attacking them with an antibody that is highly specific for tumor cells and to which a toxic molecule, such as ricin, is coupled. The approach is called a "magic bullet" technique. In principle, the toxin for the tumor would be carried to a specific site by the antibody in vivo and the attached toxin would then kill the tumor cells. Interesting results have been obtained by Krolick et al. (1980) and Gilliand et al. (1980) in this respect. However, the approach is still considered to be in the experimental stage.

(b) Monoclonal antibodies may have an important role in the correlation of neurotransmitter receptors with clinical disorders. Using conventional radioligands, several investigators have attempted to correlate disease characteristics or the extent of a therapeutic drug response with either the number or affinity of a specific receptor thought to be implicated in the disease process. For example, receptor alterations have been studied using either lymphocytes or platelets from patients with diseases such as schizophrenia, Parkinson's disease (Lefur et al., 1983), asthma (Conolley and Greenacre, 1976), and cardiovascular disorders. The validity of receptor binding data in whole lymphocytes has been questioned owing to the problem of nonspecific uptake and processing of the radioligand by lysozomes of these cells. Use of $^{125}I$-labeled monoclonal antibodies in intact lymphocyte binding studies may circumvent the above problem and more accurate quantification of receptors could be carried out in normal and disease states.

(c) Monoclonal antibodies that recognize neurotransmitter receptors may also be of potential use in developing animal models of certain disease states. This would be especially applicable to the disease states that are considered to be mediated by circulating autoantibodies to these receptors.

(d) The hybridoma technology can be used to purify an antigen by several thousandfold. The monoclonal antibody (Mab) that is obtainable in relatively large quantities can be coupled to an insoluble support, such as Sepharose, and the neurotransmitter receptor can be purified by adsorption to the bound Mab, followed by elution with high ionic strength buffer or by lowering the pH. In fact, Mabs were utilized in purification of the nicotinic cholinergic receptor (Lennon et al., 1980) and the β-adrenergic receptor (Fraser and Venter, 1980).

(e) One of the obvious and exciting applications of monoclonal antibodies to neurotransmitter receptors in the clinical field concerns the use of monoclonal antibodies as highly specific therapeutic pharmacological agents. For example, a monoclonal antibody with antagonist properties specific for kidney dopamine receptor D-2 could perhaps provide all the clinically effective advantages of neuroleptic drugs currently prescribed for reducing the renal blood flow (Cavero et al., 1982). Such antibodies may have very few or none of the side effects seen with the neuroleptic drugs. Similarly, if the antibodies could be modified to cross the blood–brain barrier (in fact some of the polyclonal antibodies do cross the barrier), then they could be used to treat schizophrenia. Conversely, Mabs possessing agonist properties could be used to treat Parkinson's disease. Other examples are β-adrenergic receptor monoclonal antibodies. A Mab specific for cardiac β-1 adrenergic receptors could offer all the benefits of β-adrenergic receptor blocking agents (such as propranolol) currently used, but with much greater specificity. Antibodies that are both specific and possess agonist properties for β-2 adrenergic receptors could be used in the treatment of asthma. In addition, because monoclonal antibodies can be isolated that bind with high specificity to single antigens expressed on only certain cell types, they can aid in directing the action of therapeutic agents to the proper target to which they have been covalently coupled.

## 10. Studies of Neurotransmitter Receptor Structure and Function Using Monoclonal and Antiidiotypic Antibodies as Probes

The neurotransmitter receptors of the cell's plasma membrane are present in very small numbers (10,000–50,000/cell). Direct radioligand binding assays have made it possible to study the structure and function of the receptors to some extent, but detailed structural and functional studies of these receptors have been hampered by the limited amounts of receptor proteins in tissue. Because of a monoclonal antibody's high affinity and specificity, the hybridoma technology is uniquely suited for receptor biochemistry. Using hybridoma technology, Gullick et al. (1982) and Gullick and Lindstrom (1983) investigated the detailed structural properties of the subunits of acetylcholine receptors (subunits of acetylcholine receptors are in the molar ratio of $\alpha 2\beta\gamma\delta$ and the genes for all the subunits have been cloned and sequenced). The results of their studies extended to myasthenia gravis, and experimental autoimmune myasthenia gravis demonstrated that autoantibodies are made against many parts of the acetylcholine receptor molecule, and there is one region that is disproportionately immunogenic. Further studies using monoclonal antibodies to identify subunits of the acetylcholine receptor expressed in cells transfected with subunit cDNAs should help in detailed structural characterization of each subunit. Schreiber et al. (1980) and Strosberg (1983), using antiidiotypic antibodies against β-adrenergic receptors, showed that a receptor and its antineurotransmitter antibody share structural configurations that may be recognized by an antireceptor or antiidiotypic antibody. Venter (1982) and Venter et al. (1984) have investigated the detailed properties of α-adrenergic, β-adrenergic, and muscarinic cholinergic receptors using monoclonal antibodies. Their studies indicate that structural similarities exist among the different types of receptors. Their studies support the concept that pharmacologically diverse types of receptors contain conserved immunogenic regions that are responsible for their interactions with the same membrane effector system.

Monoclonal antibodies have also been applied to study other cell surface receptors, such as low-density lipoprotein receptors

(Beisiegel et al., 1981), antithyroid-stimulating hormone receptors (Valente et al., 1982), and other cell surface components. Cuello et al. (1980) have produced a monoclonal antibody against substance P (a peptide acting as a neurotransmitter or a neuromodulator). The antibody was used to map the substance P pathways in the brain, for neuropathological investigations and radioimmunoassays. Hybridoma technology has not yet been applied in great detail to the central nervous system neurotransmitter receptors and the next decade should witness an exponential increase in research in this area.

## 11. Identification of Gene(s) for Neurotransmitter Receptors Using Monoclonal Antibodies

Mapping of the human genome has made enormous contributions to our understanding of various genetic diseases at the molecular level. In the future, further advances in recombinant DNA and hybridoma technology combined with a knowledge of linkage relationships among genes should facilitate early prenatal diagnosis of adverse phenotypes. Development of hybridoma technology has proven to be very useful in the area of gene mapping. Somatic cell genetic studies depend heavily upon species and not individual variations. As monoclonal antibodies are generally produced in mice they have proven exceptionally useful in detecting human/mice species differences. Table 5 shows the assignment of genes for some cell surface receptors to human chromosomes. In most of these studies, monoclonal antibodies produced against a receptor of interest or other cell surface component were used for the assignment of genes.

The identification of a human gene for a neurotransmitter receptor can be accomplished by using the same principles applied to the identification of genes for other cell surface antigens.

This procedure involves transferring human DNA to mouse cells and then identifying those that express human receptor genes by their binding of fluorescent antireceptor antibodies. Isolation of the human DNA from these mouse cells ultimately gives the DNA sequence of the human receptor. The stable integration of human DNA into the mouse genome in this DNA-mediated gene transfer method is often ensured by mixing the donor DNA with a plasmid containing a thymidine kinase (tk) gene and using tk⁻ mouse cells as recipients. Therefore with hypoxanthine, aminoterin, and thymidine (HAT) added to the medium, only cells that have acquired the tk gene survive (*see* Fig. 8). Fortuitously these cells also acquire 1000–3000 kilobases (kb) of human

TABLE 5
Summary of Gene Assignment to Human Chromosomes
Carried Out With Monoclonal Antibodies

| Cell surface receptor/component | Locus assignment chromosome no. | Reference |
|---|---|---|
| Low-density liproprotein receptors | 19 | Francke et al., (1984) |
| Interferon receptors | 21 | Slate et al., (1978) |
| Transferrin receptors | 3 | Goodfellow et al., (1982) |
| β-2 Microglobulin | 15 | Goodfellow et al., (1975) |
| Epidermal growth factor receptor | 7 | Shimizu et al., (1979) |
| HAL-A,B,C | 6 | Barnstable et al., (1978) |

DNA and express much of it. After pooling and incubating the HAT-resistant cells with flourescent anti-human receptor antibody (monoclonal, if possible), the brightest (1–2%) are isolated by flourescent-activated cell sorting techniques. These cells are then grown and resorted until a pure population of human receptor (+) mouse cells is obtained. Since the human genome contains about 2 million kb, only 1 out of 2000 tk$^+$ transfected cells contain any one particular gene. It is therefore essential to have highly selective procedures such as monoclonal antibody recognition to enable the rapid identification and isolation of the rare transfected cell that expresses the human receptor gene of interest.

After DNA from these cells is extracted, its content of human DNA is verified by hydridization with nucleotide probes for species-specific human repetitive DNA sequences. These hybridizations are generally carried out by a Southern transfer technique (Maniatis et. al., 1982). This DNA is then used in a second transfection to produce another human receptor (+) mouse cell that now contains only hundreds of kb of human DNA, yet most likely also some of the human repetitive sequences. After extraction, this cell's DNA is cut into small fragments by restriction endonucleases, joined to a plasmid, and then transferred to bacteria to create a "library." This "library" is then probed for human repetitive sequences, and DNA from positive cells tested for transfection of the human receptor (+) phenotype to mouse cells. Those that have this activity have the human receptor gene, which upon sequencing devulges the amino acid sequence of the

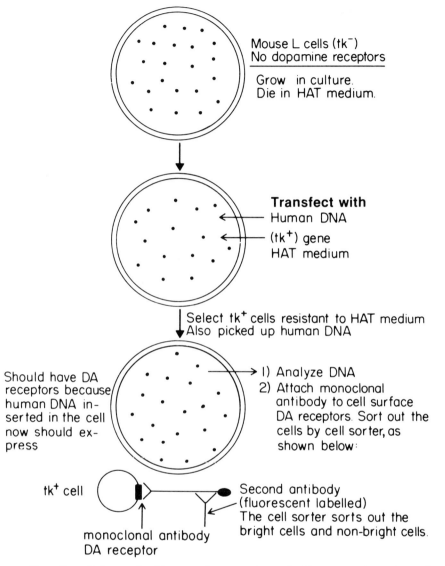

Mouse L cells (tk⁻)
No dopamine receptors

Grow in culture.
Die in HAT medium.

**Transfect with**
Human DNA

(tk⁺) gene
HAT medium

Select tk⁺ cells resistant to HAT medium
Also picked up human DNA

Should have DA
receptors because
human DNA in-
serted in the cell
now should ex-
press

1) Analyze DNA
2) Attach monoclonal
   antibody to cell surface
   DA receptors. Sort out the
   cells by cell sorter, as
   shown below:

tk⁺ cell

Second antibody
(fluorescent labelled)
The cell sorter sorts out the
bright cells and non-bright cells.

monoclonal antibody
DA receptor

Fig. 8.   Schematic diagram for gene transfection experiments and identification of human gene for neurotransmitter receptors. A hypothetical model for dopamine (DA) is used. Abbreviations: tk, thymidine kinase; H, hypoxanthine; A, aminopterin; t, thymidine.

receptor protein. The transferrin receptor has been isolated and partially sequenced using this approach. The LDL-receptor (its gene cloned and sequenced by another molecular strategy) and transferrin receptor have been assigned different chromosomes (*see* Table 5).

It should be possible using these molecular approaches to make great strides in isolating neurotransmitter receptor genes and obtaining new information concerning receptor structure, function, and evolution.

## 12. Conclusion

In this paper, a short description of the procedures used for the production of monoclonal antibodies and the application of hybridoma technology to neurotransmitter receptors is discussed. Although the technique for the preparation of monoclonal antibodies appears relatively simple, one should be prepared to commit hundreds of hours of labor to screen the clones. The screening assays should be adequately standardized even before the immunization.

The unique properties of monoclonal antibodies will help elucidate in detail the structural and functional aspects of neurotransmitter receptors. In addition, monoclonal antibodies have been used to define new cell surface antigens. These cell surface antigens have been mapped and used in hybrids as markers for specific chromosomes. Further discoveries of specific antigenic markers for chromosomes will lead to additional possibilities for chromosome manipulation of somatic cell hybrids and ultimately to a more accurate map of the human genome.

The application of monoclonal antibodies and recombinant DNA technology to the study of neurotransmitter receptors is new (e. g., cloning of the gene for acetylcholine receptors, mapping of substance P pathways using monoclonal antibodies, and so on.) The recombinant DNA and hybridoma technology should provide an extremely powerful way of studying the nervous system and will certainly provide answers to many of the outstanding problems in the field of neuroscience.

## Acknowledgments

We wish to thank Drs. J. Morgan, S. Spector, H. Yamamura, G. Rajakumar, J. Ramwani, and G. Ross for their help and constructive criticisms of the manuscript. Part of this work was supported by the Ontario Mental Health Foundation.

## References

Barnstable C. J., Bodmer W. F., Brown G., Galfre G., Milstein C., Williams A. F., and Zigler A. (1978) Production of monoclonal antibodies to Group A erythrocyte, HLA, and other human cell surface antigens—new tool for genetic research. *Cell* **14**, 9–20.

Beisiegel U., Schneider W. J., Goldstein J. L., Anderson R. G. W., and Brown M. S. (1981) Monoclonal antibodies to the low-density lipoprotein receptor as probes for study of receptor-mediated endocytosis and the genetics of familial hypercholesterolemia. *J. Biol. Chem.* **256**, 11923–11931.

Cavero I., Massingham R., and Lefevre-Berg F. (1982) Peripheral dopamine receptors, potential targets for a new class of antihypertensive agents. *Life Sci.* **31**, 1059–1069.

Conolley M. E. and Greenacre J. K. (1976) The lymphocyte β-adrenoceptor in normal subjects and patients with bronchial asthma. *J. Clin. Invest.* **51**, 1307–1316.

Cuello A. C., Galfre G., and Milstein C. (1980) Development of a monoclonal antibody against a neuroactive peptide: immunocytochemical applications. *Adv. Biochem. Psychopharmacol.* **21**, 349–363.

Fazekas De St. Groth, S. and Scheidergger D. (1980) Production of monoclonal antibodies: strategy and tactics. *J. Immunol. Methods* **35**, 1–21.

Francke U., Brown M. S., and Goldstein J. L. (1984) Assignment of the human gene for low-density lipoprotein receptor to chromosome 19: synteny of a receptor, a ligand, and a genetic disease. *Proc. Natl. Acad. Sci. USA* **81**, 2826–2830.

Fraser C. M. and Venter, J. C. (1980) Monoclonal antibodies to beta-adrenergic receptors: Use in purification and molecular characterization of beta receptors. *Proc. Natl. Acad. Sci. USA* **77**, 7034–7038.

Gilliand D. G., Steplewski Z., Collier R. J., Mitchell K. F., Chang T. H., and Koprowski H. (1980) Antibody-directed cytotoxic agents: use of monoclonal antibody directs the action of toxin A chains to colorectal carcinoma cells. *Proc. Natl. Acad. Sci. USA* **77**, 4593–4597.

Goodfellow P. N., Banting G., Sutherland R., Greaves M., Solomon E., and Povey S. (1982) Expression of human transferrin receptor is controlled by a gene on chromosome 3. Assignment using species specificity of a monoclonal antibody. *Somatic Cell Genet.* **8**, 197–206.

Goodfellow P. N., Jones E. A., VanHeyningen V., Solomon E., Bobrow M., Miggiano V., and Bodmer W. F. (1975) The β-2-microglobulin is on chromosome 15 and not in the HLA region. *Nature* (Lond.) **254**, 267–269.

Gullick W. and Lindstrom J. (1983) Comparison of the subunit structure of acetylcholine receptors from muscle and electric organ of *Electrophorus electricus*. *Biochem.* **22**, 3801–3807.

Gullick W. J., Tzartos S., and Lindstrom J. (1982) Monoclonal antibodies as probes for acetylcholine receptor structure. *Biochem.* **20**, 2173–2180.

Hempstead J. L. and Morgan J. I. (1983) Monoclonal antibodies to the rat olfactory sustentacular cell. *Brain Res.* **288**, 289–295.

Kohler G. and Milstein C. (1975) Continuous cultures of fused cells secreting antibodies of predefined specificities. *Nature* (Lond.) **256**, 495–498.

Krolic K. A. , Villemez C., Isakson P., Uhr J. W., and Vietta E. S. (1980) Selective killing of normal or neoplastic cells by antibodies coupled to the A chain of ricin. *Proc. Natl. Acad. Sci. USA* **77**, 5419–5423.

Lefur G., Zarifian E., Phan T., Cuche H., Flamier A., Bouchami F., Burgevin M. C., Loo H., Gerard A., and Uzan A. (1983) [$^3$H]-Spiroperidol binding on lymphocytes: changes in two different groups of schizophrenic patients and effect of neuroleptic treatment. *Life Sci.* **32**, 249–255.

Lennon V. A., Thompson M., and Chen J. (1980) Properties of nicotinic acetylcholine receptor isolated by affinity chromatography using monoclonal antibodies. *J. Biol. Chem.* **255**, 4395–4398.

Maniatis T., Fritsch E. F., and Sanbrook J. C. (1982) *Molecular Cloning, A Laboratory Manual*. Cold Spring Harbour Laboratory, Cold Spring Harbour, New York.

Reading C. L. (1982) *In vitro* immunization for the production of monoclonal antibodies. *J. Immunol. Methods* **53**, 261–291.

Schreiber A. B., Couraud P. O., Andre C., Vray B., and Strosberg A. D. (1980) Anti-alprenolol anti-idiotypic antibodies bind to β-adrenergic receptors and modulate catecholamine-sensitive adenylate cyclase. *Proc. Natl. Acad. Sci. USA* **77**, 7385–7389.

Shimizu N., Behzadian M. A., and Shimizu Y. (1979) Human gene mapping. Birth defects. *Orig. Art. Ser.* **15**, 201–215.

Slate D. L., Shulman, L., Lawrence M., Revel M., and Ruddle F. M. (1978) Presence of human chromosome 21 alone is sufficient for hybrid cell sensitivity to human interferon. *J. Virol.* **25**, 319–325.

Sternick J. L. and Strumer A. M. (1984) A new high-yielding immunization protocol for monoclonal antibody production against soluble receptors. *Hybridoma* **3**, 74–75.

Stocker J. W., Forster H. K., Miggiano V., Stahli C., Straiger G., Takacs B., and Staehein T. (1982) Generation of two new mouse myeloma cell lines PAI and PAI-0 for hybridoma production. *Res. Disclosure* **217**, 154–157.

Strosberg A. D. (1983) Anti-idiotype and antihormone receptor antibodies. *Springer Semin. Immunopathol.* **6**, 67–78.

Valente W. A., Vitti P., Yavin Z., Yavin E., Rotella C. M., Girolman E. F., Troccafondi R. S., and Kohn L. D. (1982) Grave's monoclonal antibodies to the thyrotropin receptor: stimulating and blocking antibodies derived from the lymphocytes of patients. *Proc. Natl. Acad. Sci. USA* **79**, 6680–6684.

Venter J. C. (1982) Monoclonal antibodies and autoantibodies in the isolation and characterization of neurotransmitter receptors: the future of receptor research, *J. Mol. Cell. Cardiol.* **14**, 687–693.

Venter J. C., Eddy B., Hall L. M., and Fraser C. M. (1984) Monoclonal antibodies detect the conservation of muscarinic cholinergic receptor structure from Drosophila to human brain and possible structural homology with α-1-adrenergic receptors. *Proc. Natl. Acad. Sci. USA* **81**, 272–276.

# Chapter 6

# Identification of Central Transmitters

## Microiontophoresis and Micropressure Techniques

### J. F. MACDONALD

## 1. Introduction

The synapse is recognized as the major structure of interneuronal communication, and a likely site of integration and information storage within networks of neurons and the central nervous system itself. Communication via these synaptic contacts involves a process of transduction of electrical signals (action potentials), to chemical signals (release of transmitter and its activation of postsynaptic receptors), and then back to electrical information (postsynaptic potentials). The chemical step in this process dictates that the pharmacology of synaptic transmission is of substantial importance in determining the functions of single neurons and their interactions with each other. Synapses within the central nervous system are not readily accessible to experimenters because of the microcosmic scale of these structures and the complexity of their anatomical relationships with other neurons, glial cells, and so on. Hence, the necessity for techniques that permit the delivery of microquantities of drugs, putative transmitters, and ions to the vicinity of synaptic contacts.

One traditional approach to transmitter identification has been simply to choose a series of criteria (Werman, 1966) whereby a possible candidate compound (exogenous or putative transmitter) can be compared to the selected (endogenous) transmitter's action. This approach requires that the exogenous substance mimic as closely as possible the action of the endogenous trans-

197

mitter on the postsynaptic membrane ("identity of action"). This is possibly the most important of the criteria (Werman, 1966) to be considered.

Such an approach demands that the exogenous substance be applied to the appropriate region (subsynaptic) of the postsynaptic neuron with kinetics that closely resemble the release of the transmitter from the presynaptic terminal. The transmitter and exogenous substance must bind to identical receptors (same affinity) and activate or inactivate identical ionic channels (same efficacy). Localized delivery of the putative transmitters to central mammalian neurons has been achieved primarily by the technique of microiontophoresis, but more recently by that of micropressure.

## 1.1. Microiontophoresis

The basic principles of iontophoresis have been clearly and simply described by Krnjević (1971). This technique takes advantage of the fact that many putative transmitters, drugs, and so on can be dissociated in water to give a charged ionic species. Thus, when a potential difference is applied across the solution, the ions will migrate in the field and effectively carry a current (iontophoretic current). Provided the net charge on the compound or ion of interest is known, ejection is achieved by applying a field of the same polarity (i.e., positive voltage to eject a positively charged ionic species). The movement of ions in solution is slow and depends strongly on interaction with molecules of water. Therefore, the conductivity of a microiontophoretic solution will depend on the mobility of its ionic species.

## 1.2. Micropressure

The use of pressure to eject solutions from micropipets (i.e., pneumatic system) into the central nervous system is not a recent development. Krnjević et al. (1963) were the first to use such a technique to apply acetylcholine (ACh) and other putative transmitters from micropipets. The top of the micropipet is attached to a source of positive pressure that simply forces the solution out the micropipet tip. However, applying drugs in this fashion suffered from several major drawbacks, including a constant and significant leakage of substance and an inability to effect rapid and reproducible onset and cessation of the pressure application. For these reasons, microiontophoresis was almost exclusively the technique of choice until the late 1970s.

There are many reasons for delivering small amounts of drugs, ions, and so on to the outer surface of neuronal membranes or to their interior: for example, the introduction of dyes such as Lucifer Yellow into individual neurons in order to correlate electrophysiological properties of the neuron with morphological properties (MacVicar and Dudek, 1982). Furthermore, the actions of many clinically neuroactive drugs (i.e., anesthetics, anticonvulsants, anxiolytics, and so on) may ultimately be the result of modifications of neuronal excitability mediated by intracellular or extracellular receptive mechanisms. The question of how these drugs activate or modify cellular mechanisms of excitability may be answered, at least in part, through experiments that utilize microdelivery (iontophoresis and pressure) techniques. The present discussion will focus on the use of microdelivery techniques to identify transmitters within the mammalian central nervous system.

## 2. Objectives of Transmitter Identification

If the major approach is to determine the identity of a particular neurotransmitter, its microdelivery is attempted initially with two basic objectives: (1) identification and characterization of postsynaptic (and in some cases presynaptic) receptors, and (2) identification and characterization of the postsynaptic effect (most often a change in ionic permeability of the postsynaptic membrane). Once these objectives have been met, equivalence of action of the exogenous substance and the transmitter must be demonstrated. The existence of receptors for exogenous compounds need not imply that synaptic or junctional receptors are involved, and receptors may be unrelated anatomically to postsynaptic specializations (i.e., extrajunctional receptors).

The effects of many compounds (i.e., local anesthetics or high concentrations of most compounds) are not mediated upon the recognition site of the receptor; instead, indirect modifications in membrane receptors (for neurotransmitters, and so on) or effectors (i.e., ion channels) may be their mechanism of action. Such is the case with a number of antagonists of acetylcholine (ACh) receptors for which the dose–response relationships demonstrate noncompetitive interactions and the site of recognition of the antagonist is likely to be the ion channel itself (Karlin, 1983).

The traditional pharmacological approach to the study of receptors entails the application of equilibrium kinetics. A mathematical model of drug–receptor interaction is compared with the results of an actual experiment. In order to employ equilibrium kinetic analysis, the assumptions underlying the particular model must be valid. For example, the effect of the agonist is measured at a time when drug receptor equilibrium has been attained. Unfortunately, microiontophoretic (and micropressure) application of drugs to mammalian central neurons has seldom provided adequate assurance that the appropriate assumptions are met.

The recognition of the agonist by the receptor and its activation of the effector is most often modeled by analogy to the law of mass action:

$$A + B \underset{k_2}{\overset{k_1}{\rightleftharpoons}} C + D \tag{1}$$

The rate at which this chemical reaction proceeds in the forward direction is proportional to the concentration of the reactants: $V$ (forward) = $k_1[A][B]$. The reverse direction is proportional to the concentrations of the products: $V$ (reverse) = $k_2[C][D]$. At equilibrium, the net formation of products (or dissociation to reactants) is zero by definition, with the forward rate equal to the reverse rate [$V$(forward) = $V$(reverse)]. This condition of equilibrium then permits calculation of the ratio of rate constants [$K = k_2/k_1 = (A)(B)/(C)(D)$].

In the specific case of an agonist (A) and its receptor (R) the interaction can be described as follows:

$$A + R \underset{k_2}{\overset{k_1}{\rightleftharpoons}} AR \tag{2}$$

The forward velocity of this reaction is therefore equal to the product of the forward rate constant ($k_1$), the concentration of unbound agonist (A–AR), and the concentration of free receptor (total number of receptors, $r_t$, minus the number of receptors with associated agonist, AR). The rate of dissociation is then the product of the reverse rate constant ($k_2$) and the concentration of agonist–receptor complex (AR). Equilibrium provides the condition in which these rates can be considered equivalent and hence the apparent dissociation constant of the complex ($K = k_2/k_1$) can be calculated from the relationship:

$$AR \text{ (at equilibrium)} = Ar_t/A + K \tag{3}$$

This equation provides a starting point in the mathematical description of the relationship between the concentration of agonist (A) and the concentration of complex (AR). For example, measurements of $K$ give some idea of how close the relationship is between an agonist and its receptor. What it does not provide is the relationship between AR and effect. Furthermore, it is usually not possible to measure the concentration of receptors present nor the concentration of complex AR. Occupational theories attempt to provide a theoretical model of this relationship and hence effect can be measured instead of AR. Receptor concentration can be related to maximal effect (classical theory) or to the fraction of receptors occupied (Stephenson's theory), giving a mathematical definition of the effectiveness of various agonists, such as intrinsic activity or efficacy. Other theories utilize the rate constants themselves rather than AR in modeling the relationship between agonist concentration and effect (Paton's rate theory). Provided the assumptions of the particular model are valid, it then becomes possible to determine $K$ from dose–response data. It should be emphasized that such models provide precise mathematical definitions for terms such as agonist, antagonist, affinity, efficacy, and so on, and therefore present a serious attempt to define the agonist–receptor interaction.

Thus, the experimenter employing microdelivery techniques to study receptors in central mammalian neurons must appreciate the constraints imposed by such modeling (equilibrium analysis). Understanding the importance and relevance of such models is the only way that aimless microiontophoresis of substances can be avoided.

What is required at even the most elementary level?

- The concentration of the agonist (or antagonist) at the receptors must be known precisely.
- The concentration must be varied to give different doses of agonist at the receptors. It is the dose-response data that will provide measurements of dissociation constants (e.g., affinity). In some models (i.e., classical theory) the dose that gives the maximum effect must also be known.
- Measurements of dose must be made at such a time that the agonist is in equilibrium with the receptor. This requires that any change in agonist concentration at the receptors must be slow in comparison with the rate constants of agonist–receptor interaction.
- Equation (3) requires an assumption in order to simplify it to this version. It is assumed that the concentration of A is much greater than the concentration of AR: Hence the term (A–AR) that indicates that the concentration of free agonist simplifies to A. Occupational theories in particular require that the total

number of receptors is small in comparison to the concentration of agonist and/or antagonist. This is likely to be achieved when drugs are applied exogenously but may not be true of agonist in the form of transmitter released from the presynaptic terminal.

## 3. Microiontophoresis

Considerable effort has gone into determining the amount of agonist released by microiontophoresis either by assaying the release in vitro or by appropriate modeling of the physical properties of solutions in micropipets. Nevertheless, in the vast majority of experiments no attempt is made at measuring the amount of release largely because of the technical difficulties inherent in making such measurements. Before briefly reviewing these attempts to quantify agonist release from micropipets, it is of some value to consider the behavior of microiontophoretic pipets based solely on empirical observations made during actual experiments:

(a) Concentrations of agonist in the molar range are ejected more readily than low concentrations in the millimolar range. Some agonists are ejected much more readily at equivalent concentration than others, regardless of the microiontophoretic dose (current). Passing currents across microiontophoretic pipets containing some compounds or ions (i.e., divalent cations, catecholamines, cAMP) can be difficult if not impossible, which further compounds the problems of ensuring release. Thus, there are wide variations in the ease with which substances can be ejected.

(b) The majority of pipets (if not all) spontaneously release agonist solution. This is most prevalent when larger diameter tip pipets are employed and/or when the concentration of agonist is high.

(c) The conventional means of countering spontaneous agonist release is to apply a retaining current (a current of the opposite polarity to the ejecting current).

(d) If the diameter of the pipet is too large, retaining currents of even high magnitude will fail to prevent spontaneous release.

(e) Retaining currents affect the behavior of agonist release by subsequent ejecting currents. This is generally observed as a reduction in neuronal response to a given iontophoretic dose when preceded by the use of a retaining current. A related phenomenon, warm-up, occurs when repeated brief applications of the same iontophoretic dose evoke progressively larger neuronal responses until some steady-state value is reached.

(f) The use of very-high-resistance pipets with tip diameters similar to those used for intracellular recording (less than 0.5 μm) greatly reduces spontaneous release to the degree that retaining currents may not be required.

(g) If the objective is to apply the agonist very rapidly, it is better to use a small diameter tip pipet rather than a large one. Small diameter pipets give rapid response times, whereas large ones are more useful for prolonged applications.

(h) A proportion of the micropipets will pass current satisfactorily, but simply will not eject agonist. These micropipets will often demonstrate signs of transient increases in resistance and difficulties in maintaining constant current.

(i) The amount of agonist released is probably linearly related to the applied dose, provided the periods of application are relatively long.

(j) Large ejecting currents may release solutions of low conductivity, but do so unreliably.

(k) There will be a wide variation between pipets with regard to the amount of agonist released.

## 3.1. Measurements of Agonist Release

### 3.1.1. Passive Release

3.1.1.1. DIFFUSION (STEADY STATE). Krnjević et al. (1963) modeled the micropipet tip as a hollow cone with an internal opening radius $(r_i)$ and an angle of tip taper $(\theta)$. All fluxes were considered to be directed radially. Under assumptions that the body of the pipet is an inexhaustible supply of agonist (in their case, ACh) with a diffusion gradient established down the pipet tip, and that the diffusion of agonist is considered to take place under steady-state conditions, a simplified equation was derived to describe diffusional flux $(Q_d)$:

$$Q_d = C_i \, D\pi \, \tan\theta r_i \tag{4}$$

where $D$ is the diffusion constant of the agonist in solution and $C_i$ is the concentration of agonist in the pipet.

Therefore, release caused by diffusion is directly proportional to internal tip radius and agonist concentration.

3.1.1.2. BULK FLOW (HYDROSTATIC PRESSURE). Under the same modeling conditions (with similar assumptions about pipet tip geometry and concentration gradients), Krnjević et al. (1963) calculated the rate of flow ($Q_{fl}$) of solution out of the pipet due to the pressure differential ($p^*$) between the tip and the top of the pipet:

$$Q_{fl} = 3\pi\tan\theta p^* \, r_i^3 \, (8\epsilon)^{-1} \tag{5}$$

where $\epsilon$ is the viscosity coefficient of the solution.

The amount of agonist released by bulk flow is therefore $Q_{fl}C_i$. This relationship demonstrates that bulk flow is strongly dependent upon the internal tip radius of the micropipet (to the third power).

3.1.1.3. TOTAL SPONTANEOUS RELEASE (STEADY-STATE). The fluxes of agonist caused by diffusion and bulk flow can be added together algebraically to give the total spontaneous flux:

$$Q = Q_d + Q_{fl} = \pi\tan\theta C_i \, [Dr_i + 3p^* r_i^3 \, (8\epsilon)^{-1}] \tag{6}$$

The total spontaneous release of agonist is therefore always dependent on bulk flow (unless of course the hydrostatic pressure of the tissue is equal to that of the column of solution).

### 3.1.2. Methods of Reducing Spontaneous Release

3.1.2.1. RETAINING CURRENTS. Applications of small currents (relative to the ejection current and of opposite polarity) have traditionally been used to reduce spontaneous release of agonist (Del Castillo and Katz, 1955). Retaining currents will establish a reverse concentration gradient for the agonist along the micropipet tip and effectively reduce the concentration in the tip available for iontophoretic release (Kelly, 1975). Purves (1979) confirmed this reverse concentration gradient under direct observation of iontophoretic pipets and demonstrated the subsequent reduction in efficacy of iontophoretic ejections. Thus, it takes appreciable time for reestablishment of the agonist concentration in the tip after exposure to a retaining current. Bradshaw et al. (1973) observed essentially the same phenomenon when retaining currents were used to reduce the spontaneous release of labeled noradrenaline.

Retaining currents (and the reverse concentration gradients they establish) have a profound effect on the kinetics of agonist release. In particular they greatly potentiate the disparity between the kinetics of onset and offset of agonist release (Purves, 1979).

For example, the apparent long delay in responses to microionto-phoretically applied substance P can be attributed at least in part to the actions of the retaining current upon subsequent microiontophoretic ejection (Guyenet et al., 1979).

When iontophoretic applications of agonists are brief and re-peated at a constant frequency, the evoked neuronal responses increase in magnitude until a steady-state value is reached. This warm-up of iontophoretic pipets is commonly observed and likely caused by a similar depletion of the pipet tip by spontaneous re-lease (Kelly, 1975; Purves, 1979).

3.1.2.2. LOWER AGONIST CONCENTRATION. One method of reducing the spontaneous release of agonist is simply to reduce its concentration in the pipet (Bradshaw and Szabadi, 1974). This will present several major disadvantages. If the concentration is reduced to below 0.1$M$, the contribution of electroosmosis to re-lease (see below) becomes significant (Krnjević, 1971). Further-more, a reduction in the ionic strength of the agonist will mean that other charged species rather than the agonist are likely to carry the iontophoretic current. Nevertheless, if concentrations of above 0.1$M$ are used, lower concentrations are likely to give less spontaneous release than higher concentrations. This has proven true empirically as well as theoretically (Purves, 1980).

3.1.2.3. NEGATIVE PRESSURE. One method of reducing spontane-ous release from the pipet would be to apply a negative pressure to the back of iontophoretic pipets (Kelly, 1975). Bulk flow should stop provided the negative pressure is just equal to the difference between the hydrostatic pressure of the pipet and that of the tis-sue. Diffusion would then be the only source of spontaneous re-lease. This is unlikely to be a useful approach because in practice neither the hydrostatic pressure of the pipet nor the tissue is known.

*3.1.3. Active Release*

3.1.3.1. IONTOPHORESIS. The release of charged agonist has been described using "Faraday's Law" (Krnjević et al., 1963):

$$Q_i = tI/zF \tag{7}$$

where $Q_i$ is the iontophoretic flux of agonist, $I$ is the applied cur-rent, $t$ is the transport number (see below) of the particular ago-nist in solution, $F$ is the Faraday constant (in coulombs), and $z$ is the valence number of the agonist ion.

Unfortunately, not all of the charge carried by the iontophor-etic current will represent transfer of the charged agonist. Because of interactions with the water molecules, they can differ in relative mobilities and therefore the proportion of the iontophoretic current carried by ions can differ significantly from the ideal ($t = 0.5$ for monovalent cation or anion). Therefore, transport numbers must be determined empirically from the ratio of the total assayed amount of agonist released ($n$) to the total charge ($Q$) passed through the pipet to achieve that release ($t = nF/Q$). The procedure of adding a relatively mobile salt (i.e., NaCl) to the agonist solution with the objective of increasing the efficiency of the iontophoretic current (presumably to increase electroosmotic release of agonist) will result in a significant proportion of the iontophoretic current being carried by the contaminating salt. This can substantially reduce the amount of agonist released for a given iontophoretic current (Guyenet et al., 1979).

The relationship given in Eq. (7) is more accurately described as Hittorf's Law (Purves, 1977, 1979) and cannot be added algebraically to the diffusion term (for spontaneous release), since iontophoresis and diffusion interact. Purves (1977, 1979) has advocated the use of the Nernst–Planck equation that takes this interaction into account. Consequently, the amount of agonist release predicted by this relationship differs from that of Eq. (7) in several features. For example, the amount of agonist released is linearly related to the applied current only when the iontophoretic current exceeds a given value (applied voltage drop across the pipet exceeds 100 mV). Purves (1979) measured the release of a fluorescent compound (quinacrine HCl) and was able to confirm this predicted region of nonlinear release. In practice iontophoretic currents usually exceed this criterion except when potent agonists are applied from large-tipped (low-resistance) pipets (Purves, 1979). Direct examinations of the relationship between release of labeled substances and applied iontophoretic current have generally confirmed this direct proportionality (Krnjević et al., 1963; Kelly, 1975; Guyenet et al., 1979; Dray et al., 1983; Hicks, 1984).

3.1.3.2. BULK FLOW (ELECTROOSMOSIS). A complicating phenomenon can arise when iontophoresis of solutions of low ionic strength (i.e., peptides) is attempted. A boundary potential occurs at the interface of the relatively negatively charged glass of the pipet and the solution. Application of a potential gradient down the pipet (i.e., during application of an iontophoretic current) will result in the bulk flow of solution in the direction determined by the polarity of the iontophoretic current. Electroosmotic flow cannot easily be measured, but Krnjević et al. (1963) made an estimate from the migration of cylindrical glass particles in solutions contained within micropipets. Using this technique, Krnjević and Whittaker (1965) found that large outward (positive) iontophoretic currents were able to eject substances of low conductivity from sucrose solutions. This presented the opportunity to eject poorly charged agonists using large positive currents. The volume of ejected solution can be considered to be approximately proportional to the magnitude of the current, but not reliably so (Krnjević and Whittaker, 1965). The appropriate polarity of the iontophoretic current for electroosmotic release must be determined for an individual solution by empirical means (i.e., observing the migration of glass particles) (Krnjević, 1971).

3.1.3.3. PIPET VARIABILITY AND ROGUES.   An unfortunate finding with microiontophoresis is that the amount of agonist released is strongly dependent on unknown properties of individual pipets and not on resistance or estimates of tip diameter (Kelly, 1975). This is particularly obvious in the case of so-called rogue pipets (Kelly, 1975). With these pipets the application of the iontophoretic current is associated with a sudden increase in pipet resistance (perhaps because of blocking of the electrode by contaminants in the agonist solution and during in vivo recording by plugging of the tip with cellular membranes or debris) and by a dramatic drop in the amount of agonist released even though iontophoretic current is maintained (Purves, 1979). Such pipets are usually "noisy" during current passage and therefore can be discarded (Hoffer et al., 1971; Kelly, 1975). However, blockade of the pipet tip need not result in an infinite tip resistance and the current may flow along the surface of the glass (Krnjević et al., 1963; Krnjević and Whittaker, 1965), making it difficult to identify pipets that pass current but fail to eject agonist. Little can be done to reduce this interpipet variability of agonist release.

3.1.3.4. IMPORTANCE OF SMALL TIP DIAMETER.   Since the initial detailed description of iontophoretic release of ACh from micropipets (Krnjević et al., 1963) it was clear that the diameter of pipet tips (or more correctly the internal radius $r_i$) should be limited to a maximal value. Bulk flow due to the hydrostatic pressure differential of the micropipet is dependent upon the cube of $r_i$. It is therefore possible to calculate (for a particular agonist concentration) the tip size where bulk flow predominates over diffusion (i.e., 0.125 μm internal radius for a 3$M$ AChCl pipet and a corresponding resistance of about 25–47 mΩ). Once the internal tip radius exceeds about 0.25 μm (about 1 μm external tip diameter), it may be impractical to employ retaining currents to prevent spontaneous release because the currents required would be much greater than those used to eject the agonist (Krnjević et al., 1963).

Purves (1979) has also demonstrated that the kinetics of the onset and offset of iontophoretic release is highly dependent on tip diameter. Ideally, the application of the iontophoretic current should result in an instantaneous increase of agonist release to the steady-state value, followed by an instantaneous drop to zero upon cessation of the current. However, calculations based upon the Nernst–Planck equation show that the time constants of onset and offset of release depend strongly upon $r_i$. In comparing a pipet with an $r_i$ equal to 0.1 µm with one of 1.1 µm, the time to reach the steady-state amount of agonist ejection slows by a factor of a thousand (from several milliseconds to several seconds). Direct observations of quinacrine release were seen to be faster with high-resistance (small-tip-diameter) pipets than they were with low-resistance (large-tip-diameter) pipets (Purves, 1979).

The linearity of agonist release also depends on the tip diameter ($r_i$) of the pipet. If small currents are used to eject a particularly potent agonist from large tip diameters, it is possible that the 100 mV criterion of potential drop across the iontophoretic electrode will fail to be satisfied. This would greatly confound certain types of analysis that depend strongly on the accuracy of release of small amounts of agonist (i.e., the use of the limiting slopes of log-log dose–response curves)(Werman, 1969; Dudel, 1975).

Microiontophoresis of agonists should be restricted to external tip diameters of 1 µm or less and high-resistance micropipets should be considered a necessity if the receptor is susceptible to desensitization (Dreyer and Peper, 1974; Dudel, 1975).

## 4. Micropressure

Micropressure application of agonist simply increases the bulk flow caused by hydrostatic pressure ($p^*$). The internal radius $r_i$ is therefore of critical importance in determining the efflux of solution. Unfortunately, it is difficult to measure $r_i$ and Krnjević et al. (1963) were forced to calculate it indirectly from measurements of electrode resistance. Given AChCl as an agonist at a concentration of 3$M$ they determined that bulk flow would account for 90% of the release when $r_i$ was equal to 0.4 µm (corresponding resistance of about 8 m$\Omega$).

This technique would seem to have considerable advantage over iontophoresis as a means of microdelivery. The release of an agonist would not be dependent on its ionic strength in solution (hence uncharged agonists could be applied) and electroosmosis would obviously not be a factor. Furthermore, the amount of agonist released can be accurately calculated as the sum of the diffusion and bulk flow terms as given above. Krnjević et al. (1963) found a reasonably good correspondence between the calculated amount of AChCl released and the applied pressure. However, microiontophoresis continued to be the preferred technique of drug delivery for several reasons (Krnjević et al., 1963):

1. The onset and cessation of drug release by micropressure was found to be slow relative to comparable iontophoretic applications. It was suggested that the lag resulted in part from friction of the solution in the lumen of the pipet. A delay would also be introduced as a consequence of the time required to establish a pressure gradient across the tubing and valves connecting the pipet and the source of positive pressure.
2. It was difficult to control the rate of flow of solution from pipets because small changes in internal tip radius cause very large changes in bulk flow. *In situ* placement of the micropipet would be exceptionally difficult because plugging of the tip with tissue (or breakage of the tip, enlarging its diameter) would result in a wide variability of flow at the same application pressure.
3. Attempts to apply agonists via bulk flow utilized the same pipets as used for iontophoresis. Hence the high concentrations of agonist (molar range) employed, and their spontaneous leakage resulted in a significant and uncontrolled release of agonist.

Many of these initial concerns have been overcome. For example, electronic valves have been developed that permit applications with response times as short as 2 ms (McCaman et al., 1977) and that automatically vent the pressure head upon cessation of the application. Thus, rapid ejections of small volumes (usually picoliters) from pipets (single-barrel tip diameters, 0.6–5 $\mu$m) can be achieved reliably using pressure both *in situ* and in vivo (McCaman et al., 1977; Sakai et al., 1979; Palmer et al., 1980; Dray et al., 1983). Furthermore, for a given micropipet the amount of labeled substance released or the volume of ejected solution can be linearly related either to the duration of the applied pressure pulse (pressure constant) or to the magnitude of the pressure (duration constant) (McCaman et al., 1977; Sakai et al., 1979; Palmer et al., 1980; Dray et al., 1983).

Unlike iontophoresis, the volume of ejected solution can be measured with the pipet in vitro and therefore the amount of agonist released can be calculated with relative ease. However, since the calibration is done in vitro, it will not necessarily follow that a pulse of given magnitude and duration will release the same volume (and therefore amount) *in situ*. Changes in $r_i$ that occur during penetration of the tissue (and the hydrostatic pressure of the tissue itself) will alter the release characteristics *in situ*.

McCaman et al. (1977) found that pressure and iontophoretic applications of ACh or serotonin to Aplysia neurons evoked responses with identical rise and decay times. Identical amounts (calculated) of agonist released either by microiontophoresis or pressure application gave responses of similar magnitude. However, in the case of agonists that have low and variable iontophoretic transport numbers, micropressure has been shown to be a more reliable method of ejection (Dray et al., 1983).

## 4.1. Control of Spontaneous Release

Palmer et al. (1980) reported that they could see no indication of warmup during pressure applications. Nevertheless, the same concentration gradient (caused by diffusion) will develop in the pipets used for pressure applications as those for microiontophoresis. The major difference would be the much lower concentrations of agonist employed for micropressure, and thus much less of a gradient would be present. This may not be the only factor "denuding" the pipet, however, and some backfilling of the electrode upon cessation of the pressure pulse has been observed when dyes were included in the pipet (McCaman et al., 1977). Such factors would limit the usefulness of very brief applications of agonist by pressure.

Spontaneous leakage will occur because of diffusion and bulk flow no matter how low the agonist concentration in the pipet. Therefore the argument that micropressure applications will not desensitize liable receptors (McCaman et al., 1977) is not necessarily valid. Rather, the concentration of agonist reaching the receptors is likely to be subthreshold. A small negative pressure (backflow) on the pipet would eliminate leakage but reduce the reproducibility of subsequent agonist ejections. This can be overcome to some extent by placing the pipet distant from the neuron (at least in the case of in vitro preparations).

## 4.2. Pipet Variability

Most authors report that a proportion of the agonist pipets utilized for micropressure fail to pass any solution regardless of the magnitude of the applied pressure. Furthermore, the amount of agonist or the volume ejected can vary significantly between pipets even when attempts are made to keep tip diameter similar. Micropipets with tip diameters much below 0.6 μm usually fail to eject solution with any consistency (Sakai et al., 1979).

Rogue pressure pipets are in fact considerably more common than rogue iontophoretic pipets. Micropressure pipets that fail to eject solution can also have remarkably large tip diameters (5–15 μm) (Choi and Fischbach, 1981; Heyer and Macdonald, 1982) and need not be exposed to any biological tissue. Even though solutions are filtered and glass for pipets kept clean, plugging of micropressure pipets occurs often (perhaps by bubbles of air). A direct visual observation of the patency of the micropipet is possible in vitro that permits rejection of many of these rogue pipets. However, even during experiments in which the pipet does not contact the tissue, the tip can become plugged or will demonstrate bizarre characteristics of release such as failure to release during the pressure pulse with a transient release upon cessation of the pulse (*see* Fig. 1). In using both microiontophoresis (high-resistance pipets) and pressure to apply excitatory amino acids to cultured spinal cord neurons, a much greater reliability and reproducibility between individual pipets has been observed with microiontophoresis (MacDonald, unpublished observations).

## 4.3. Pressure Microperfusion

Certain in vitro preparations of central mammalian neurons present a unique opportunity for the use of pressure techniques. For example, neurons can be grown in dissociated tissue cultures and because individual neurons can be observed directly under phase-contrast microscopy, it becomes possible to perfuse the region of the soma (and at least a portion of the dendrites) while recording intracellularly. During relatively prolonged pressure applications the concentration of agonist reaching the receptors should approach, and likely reaches, the concentration in the micropipet provided the volume of ejected agonist solution is sufficient to flood the local region of the neuron. Hence the term pressure microperfusion.

Microperfusion requires that relatively large volumes (nanolitres, not picolitres) of agonist solution be ejected and that the ejection last long enough to allow the agonist concentration to reach a steady state (seconds, not milliseconds). This is most easily achieved with large diameter tip pipets that maximize bulk flow. However, once the tip opening exceeds a certain diameter, the hydrostatic pressure of the column of solution alone can cause a

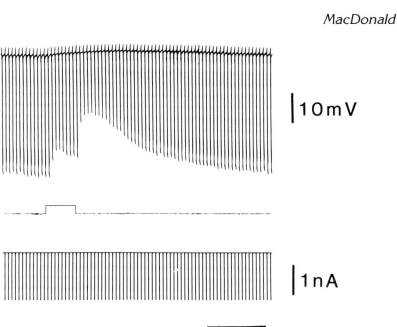

|10mV

|1nA

10s

Fig. 1.    The pressure application of 500 μm L-aspartic acid to a cultured spinal cord neuron demonstrates an unexpected pattern of agonist release.

The upper trace is a recording of membrane potential set, by current injection, at the reversal potential for L-aspartic acid. Hypolarizing current pulses (50 ms duration, lower trace) were passed via a bridge circuit in order to assess membrane input conductance. The middle trace gives an indicator of the pressure application (300 kPa) that is associated with a small increase of conductance in the neuron (a reduction of the voltage deflections recorded in the upper trace). Cessation of the pressure application in contrast resulted in a much larger response. Visual observation of this pipet outside of the bathing solution confirmed that most of the solution was released upon cessation of the pressure pulse.

flow rate sufficient to activate the "flow artifact" (see below). Such very large pipets (in excess of 10 μm) have been used to supply solutions to neurons in culture (Barker and Ransom, 1978; MacDonald and Wojtowicz, 1982), but they provide little or no control over flow rate. Pipet tips should be about 1–5 μm in diameter and pressures should be minimal (i.e., 5–15 kPa)(Choi and Fischbach, 1981; Heyer and Macdonald, 1982; MacDonald and Wojtowicz, 1982) in order to achieve a large ejection volume, but with a flow rate sufficiently low that the flow artifact does not confound the recording. Greater consistency can also be achieved by careful measurement of tip diameter (Choi and Fischbach, 1981).

# 5. Artifacts (Is the Agonist the Only Stimulus Delivered?)

Both microiontophoretic and micropressure techniques have some inherent disadvantages with the particular vehicle of delivery of substances.

## 5.1. Microiontophoresis

Microiontophoretic currents are usually small (1–1000 nA), but the current is applied across the electrode tip through the preparation to ground. If the current is strictly applied across the extracellular space it is unlikely to have much effect upon neuronal excitability (provided the extracellular space provides a relatively low resistance pathway to ground). However, when iontophoretic pipets are used in vivo the extracellular resistance may be sufficient that changes in neuronal excitability occur (Krnjević and Phillis, 1963).

### 5.1.1. Extracellular Recording

A relatively weak excitation occurs with negative iontophoretic currents, whereas a weak inhibition may be produced with positive currents (Krnjević, 1971). In contrast, more profound and different actions of microiontophoretic current are likely to occur when the pipet tip actually comes in close contact with neuronal membranes. In this situation the current, applied directly across the membrane, is analogous to an intracellular application of current. Hence the relationship between excitation or inhibition and polarity of the microiontophoretic current is reversed (Krnjević and Phillis, 1963; Krnjević, 1971). This current may directly affect the neuron activity being recorded, but may also change the excitability of this neuron by acting upon other neuronal somata or presynaptic terminals.

Direct actions by current can usually be recognized by their immediate onset and offset and by the reversal of excitation to inhibition (or vice versa) with reversal of current polarity. A possible solution to this problem is to provide current balancing or neutralization via an alternate barrel of the multibarreled pipet. A current of equal but opposite polarity to the microiontophoretic current is applied simultaneously to an indifferent barrel (i.e., NaCl), effectively neutralizing the net current flow in the region of the electrode tip. Such neutralization can be highly effective, but only provided the individual current flows are between pipet barrels and not across some membrane structure (i.e., the neuronal membrane).

## 5.1.2. Intracellular Recording

The problem of direct current actions upon the excitability of neurons is much the same for intracellular as for extracellular recording. However, contact with the neuronal membrane by the iontophoretic pipet is readily recognized by large and abrupt changes in membrane potential that are reversed with a change in the polarity of the iontophoretic current. Indirect effects mediated by other neurons or presynaptic terminals may also be detected by watching for increased synaptic activity coincident with the microiontophoretic application. The standard practice (at least in the case of in vitro preparations) is to lower bath concentrations of $Ca^{2+}$ and elevate $Mg^{2+}$ in order to suppress synaptic transmission. This technique presents some problem for agonists, such as excitatory amino acids, whose responses demonstrate strong dependence on the presence of divalent cations (MacDonald and Wojtowicz, 1982; MacDonald and Schneiderman, 1984). An alternate and often more successful approach is to employ tetrodotoxin (TTX) to reduce interneuronal communication (*see* Fig. 2) by blocking $Na^+$ action potentials (Choi and Fischbach, 1981; MacDonald and Wojtowicz, 1982; Brown and Griffith, 1983). Although TTX does not ensure elimination of presynaptic actions (i.e., Choi and Fischbach, 1981; MacDonald, 1984), the use of this drug greatly facilitates pharmacological experiments on central mammalian neurons.

Another problem arises when intracellular recordings are used. A coupling develops between the current applied across the iontophoretic pipet and the recording electrode. This is detected by the intracellular electrode as a shift of membrane potential (*see* Fig. 3); but it does not usually result in a flow of current across the membrane, and consequently, the activity of the cell (its excitability) is not altered. If the microiontophoretic pipet is actually inside the neuron (i.e., for the purpose of injecting substance) this is not the case and the entire iontophoretic current is applied across the neuronal membrane. Both the passive shift of potential caused by coupling and the active current injection that occur during intracellular microiontophoresis are highly undesirable. The passive shift introduces a significant error in measurements of membrane potential. In addition the large and active shifts that occur during intracellular iontophoresis can be sufficient to physically break down the membrane and at best obscure any change in excitability induced by the test substance.

Current neutralization systems have been developed to deal with these problems (Krnjević, 1971; Kelly, 1975; Hicks, 1984). An independent micropipet, an alternate barrel of a multi-barreled

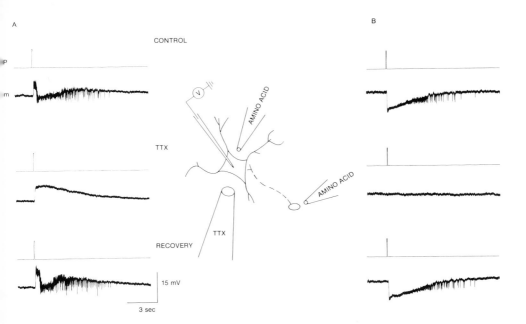

Fig. 2. Tetrodotoxin (TTX) blocks synaptic interaction between two spinal cord neurons. A recording electrode (1*M* potassium methylsulfate) was inserted into the central neuron and kainate (100 μ*M*) applied by pressure (*P*) either close to the recorded neuron (A) or at the soma of a second neuron (B) (electrodes and neurons diagramatically represented). The response to this amino acid is shown before, during, and after the positioning of a blunt-diffusion pipet containing 1 μ*M* TTX near the recorded neuron. When the amino acid was applied to the soma of the recorded neuron, depolarization of the membrane ($E_m$) was interrupted by a barrage of inhibitory postsynaptic potentials probably originating from a small cell close to the recorded neuron (not shown). Diffusion of TTX eliminated these synaptic potentials (A). Similarly, when kainate was applied to a second soma about 150 μm from the first, inhibitory postsynaptic potentials were evoked and TTX readily and reversibly abolished the response (B) (Fig. from MacDonald and Wojtowicz, 1982).

pipet, or the bath (in vitro preparations) can be used to provide a simultaneous and equal but opposite balancing current (in principle identical with balancing during extracellular recording). Figure 3 demonstrates that during extracellular iontophoresis quite large passive shifts in membrane potential can be neutralized using such a system. Krnjević et al. (1975, 1978) were the first to routinely use such a current neutralization system to inject a variety of substances into mammalian central neurons in vivo (cat spinal cord motoneurons). This meant that sensitivity parameters of neuronal excitability, such as membrane potential, input conductance, and size of spike after-potentials could be measured before,

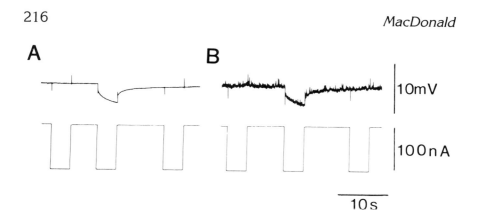

Fig. 3.    Recording from an intracellular electrode outside (A) and inside (B) a cultured spinal cord neuron (upper traces). A series of three negative iontophoretic current pulses was applied in both A and B, but the second pulse in each series was not balanced (see text). Without balancing the iontophoretic current, a hyperpolarizing artifact resulting from coupling between the recording electrode and the iontophoretic pipet is observed. This artifact is of identical magnitude inside or outside the neuron and outlasts termination of the iontophoretic current. It is virtually eliminated by balancing the current of application, and only rapid transient changes of potential are seen with onset and offset of the iontophoretic current (outer responses of each group).

during, and after intracellular microiontophoretic injections of transmitters, drugs, and ions.

## 5.2. Micropressure

### 5.2.1. The Flow Artifact

The pressure application of transmitters, drugs, or ions to central mammalian neurons side-steps some of the problems associated with microiontophoresis. However, it introduces a set of mechanical problems associated with the flow of solution. Both in vitro and in vivo applications of control solution (i.e., the bathing solution in the case of in vitro preparations or an osmotically suitable solution for in vivo preparations) can inhibit the firing of central neurons (Poulain and Carette, 1981; MacDonald and Wojtowicz, 1982). This inhibition of firing is proportional to the magnitude of the flow from the micropipet (MacDonald and Wojtowicz, 1982).

Figure 4 illustrates the response of a spinal cord neuron (grown dissociated in tissue culture) to applications of the bathing solution (control solution). This pipet had a tip diameter of about 10 μm and is larger than those usually used to apply agonists to these neurons. At resting membrane potential the response was

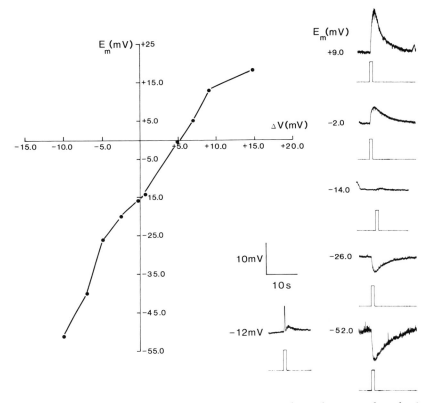

Fig. 4. Demonstration of the flow artifact that results during pressure applications of solutions from relatively large diameter pipets (see text).

An intracellular recording was made from a cultured spinal cord neuron and the bathing solution alone (control solution) was applied from a pipet with a tip diameter of about 10 μm (200 kPa) and chosen for its high flow rate. The bathing solution contained 5 m$M$ calcium and 25 m$M$ tetraethylammonium in addition to TTX. The pressure application hyperpolarized the neuron by about 5 mV at resting potential. When direct current was injected through the intracellular electrode in order to adjust membrane potential ($E_m$) to various values, a reversal potential for this response could be clearly demonstrated (sample responses and pressure indicators given on the right side with values of $E_m$ shown). The change in potential (ΔV) evoked by the application is plotted on the axis versus $E_m$ on the ordinate and the value of the reversal potential was found to be about −16 mV.

Note that an application of solution at the potential of −12 mV depolarized the neuron and evoked an action potential (largely a calcium spike under the given recording conditions). These spikes inactivate at more depolarized potentials.

hyperpolarizing and is similar to the flow artifact described by MacDonald and Wojtowicz (1982) and by Heyer and Macdonald (1982). A change in input conductance can be characteristic of this artifact (Fig. 5). It is also possible to demonstrate a reversal potential (about −15 mV in Fig. 4) and measure the change of input conductance at this potential when direct current is passed through the intracellular electrode (Fig. 5). In some cases the reversed (depolarizing) response of the cell was capable of triggering a spike (*see* insert, Fig. 4), demonstrating the capacity of this response to influence neuronal excitability. That this observed change of input conductance and potential is not exclusively a response of the neuronal membrane is strongly suggested by the observation that a similar response can be evoked from the recording microelectrode after it has been withdrawn from the neuron. With no current being injected through the intracellular electrode, a membrane hyperpolarization of similar time course but much smaller magnitude is still elicited by the pressure microperfusion (Fig. 6). A small increase of the resistance of the

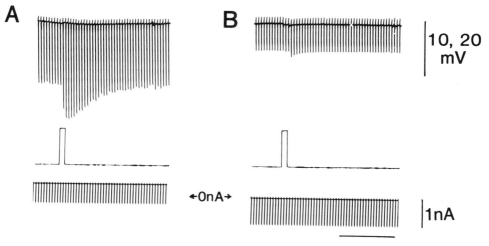

Fig. 5. Recording of input conductance from the same neuron shown in Figs. 4 and 6. Current was injected through the recording electrode in order to adjust membrane potential (top trace) to the reversal potential for the flow artifact (A). Hyperpolarizing pulses (50 ms, lower trace) were passed to assess input conductance. The pressure application of control solution decreased input conductance (increased the magnitude of hyperpolarizing voltage deflections). After the electrode was removed from the neuron, the pressure was applied again with the bridge slightly unbalanced (B) and a change of electrode resistance was observed. The voltage calibration was 10 mV for A and 20 mV for B.

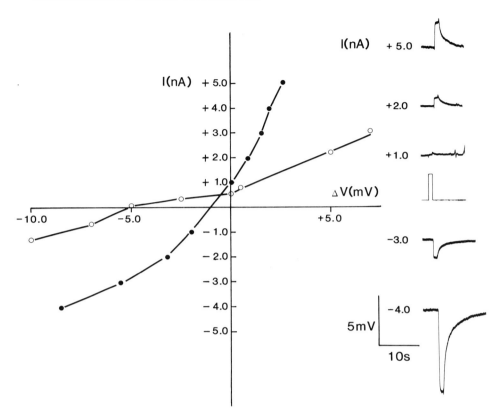

Fig. 6. The flow artifact can be observed as a change in the voltage measured by the recording electrode when either inside or outside the neuron. Sample records are given to the right of the figure that illustrate the response of the electrode itself to the pressure application of control solution (only a single pressure indicator given). The currents injected through the electrode inside (open circles) or after the electrode was withdrawn from the cell (closed circles) are plotted versus the change in voltage (ΔV) evoked by the same pressure application of control solution (200 kPa). In either case a clear reversal potential is demonstrated. Note that the response of the electrode cannot be attributed to fragments of neuronal membrane covering the electrode tip because the same response could be elicited when a second recording electrode was located in the same position just above (outside) the neuron.

recording electrode is also observed (Fig. 5) and with current injection a reversal potential is demonstrated as well (Fig. 6).

This artifactual response of the recording electrode is directly dependent on the flow of solution. (It is also accentuated by osmotic differences between the applied solution and the bathing solution.) A likely explanation for the effect is that the flow of solution across the charged glass surface of the recording electrode (and perhaps the neuronal membrane itself) generates a streaming current (Rutgers, 1940). The surface layer of ions (cations) attracted to the glass surface will flow against the charge of the glass surface, generating an electrokinetic current. The current injected just adequate to null this putative streaming current presumably is responsible for the apparent reversal potential (i.e., about +0.6 nA, *see* Fig. 6).

The use of a second control pipet (or alternate barrel of a multibarreled pipet) in conjunction with the test pipet is not an adequate control for this artifact unless the flow rate (and the direction of flow) in both control and test pipets can be shown to be identical. The large variability in flow rates between pipets makes this difficult to achieve and it is not possible to ensure equal flow rates for different pipets (or barrels of the same pipet) when they are inserted into tissue. Therefore, the best solution seems to be to reduce the flow rate to as low a value as possible or to use very brief periods of application and wait an interval after the application in order to measure the response to the test substance (Choi and Fischbach, 1981; MacDonald and Wojtowicz, 1982; Heyer and Macdonald, 1982). The latter of these solutions is not amenable with the requirements of equilibrium analysis. The most satisfactory solution is to replicate the application of agonist by microiontophoresis (Fig. 7).

## 5.3. Mechanical Artifacts

The application of a microiontophoretic current to a glass micropipet causes a sudden vibration of its tip. In fact, penetration of neurons by intracellular electrodes is often achieved by passing a brief (i.e., 50 ms) current pulse through the electrode. The tip presumably acts like a piezoelectric crystal, with a response that drives the tip through the neuronal membrane. This kinetic response of the micropipet is probably of little significance as a possible source of a stimulus to the neuron provided the micropipet tip is relatively far from the membrane. However, when a high-resistance micropipet with small-tip diameter is used close to the neuronal membrane, with the objective of rapidly applying the test substance (either by iontophoresis or pressure), a standing pressure wave may be directed against the neuron, causing deformation and stretching of the membrane. Applications of test solutions conforming to this description are characterized by a depolarizing (excitatory) neuronal response (*see* Fig. 8). This response peaks in a period of milliseconds and then rapidly declines whether or not the microiontophoretic current or pressure

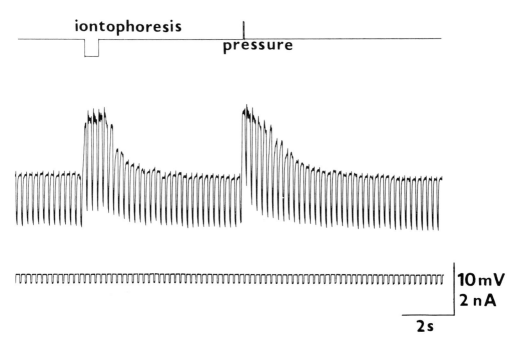

Fig. 7. Response of a cultured spinal cord neuron to L-aspartic acid applied either by iontophoresis or by pressure. An indicator for the time of agonist application is given above the voltage trace. Hyperpolarizing current pulses (lower trace) were passed via a bridge circuit as previously described in order to measure input conductance. This amino acid evoked an apparent and voltage-dependent reduction of input conductance (increase in the size of voltage deflections) regardless of the method of agonist application (MacDonald and Wojtowicz, 1982). The concentration of L-aspartic acid was 500 μM for the pressure application with a pH (7.4) exactly matching that of the bathing solution. Flow was kept minimal to avoid the resulting artifact and limited by making the duration of the application relatively short. A much higher concentration of L-aspartic acid (0.5M, pH 8.4) was used for microiontophoresis.

application is maintained. Brief repetitive presentations are also characterized by an apparent desensitization. This fast desensitizing response is generated by an inward current and may be carried primarily by $Na^+$ ions (Gruol et al., 1980; Krishtal and Pidoplichko, 1980). It is not clear whether or not this response is generated by $H^+$ ions, drugs, subtle changes in the osmolarity of the solution, or by the movement of the micropipet tip itself. However, extreme caution needs to be exercised in the interpretation of such responses, since the matter is not yet resolved.

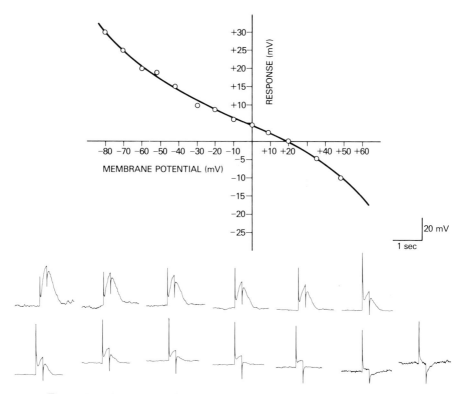

Fig. 8. An example of a rapid depolarization evoked in a neuroblastoma cell (108CC15) by microiontophoretic ejections of met-enkephalin (pH 4.0). This response demonstrated characteristics similar to a rapid desensitizing response in cultured spinal cord neurons described by Gruol et al. (1980). When the membrane was adjusted to various membrane potentials a reversal potential could also be shown. Sample responses to a standard iontophoretic ejection (positive current) are given below plotted values (MacDonald and Barker, unpublished results). Transient artifacts from the onset and offset of the iontophoretic applications are found in each response.

## 5.4. Artifacts of Acidification

Solutions applied by micropressure should be maintained at physiological pH and therefore there is little reason to expect local changes in pH at the injection site. Caution should be exercised, particularly with strong acids, to see that the solution contains adequate buffering power to maintain a neutral pH. In contrast, the microiontophoretic application of many compounds, and weak bases in particular, requires acidification of the solution in order to ensure adequate charging of the drug molecules. Although the concentration of $H^+$ ions is usually very much less than that of the test substance, the mobility of these ions is high. Consequently, the flow of current during positive iontophoretic currents (particularly when used to eject substances that have low transport numbers) is likely to be carried in part by $H^+$ ions. This will result in a decrease of pH near the tip of the iontophoretic pipet. Setting a lower limit for the pH of a test solution does not ensure that acidification will not occur as a result of the variability of transport between pipets (Gruol et al., 1980). A control application (i.e., same ejection current from a NaCl solution acidified to the same pH) will not necessarily release the same amount of $H^+$ ions. When responses to $H^+$ ions are suspected, it is best to confirm the results using micropressure (*see* Fig. 9).

Excitatory responses to $H^+$ ions were initially noted during in vivo recordings of extracellular spike activity (Krnjević and Phillis, 1963). More recent intracellular recordings from spinal cord neurons in vivo (Marshall and Engberg, 1980) and in vitro (Gruol et al., 1980) have shown that $H^+$ ions cause inhibition (*see* Fig. 9) as well as excitation of central neurons. Acidification causes a complex pattern of membrane potential and conductance changes, including a membrane hyperpolarization (decreased conductance) followed by depolarization (increased conductance). Similar results have been observed in invertebrate neurons (Barker and Levitan, 1972).

20mV

1000nA

20s

Fig. 9.    An intracellular recording from an unidentified midbrain neuron grown in tissue culture as an explant slice. TTX was not included in the bathing solution and substantial spontaneous activity was present (action potentials are truncated by the pen recorder). Increasing microiontophoretic doses of dopamine (pH 5.0), as indicated by the lower trace, were associated with inhibition and membrane hyperpolarization. This inhibition could not be reproduced with pressure applications of dopamine (1 m$M$), and it is likely that the response to microiontophoretic application can be attributed to ejection of $H^+$.

Numerous other actions of acidification upon membrane excitability have also been noted. The inhibition of $Na^+$ channels by $H^+$ ions has been well characterized (Begenisich and Danko, 1983) and probably accounts for the observed increase in spike threshold during microiontophoresis of $H^+$ ions (Gruol et al., 1980). Selective blockade of $Ca^{2+}$ vs $Na^+$ action potentials has also been reported (Spitzer, 1979).

## 6. Have the Objectives of Equilibrium Analysis Been Met?

The use of equilibrium analysis requires, in addition to the attainment of drug–receptor equilibrium, a precise knowledge of the agonist concentration at the receptors. In most experiments the distribution of agonist is not uniform and constant within the tissue volume, nor is it constant with respect to time. Therefore, an unknown amount is released into an unknown volume to reach receptors in an unknown distribution within this volume. Even an absolute measure of the amount of agonist released (determined by direct assay or by theoretical calculation) is of limited value. In the case of in vivo preparations, the agonist is ejected into a tortuous extracellular space defined by large numbers of neuronal and nonneuronal (glia) elements. Furthermore, these elements are far from passive and may contain (and most certainly do, in many instances) mechanisms for sequestering or enzymatically degrading the agonist. Therefore, microiontophoresis has not been and is not likely to be a successful technique for equilibrium analysis of agonist responses in the central nervous system *in situ*.

Calibrated micropressure applications for which a linear relationship is shown between time of application (or the magnitude of the applied pressure) and amount of agonist released do not provide any advantage over microiontophoresis in this respect. Figure 10 is an example of dose–response curves (uncalibrated) for kainic and glutamic acids compared in this fashion. Knowing the amount of agonist injected into an unknown volume, of a nonhomogeneous and nonpassive medium, and with unknown receptor geometric arrangement and density, is basically the same problem. The only advantage gained is the ease in calculating the amount of agonist released (assuming of course that it is the same *in situ* as it is when the calibration is done in vitro).

Microiontophoresis of cholinergic agonist at the neuromuscular junction is the only example of vertebrate studies in which

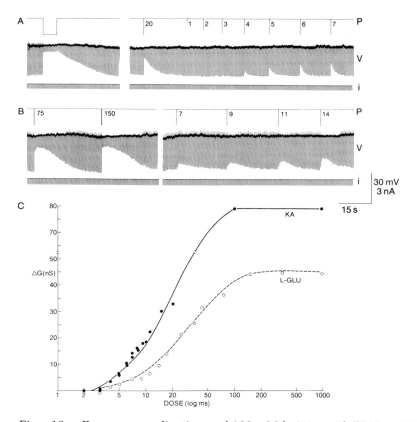

Fig. 10. Pressure applications of 100 $\mu M$ kainic acid (KA) or 500 $\mu M$ L-glutamic acid (L-GLU) to a cultured spinal cord neuron. Current was injected to bring the membrane potential to the reversal potential for the amino acids and the dose of amino acid varied by changing the duration of each application. Dose (duration in milliseconds)–response ($\Delta G$, input conductance) are plotted (C) below sample intracellular recordings (A and B). Pressure indicators (P), membrane voltage (V), and injected current (i) are shown.

The slope of the dose–response curve for kainic acid was significantly greater than that for L-glutamic acid, which confirms a similar relationship seen for in vivo spinal cord neurons (extracellular recordings, MacDonald and Nistri, 1978). Nevertheless, this technique cannot ensure an adequate knowledge of the concentration of agonist at the receptors and is therefore subject to error (see text).

equilibrium analysis has met with some degree of success. In this preparation the geometric relationship of the receptors relative to the microiontophoretic pipet has been determined. The concentration of agonist at the receptors can be calculated using diffusion equations (Dreyer et al., 1978). Dionne et al. (1978) were, in addition, able to employ agonist-sensitive electrodes to assist in the calculation of concentrations at the receptors. Coupled with this analysis was the use of high-resistance iontophoretic micropipets that reduced the possibility of desensitization. In this preparation, microiontophoresis has permitted a detailed examination of the relationship between agonist concentration and effect. Thus the nonsigmoidal relationship of the dose–response relationship and apparent cooperativity of this cholinergic receptor has been convincingly demonstrated.

Comparable in vitro preparations of the central neurons are now available (i.e., dissociated neurons in culture, acutely isolated individual adult neurons). However, the distribution of relevant receptors over the surface of these neurons is not known. This is because of the likelihood that receptor distributions differ from one neuron to the next. These difficulties are likely to be overcome by persistence and refinement of techniques (i.e., use of agonist-sensitive electrodes and methods for locating receptors on living neurons).

Pressure microperfusion of agonist solution to neurons in vitro for the purpose of constructing dose–response relationships for agonists has not been extensive. This is largely because of the difficulties in applying a large number of different doses of agonists without dislodging the recording electrode. The simplest approach is to have many pipets, each containing a different concentration of agonist (*see* Fig. 11) and apply them under the constraints previously described. The major drawback of this technique is the inconsistency of flow between pipets. Flow artifacts may result with some applications and not with others and the region of local perfusion may differ between pipets. Nevertheless, such a technique provides evidence about threshold doses of agonists as well as putative quantitative information about the dose–response relationships (Dichter, 1980; Choi and Fischbach, 1981; MacDonald and Wojtowicz, 1982; Heyer and Macdonald, 1982; MacDonald and Barker, 1982).

## 7. Antagonists and Transmitter Identification

One of the strongest criteria for transmitter identification is the coincidental blockade of appropriate synaptic potential (or synaptic current) and the exogenously applied agonist (the putative transmitter) by a competitive antagonist (Werman, 1966). Competitive antagonism is established by pharmacological criteria derived from equilibrium analysis and is therefore subject to the va-

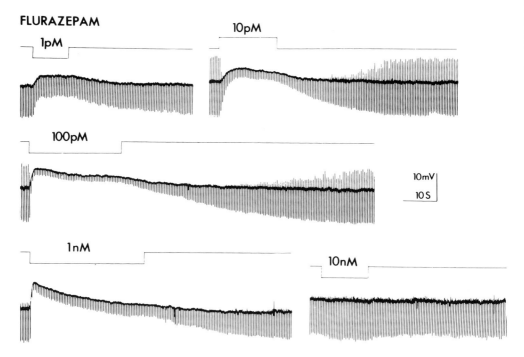

Fig. 11. The increase in membrane conductance evoked by flurazepam fades with increasing dose. Pressure applications of flurazepam from five different pipets, with concentrations ranging from 1 pM to 10 nM, are illustrated. The response to each concentration is shown below the pressure indicator. Doses of flurazepam were applied randomly, and this figure is a summary selected from results of over 30 individual applications. Membrane conductance was assayed by passing constant current pulses (50 ms) and measuring the resulting voltage excursions. Flurazepam at a concentration of 1 pM depolarized the membrane and increased conductance (decrease in voltage deflections). Increasing the concentration to 100 pM maximized this response, but a significant fading developed with 1 nM flurazepam. Applications of 10 nM consistently failed to activate this response even though visual checks of drug ejection demonstrated perfusion of the neuron. Resting potential: −51 mV. (Reproduced from MacDonald and Barker, 1982.)

lidity of assumptions inherent in such an analysis. For example, the number of agonist and antagonist molecules should be large relative to the number of receptors. This is the case when the agonist is applied exogenously, but is it true for agonist (transmitter) release from the presynaptic terminal?

Quastel and Pennefather (1983) have shown that at the neuromuscular junction the number of receptors is likely to exceed significantly the number of molecules of ACh released during synaptic activation. Therefore, manipulations that resulted in large reductions in the number of receptors, such as the presence of the competitive antagonist ( + )-tubocurarine, failed to reduce the synaptic event (miniature endplate current) even though responses to an agonist were greatly depressed (as predicted from equilibrium analysis). They concluded that synaptic transmission at the neuromuscular junction is relatively insensitive to blockade by agents that reduce receptor number (i.e., competitive antagonists). If this finding can be generalized to other transmitters and to synapses within the central nervous system, it suggests that using an antagonist to identify the endogenous transmitter is subject to the error of rejecting the exogenous agonist as a transmitter candidate when it is in fact the endogenous transmitter. Ironically, Quastel and Pennefather (1983) further demonstrated that if receptor affinity for the agonist remains high but ionic conductance is reduced (e.g., during channel blockade by local anesthetics), then the synaptic event may be more sensitive to depression by the local anesthetic than is the response to the exogenously applied agonist. This presents the possibility of mistaking a channel blocker (or agent that reduces the ionic conductance of the agonist) for an antagonist of the endogenous transmitter.

## 8. Single Channel Analysis: Identification of Transmitters

Recently, techniques have been developed and utilized to characterize the average behavior of single ionic channels activated by agonists in mammalian central neurons. These include fluctuation analysis, relaxation analysis, and direct measurements of single-channel properties (i.e., conductances and open times) using the patch clamp technique. In some preparations the time constant of decay of a particular synaptic current has been found to be similar, if not identical, to the mean single-channel open time of the most likely transmitter candidate (Mathers and Barker, 1982). Thus, it may be possible to select the actual transmitter from a series of agonists (all of which have affinity and efficacy for the same postsynaptic receptor) simply by matching the time constant of decay of the synaptic current with the mean open time of the exogenously applied candidate.

Such an approach has been used by Barker and coworkers to identify gamma-aminobutyric acid (GABA)-mediated synaptic currents in cultured neurons from the spinal cord and hippocampus (Barker and McBurney, 1979; Barker et al., 1982; Segal and Barker, 1984). For example, the mean open times for GABA-activated channels (GABA applied either by microiontophoresis or pressure) have been measured as well as the decay constants of certain inhibitory synaptic currents and a close correspondence found. Further support for measurements of these parameters being important in the identification of the transmitter was provided by the parallel modulation of single channel open time and decay of the inhibitory synaptic current by neuroactive compounds such as pentobarbital and not by antagonists, such as picrotoxin and bicuculline, that may simply change the kinetics of channel activation or reduce the number of activated channels (Barker and McBurney, 1979; Study and Barker, 1981; Barker et al., 1983; Segal and Barker, 1984).

Thus, the "identity of action" criterion (Werman, 1966), coupled with techniques of single channel analysis, has substantial potential for success in the identification of transmitters in the central mammalian nervous system. Microiontophoresis and micropressure will likely continue to be useful techniques for microdelivery of putative transmitters.

## Acknowledgments

The author gratefully acknowledges the support of the Medical Research Council of Canada, the assistance of Ms. Brenda Chin Choy in preparation of the manuscript, and Dr. M.E. Morris for reading the manuscript.

## References

Barker J. L. and Levitan H. (1972) The antagonism between salicylate-induced and pH-induced changes in the membrane conductance of molluscan neurons. *Biochim. Biophys. Acta.* **274,** 638–643.

Barker J. L. and McBurney R. N. (1979) Phenobarbitone modulation of postsynaptic GABA receptor function on cultured mammalian neurons. *Proc. Roy. Soc. Lond. B.* **206,** 319–327.

Barker J. L., McBurney R. N., and MacDonald J.F. (1982) Fluctuation analysis of neutral amino acid responses in cultured mouse spinal cord neurons studied under voltage-clamp. *J. Physiol. (London)* **322,** 365–387.

Barker J. L., McBurney R. N., and Mathers D. A. (1983) Convulsant-induced depression of amino acid responses in cultured mouse spinal neurons studied under voltage-clamp. *Brit. J. Pharmacol.* **80,** 619–629.

Barker J. L. and Ransom B. R. (1978) Amino acid pharmocology of mammalian central neurons grown in tissue culture. *J. Physiol. (London)* **280,** 331–354.

Begnisich T. and Danko M. (1983) Hydrogen block of the sodium pore in squid giant axon. *J. Gen. Physiol.* **82,** 599–618.

Bradshaw C. M., Roberts M. H. T., and Szabadi E. (1973) Kinetics of the release of noradrenaline from micropipettes: interaction between ejecting and retaining currents. *Brit. J. Pharmacol.* **49,** 667–677.

Bradshaw C. M. and Szabadi E. (1974) The measurement of dose in microelectrophoresis experiments. *Neuropharmacology* **13,** 407–415.

Brown D. A. and Griffith W. H. (1983) Persistent slow inward calcium current in voltage-clamped hippocampal neurons of the guinea-pig. *J. Physiol. (London)* **337,** 303–320.

Choi D. W. and Fischbach G. D. (1981) GABA conductance of chick spinal cord and dorsal root ganglion neurons in cell culture. *J. Neurophysiol.* **45,** 605–620.

Del Castillo J. and Katz B. (1955) On the localization of acetylcholine receptors. *J. Physiol. (London)* **128,** 157–181.

Dichter M. A. (1980) Physiological identification of GABA as the inhibitory transmitter for mammalian cortical neurons in cell culture. *Brain Res.* **190,** 111–121.

Dionne V. E., Steinbach J. H., and Steven C. F. (1978) An analysis of the dose–response relationship at voltage-clamped frog neuromuscular junction. *J. Physiol. (London)* **281,** 421–444.

Dray A., Hanley M. R., Pinncock R. D., and Sandberg B. E. B. (1983) A comparison of the release of substance P and some synthetic analogues from micropipets by microiontophoresis or pressure. *Neuropharmocology* **22,** 859–863.

Dreyer F. and Peper K. (1974) Iontophoretic application of acetylcholine: advantages of high resistance micropipets in connection with an electronic current pump. *Pflugers Arch.* **348,** 263–272.

Dreyer F., Peper K., and Sterz R. (1978) Determination of dose–response curves by quantitative iontophoresis at the frog neuromuscular junction. *J. Physiol. (London)* **28,** 395–419.

Dudel J. (1975) Kinetics of postsynaptic action of glutamate pulses applied iontophoretically through high-resistance micropipets. *Pflugers Arch.* **356,** 329–346.

Gruol D. L., Barker J. L., Huang M. L., MacDonald J. F., and Smith T. G. (1980) Hydrogen ions have multiple effects on the excitability of cultured mammalian neurons. *Brain Res.* **183,** 247–252.

Guyenet P. G., Mroz E. A., Aghajanian G. K., and Leeman S. E. (1979) Delayed iontophoretic ejection of substance P from glass micropipets: correlation with time-course of neuronal excitation in vivo. *Neuropharmacology* **18,** 553–558.

Heyer E. J., and Macdonald R. L. (1982) Calcium- and sodium-dependent action potentials of mouse spinal cord and dorsal root ganglion neurons in cell culture. *J. Neurophysiol.* **47,** 641–655.

Hicks T. P. (1984) The history and development of microiontophoresis in experimental neurobiology. *Prog. Neurobiol.* **22,** 185–240.

Hoffer B. J., Neff N. H., and Siggins G.R. (1971) Microiontophoretic release of norepinephrine from micropipets. *Neuropharmacology* **10,** 175–180.

Karlin A. (1983) Anatomy of a receptor. *Neurosci. Commentaries* **1,** 111–123.

Kelly J. S. (1975) Microiontophoretic Application of Drugs Onto Single Neurons, in *Handbook of Psychopharmacology,* vol. 2 (Iversen L.L., Iversen S.D., and Snyder S.H., eds.), pp. 29–67. Plenum, New York.

Krishtal O. A. and Pidoplichko V. I. (1980) A receptor for protons in the nerve cell membrane. *Neurosci.* **5,** 2325–2327.

Krnjević K. (1971) Microiontophoresis, in *Methods of Neurochemistry,* vol. 1 (Fried R., ed.). pp. 130–172. Marcel Dekker, New York.

Krnjević K., Puil E., and Werman R. (1975) Evidence for $Ca^{2+}$-activated $K^+$ conductance in cat spinal motoneurons from intracellular EGTA injections. *Can. J. Physiol. Pharmacol.* **53,** 1214–1218.

Krnjević K., Mitchell J. F., and Szerb J. C. (1963) Determination of iontophoretic release of acetylcholine from micropipets. *J. Physiol. (London)* **165,** 421–436.

Krnjević K., and Phillis J. W. (1963) Iontophoretic studies of neurons in mammalian cerebral cortex. *J. Physiol. (London)* **165,** 274–304.

Krnjević K., Puil E., and Werman R. (1978) EGTA and motoneuronal after-potentials. *J. Physiol. (London)* **275,** 199–223.

Krnjević K. and Whittaker V. P. (1965) Excitation and depression of cortical neurons by brain fractions released from micropipets. *J. Physiol. (London)* **179,** 298–322.

MacDonald J. F. (1984) Substitution of extracellular sodium ions blocks the voltage-dependent decrease of input conductance evoked by L-aspartate. *Can. J. Physiol. Pharmacol.* **62,** 109–115.

MacDonald J. F. and Barker J. L. (1982) Multiple actions of picomolar concentrations of flurazepam on the excitability of cultured mouse spinal neurons. *Brain Res.* **246,** 257–264.

MacDonald J. F., and Nistri A. (1978) A comparison of the action of glutamate, ibotenate, and other related amino acids on feline spinal interneurons. *J. Physiol. (London)* **275,** 449–465.

MacDonald J. F., and Schneiderman J. H. (1984) L-aspartic acid potentiates 'slow' inward current in cultured spinal cord neurons. *Brain Res.* **296,** 350–355.

MacDonald J. F., and Wojtowicz J. M. (1982) The effects of L-glutamate and its analogues upon the membrane conductance of central murine neurons in culture. *Can. J. Physiol. Pharmacol.* **60,** 282–296.

MacVicar B. A., and Dudek F. E. (1982) Electrotonic coupling between granule cells of rat dentate gyrus: physiological anatomical evidence. *J. Neurophysiol.* **47,** 579–592.

McCaman R. E., McKenna D. G., and Ono J. K. (1977) A pressure system for intracellular and extracellular ejections of picoliter volumes. *Brain Res.* **136**, 141–147.

Marshall K. C., and Engberg I. (1980) The effects of hydrogen ion on spinal neurons. *Can. J. Physiol. Pharmacol.* **58**, 650–655.

Mathers, D. A., and Barker J. L. (1982) Chemically induced ion-channels in nerve cell membranes. *Int. Rev. Neurobiol.* **23**, 1–34.

Palmer M. R., Wuerthele S. M., and Hoffer B. J. (1980) Physical and physiological characteristics of micropressure ejection of drugs from multibarreled pipets. *Neuropharmacology* **19**, 931–938.

Poulain P. and Carette B. (1981) Pressure ejection of drugs on single neurons *in vivo:* Technical considerations and application to the study of estradiol effects. *Brain Res. Bull.* **7**, 33–40.

Purves R. D. (1977) The release of drugs from iontophoretic pipets. *J. Theor. Biol.* **67**, 789–798.

Purves R. D. (1979) The physics of iontophoretic pipets. *J. Neurosci. Meth.* **1**, 165–178.

Purves R. D. (1980) Effect of drug concentration on release from iontophoretic pipets. *J.Physiol. (London)* **300**, 72P–73P.

Quastel D. M. J., and Pennefather P. (1983) Receptor blockade and synaptic function. *J. Neural Transm., Suppl.* **18**, 61–81.

Rutgers A. J. (1940) Streaming potentials and surface conductance. *Trans. Faraday Soc.* **36**, 69–80.

Sakai M., Swartz B. E., and Woody C. D. (1979) Controlled microrelease of pharmacological agents: measurements of volume ejected *in vitro* through fine tipped glass microelectrodes by pressure. *Neuropharmacology* **18**, 209–213.

Segal M., and Barker J. L. (1984) Rat hippocampal neurons in culture: voltage clamp analysis of inhibitory synaptic connections. *J. Neurophysiol.* **52**, 469–487.

Spitzer N. C. (1979) Low pH selectively blocks calcium action potentials in amphibian neurons developing in culture. *Brain Res.* **161**, 555–559.

Study R. E., and Barker J. L. (1981) Diazepam and ( − ) pentobarbital: fluctuation analysis reveals different mechanisms for potentiation of GABA responses in cultured central neurons. *Proc. Nat. Acad. Sci. USA* **78**, 7180–7184.

Werman R. (1966) Criteria for identification of a central nervous system transmitter. *Comp. Biochem. Physiol.* **18**, 745–766.

Werman R. (1969) Electrophysiological approach to drug-receptor mechanisms. *Comp. Biochem. Physiol.* **30**, 997–1017.

# Chapter 7

# Electrical and Chemical Stimulation of Brain Tissue In Vivo

## A. J. Greenshaw

## 1. The Intracranial Approach

The use of general systemic manipulations to alter brain activity, often coupled with ex vivo biochemical analysis, has led to significant advances in our understanding of central nervous system (CNS) activity. Nevertheless, the search for a clear understanding of the function of separate brain regions has necessitated the use of intracranial preparations. The advantages of this approach are twofold: It allows stimulation of localized areas of brain tissue, which is important in the analysis of functions of separate nuclei and neural pathways; furthermore the intracranial approach enables us to apply compounds to the CNS that do not readily cross the blood–brain barrier (*see* Oldendorf, 1971). It is notable that the majority of substances believed to be neurotransmitters fall into this category of compounds.

An attempt is made in the present chapter to provide a useful general description of methodology in this area. This account is divided into two sections: one dealing with electrical and one with chemical stimulation. It is important for the reader to appreciate that quite different problems are associated with each area. Electrical pulses represent a form of stimulation that is easy to apply and infinitely variable. Chemical stimuli are more difficult to handle in terms, for example, of concentration, volume, and problems of chemical stability.

With electrical stimuli the effects of stimulation are often clear enough at the level of the whole organism, but their interpretation is often unclear at the cellular level. This interpretative problem is less evident with chemical stimuli. Within certain limits the pharmacological profiles of chemical stimuli allow an interpretation of effects at the cellular level (i.e., in terms of receptor activation or blockade, effects on uptake or release mechanisms, and so on).

Although pharmacology is often adopted to assess effects of electrical stimulation, the analysis is often unsatisfactory for determining the direct effects of this form of stimulation. The alternative strategy is purely electrophysiological and many researchers have adopted indirect methods (with attendant problems of interpretation). Direct electrophysiological methods involving the recording of individual cellular responses are more demanding, particularly in freely moving animals. Thus, chemical stimuli are somewhat difficult to apply, but their effects are often well-defined at the cellular level. Electrical stimulation is relatively easy to apply, but its effects are often difficult to interpret at the cellular level. In the latter case a complex technology has arisen, incorporating both direct and indirect electrophysiological techniques.

As a consequence of these contrasting areas of difficulty there is a different emphasis within each section of this chapter. The section dealing with electrical stimulation provides a limited outline of technique. This information covers the use of basic procedures. The reader is, however, urged to explore the cited references for a more detailed view of interpretative issues. At the end of this section a few brief examples of applications of electrical stimulation are outlined in a general sense. In the section dealing with chemical stimulation the limited variety of general methods is outlined in detail, together with a more analytical description of illustrative applications. In this case the reader may feel better informed, but will undoubtedly face a few more problems in terms of the practice of basic technique *per se*. Nevertheless, using these techniques for electrical and chemical stimulation is not difficult. Using them effectively, with the possibility of a clear interpretation of the consequences, yields both interesting and rewarding questions.

## *1.1. Stereotaxic Surgery*

The ability to localize specific brain regions in vivo is central to techniques for electrical and chemical stimulation of the brain. The precise localization of electrodes and injection cannulae into specific brain areas is, of course, achieved with the use of a stereotaxic instrument. A number of brain atlases are commercially available that form an accurate description of brain structures in a variety of species. Although a variety of species are used in neuroscience research, much of this work focuses on the laboratory rat, and for this reason the present chapter will deal almost exclusively with techniques as they are applied to this species. Nevertheless, the principles and procedures described here are equally applicable to other species. Various atlases of the rat brain are available (De Groot, 1959; Fifková and Maršala, 1967; Hurt et al., 1971; Paxinos and Watson, 1982; König and Klippel, 1974; Pellegrino et al., 1979).

Each atlas provides coordinates in three planes—anterior–posterior, lateral, and dorsoventral (or horizontal)—in relation to a particular reference point. Two reference points are used in stereotaxic surgery, one based on bregma (the point of intersection of the frontoparietal and sagittal sutures); the other is the interaural zero (i.e., the central point between the tips of the ear bars on the stereotaxic frame). An important point about choice and use of different brain atlases is that some atlases provide two sets of anterior–posterior coordinates (i.e., one based on bregma and the other on the interaural zero); furthermore, atlases vary in terms of the orientation of the skull. For example, König and Klippel (1967) use a flat skull orientation at which the incisor bar on the stereotaxic instrument is set at 2.4 mm below the interaural zero, whereas Pellegrino et al. (1979) adopt a skull orientation whereby the incisor bar is set at 5.0 mm above the interaural zero. Skull orientation is particularly important when preparing brains for histology and assessing areas that are chosen for manipulation in relation to plates in the atlas (*see* Bureš et al., 1983). A number of full accounts of techniques for stereotaxic surgery are available (*see* Cooley and Vanderwolf, 1978; Bureš et al., 1983).

## 1.2. Anesthesia

An important consideration in the use of stereotaxic techniques is the need for effective anesthesia. The anesthetic of choice depends very much on the type of preparation used in the laboratory. For acute preparations a long-acting anesthetic, such as urethane, is preferable to a shorter-acting compound; nevertheless, when postoperative recovery is required, urethane is not suitable because the effects of this anesthetic may, at sublethal doses, last for many hours. For experiments in which chronic preparations are used, and in which animals are allowed to regain consciousness, barbiturate anesthesia is most commonly used (e.g., 40–50 mg/kg of sodium-pentobarbitol for the rat). Nevertheless, there are some problems associated with anesthesia, particularly in terms of respiratory failure. With this consideration in mind it should be noted that an inhalation anesthetic, such as halothane

($2\%$ in $O_2$), is far superior to other forms of anesthesia because animals can be revived within minutes of terminating the anesthetic. In relation to respiratory failure, pretreatment with atropine sulfate (0.1 mg/kg subcutaneously) is often used as a preventative measure: In most cases, however, this is unnecessary if healthy rats are used for experimentation.

### 1.3. Suitable Connection Systems for Electrical and/or Chemical Stimulation of CNS in Unrestrained Animals

Acute preparations require very little in terms of specialized hardware for electrical or chemical manipulations. A simple electrode or microinjection syringe mounted on a stereotaxic micromanipulator will suffice. However, for chronic preparations, skull assemblies are necessary.

A variety of techniques are commonly used for anchoring electrodes or cannulae to the cranial bones for long periods. The most commonly used technique involves the insertion of small stainless steel screws into the skull and the use of dental acrylic to form a solid skull cap that will securely anchor the electrodes or cannulae in place. In some conditions, and with some species, more drastic measures have been employed, such as the use of anchoring bolts or threading wire under the cranial bone, to form a more durable implant. With laboratory rats this is usually unnecessary.

The administration of electrical stimuli or the application of chemical stimuli to freely moving animals requires the use of efficient connection systems. In the case of electrical stimulation this means the use of flexible connectors with good low-resistance contacts that allow free movement. There are a number of possibilities for the laboratory construction of slip rings that have been described previously (for example, *see* Bureš et al., 1983). Typically, under these conditions, mercury-track slip rings are employed. Nevertheless, gold-track slip rings should be used whenever possible. There are two reasons for this: (1) Mercury is highly toxic and its presence in laboratories undesirable, particularly with applications in which spillage may occur. (2) Mercury-track slip rings require a good deal of maintenance because they have a tendency to become noisy with the formation of oxide when the mercury is inevitably exposed to air, as is the case with practically all systems built in the laboratory. Gold-track slip rings, although moderately expensive, are extremely reliable and require only occasional maintenance. It is notable that for some applications, i.e., particularly those involving electrophysiological recordings, the use of either gold-track or mercury-track slip rings is not desirable because even quite low resistance changes can affect the nature of recordings. Thus, in electrophysiological experiments, some researchers prefer to have a direct connection between the electrode, or electrodes, and the subject; this causes problems when animals move freely and twist the cable. However, it is apparent that for short duration recordings, manual disentangling of connector wires may be preferable to some of the problems encountered with noisy slip rings.

With direct chemical stimulation of the CNS, it is, in many cases, possible to administer microinjections to animals prior to experimental testing. In this "off the baseline" approach no fluid swivel is necessary. Nevertheless for some applications the use of a fluid-tight, flow-through swivel may be necessary. The most satisfactory systems are those that are commercially available, particularly in the case in which electrical and chemical stimulation of the brain are used simultaneously (this is achieved with so-called electrocannula swivels). There are, however, a number of reports that describe the simple construction of fluid swivels for animal research (Brown et al., 1976; Pickens and Thompson, 1975; Blair et al., 1980). Most are single-channel fluid swivels. In some experiments it is necessary to have more than one fluid channel; this application is relatively rare in the case of direct chemical stimulation of the CNS. Nevertheless, when push–pull perfusion is used as a technique for the application of drugs (Redgrave, 1978), a double-channel swivel is needed. A number of double-channel swivel designs have been described (Nicolaidis et al., 1974; Blair et al., 1980). In this case, the latter (Blair et al., 1980) is an inexpensive double-channel, low-torque swivel that is easily assembled with readily available materials and common laboratory tools. The reader is recommended to consult one or more of these publications for details concerning in-laboratory construction of these devices.

A critical feature of systems for electrical or chemical stimulation of the brain in freely moving animals is the coupling between the electrode or cannula at the skull with the flexible connection cable or tubing.

Various approaches to this problem have been successfully adopted, and a number of designs for connectors are available. Screw-on collar fittings, such as those provided by Plastic Products Inc., are perhaps the best solution (*see* Fig. 1), although the use of transistor sockets for electrical contacts or for fluid, luer lock connectors provides alternative strategies.

It is notable that a coiled spring covering provides an effective measure against damage to cable or tubing caused by biting, which is a frequent problem.

A.                                                                  B.

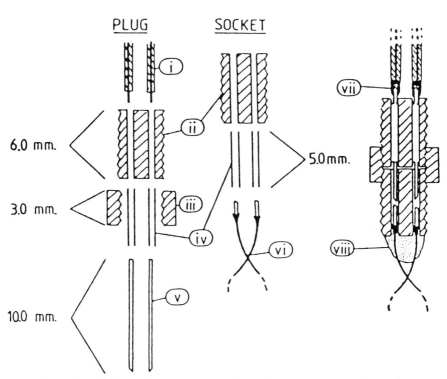

Fig. 1. Schematic representation of a screw-on electrode connector assembly. (A) Exploded cross-sectional assembly. (B) Assembled cross-sectional view. The dimensions are suggested, but may easily be adapted to suit the experimental conditions.

    (i) Insulated connector wire—to stimulator
   (ii) Nylon bolts—cut and drilled
  (iii) Nylon nut
  (iv) Steel hypodermic tubing
   (v) Steel pins
  (vi) Insulated electrode wire (≤300 μm bare diameter)
 (vii) Suitable solder for stainless steel
(viii) Epoxy resin

## 2. Electrical Stimulation

Since the pioneering works of Simonoff (1866) and Hess (1928), using electrical brain stimulation in freely-moving animals, and of Fritsch and Hitzig (1870), who first demonstrated electrical excitability of cortex in man, the electrical stimulation of brain tissue, in vivo, has been a technique that has been usefully applied to the analysis of brain functions. There are a number of in-depth reviews of this technique in literature (Sheer, 1961; Delgado, 1964; Ranck, 1975; Patterson and Kesner, 1981), as well as other more basic descriptions, such as that provided by Bureš et al. (1983). In the latter source there is a section dealing with basic concepts and electrical circuits that will be extremely useful to those who may be unfamiliar with these issues.

To usefully employ electrical stimulation techniques to CNS preparations, a fundamental understanding of neural activity is necessary, i.e., some understanding of membrane processes in relation to action potentials and their regulation in brain tissue. A clear exposition of some fundamental aspects of these mechanisms has been provided by Katz (1966) (for an overview, *see* Kuffler et al., 1984).

The flow of electrical current across neural membranes may change the permeability of these membranes to certain ions and result in the triggering of nerve impulses or action potentials. Under physiological conditions, this transmembrane current is generated by the activity of neurons whose effects may be mimicked by electrical fields that are artificially induced by stimulation. Stimulation may be either external to the membrane or intracellular, in certain cases. Intracellular stimulation is not dealt with in the present chapter because this technique is not applied to the analysis of the system as a whole, but rather to the behavior of single cells. An excellent basic account of intracellular stimulation technique is available (Byrne, 1981).

Extracellularly applied current flows mainly through the extracellular space, a small proportion entering the intracellular space. With a cathodal (i.e., negative) stimulus, a reduction of the membrane potential (depolarization), caused by outward current, induces neural stimulation (the outward current serves to carry positive charges from the inside to the outside of the neuron). This effect is balanced by a reentry of the same amount of inward current (serving to increase the membrane potential: hyperpolarization) in other parts of the neuron (*see* Fig. 2). Under certain circumstances, this hyperpolarization may block impulse generation (*see* section 2.8.4).

A number of factors determine the efficacy of an electrical stimulus in relation to neural activity, the main determinants being the configuration of the electrodes (which will determine the nature of the stimulus field) and the intensity and duration of stimulation. Duration of stimulation is particularly important be-

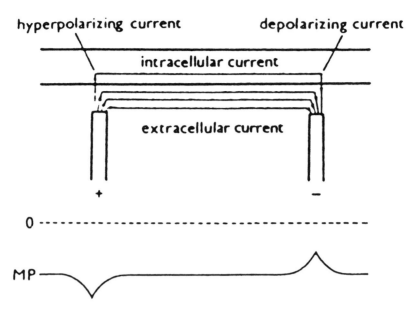

Fig. 2. The effect of electrical current on a nerve fiber. (Above) Scheme of the electrode arrangement. (Below) Changes in membrane potential (MP). 0, level of extracellular potential (from Bureš et al., 1983, reproduced with permission of the publisher and authors).

cause excessively long stimulation is made ineffective by the process of depolarization that blocks impulse conduction, or by accommodation (a gradual increase of threshold).

## 2.1. Electrode Configuration

A full discussion of suitable metals for electrodes may be found in the article by Delgado (1964). For practical purposes, stainless steel (typically 100–300 μm bore diameter) is a suitable substance; platinum/radium or pure iridium are far superior in terms of resistance to corrosion, but represent a relatively expensive choice. At least two electrodes are required to form any circuit. There are three basic forms for electrode arrangement; these are monopolar, twisted bipolar, and concentric bipolar. With monopolar stimulation, one electrode is located at the site of stimulation and a larger electrode (usually a reusable silver screw) is located at some distal region outside the brain, such as the surface of the skull. In the twisted bipolar configuration, the electrode tips are located close to each other at the site of stimulation. With concentric bipolar electrodes, both electrode poles are located at the stimulation site.

Interelectrode distance is smallest with the concentric bipolar configuration and, of course, greatest in the monopolar case. Interelectrode distance will affect the geometry of the stimulus field in relation to the target site. With small interelectrode distance, current density is highest at the site of stimulation between electrodes, and with large interelectrode distance, current density is highest at the active electrode–tissue contact. In the monopolar case, there is an approximately spherical field of current, the density of which decreases in proportion to the square of the distance from the active electrode. These electrode configurations are illustrated by Fig. 3A. For some mapping studies, movable unipolar electrode assemblies are used. [*See* Miliaressis (1981) and Miliaressis and Phillipe (1984) for good examples of such systems.]

Electrodes should be insulated, except for a cross-sectional area at the tip or for up to 0.5 mm above the bare tip. Insulation should be checked with a suitable Ohmmeter prior to implantation.

## 2.2. Wave Forms

The simplest stimulation wave form consists of pulses of direct current (dc). Parameters that are varied are: pulse duration that typically varies from the $\mu s$ to the ms range, up to a maximum of around 2 ms for CNS applications; pulse amplitude; and frequency of stimulation. The duration of each train of stimulation is

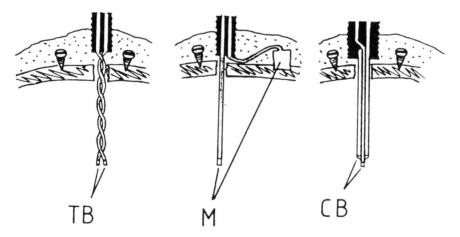

Fig. 3A.    Electrode configurations: A schematic view of implanted electrodes (the lines point to the two electrical poles in each case). Abbreviations: TB, "twisted bipolar"; M, "monopolar"; CB, "concentric bipolar."

usually fixed. The rise and fall time of the leading and trailing pulse edges, respectively, are also important; these should be minimized because slow changes induced by gradual rise and fall time may greatly affect neural responsiveness.

Sinusoidal wave forms are also used in this context. These alternating current (ac) wave forms are characterized by peak-to-peak amplitude, or by the equivalent root mean square (rms) values, and by frequency. The rms values for ac wave forms correspond to dc stimuli of equivalent power. These rms values for ac stimulation may be obtained by dividing the peak-to-peak amplitude by $2 \times (2)^{1/2}$. Bidirectional square waves of alternating polarity may also be used to prevent electrode polarization that may be induced by unidirectional dc stimulation (*see* Doty and Bartlett, 1981). These wave forms are schematically represented in Fig. 3B.

## 2.3. Stimulators

There are basically two forms of stimulators that are used in this context. These are either constant-voltage or constant-current stimulators. Constant-voltage stimulators maintain a constant potential difference at the electrodes, whereas constant-current stimulators maintain a passage of constant current through the tissue. Constant-voltage stimulators effectively have a low internal resistance, therefore the output voltage is not dependent upon variation of the output current. Constant-current stimulators, on the other hand, effectively have a high internal resistance, and output current is, to within certain limits, independent of tissue-resistance variability. For most purposes, constant-current stimulation is preferable, i.e., when there is high variability of electrode-tissue resistance, possibly because of poor contact or electrode polarization, combined with fairly constant resistivity of the stimulated tissue. Under these conditions, constant-current stimulation yields more reproducible results than constant-voltage stimulation.

There are a number of circuits in the literature that may be used to build stimulators. Simple stimulators may, however, be derived from equipment that is generally available in laboratories. The simplest stimulator may be derived by using a Variac, which is a constant voltage source, connected to a 0.5 m$\Omega$ resistor in series with its output to convert it to a relatively constant current source. This type of device will deliver sinusoidal stimulation at main supply frequency.

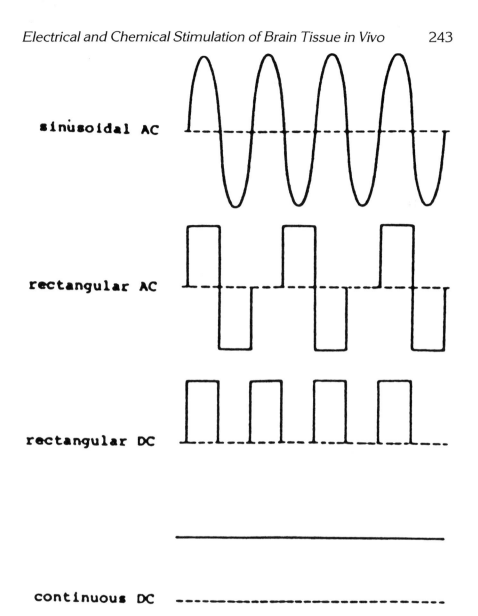

sinusoidal AC

rectangular AC

rectangular DC

continuous DC

Fig. 3B.   Types of electrical stimuli (from Bureš et al., 1983, reproduced with permission of publisher and authors).

## 2.4. Monitoring Electrical Stimulation

It is particularly important to monitor electrical stimulation during experiments. This may be done by measuring the voltage drop across a known resistor placed in series with the animal (*see* Fig.

4). By applying Ohm's law, a measure of current may be derived from this voltage drop:

Current (I)(in amperes) = Voltage (E)(in volts)/Resistance (R)(in ohms)

Thus, if a 10 kΩ resistor is used, a voltage drop of 1 V pk-pk (i.e., peak-to-peak) on the oscilloscope is equivalent to 100 μA for dc stimulation. For ac stimulation, the current should be derived by dividing the peak-to-peak voltage drop by $2 \times (2)^{1/2}$ (i.e., 1 V pk-pk across a 10 kΩ resistor would be equivalent to approximately 35.4 μA rms). With a known current and voltage source it is possible to use Ohm's law to calculate resistance between the electrode tips in brain or, in the monopolar case, between the active electrode and the distal electrode. It is notable that the resistance across the electrode tips may change with variation in stimulus parameters and also in relation to the age of the implanted electrode. Thus, it is clear that to maintain a constant current with a current source that may have only an approximate constant-current output, small adjustments may have to be made through the course of the experiment to offset this change in resistance or impedence across electrode tips.

Sinusoidal stimulation or the use of bidirectional rectangular pulses minimizes the risk of electrolytic damage to tissue that may be a problem with unidirectional dc current pulses. Nevertheless, very short unidirectional dc pulses may be used for electrical

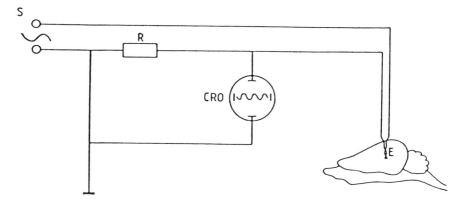

Fig. 4.   Circuit for measuring current delivered through an intra-cranial electrode assembly (E). Current from the stimulator (S) is calculated from the voltage drop on the cathode ray oscilloscope (CRO) across a 10 kΩ precision resistor (R).

brain stimulation studies without undue problems of electrode polarization or electrolytic damage (but, *see* section 2.8.5).

## 2.5. Choice of Stimulation Parameters

Frequency of stimulation may be varied systematically for brain stimulation experiments or fixed at one value. For ac stimulation, frequencies between 30 and 100 Hz may be effective. For dc stimulation, a frequency range of up to 200 Hz may be effective. However, maximum effective frequency (and interpulse interval if pulse pairs are used) will be determined by the refractory period of the population of neurons that is primarily activated by the stimulus (e.g., *see* Deutsch, 1964 and Gallistel et al., 1969).

Currents of up to 2000 μA may be necessary when very short dc pulses are used; with sinusoidal stimulation, however, 200–300 μA should be regarded as an approximate upper limit. Nevertheless, maximum current will depend on the diameter of electrode wire and on the particular preparation studied and should be determined by particular experimental conditions. Durations of pulse trains are generally between 0.1 and 1 s for most purposes. For certain applications, however, a continuous train of pulsatile stimulation has been delivered for periods up to 30–60 min (e.g., Jones et al., 1981). In the latter case, the current and frequency are typically much lower than those that would be used for short trains of stimulation.

## 2.6. Multiple Electrode Use and Cross-Talk: Stimulus Isolation Units

In cases where stimulation is combined with the recording of electrical brain activity (*see* Vanderwolf and Leung, this volume) or when two electrical stimuli are used, it is necessary to isolate the stimulus from ground to prevent interference in the recording case and to prevent cross-talk between the poles of different electrodes in the multiple-stimulation case. For this purpose, stimulus-isolation units are employed. In most cases with stimulus-isolation, each stimulus is provided from a separate battery-operated source; this is particularly common in electrophysiological experiments (e.g., *see* Moroz and Bureš, 1983). Typically, a conventional stimulator is used to generate pulses, the duration and frequency of which control current generated from the stimulus-isolation unit. Usually this is achieved with photoelectric coupling. The interference of stimulating current with recording and the interaction between multiple-stimulation sites are represented schematically in Fig. 5. In the case of battery-

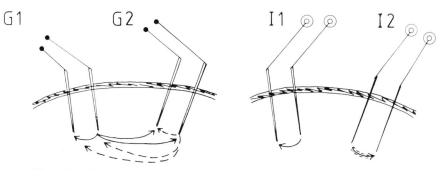

Fig. 5.   Schematic representation of multiple pairs of electrodes in the same brain: An illustration of the function of stimulus isolation units. Arrows represent interaction between electrode poles. G1 and G2 are grounded stimulators: When simultaneously activated considerable electrical interaction (cross-talk) is evident. I1 and I2 represent "ideal" stimulus isolation units: When activated simultaneously each circuit is separate and no "cross-talk" is observed.

operated stimulators, there are few problems because the stimulators can easily be disconnected from ground. Mains-supplied stimulators are grounded devices and, therefore, must be connected through some form of stimulus-isolation circuit. Under these conditions, as Bureš et al. (1983) have pointed out, although resistive isolation of the stimulator from ground is easily accomplished, capacitive isolation is virtually impossible. The latter is dependent on position and length of connecting leads and on the position and construction of the stimulator, i.e., the nature of coupling between the stimulator and the stimulus-isolation unit. Nevertheless the capacitive component of stimulus artifacts can be minimized by using short pulses. It is important to note that with stimulus isolation, unless a battery-operated measurement device is used, stimulation parameters should be set up and checked (preferably with an oscilloscope) only prior to each experimental session.

## 2.7. Circuits

As stated earlier, there are a number of circuits in the literature that may be used for the construction of various types of stimulators. The construction of most of these devices uses transistor–transistor logic (TTL) circuitry. The use of TTL to program stimulation is invaluable in this technique. A basic working knowledge of TTL programming is, therefore, desirable for researchers who are interested in using any but the most simple of stimulator applications. An excellent and lucid account of basic TTL programming has been provided by Bureš et al. (1983), and this source of information may prove invaluable to the researcher with no experience in this area. An appropriate circuit for the construction of dc stimulators has been provided by Doty and Bartlett (1981).

## 2.8. Practical Considerations

In a recent, excellent exposition of effects of extracellular stimulation at the "cellular level," Ranck (1981) has listed a number of practical suggestions that should be taken into account when extracellular stimulation is used. His main points are listed briefly below.

### 2.8.1. Current Sources and Electrode Configuration

Constant-current, and not-constant-voltage sources, should be used whenever possible. The rationale for this in terms of reproducibility of results has been outlined earlier.

The use of monopolar cathodal stimulation is recommended in preference to bipolar stimulation; this electrode configuration represents the simplest field of stimulation in the geometrical sense and allows clearer interpretation of the data in terms of cellular responses. Ranck has suggested that bipolar electrodes should only be used when it is necessary to eliminate or minimize stimulus artifacts. Bipolar arrangements are, however, frequently adopted; the rationale for this is related primarily to convenience and the widespread use of sinusoidal stimulation.

### 2.8.2. Strength Duration Analysis

By recording a given response from the system under investigation and varying the strength (i.e., intensity) and duration of constant-current pulses, it is possible to plot strength vs duration of the pulse to give a constant response. Data from such an analysis are invaluable, particularly in terms of choosing stimulation parameters that are optimal for detecting response changes. Strength duration curves obtained in this way may fit a hyperbolic or exponential family of curves (*see* White, 1976). Many of these functions fit the equation $I = I_r (1 + C/t)$. I represents the current, $I_r$ is the threshold or rheobase current, $t$ is the time, and C the chronaxie. The chronaxie is the pulse duration on this function at twice the threshold current. Different fiber populations have different chronaxies (*see* Ranck, 1975). All CNS myelinated fibers have chronaxies of 50–100 µs; thus, to stimulate myelinated fibers, pulse durations of 50–100 µs are adequate when dc stimulation is used. A typical strength–duration curve is illustrated in Fig. 6.

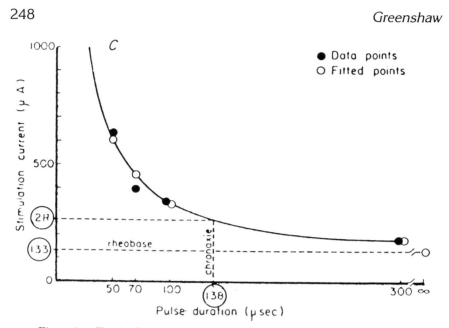

Fig. 6. Typical representation of a strength–duration function: Both observed values and theoretical values are displayed. Note the rheobase is the threshold current. The chronaxie is represented by the intercept of twice the rheobase with the curve relation to pulse duration (from White, 1976, reproduced with permission of the publisher and author).

### 2.8.3. Current–Distance Relationships

In predicting the responsiveness of nerve cells to extracellular stimulation it is necessary to consider current–distance relationships. Ranck (1975) has reviewed the literature on current-distance relations in the CNS. This author has suggested that inadequate attention is paid to these considerations and has given detailed instructions on the possible use of available data. Quite simply, the sensitivity of axons to extracellular stimulation will be related to (a) the diameter of the axon, and (b) the distance of the axon from the stimulating electrode. Two axons of identical diameter will exhibit responses that vary as a function of distance from the stimulating electrode. With axons of different diameter, however, the larger diameter axon will show greater sensitivity, i.e., be stimulated by a lower current than a smaller diameter axon at the same distance from the electrode. Thus, information related to the relative diameter of the axons that are to be stimulated and their spatial relation to the stimulating electrode may allow useful predictions to be made concerning the population of axons stimulated by a given electrode.

## 2.8.4. Anodal-Surround Effect

A number of problems associated with extracellular stimulation, some of which are less well known than they should be, should be taken into account. One such factor is the anodal-surround effect. As outlined earlier, when cathodal extracellular stimulation is applied, only a small quantity of current enters the intracellular space. Thus, extracellularly applied current flows mainly through the extracellular space. Reduction of the membrane potential caused by outward current induces neural stimulation and this effect is balanced by a reentry of the same amount of inward current in other parts of the neuron. As the neuron is depolarized increasingly at one site, hyperpolarization is increased at some other site. The consequence of this is what has been called "anodal-surround blocking." Large stimulating currents induce sufficient hyperpolarization to block the propagation of action potentials, an effect that is not seen at smaller currents. With a monopolar cathode, therefore, there is a spherical field of stimulation, the inner core of which may not result in an increase of neural activity, but rather will serve to block the propagation of action potentials. This phenomenon is particulary important when one is trying to relate the effects of stimulation to electrode site. The anodal-surround effect is discussed in detail by Ranck (1975, 1981) and the reader is recommended to consult these excellent reviews for further information. The interested reader should consider the relevance of current–distance relationships for anodal-surround blocking (*see* Ranck, 1981).

At this point it should be noted that anodal (i.e., positive) stimulation may also be effective in facilitating neural activity. Here, the converse of the cathodal stimulation case applies: Hyperpolarization occurs at the anode, and at some other part of the neuron, depolarization occurs. In most cases, far more anodal current than cathodal current is required to stimulate neurons. An exception to this is when cell bodies and dendrites are being stimulated (*see* Ranck, 1975).

## 2.8.5. Electrode Size

Small electrode tips may be desirable for localized stimulation. Although it is clear that there are advantages to this in terms of differentiating between closely related groups of cells, there is a trade-off between localization and the electrolytic tissue damage caused by cathodal stimulation with very small electrodes (at which extremely high-current density may be achieved). The choice between tissue damage and greater localization of stimulation sites should be assessed in relation to the particular application of interest. When stimulation is applied repeatedly and chronically, as in some behavioral experiments, electrolytic damage induced by very small electrodes may be prohibitive (*see* Asanuma and Arnold, 1975).

### 2.8.6. Electrode Orientation

The orientation of electrode tips when bipolar electrodes are used is an important factor: When stimulating axons, the orientation should ideally be parallel to that of the axonal projection. Indeed, when dc pulses are used, it is extremely important to maintain the polarity of stimulation, i.e., for most applications the active electrode should always be the cathode (this is particularly important when using bipolar electrodes under conditions in which connectors can be coupled to the electrode in either direction). Details concerning the size and shape of the electrode tips should be specified, i.e., whether the electrodes are bared for any distance or are insulated except for a cross-sectional area at the tip.

## 2.9. Typical Applications

Electrical stimulation of the brain in vivo is a procedure that has been adopted for a variety of purposes. An excellent review that outlines many aspects of the application of this technique has been written by Doty (1969). Overt behavioral responses to 'electrical brain stimulation in a variety of species represent perhaps the highest level of analysis of stimulation-induced changes in brain activity. There are many aspects of the application of brain stimulation in neuroscience and for practical reasons only a limited illustration is possible in the present chapter. The ultimate goal of these applications is to understand the relationship between the activity of specific neural systems and behavior. Two levels of analysis have been chosen for a brief description. On the one hand, experiments are briefly mentioned that attempt to correlate electrical stimulation of specific groups of neurons to neurochemical changes in different brain regions. At another level, a behavioral analysis of the responsiveness of stimulation sites is briefly outlined.

In this context perhaps the most intense research area is currently related to the analysis of neurochemical activity that correlates with the activiation of intracranial reward or reinforcement sites. A number of authors have analyzed changes in central catecholamine metabolism in response to electrical stimulation of different brain areas, in particular those sites that may act as positive reward sites in the rat. Garrigues and Cazala (1983) have recently analyzed rates of intracranial self-stimulation responding in two inbred strains of mice selected for high and low catecholamine activity, respectively. These authors have observed that the strain with higher catecholamine activity exhibits higher self-stimulation response rates with equivalent electrode placements, relative to the strain with low catecholamine metabolism. These authors then assessed the effects of rewarding stimulation on catecholamine metabolism with electrodes implanted in either the dorsal or ventral part of the lateral hypothalamus. Significant enhancement of catecholamine turnover was noted in nerve terminal regions in hippocampus, cortex, and nucleus accumbens in these animals. These results were interpreted as evidence for the involvement of the dorsal noradrenergic and mesolimbic dopaminergic systems in lateral hypothalamic self-stimulation. Using a similar approach, van Heuven-Nolsen et al. (1983) have investigated the effects of electrical stimulation of the ventral-tegmental area on catecholamine metabolism in discrete regions of rat brain. These authors, adopting a strategy of stimulating cell bodies in, and lateral to, the ventral-tegmental area, report different patterns of dopamine or noradrenaline utilization in terminal regions of the axonal projections of these cell bodies and in terminal regions of the dorsal-noradrenergic bundle, respectively. At the sites of the electrodes an enhanced turnover of noradrenaline was also observed. These authors, in agreement with the previous study, concluded that activation of parts of mesocorticolimbic-dopaminergic projections and of the dorsal-noradrenergic projection is correlated with intracranial self-stimulation. Using a more global approach, Yadin et al. (1983) have assessed the possibility that a common neural substrate for the effects of medial forebrain bundle self-stimulation and extra-diencephalic self-stimulation may be revealed by an analysis of neural activity involving 2-deoxyglucose autoradiography. In this study, rats with electrodes in posterior or anterior medial forebrain bundle, the medial prefrontal cortex, or the locus ceruleus were allowed to self-stimulate for a 45-min period following the injection of $[C^{14}]$-2-deoxyglucose. They were then killed and their brains assayed by an autoradiographic technique. No overlap was reported between activity induced by medial forebrain bundle stimulation and that induced by stimulation at sites outside the diencephalon. Thus, Yadin et al. propose that it is unlikely that there is a common substrate to the effects of medial forebrain bundle stimulation and stimulation of extradiencephalic sites. A similar form of analysis, also using an autoradiographic analysis with labeled 2-deoxyglucose, has recently been adopted by Porrino and colleagues (1984).

On another level, behavioral responses have been monitored in relation to changing specific parameters of brain stimulation. The effectiveness of stimulation is related to changes in behavior. An elegant example of this kind of work is to be found in White's strength-duration analysis of reinforcement pathways in the medial forebrain bundle of rats (White, 1976). In this study, White carried out strength-duration analysis as described earlier, i.e., plotted the relationship between current and pulse width (for the cathodal component of biphasic dc stimulation) in relation to different behavioral tests. In these experiments White assessed the relative efficacy of reinforcing stimulation when animals were required to emit different responses contingent on delivery of medial forebrain bundle stimulation. White compared chronaxie estimates in these different situations and reported that there were reliably different chronaxies for the strength-duration analysis comparing self-stimulation in a situation requiring a tail movement response with that requiring a runway response. There were no reliable differences between the chronaxies derived from other behavioral responses. White has suggested that because chronaxies are proportional to time-constants of excitation of stimulated neurons and to current–distance relations (a consistent relationship between stimulation points and relevant neurons being unlikely), this chronaxie difference indicates that the reinforcement of tail movement and alley running responses, respectively, involve independent sets of neurons that are activated by the same electrodes. Although there are certain assumptions implicit in this analysis, it is indicative of possible interpretations of data from experiments that attempt to assess the nature of the neuronal population involved in a particular behavioral response, by varying certain parameters of the stimulation that is used. Further examples of this kind of work are to be found in studies published by Bodnar and his colleagues (1982) using pulse pairs (*see* Deutsch, 1984 and Gallistel et al., 1969) and multiple stimulation sites, and by the studies of Bielajew and her colleague employing a behavioral analog of classical collision testing (*see* Bielajew and Shizgal, 1982). Although some researchers may find the assumptions implicit in this form of analysis too far-reaching, these studies do illustrate the rather challenging problems that are being analyzed with the elegant approach of varying parameters of stimulation in relation to the measurement of simple behavioral responses.

# 3. Chemical Stimulation

## 3.1. General Appraisal

The use of techniques that allow the application of compounds directly into brain tissue is central to the study of CNS function. The reasons for this are twofold; first a number of compounds of interest do not pass the blood–brain barrier and, therefore, without applying these substances intracranially it is impossible to assess their influence on brain activity. Furthermore, whereas the systemic administration of pharmacologically active compounds allows an analysis of their effects on the system as a whole, it is difficult to analyze those components of brain activity that may underlie the syndrome of change induced by peripheral drug application. The use of intracranial chemical stimulation circumvents both of these problems to a large extent. This approach can, therefore, be viewed at two levels. In the broad sense, direct application of chemicals into brain tissue or into the cerebroventricular system provides a means of assessing the central effects of a wide range of compounds. At a more sophisticated level, the objective of chemical brain stimulation techniques is to assess the functional role of different chemicals in the CNS. In the latter case we are confronted with a number of problems related to the establishment of a physiological role for pharmacologically active substances. Although the techniques described here do play a significant role in this attempt, there are a number of fundamental problems related to establishing physiological functions of centrally applied compounds. Nevertheless, when the results of experiments involving chemical stimulation of brain tissue using the present techniques are compared with the related body of literature in neuroscience, evidence gleaned from the present approach does contribute significantly to our view of the CNS function of various endogenous substances, particularly at the level of their effects on behavior.

Myers (1974) has discussed many of the problems of assessing the effects of chemical stimulation of the brain. The reader is directed to his volume, *Handbook of Drug and Chemical Stimulation of the Brain,* for an alternative and detailed discussion of these problems and also for a source of experiments dealing with these techniques up until 1973. As Myers has pointed out, any compound may excite or depress neuronal activity in the brain or affect temporal characteristics of neural firing. These effects may be related to normal processes underlying neural regulation or possibly to local anesthetic or other effects. Here it is interesting to note that one largely unexplored phenomenon that may contribute to microinjection effects is the induction of spreading waves of neuronal depolarization (i.e., spreading depression: Huston and Jakobartl, 1977; Sprick et al., 1981; *see* Oitzl and Huston, 1984 for a detailed analysis of this problem). This question of mode of action is made more complex by the now evident plethora of mechanisms for controlling neural activity (Dismukes, 1979). The issue of whether a substance is acting on some feature of synaptic function is usually resolved by assessment of whether the substance in question occurs endogenously, whether it has effects on membrane potential, and whether or not the substance is stored or released in classical ways (*see* Florey, 1967). These considerations are of primary importance in attempting to establish some physiological response to chemical stimulation of the brain.

At this level it is possible to assess the effects of intracranial chemical manipulations in relation to the known effects of general systemic applications of compounds, i.e., either substances in question themselves or drugs known to influence the system that is being investigated. At the present time the large body of accumulating data relating chemical action to specific receptor sites and particular brain pathways from immunohistofluorescence studies has been invaluable in helping us to relate the effects of central injections to endogenous neurotransmitter systems. For example, the use of a relatively specific antagonist for a defined neurochemical system (such as atropine or scopolamine in the case of muscarinic cholinergic systems) may allow us to demonstrate, with the systemic administration of an antagonist, a reversal of a centrally induced effect. Alternatively, an alteration in known neural systems may be used to modify responses. For example, lesions of cell bodies may be used to modify the response to substances injected into the terminal fields of their axonal projections. Current knowledge about the development of receptor supersensitivity allows us to predict changes in the response to compounds that are known agonists at these receptors in the lesion vs nonlesion conditions. Similarly the demonstration of anatomically specific effects of intracranial injections allows us to analyze the relationship between known neuranatomical projections associated with certain neuractive substances and the response to centrally applied compounds. In the latter case there are problems related to the diffusion of the injected compound through brain tissue; this issue of lipid solubility will be dealt with later.

## 3.2. Microinjection Systems: Basic Components

The origins of microinjection systems have been attributed to Hashimoto (1915) who described the first contemporary cannula system that he used for injections into the diencephalon of the unanesthetized rabbit. The basic approach that Hashimoto adopted has not changed in principle, although it has been greatly refined in recent years. His approach was to secure a metal needle to the skull so that its tip was located within the structure of interest. This could be coupled to a syringe with a smaller diameter injection needle that could be inserted into the needle (guide cannula) that was implanted in the animal's head. In this way Hashimoto was able to apply drugs repeatedly into rabbit diencephalic structures without anesthesia. The model cannula system on which most cannulae are currently based was adapted from that of Hashimoto by von Euler and Holmgren (1956). The basic design of this cannula system is illustrated in Fig. 7. The guide cannula is implanted stereotaxically and, after recovery, the microinjection cannula is inserted to a predetermined depth below the tip of the guide cannula and a known volume of drug solution is slowly infused into the injection site using a microliter syringe.

### 3.2.1. Cannulae

Microinjection cannulae are commercially available. These commercial systems are both convenient and efficient, although it is possible to construct microinjection cannulae from stainless steel tubing derived either from hypodermic needles or from a supply of fine gage stainless steel tubing. Typically, for guide cannulae, tubing of 26 or 28 gage is used, with internal cannulae being constructed from 28- or 30-gage tubing, respectively. The length of the guide cannula will be determined by the species and depth of microinjection; for the rat, 15-mm guide cannulae are adequate for all applications. The guide cannula is implanted stereotaxically into the area of interest and held in place, as in the case of electrode systems, with dental acrylic anchored to the skull with stainless steel screws. One important point about the construction of guide cannulae in the laboratory is that these devices should have some kind of enlargement at the top end to provide

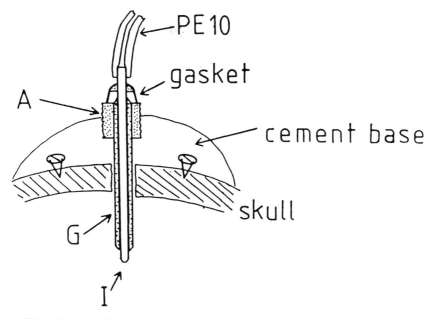

Fig. 7. (A) Schematic representation of a microinjection assembly
*in situ*. Abbreviations: G, guide cannula (23 g); A, enlargement on guide
cannula to provide firm anchorage to dental cement; I, internal or
microinjection cannula (30 g). The PE 10 tubing connects I to the
microsyringe as shown in (B).

firm anchoring within the dental acrylic. The reason for this is
self-evident: smooth stainless steel tubing will not reliably adhere
to dental acrylic and any downward pressure will cause displacement of the cannula in a downward direction, thus rendering the
cannula useless and damaging the microinjection area and lower
sites. This may be overcome by constructing the guide cannula
from a hypodermic needle and leaving part of the leur-lock fitting
at the top of the needle as a head that may be held firmly by the
acrylic head cap. Alternatively, the stainless steel guide may be
roughened along the upper sides and either epoxy resin or a suitable silver solder may be used to form a ridge that will prevent
movement of the guide cannula against the acrylic head cap. It
should be noted that when constructing guide cannulae from
stainless steel tubing, the edges of the cannulae should be ground
finely and examined under a microscope to ensure the removal of
steel splinters and jagged edges. After implantation of the guide
cannula a protective mandrel or stylet of stainless steel should be
inserted (to a level equal to the tip of the guide) to prevent guide
blockage and necrosis caused by exposure of the cerebral tissue.

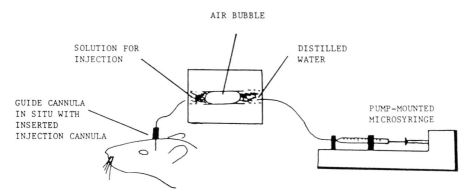

Fig. 7 (*Cont.*)　(B) Schematic view of the overall microinjection system. Note the enlarged view of the PE10 tubing represented in the figure: The air bubble serves to separate the drug-solution from the distilled water filling the "syringe-side" of the PE10 line. This bubble also serves as a useful indicator of fluid movement and pressure changes in the system.

The microinjection cannulae may be constructed similarly; the length of these will be determined by the procedure adopted by the experimenter. It is conventional to use a microinjection cannula that is either the same length as the guide cannula or projects 0.5–1.0 mm below the tip of the guide cannula. Although Routtenburg (1972) suggests that equal length of guide and microinjection cannulae is desirable, this does cause problems related to the diffusion of injected fluid back up the space between the guide cannula and the microinjection cannula. There is, of course, a trade-off between tissue disruption induced by a long microinjection cannula vs possible backflow of the injected solution. It should be noted at this point that backflow may also occur along the track of the guide cannula outside the shaft: In relation to this problem, the use of dorsal placements is useful as a control procedure.

### 3.2.2. Tubing

There is a variety of material available for tubing that forms a flexible connection between the microinjection cannula and the microinjection syringe. Any biologically inert flexible tubing is suitable for this application; however, the internal diameter of the tubing is important. PE10 (internal diameter, 0.38 mm) is commonly used under these conditions. Slightly smaller bore tubing may also be employed.

### 3.2.3. Syringes

A variety of small volume syringes are commercially available. The overall volume of the syringe will be determined by the required injection volume and the number of injections used at any one time. This actually relates to whether intracerebroventricular application or intracerebral application of solutions is being considered. In the intracerebroventricular case, quite large volumes, i.e., up to 10 μL may be used, whereas in the intracerebral case a maximum value of 1 μL should be considered, although some researchers have used larger injection volumes. Intracerebral applications, using microliter syringes with an injection volume of 0.5–1 μL, are considered conventional; 0.5 μL is a preferable volume when tissue damage caused by displacement is considered. Nevertheless, under certain conditions it is necessary to use 1 μL to achieve a large enough quantity of injected compound.

### 3.2.4. The Basic System

In considering all of these components together the system consists of a microliter syringe connected to flexible tubing that is linked to a microinjection cannula. The microinjection cannulae sits tightly in the guide cannula and may project usually 0.5–1 mm below the tip of the guide cannula into brain tissue. Obviously, it makes no sense to fill the entire system with drug solution that you wish to inject, particularly when dealing with compounds that may be extremely costly or difficult to obtain, such as peptides. To avoid the need for this, a procedure is adopted in which a small bubble of air is used as an interface between the distilled water that fills both the microinjection syringe and the injection line and the drug that is located in the microinjection cannula (*see* Fig. 7B). An added advantage of this approach is that the movement of the small air bubble that serves as the interface between the two fluids may be used as an indication of the flow of the infused solution from the microinjection cannula into the brain tissue. This air bubble, which is usually between 0.5 and 1 cm in tubing length, also serves as a pressure buffer that absorbs some of the large pressure changes that may be associated with free-hand injections, although this is not a problem when automatic injection devices are used. It should be stressed that leakage may occur even with commercial microinjection systems. This may be easily overcome by testing the entire system prior to use. Small Teflon gaskets and the use of a suitable adhesive or sealant are simple solutions to this possible problem.

## 3.3. Automatic Pumps and Free-Hand Injections

Automatic pumps are used in microinjection applications because tissue damage may be minimized by the use of a long injection time.

For some applications free-hand injections are also employed. In this case, the microliter syringe is simply controlled by the experimenter and the injection rate is controlled approximately over a set time. Typically, for the injection of compounds into brain tissue, an injection time of 3 min is employed. Some researchers use shorter periods; however, this seems largely unwarranted. Problems of variability in infusion rate that may effect the degree of tissue damage and the rate of diffusion of applied substances from the tip of the injection cannula may be minimized by the use of an automatic delivery system. The automatic approach is particularly important when longer-term infusions are considered. Such situations may arise with the use of prolonged infusions of substances into the cerebroventricular system; these applications are virtually impossible without an automatic pump system. The rate of delivery, which will vary depending upon injection volume, is controlled by electrical or mechanical means. It should be noted that mechanically controlled systems are usually more accurate than electrically controlled systems. Pumps with mechanical gears are therefore preferable for these applications. It is, of course, possible to use larger capacity syringes with automatic pump systems; however, since the accuracy of the pump is proportional to the distance traveled by the syringe plunger, it is preferable to adapt these pump systems for microinjection syringes. There are a number of automatic pumps available that are suitable for microinjection syringes; nevertheless, in most laboratories general-purpose pumps are more readily available and these machines can be easily adapted to hold microliter syringes. The volume of solution infused with a pump in unit time should be determined in preexperimental tests by measuring the displacement of a known quantity of fluid from a microliter syringe over time; this may be achieved by measuring the distance traveled by the syringe plunger. With automatic pumps there is an intrinsic error related to the lag-time with which the pump operates. Under most circumstances this error will be negligible because a relatively long infusion time is required for a relatively small volume; thus, over a period of 180 s a lag-time error for switching a pump on or off in the order of up to even 2–3 s will not introduce much error in the infusion volume. With respect to this kind of variation, it is evident that the internal diameter of flexible tubing that connects the syringe to the microinjection needle may play a certain role, i.e., if the tubing is bent with a wide internal diameter, this may cause some variation in volume displacement at the microinjection tip. Nevertheless, with tubing of small internal diameter, this problem is minimized. After the microinjection is complete the injection needle should remain in place for approximately 1 min to minimize backflow into the guide cannula. After removal of the injection needle, the protective stylet should be replaced.

### 3.3.1. An Extra-Fine Assembly for Intracerebral Microinjections

As outlined earlier, the size of the internal or microinjection cannula may be determined by the experimental situation. A 30-gage microinjection cannula will cause less damage from displacement than a 28-gage microinjection cannula; nevertheless, with a 30-gage microinjection cannula there will be more problems related to pressure damage resulting from the injection. In certain situations the size of the microinjection cannula may be critical, particularly when it is necessary to distinguish between closely related microinjection sites. For this purpose the use of very fine glass microinjection needles may be particularly advantageous.

Glass microcapillaries have rarely been used for applications of chemicals or drugs into brain tissue of conscious animals. The main reason for this is that fine glass microinjection needles are extremely fragile and, considering problems of movement and microinjections, in freely moving animals this is usually prohibitive. An attempt to circumvent the problem of breakage has been reported by Aghajanian and Davis (1975). In this study, animals were implanted with filled microcapillaries for iontophoretic application of carbachol. Thus, tissue damage was minimized by the use of a fine tip and problems of removing and replacing microinjection cannulae were apparently circumvented by the iontophoretic application of the chosen compounds. Although the authors of this report clearly demonstrated a reliable behavioral response to iontophoretically applied carbachol, this procedure has not been successfully adopted by other laboratories. Indeed, there are a number of problems associated with possible use of iontophoresis in freely moving animals. These are beyond the scope of the present chapter, but many of the issues (e.g., effective use of retaining current) will be self-evident in an account of iontophoresis (*see* MacDonald in this volume).

Problems of breakage have, nevertheless, been largely circumvented by a procedure reported by Azami et al. (1980). As described later (*see* illustrative applications), these researchers wished to distinguish between the effects of microinjections into two very closely related nuclei in the brain stem of the rat. To achieve this, Amazi et al. used a very fine glass microinjection assembly. This assembly is illustrated schematically in Fig. 8. Briefly, a guide cannula is made from a 23-gage disposable hypodermic needle, the luer-lock fitting of which is converted to a

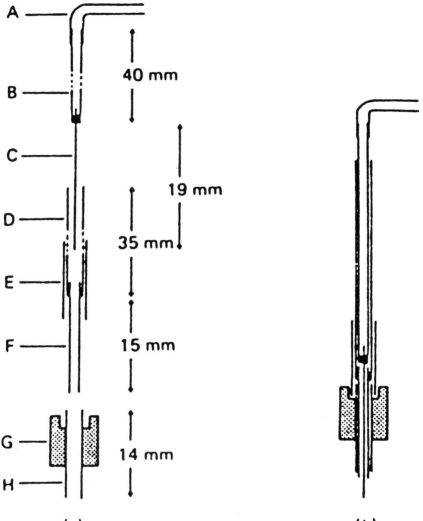

Fig. 8.   The microinjection assembly. (a) An exploded view: A, Portex tubing (PE10) connected to a microinjection system; B, 30-gage steel tube; C, glass capillary; D, 23-gage steel tube; E, 19-gage steel tube; F, 30-gage steel tube; G and H, guide cannula; (b) the assembly in position for microinjection, total length 60 mm (from Azami et al., 1980, reproduced with permission of publisher and authors).

head that serves to limit the length of insertion of the microinjection assembly (G). The guide cannula is chronically implanted so that the tip is 4 mm above the target site; a 30-gage stylet is used to keep the cannula free of blockage. The microinjection cannula is constructed from a 10-μL glass pipet heated and pulled to an outer diameter of 70–90 μm. This fine needle is fixed into a 30-gage steel tube and its length adjusted so it will project from the tip of the guide cannula by 4 mm (i.e., to the target site). Before microinjection, the glass capillary is inserted into the protective assembly labeled (DF) in the diagram and retracted until it is protected by (F). The limiting collar (E) is pressed into the groove of the head of the guide cannula (G) and the capillary is lowered into position. (D), (F), and (E) prevent lateral movement of the capillary and thus prevent breakage of the fine tip. Azami and his colleagues report that a 0.5-μL injection of 0.5% pontamine sky blue resulted in a visible spot 0.3 mm in diameter, suggesting that the application of fluid from this system is very localized.

### 3.4. Repeated Intracerebral Injections: The Electrolytic Microinfusion Transducer (EMIT) System

There are a number of problems associated with microinjection simply in terms of mechanical considerations. Apart from pump accuracy and displacement caused by the movement of connecting tubing, repeated discrete infusions of compounds with relatively large volume (0.5–1 μL) induce progressive damage to neural tissue. This effect combined with repeated displacement of tissue by insertion of the microinjection cannula itself and replacement of the protective stylet that is used to keep the guide cannula patent may lead to decreased sensitivity of the microinjection sites to pharmacological treatments. Some researchers have attempted to control for this by repeatedly administering a standard concentration of a compound with known effects at the particular site. In this way, by demonstrating a repeated response to a drug throughout a series of microinjections over a period of weeks, some researchers have demonstrated that their microinjection procedures do not disrupt the site of stimulation. Nevertheless, it would seem preferable to circumvent these problems by the use of finer intracerebral injection techniques. Recently, one such technique has been exploited quite successfully. This procedure uses the EMIT or electrolytic microinfusion transducer system.

The EMIT was first reported by Criswell (1977). Bozarth and Wise (1980, 1981) have exploited and developed this idea, illustrating its utility for CNS microinjections. The system is schematically illustrated by Fig. 9. The microinjection needle is connected to a reservoir that contains a certain volume of solution to be injected. A silver anode and platinum cathode project into this

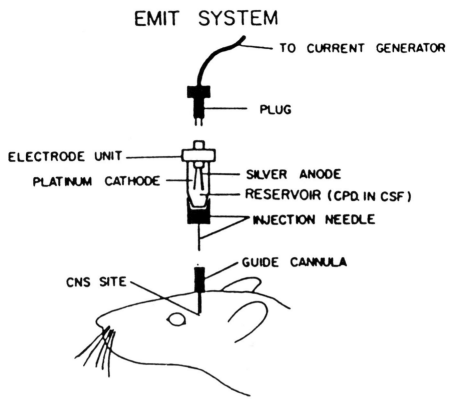

Fig. 9. The electrolytic microinjection transducer system. Micro-infusions are produced by passing a direct current (200 μA) between the silver anode and platinum cathode, with the resulting evolution of hydrogen gas forcing reproducible volumes of compound solution (CPD in CSF) through the injection cannula (from Goeders et al., 1984, reproduced with permission of publisher and authors).

reservoir, forming an electrode unit that is connected to a constant current source. Microinfusions are produced by passing 200 μA dc between the silver anode and the platinum cathode contained in this air-tight drug reservoir. The resulting evolution of hydrogen gas forces a reproducible amount of the drug solution out of the injection cannula into the microinjection site. With this procedure, a small current (6 μA dc) is used to prevent the redissolution of hydrogen that has been evolved in previous infusions.

Recently Goeders and his colleagues (1984) have calibrated this system with a radioactive procedure and report the reliable delivery of 100 ± 7 nL/microinjection. This system seems to hold great promise for repeated microinjections into brain tissue and it is notable that the design is now commercially available from

Plastic Products Inc. (Roanoke, VA, USA). Nevertheless, there may be some limitations to the use of the EMIT system. Although Goeders et al. have established that for certain substances the electrolytic process for infusion does not result in a breakdown of the compound of interest, it is possible that this procedure may result in the oxidation of certain compounds, particularly catecholic substances, such as dopamine, noradrenaline and some related compounds, and 5-HT (serotonin). This is evident because oxidation induced in a high-performance liquid chromatography unit with an electrochemical detector uses currents within this range. Thus, it would seem necessary to ensure, prior to using the technique, that the electrolytic process involved does not lead to a breakdown of the compound of interest.

Because the procedure is a fairly new one, few data are available related to this particular problem. Nevertheless, in terms of accuracy for small injection volumes this system is definitely worthy of further assessment and application in the CNS. Indeed, it is evident that if the EMIT system could be coupled to a fine microinjection system such as that described by Azami et al. (1980), it may be possible to provide maximal localization with minimum mechanical damage to tissue. A reliable low-volume pump system built along more conventional lines has recently been reported for volumes of 12.5–50 nL/infusion (Iwamoto et al., 1984). For compounds that will not resist the electrolytic process in the EMIT system, this may prove a viable alternative (*see* also section 3.6). In relation to repeated chemical stimulation of the CNS, it is interesting to note the successful application of miniature osmotic pumps for chronic microperfusions of approximately 0.5 μL/h over periods of up to 14 d (*see* Urquhart et al., 1984, pp. 207–210, for a recent review).

### 3.5. Necessary Considerations for Solutions for Microinjection

There are a number of factors related to microinjected solutions that should be taken into account when considering the specificity of experimental facts. The first of these, *volume*, has been briefly discussed above. It is evident that the volume of microinjections should be kept relatively small to avoid tissue damage. Nevertheless, there is a difficult problem in terms of the trade-off between volume and concentration. A large volume will cause mechanical disruption of brain tissue; alternatively, extremely high concentrations may cause damage related to osmotic gradients.

It may be argued that since compounds are being applied directly into brain tissue, one might expect the necessity for only low concentrations, particularly because some of the substances of interest occur in relatively small quantities within the brain. Nevertheless this is clearly not the case. As Myers (1974) has previously pointed out, quite high (in some cases what reasonably could be considered excessive) concentrations are required to achieve experimental effects. Myers (1974) has suggested that this is not surprising in view of the compartmentalization of the CNS and the fact that the synaptic cleft must, to a large extent, be a functionally discrete area. Whatever the theoretical position may be, it is evident that large quantities of compounds are required to achieve central effects in many cases. In view of this we are confronted with the problem of choosing volume or concentration. The *osmolality* of the injected solution is therefore of paramount interest to us. It is, however, extremely difficult to find appropriate controls for osmotic pressure. As suggested earlier, most substances may influence neural firing and it is extremely difficult to assess to what extent these substances affect specific processes related to neural regulation and to what extent they lead to some nonspecific disruption of membrane integrity *per se*. High concentrations of sucrose have been used to control concentration effects. At first sight it may seem appropriate to simply use a hypertonic solution that is balanced in ionic composition relative to extracellular fluid [i.e., hypertonic artificial cerebrospinal fluid (csf) preparations]. Nevertheless, high concentrations of extracellular potassium that would be achieved with such manipulations may result in irreversible cell damage (*see* Bourke et al., 1980, pp. 100–102). Furthermore, other ion imbalances, of course, have marked consequences for integrated CNS function (*see* Myers, 1974, pp. 263, 449). This is perhaps the most difficult problem in considering data obtained through the use of microinjection procedures, and one for which there is no clear solution. It may be noted, however, that in the case of active isomers of a compound (e.g., α- and β-flupenthixol, respectively, in relation to dopamine receptors), the inactive isomer is an effective agent for an assessment of concentration effects.

The pH of injected solutions is something that may be more easily controlled by the use of buffer systems; indeed, artificial csf has significant buffering capacity. Microinjection of vehicle solutions, of course, allows an exclusion of the effects of their constituents as determinants of the experimental effects.

*Lipid solubility* of injected compounds is of paramount importance in the present context. This factor is important, of course, because of attempts to localize the site of action of different compounds at the area around the microinjection cannula. Highly lipophilic compounds will tend to spread throughout brain tissue fairly rapidly and may therefore, to some degree, preclude the demonstration of site-specific effects. It is notable that the use of a hydrophilic antagonist has been employed to demonstrate that a more lipophilic agonist may be acting at a particular site (Britt and Wise, 1983). The degree of spread of compounds from an injection site may be estimated by ex vivo analysis, typically using a spread of radiolabeled substance as an index of the degree to which a substance spreads through brain tissue. This has, to a certain extent, been assessed for a number of compounds (e.g., *see* Iwamoto et al., 1984; Cox et al., 1983; Evans et al., 1975; Goeders and Smith, 1984; Myers and Hoch, 1978).

It should be noted, at this point, that crystalline application of compounds has been employed in this context. Nevertheless, because of interpretative problems both of nonspecific effects and of the quantitation of applied compounds using this procedure, crystalline application has become less popular and is now rarely used. In relation to this point, the relative solubility of compounds is an important determinant of their suitability for microinjection. As a final point in this section the *chemical stability* of the applied stimulus is of paramount importance. This is a fundamental issue in pharmacology and is, of course, usually dealt with by control of pH and temperature.

### 3.6. A Note on Intracerebral Dialysis

Recently, dialysis has been proposed as an innovative technique for chemical stimulation of local areas of brain tissue. The basic idea is that a small diameter dialysis tube is inserted into brain tissue: By pumping artificial extracellular fluid through it, the collection of neurochemicals from the outflow stage and the application of chemicals and drugs at the inflow stage, is possible. The recent "renaissance" of this technique was initially reported by Tossman and Ungerstedt (1981). The dialysis method has subsequently been adopted by other researchers for the analysis of neurochemical release and effects of chemical stimuli contained in the artificial extracellular fluid (Johnson and Justice, 1983; Imperato and DiChiara, 1984). It must be clearly acknowledged, however, that this is not a new technique. The implementation of a microdialysis system for the injection and collection of chemicals in brain was first reported by Delgado et al. (1972). These authors pointed out that the dialysis approach offers a minimization of the risk of infection and helps to avoid the possible blockage of cannulae and mechanical damage from liquid injection.

## 3.7. Intracerebroventricular Injections

Intracerebroventricular application of drugs or chemicals is a particularly useful procedure because it allows a general analysis of the effects of substances that may not cross the blood–brain barrier or an assessment of central effects of certain compounds. With these intracerebroventricular applications, tissue damage is less of a problem than in the case of repeated injections into brain tissue. Nevertheless, the correct localization of ventricular cannulae is of paramount importance. A technique that is reliable and elegant in its simplicity has been proposed by Walls and Wishart (1977); this technique is an extension of a previous method proposed by Goodrich et al. (1969). The original procedure employed infusion of liquid under pressure such that the pressure change occurring at the time of ventricular penetration indicated a successful puncture. Walls and Wishart have pointed out that such a method requires the use of sensitive pressure measurement devices and that this pressure method may also create problems of tissue damage.

With the procedure of Walls and Wishart, a 24-gage guide cannula and a 30-gage internal cannula are used. The small size of the internal cannula ensures that should the cannula tip be imbedded in tissue rather than in the ventricular space, tissue damage will be minimal since little aspiration is likely to occur with a 30-gage internal cannula. The cannula tips are beveled at an angle of 45–55° and prepared so that the internal cannula protrudes approximately 0.2 mm beyond the guide cannula. The beveled angle of the cannula tips is recommended to facilitate puncturing of the ventricular wall. The original procedure is described for implantation of cannulae into the third ventricle; however, this procedure may be employed for the lateral ventricle with ease. For the third ventricle the guide and internal cannula are positioned together in a stereotaxic frame, the arm of which is angled 10° laterally from the vertical, with the tip of the beveled guide pointing toward the midline. A short length of tubing (PE20 was recommended in the original method, but the present author has found PE10 to be more effective) is filled with distilled water, except for the space that represents the length of the internal cannula. This tubing is connected at the other end to a 10-μL syringe. The cannula assembly is lowered 9 mm from the skull's surface (1 mm posterior to bregma and 1.5 mm lateral to the sagital suture, for the third ventricle) and then retracted 0.3 mm. The syringe plunger is withdrawn by approximately 3 μL. Cerebrospinal fluid should become visible in the tubing at the junction with the internal cannula. If fluid is not visible in the tube, the plunger is returned to its original position and again withdrawn approximately 3 μL. With accurate placing of the cannula into the ventricular space this procedure usually yields csf on the first or second try. If this attempt fails, the withdrawal procedure is repeated. When csf is being drawn up into the internal cannula, the distilled water in the PE10 tube moves smoothly and csf may be drawn into the tubing and pushed back into the ventricular space with ease.

This procedure represents a very simple and effective means of accurate localization of ventricular cannulae that should be employed wherever possible.

One advantage of intracerebroventricular administration is that a fairly large volume (up to 10 μL) may be delivered within a period of 3 min. This enables the administration of relatively insoluble chemicals or compounds that may require high dosages for effective treatment. As outlined earlier, the trade-off between volume and concentration is a particular problem when we consider intracerebral applications of chemicals and drugs.

## 3.8. Typical Applications

The utility of microinjection procedures for an analysis of the effects of chemical stimulation of the brain is abundantly clear from an examination of Myers' (1974) authoritative handbook *Drug and Chemical Stimulation of the Brain*. The experiments, which are briefly outlined in this section of the chapter, are chosen because they are good examples of certain features of microinjection applications in neuroscience.

By far the most challenging problems encountered with microinjection techniques are, on the one hand, the specificity of the response to chemical stimulation in pharmacological terms and, on the other, the specificity of the response in anatomical terms. A recent study investigating a possible physiological role of indoleamines in thermoregulation provides an illustration of some aspects of these problems. Cox et al. (1983) have investigated the effects of intrahypothalamic injection of either 5-hydroxytryptamine (serotonin, 5-HT) or tryptamine (T). In this study rats were stereotaxically implanted with stainless steel guide cannulae (0.5 mm external diameter) into the hypothalamus. The tip of the guide cannula was placed 3 mm above the desired injection site. Drug injections were made 7 d after recovery via a microinjection cannula that extended 3 mm from the guide cannula. The volume of each microinjection was 1 μL injected over a period of 45 s; this is much shorter than the recommended injection time. On completion of the experiments, 1 μL of Indian ink was injected to facilitate the identification of injection sites by histological examination.

Both 5-HT- and T-sensitive sites were located within the same region of the preoptic area of the hypothalamus. When rats were tested at different ambient temperatures, intrahypothalamic application of 5-HT induced a decrease in core temperature in rats maintained at 4°C; smaller responses were obtained at higher temperatures. Tryptamine induced a hyperthermic response in animals kept at 20°C, but had no effect in rats maintained at 4 or 29°C. The hyperthermic effect of 5-HT was antagonized by systemic pretreatment with cyproheptadine (2.5 mg/kg), but not by methergoline (0.625 mg/kg) or methysergide (0.2 mg/kg). The hyperthermic effect of T was blocked by methergoline and methysergide, but was unaffected by cyproheptadine. These data were interpreted as evidence for a serotonergic pathway mediating heat loss and a nonserotonergic pathway mediating heat gain.

Under the conditions described in this experiment, it is evident that T and 5-HT exerted opposite influences on core body temperature. That these effects were pharmacologically specific is indicated by the differential efficacy of the selected antagonists—methysergide, methergoline, and cyproheptadine. In addition to the demonstration of pharmacological specificity, these authors have addressed the question of the extent to which the applied compounds spread through the brain tissue. Apart from the usual identification of effective microinjection sites and the analysis of these data for systematic differences in sensitivity relative to anterior/posterior lateral and horizontal localization, Cox et al. (1983) have analyzed the degree to which radiolabeled 5-HT and radiolabeled T respectively diffuse through the preoptic area. More than 75% of radioactive material was observed to be within the range 0.3 mm anterior and 0.3 mm posterior to the center of the injection, reaching negligible amounts at 0.8 mm.

Minimization of tissue damage and a discrete analysis of the anatomical specificity of microinjection effects are well illustrated by recent studies conducted by Azami et al. (1982) and Llewelyn et al. (1983).

In attempting to assess the possible differential involvement in responses to nociceptive stimulation of two adjacent brain stem nuclei, nucleus raphe magnus (NRM) and nucleus reticularis paragigantocellularis (NRPG), Azami et al. have employed an extra-fine microinjection technique (as described earlier). In the rat, NRM varies in width from 0.2 to 0.5 mm. To analyze the respective involvement of these two nuclei in nociceptive responses, Azami et al. have employed a glass capillary microinjection system with a tip diameter of 70–90 μm. These authors have reported the relative response to microinjections of naloxone into these two brain stem nuclei as a function of laterality, using the centrally located NRM as a reference point. In this study, microinjections were administered in a volume of 0.5 μL over a period of 3 min. Since naloxone is a rather lipophilic compound, the reported change in efficacy of this dose of naloxone in relation to distance from NRM is a good indication of the anatomical specificity of the effect since a fairly marked change in naloxone concentration after injection would be expected.

Using the same microinjection procedure, Llewelyn et al. (1983) have assessed the relative effects of 5-HT and morphine injected into NRM and NRPG. The injection protocol was equivalent to that of the previous study. In this study a significant analgesic response was observed after a microinjection of 5 μg of 5-HT into NRM but not into NRPG, although morphine at 5 μg, injected into either NRM or NRPG, produced a clear analgesic response. Furthermore, in this study pretreatment with the 5-HT receptor antagonist cinanserine (5 mg/kg ip) was observed to result in an antagonism of the analgesic effects of 5-HT injected into NRM. When animals were pretreated with the opiate receptor antagonist naloxone (1 mg/kg ip), no clear antagonism of the analgesic effect of 5-HT into NRM was observed.

These studies of effects of microinjections into NRM and NRPG on analgesic responses in the rat, taken together, represent interesting evidence for a differential involvement of these two nuclei in the regulation of nociceptive responses. Here again, the authors have attempted to demonstrate anatomical specificity of their microinjection effects by systematically assessing the response to microinjections in relation to laterality from NRM. Furthermore, two different receptor antagonists, of 5-HT and opioid systems, respectively, have been used in an attempt to distinguish between the pharmacological responses of these two injection sites. In this case, it is evident that the use of a very fine microinjection technique in these studies may have contributed significantly to this analysis.

It is clear from the earlier discussion that a major problem with microinjections is that of tissue damage induced by repeated administration of compounds with this technique. One paradigm, in which this problem is of paramount significance, is that of intracerebral self-administration of drugs. It has been known for a number of years that animals may be trained to self-inject certain compounds, typically drugs of abuse, usually by the intravenous route (Pickens and Thompson, 1975). The logical extension of this paradigm is to analyze compounds that will be self-administered directly into different brain areas. The aim of this research is to elucidate pharmacological substrates underlying reward or reinforcement in the CNS. Although some researchers have successfully investigated intracerebral self-administration of drugs using conventional microinjection techniques (Olds, 1982), this approach has been problematic in relation to achieving accurate repeatable administration of fixed doses of compounds into discrete brain areas without significant tissue damage. It is in this area that the EMIT system described earlier has had most impact. In this context, the EMIT system has been used to analyze the effects of a variety of compounds, including morphine and cocaine. A recent study investigating intracranial self-administration of cocaine into medial prefrontal cortex will serve to illustrate the utility of this approach.

Goeders and Smith (1983) have reported that cocaine is reliably self-administered into the medial prefrontal cortex, but not into the nucleus accumbens or the ventral-tegmental area in rats. In these experiments animals are given access to a lever, depression of which will result in the delivery of a fixed quantity of cocaine directly into the area of interest via an EMIT system. Goeders and Smith (1984) have assessed the self-administration of cocaine into the medial prefrontal cortex in relation to (1) the extent of spread of tritiated cocaine through medial prefrontal cortex and surrounding brain areas when the drug was given on a response-independent basis at a dose and schedule equivalent to the maximum rate of self-administration, and (2) to determine the role of cholinergic, dopaminergic, and noradrenergic neurons in the maintenance of intracranial cocaine self-administration into the medial prefrontal cortex.

To determine the radioactive spread of tritiated cocaine, rats were either given 10, 20, or 40 response-independent infusions of cocaine into the medial prefrontal cortex. Each infusion contained 100 pmol of cocaine. Immediately following the last infusion animals were frozen in liquid nitrogen. After warming to $-20°C$ the brains were removed and cut into 1-mm coronal sections and each section cut into 1-mm cubes. Each tissue cube was assayed for radioactivity. This analysis indicated that the infused cocaine was localized in the medial prefrontal cortex. After the maximum number of injections and the longest time, 67% of total recovered radioactivity was located in the 1-mm cube containing the injection cannula tip. Ninety-three percent of the total recovered radioactivity was located within a 1-mm radius of this site. These data indicate quite strongly that self-administered cocaine exerted its effects in the area immediately around the injection site and did not diffuse to other brain regions.

Tests with receptor antagonists added to the cocaine injection solution indicated that $D_2$-dopaminergic receptors may be directly involved in the reinforcing or rewarding properties of cocaine at this site, whereas muscarinic cholinergic, $\alpha$- and $\beta$-nonadrenergic, and $D_1$-dopaminergic receptors are not.

This latter analysis involving 5-s microinjections of 100 nL of cocaine hydrochloride solution provides impressive evidence for the efficacy and reliability of the EMIT system. Indeed the fact that intracranial self-administration under these conditions has been observed in some animals during more than 60 consecutive experimental sessions (Goeders and Smith, 1983) represents impressive evidence for relatively little tissue damage in response to these treatments.

## 4. A Cautionary Note on Controls for Damage to CNS Tissue

Damage to CNS tissue as a consequence of the implantation of electrodes and cannulae is an important issue in the interpretation of experimental data in this field. Wyss and Goldstein (1976) and Isaacson (1981) have discussed this issue for electrode implantation, and the general issues are equally applicable to the implantation of cannulae (*see* Myers, 1974, pp. 68–70). Examples of altered behavioral responses induced as a consequence of electrode or cannula implantation may be found in studies by Martin and Hammond (1983) and by Greenshaw and Burešova (1982), respectively. In the former study, electrode implantation was observed to enhance sodium appetite, thus disrupting generalization of a lithium chloride based conditioned taste aversion to sodium chloride. In the study of Greenshaw and Burešova (1982), it was observed that cannulation and infusion of saline into the third cerebral ventricle attenuated the potency of intraperitoneally injected *d,l*-amphetamine in a classical conditioning paradigm. There are many such examples in the literature and the reader is cautioned to consider these possibilities when interpreting data from experiments involving direct CNS stimulation. An awareness of this issue and the design of experiments with suitable control groups is the simple, practical answer to these possible interpretative problems.

## References

Aghajanian G. K. and Davis M. (1975) A method of direct chemical brain stimulation in behavioral studies using microiontophoresis. *Pharmacol. Biochem. Behav.* **3,** 127–131.

Asanuma H. and Arnold A. P. (1975) Noxious effects of excessive currents used for intracortical microstimulation. *Brain Res.* **96,** 103–107.

Azami J., Llewlyn M. B., and Roberts M. H. T. (1980) An extra-fine assembly for intracerebral microinjection. *J. Physiol.* (Lond.), **305,** 18P–19P.

Azami J., Llewelyn M. B., and Roberts M. H. T. (1982) The contribution of nucleus ventricularis paragigantocellularis and nucleus raphe magnus to the analgesia produced by systemically administered morphine, investigated with the microinjection technique. *Pain* **12,** 229–246.

Bielajew C. and Shizgal P. (1982) Behaviourally derived measures of conduction velocity in the substrate for rewarding medial forebrain bundle stimulation. *Brain Res.* **237,** 107–119.

Blair R., Fishman B., Amit Z., and Weeks J. R. (1980) A simple double channel swivel for infusions of fluids into unrestrained animals. *Pharmacol. Biochem. Behav.* **12,** 463–466.

Bodnar R. J., Ellman S. J., Steiner S. S., Ackerman R. F., and Coons, E. E. (1982) Intracranial self-stimulation: Temporal interactions among mesencephalic and diencephalic sites. *Physiol. and Behav.* **28,** 473–482.

Bourke R. S., Kimelberg H. F., Nelson L. R., Barron K. D., Anen E. L., Popp A. J., and Waldman J. B. (1980) Biology of Glial Swelling in Experimental Brain Edema, in *Advances in Neurology,* Vol. 28 (Cervos-Navarro J. and Ferszt R., eds.), Raven, New York.

Bozarth M. A. and Wise R. A. (1980) Electrolytic microinfusion transducer: An alternative method of intracranial drug application. *J. Neurosci. Methods* **2,** 273–275.

Bozarth M. A. and Wise R. A. (1981) Intracranial self-administration of morphine into the ventral tegmental area in rats. *Life Sci.* **28,** 551–555.

Britt M. D. and Wise R. A. (1983) Ventral tegmental site of opiate reward. *Brain Res.* **258,** 105–108.

Brown Z. W., Amit Z., and Weeks J. R. (1976) Simple flo-thru swivel for infusions into unrestrained animals. *Pharmacol. Biochem. Behav.* **5,** 363–365.

Bureš J., Burešová O., and Huston J. P. (1983) *Techniques and Basic Experiments for the Study of Brain and Behaviour,* Elsevier, Amsterdam.

Byrne J. H. (1981) Intracellular Stimulation, in *Electrical Stimulation Research Techniques* (Patterson M. M. and Kesner R. P., eds.), pp. 37–59, Academic, New York.

Cooley R. K. and Vanderwolf C. H. (1978) *Stereotaxic Surgery in the Rat: A Photographic Series.* A. J. Kirby, London, Ontario, Canada.

Cox B., Davis A., Juxton V., Lee T. F., and Martin D. (1983) A role for an indoleamine other than 5-hydroxytryptamine in the hypothalamic thermoregulatory pathways of the rat. *J. Physiol.* (Lond.) **337,** 441–450.

Criswell H. E. (1977) A simple chronic microinjection system for use with chemitrodes. *Pharmacol. Biochem. Behav.* **6,** 237–238.

De Groot J. (1959) *The Rat Forebrain in Stereotaxic Coordinates.* North-Holland, Amsterdam.

Delgado J. M. R. (1964) Electrodes for Extracellular Recording and Stimulation, in *Physical Techniques in Biological Research*, Vol. 5, (Nastuk W. L., ed.), pp. 88–143, Academic, New York.

Delgado J. M. R., De Feudis F. V., Roth R. H., Ryugo D. K., and Mitruka B. M. (1972) Dialytrode for long term intracerebral infusion in awake monkeys. *Arch. Int. Pharmacodyn. Ther.* **198**, 9–21.

Deutsch J. A. (1964) Behavioral measurement of the neural refractory period and its application to intracranial self-stimulation. *Comp. Physiol. Psychol.* **58**, 1–9.

Dismukes R. K. (1979) New concepts of molecular communication among neurons. *Behav. Brain Sci.* **2**, 409–448.

Doty R. W. (1969) Electrical stimulation of the brain in behavioral context. *Ann. Rev. Psychol.* **20**, 289–320.

Doty R. W. and Bartlett J. R. (1981) Stimulation of the Brain Via Metallic Electrodes, in *Electrical Stimulation Research Techhniques* (Patterson M. M. and Kesner R. P., eds.), pp. 71–103, Academic, New York.

Evans B. K., Armstrong S., Singer G., Cooke R. D., and Burnstock G. (1975) Intracranial injection of drugs: Comparison of diffusion of 6-OHDA and Guanethidine. *Pharmacol. Biochem. Behav.* **3**, 205–217.

Fifková E. and Maršala J. (1967) Stereotaxic Atlases for the Cat, Rabbit and Rat, in *Electrophysiological Methods in Biological Research* (Bureš J., Petran M., and Zachar J., eds.) pp. 653–731. Academic, New York.

Florey E. (1967) Neurotransmitters and modulators in the animal kingdom. *Fed. Proc.* **26**, 1164–1178.

Fritsch G. and Hitzig E. (1870) Uber die elektrische Erregbarkeit des Grosshirns. *Arch. Anat. Physiol.* **37**, 300–332.

Gallistel C. R., Rolls E. T., and Greene D. (1969) Neuron function inferred from behavioral and electrophysiological estimates of refractory period. *Science* **166**, 1028–1030.

Garrigues A. M. and Cazala P. (1983) Central catecholamine metabolism and hypothalamic self-stimulation behaviour in two inbred strains of mice. *Brain Res.* **265**, 265–271.

Goeders N. E. and Smith J. E. (1983) Cortical involvement in cocaine reinforcement. *Science* **221**, 773–775.

Goeders N. E. and Smith J. E. (1985) Parameters of Intracranial Self-Administration of Cocaine Into the Pre-Frontal Cortex, in *Problems of Drug Dependence, NIDA Research Monographs* (in press.)

Goeders N. E., Lane J. D., and Smith J. E. (1984) Self-administration of methionine enkephalin into the nucleus accumbens. *Pharmacol. Biochem. Behav.* **20**, 451–455.

Goodrich C. A., Greehey B., Miller T. B., and Pappenheimer J. R. (1969) Cerebral ventricular infusions in unrestrained rat. *J. Appl. Physiol.* **26**, 137–140.

Greenshaw A. J. and Burešová O. (1982) Learned taste aversion to saccharin following intraventricular or intraperitoneal administration of d,l-amphetmine. *Pharmacol. Biochem. Behav.* **17**, 1129–1133.

Hashimoto M. (1915a) Fieberstudien. I Mitteilung: Über die spezifische Überempfindlichkeit des Wärmzentrums an sensibilisierten tieren. *Arch Exper. Pathol. Pharmakol.* **70**, 370–393.

Hashimoto M. (1915b) Fieberstudien. II Mitteilung: Über den Einfluss unmitelbaver Erwärmung und Abkühlung des Wärmzentrums auf die Temperaturwirkungen von verschiedenen pyrogenen und antiphyretischen Substanzen. *Arch Exper. Pathol. Pharmakol.* **70**, 394–425.

Hess W. R. (1928) Hirnreizversuche uber den mechanismus des schlafes. *Arch. Psychiatr.* **86**, 287–292.

Hurt E. A., Hanaway J., and Netsky M. G. (1971) Stereotaxic atlas of the mesencephalon in the albino rat. *Confin. Neurol.* (Basel) **33**, 93–115.

Huston J. P. and Jakobartl L. (1977) Evidence for selective susceptibility of hippocampus to spreading depression induced by vasopressin. *Neurosci. Lett.* **1**, 291–296.

Imperato A. and DiChiara G. (1984) Trans-striatal dialysis coupled to reverse phase high performance liquid chromatography: A new method for the study of the *in vivo* release of endogenous dopamine and matabolites. *J. Neurosci.* **4**, 966–977.

Isaacson R. L. (1981) Brain Stimulation Effects Related to Those of Lesions, in, *Electrical Stimulation Research Techniques* (Patterson M. M. and Kesner R. P., eds.), pp. 205–217.

Iwamoto E. T., Williamson E. C., Wash C., and Hancock R. (1984) An improved drug infusion pump for injecting nanoliter volumes subcortically in awake rats. *Pharmacol. Biochem. Behav.* **20**, 959–963.

Johnson R. D. and Justice J. B. (1983) Model studies for brain dialysis. *Brain Res. Bull.* **10**, 567–571.

Jones R. S. G., Juorio A. V., and Boulton A. A. (1981) Changes in levels of dopamine and tyramine in the rat caudate nucleus following alterations of impulse flow in the nigro-striatal pathway. *J. Neurochem.* **40**, 396–401.

Katz J. (1966) *Nerve Muscle and Synapse.* McGraw-Hill, New York.

König J. F. R. and Klippel R. A. (1974) *The Rat Brain: A Stereotaxic Atlas of the Forebrain and Lower Parts of the Brain Stem.* Krieger, New York.

Kuffler S. W., Nicholls J. G., and Martin A. A. (1984) *From Neuron to Brain,* 2nd Ed., Sinauer Associates, Sunderland, Massachusetts.

Llewelyn M. B., Azami J., and Roberts M. H. T. (1983) Effects of 5-hydroxytryptamine applied into nucleus raphe magnus on nociceptive thresholds and neuron firing rate. *Brain Res.* **258**, 59–68.

Martin R. L. and Hammond G. R. (1983) Lateral hypothalamic electrode implantation disrupts lithium chloride based generalized aversion to sodium chloride by enhancing sodium appetite. *Physiol. Psychol.* **11**, 63–72.

Miliaressis E. (1981) A miniature moveable electrode for brain stimulation in small animals. *Brain Res. Bull.* **7**, 715–718.

Miliaressis E. and Phillipe L. (1984) The pontine substrate of circling. *Brain Res.* **293**, 143–152.

Moroz V. M. and Bureš J. (1982) Cerebellar unit activity and the movement disruption induced by caudate stimulation in rats. *Gen. Physiol. Biophys.* **1**, 53–70.

Myers R. D. (1974) *Handbook of Drug and Chemical Stimulation of the Brain.* Van Nostrand Rheinold, New York.

Myers R. D. and Hoch D. B. (1978) $^{14}$C-Dopamine microinjected into the brainstem of the rat: Dispersion kinetics, site content and functional dose. *Brain Res. Bull.* **3**, 601–609.

Nicolaidis S., Rowland N., Meile M.-J., Marfaing-Jallat P., and Pesez A. (1974) A flexible technique for long term infusions in unrestrained animals. *Pharmacol. Biochem. Behav.* **2**, 131–136.

Oitzl M. S. and Huston J. P. (1984) Electroencepholographic spreading depression and concomitant behavioral changes induced by intrahippocampal injections of ACTH1-24 and D-Ala$^2$-Met enkephalinamide in the rat. *Brain Res.* **308**, 33–42.

Oldendorf W. H. (1971) Brain uptake of radiolabelled amino-acids, amines and hexoses after arterial injection. *Am. J. Physiol.* **221**, 1629–1639.

Olds M. E. (1982) Reinforcing effects of morphine in the nucleus accumbens. *Brain Res.* **237**, 429–440.

Patterson M. M. and Kesner R. P. (Eds.) (1981) *Electrical Stimulation Research Techniques.* Academic, New York.

Paxinos G. and Watson C. (1982) *The Rat Brain in Stereotaxic Co-ordinates.* Academic, New York.

Pellegrino L. J., Pellegrino A. S., and Cushman A. J. (1979) *A Stereotaxic Atlas of the Rat Brain.* Appleton-Century-Crofts, New York.

Pickens R. and Thompson T. (1975) Intravenous preparation for self-administration of drugs by animals. *Am. Psychologist* **30**, 274–275.

Porrino L. J., Esposito R. N., Seeger T. F., Crane A. M., Pert A., and Sokoloff L. (1984) Metabolic mapping of the brain during rewarding self-stimulation. *Science* **224**, 306–309.

Ranck J. B. Jr. (1981) Extracellular Stimulation, in *Electrical Stimulation Research Techniques.* (M. M. Patterson and R. P. Kesner, eds.), pp. 1–36, Academic, New York.

Ranck J. B. Jr. (1975) Which elements are excited in electrical stimulation of mammalian central nervous system? *Brain Res.* **98**, 417–440.

Redgrave P. (1978) Modulation of intracranial self-stimulation behaviour by local perfusions of dopamine, noradrenaline and serotonin within the caudate nucleus and nucleus accumbens. *Brain Res.* **155**, 277–295.

Routtenberg A. (1972) Intracranial chemical injection and behavior: A critical review. *Behav. Biol.* **7**, 601–641.

Sheer D. E. (Ed.) (1961) *Electrical Stimulation of the Brain.* University of Texas, Austin, Texas.

Simonoff L. N. (1866) Die hemmungs mechanismen der Sangethiere experimentell bemeisen. *Arch. Anat. Physiol.,* Leipzig **33**, 545–564.

Sprick U., Oitzl M.-S., Ornstein K., and Huston J. P. (1981) Spreading depression induced by microinjection of enkephalins into the hippocampus and neocortex. *Brain Res.* **210**, 243–252.

Tossman U. and Ungerstedt U. (1981) Neuroleptic action on putative amino-acid neurotransmitters in the brain studied with a new technique of dialysis. *Neurosci. Lett.* Suppl 7, S749.

Urquhart J., Fara J. W., and Willis K. L. (1984) Rate-controlled delivery systems in drug and hormone research. *Ann. Rev. Pharmacol. Toxicol.* **24,** 199–236.

von Euler C. and Holmgren B. (1956) The thyroxine 'receptor' of the thyroid-pituitary system. *J. Physiol.* (Lond.) **131,** 125–136.

van Heuven-Nolsen D., van Wolfswinkel L., van Ree J., and Versteeg D. H. G. (1983). Electrical stimulation of the ventral tegmental area and catecholamine metabolism in discrete regions of the rat brain. *Brain Res.* **268,** 362–366.

Walls E. K. and Wishart T. B. (1977) Reliable method for cannulation of the third ventricle of the rat. *Physiol. Behav.* **19,** 171–173.

White N. (1976) Strength-duration analysis of the organisation of reinforcement pathways in the medial forebrain bundle of rats. *Brain Res.* **110,** 575–591.

Wyss J. M. and Goldstein R. (1976) Lesion artifact in brain stimulation experiments. *Physiol. Behav.* **16,** 387–389.

Yadin E., Guarini V., and Gallistel C. R. (1983) Unilaterally activated systems in rats self-stimulating at sites in the medial forebrain bundle, medial prefrontal cortex or locus coeruleus. *Brain Res.* **266,** 39–50.

# Chapter 8

# Perfusion Techniques for Neural Tissue

Q. J. PITTMAN, J. DISTURNAL, C. RIPHAGEN, W. L. VEALE, AND L. BAUCE

## 1. Introduction

The physiological and pharmacological examination of the nervous system enjoyed enormous popularity over the last several decades. It is now possible to record the electrical currents generated by ions passing through individual membrane channels and to define precisely the molecular characteristics of individual receptor aggregates. Success in these endeavors has arisen, in part, from the use of sophisticated biochemical, electrophysiological, and pharmacological techniques, often carried out in isolated neural tissue. Despite the wealth of information generated by this approach to brain function, our knowledge of how the brain works as a whole to integrate and direct a basic physiological or behavioral drive is still scanty. The brain is well-endowed with putative neurotransmitter molecules whose functions are still unknown. The utility of perfusion technology as a potent tool for addressing this problem is outlined in Table 1 and also has been pointed out previously (Myers, 1972). In this chapter, we will examine critically some of the theoretical and methodological issues raised in perfusing cerebral structures, and give descriptions of the methodology involved. In view of space limitations, frequent reference to the literature will be made to assist the reader in obtaining precise descriptions of some of the techniques under discussion. Of particular value is the 1972 chapter by Myers.

279

TABLE 1
Uses of Perfusion Techniques

1.  Survey molecules for suspected neurotransmitter action
2.  Correlate pharmacological actions with specific behaviors and phys-
    iological responses
3.  Determine site(s) of action for suspected transmitters and drugs
4.  Establish pharmacological profiles
5.  Detect release of endogenous compounds in association with
    specific functions.
6.  Perform dynamic studies on neurotransmitter turnover
7.  Work with unanesthetized, unstressed preparations
8.  Manipulate brain extracellular environment

## 2. Methodological Considerations

### 2.1. Perfusion Media

A variety of perfusion media have been successfully utilized in
perfusion studies. The primary requirement appears to be that
they are isotonic with respect to the brain extracellular space. The
ionic composition of cerebrospinal fluid (CSF) is known (Wood,
1980), and any number of artificial CSFs should prove suitable for
perfusion studies, particularly if it can be shown that introduction
of the vehicle alone provides no untoward signs on the part of the
animal or system under study. Despite publications reporting use
of inert perfusion media, e.g., sucrose only (Cooper et al., 1979),
the inclusion of ions essential for maintenance of neuronal resting
potential and transmitter release is to be encouraged. As an example, synaptic release of neurotransmitters requires calcium and
most artificial CSF recipes contain calcium. Nevertheless, the
ionic compositions of artificial CSFs should be examined carefully, since some (e.g., Yamamoto, 1972) contain relatively high
levels of calcium that can suppress cellular activity (Pittman et al.,
1981). Recent studies using ion-sensitive electrodes indicate that
*unbound* extracellular calcium is approximately 1–1.5 mM
(Heinemann et al., 1977) and a level approximating this should be
used.

When using bicarbonate-buffered physiological solutions, it
is necessary to bubble the solution with 5% $CO_2$/95% $O_2$ to maintain an appropriate pH (7.3–7.4). Whether or not the presence of
elevated oxygen tension in the perfusate enhances the physiological response of brain tissue is open to question, and, in our
hands, effects on physiology or on release of neurotransmitters
do not seem to differ when using an oxygenated medium. In con-

trast to the situation with in vitro preparations, a perfused, intact brain retains its own blood supply so that tissue oxygenation should proceed normally. In isolated CNS tissue, however, maintenance of high oxygen tension is extremely important, and some additional success has been reported with the use of oxygen-carrying hydrocarbons or of small amounts of $H_2O_2$ as a means of supplying oxygen to the tissue (e.g., Llinas et al., 1981).

In chronic animal preparations, it is well recognized that perfusates introduced into the brain must be sterilized to avoid bacterial tissue contamination. What is important to remember also is that acute perfusions with nonsterile fluids may invite leukocyte infiltrations of the tissue, leading to release and possible collection of leukocytic mediators. Therefore, all solutions (as well as cannulae and syringes) should be sterilized for perfusion studies. This can be carried out by filtering the fluid through a submicron millipore filter and by autoclaving all equipment used.

The other major contaminant that may be present even in a sterilized solution is endotoxin, a fever-producing (pyrogen) molecule that can pass through even a 0.22 μm filter. When possible pyrogenicity of the solution is an important consideration (i. e., during experiments on thermoregulation or on certain behaviors that may be altered by a febrile episode), the medium and glassware may be heated to 180°C for 2 h to destroy the pyrogen molecules. The nonpyrogenicity of the solution may be monitored by recording the animal's rectal temperature or by carrying out an in vitro pyrogen assay (Cooper et al., 1972).

Prewarming the perfusate to 37°C does not appear to be required at the generally low perfusion rates utilized in cerebral ventricular and push-pull perfusions. With the use of stainless steel cannulae and guide tubes in push-pull perfusion, the countercurrent flow appears to warm the perfusates to brain temperature (unpublished observations). However, perfusions of isolated neural tissue require warming, and maintaining these tissues at 34–35°C rather than the normal body temperature of 37–38°C often appears to enhance viability of the tissue (Dingledine et al., 1980). However, electrophysiological properties of cells can be altered by reductions of several degrees in body temperature (Pierau et al., 1976), and appropriate controls should be carried out at 37°C.

## 2.2. Local Tissue Alterations

Several authors have commented that perfusion studies are of doubtful value because of local tissue damage, particularly with respect to push-pull perfusion (Bloom and Giarman, 1968; Chase and Kopin, 1968; Izquierdo and Izquierdo, 1971). Although some of the more recent methodological innovations and careful attention to detail will limit the amount of damage, it is unavoidable that introduction of the cannulae into the brain will damage brain tissue. Examination for damaged neurons at the termination of the experiment indicates that a shell of neurons approximately 0.2–0.6 mm thick is damaged around the cannula shaft (Yaksh and Yamamura, 1974). At the perfusion site, when isotonic solutions are perfused, widespread destruction of local tissue does not occur (Bartholini et al., 1976; Yaksh and Yamamura, 1974). One way to detect neuronal and glial damage is to measure the quantities of certain intracellular markers, such as lactate dehydrogenase, or neuron- and glial-specific proteins in collected perfusates. It would appear that, except for the initial trauma resulting from lowering of the cannula or from alterations in flow, the level of intracellular markers remains low throughout a perfusion (Honchar et al., 1979).

Some push-pull perfusion penetrations or ventricular perfusions will reveal blood in the perfusate. Often this disappears after several minutes, but results from this period of the experiment should be discarded since blood can contribute significant quantities of 5-hydroxytryptamine, (5-HT; serotonin), prostaglandins, and other circulating substances to the area. For example, we have found that as little as 1 $\mu$L of blood in 1 mL of push-pull perfusate will produce detectable levels of serotonin (Fig. 1).

Honchar et al. (1979) have examined the patency of the blood–brain barrier during push-pull perfusions and have calculated that approximately 2 nL of serum/min penetrated the vascular barrier into the perfused site. This, however, may not be a function of the perfusion, since Yaksh and Yamamura (1974) noted similar quantities of a vascular label in perfused and unperfused brain areas. Nevertheless, these observations underline the importance of carrying out appropriate controls to insure that substances appearing in a perfusate are of local rather than vascular origin.

A consequence of tissue damage in brain is the establishment of a glial scar around the cannula tract; this feature has prompted some authors to advocate perfusion of fresh brain sites with removable push-pull cannulae lowered through guide tubes rather than with chronically implanted cannulae. Although some investigators have apparently achieved good results with chronic preparations, it appears that a period of three successive push-pull perfusions, each 1 wk apart, in one locus can result in decline of releasable substances to zero (Bayon et al., 1981). This has been attributed to glial invasion and isolation of the perfusion site from neurally active tissue. However, a chronically implanted cannula

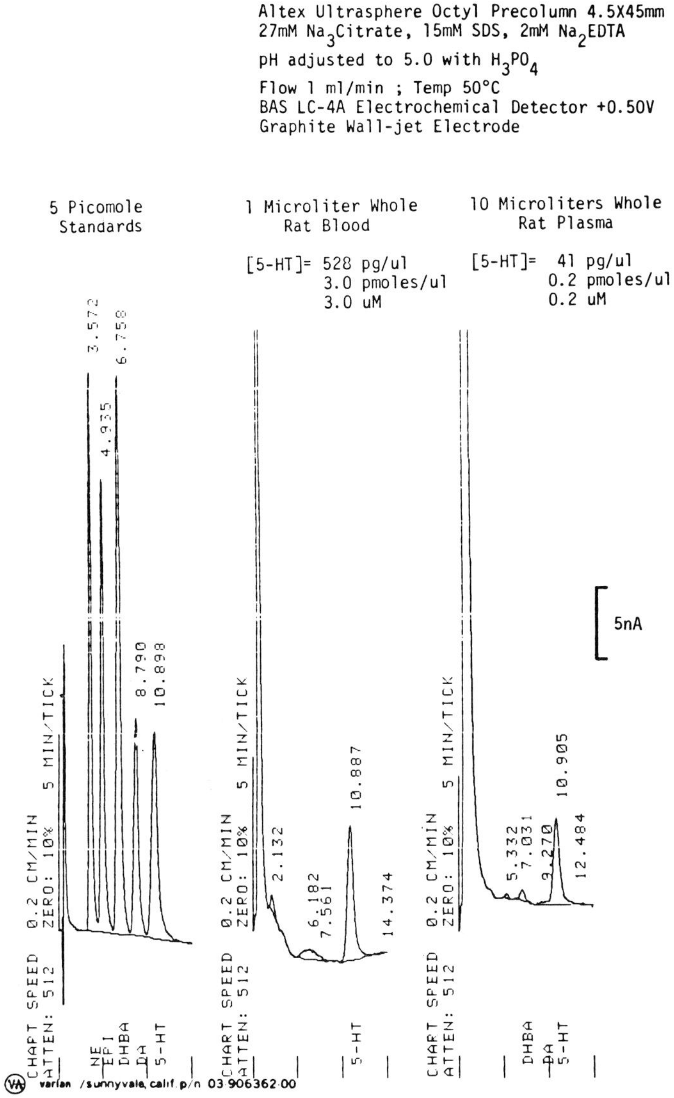

Fig. 1. High-pressure liquid chromatogram (left) for 5 pmol standards of norepinephrine (nonadrenaline; NE), epinephrine (adrenaline; $EP_1$), dihydroxybenzylamine (DHBA), dopamine (DA), and serotonin (5-HT). The middle chromatogram illustrates the quantities of the above amines in 1 µL of whole rat blood. The right chromatogram illustrates the content in 10 µL of whole rat plasma. Note the quantity of 5-HT in rat blood and plasma.

will allow the blood–brain barrier time to repair itself, thereby eliminating many of the problems of bleeding and vascular invasion of the perfusate.

## 2.3. Application of Exogenous Substances

Perfusion methods have proven extremely popular for application of substances to neural tissue. They offer a distinct advantage over single or intermittent application; in particular, perfusion can be used to apply a defined quantity of a drug or putative neurotransmitter over a relatively defined period of time. Furthermore, the dynamic aspect of the perfusion allows a more rapid exchange of different concentrations and a better carrying out of dose–response curves than can be achieved by a single application. They also permit application of substances concurrently with collection of released substances, thus allowing a neurochemical analysis of a local area of neural tissue. These considerations make perfusion a powerful tool for drug application; however, a number of considerations must be addressed in order to make interpretation of data meaningful.

It is thought that molecules such as neuropeptides do not easily enter the central nervous system, but it is possible that, with the local trauma associated with establishing the in vivo perfusion, the blood–brain barrier may open sufficiently to allow passage of even these hydrophilic molecules. Of course, there are areas of the brain that are devoid of a classical blood–brain barrier and therefore easily accessible to substances in the vascular system.

Because of the well-known fact of the relative inaccessibility of many circulating neural substances into the brain (e.g., peptides and other large molecules; Davson and Welch, 1971), it has long been taken for granted that these compounds do not cross out of the brain with facility. However, substances that enter or are perfused through the cerebral ventricular system can reach the periphery in a surprisingly short period of time via bulk flow through the arachnoid villi. For example, neurohypophyseal peptides perfused in low quantities throughout the lateral ventricular system of the guinea pig appear in the blood in sufficient quantities to exert a biological action (Robinson, 1983). We have found also (unpublished observations) that tritiated vasopressin injected intrathecally can be detected in the aortic circulation within 1 min of application of as little as 3 pmol of the substance. Therefore, particularly when one is dealing with substances that also exert potent effects in the periphery, it is necessary to verify that any observed actions are indeed central in origin (c.f., Pittman et al.,

1982). Appropriate controls for this include the administration of identical or even greater quantities of a substance into the circulation, or the utilization of a radioactive label to trace the passage of the molecule into the periphery.

The extent of the passage of substances throughout brain tissue will be influenced by a number of factors, including the amount of extracellular space, the presence of degradative enzymes and uptake mechanisms, and possibly the intracranial pressure. In contrast to the relative impermeability of the brain–vascular barrier, the ependyma between the ventricular space and brain tissue is more easily penetrated by molecules. It is this feature that permits ventricular perfusion to be used to apply neuroactive agents to brain. Nevertheless, it is likely that substances diffuse a distance of less than 1 mm from the ventricular space into tissue (Fuxe and Ungerstedt, 1968).

Neurotransmitter contents of a variety of brain areas, as well as of cerebrospinal fluid, have now been calculated. Though there are often marked regional variations, particularly in tissue concentrations, quantities of many presumptive neurotransmitter agents within brain tissue can be very low. The fact that brain content of a neurotransmitter may be low should not detract from the potential physiological importance of perfusion studies in which larger quantities of peptide are applied to the brain than may, in fact, be found there normally. This is of particular importance when discussing the action of certain substances that are found to circulate within the bloodstream as well as to have a putative neurotransmitter action within the brain. Many receptors for circulating substances are indeed sensitive to low levels (e.g., femtomolar amounts) of circulating hormones. A few studies using electrophysiological techniques on brain tissue slices indicate that peptides such as vasopressin may be active at $\mu M$ to $nM$ concentrations (Muhlethaler et al., 1983; Pittman et al., 1980), whereas in the kidney their physiological actions are present at much lower concentrations. Therefore, one must consider that the central receptors for such substances may be of much lower affinity than those in the periphery, necessitating application of higher doses to exert physiological effects. A second consideration in establishing an appropriate concentration of the perfused substances is that homeostatic processes under central nervous system control are by their very nature designed to resist change. Therefore, a literal flooding of every receptor with supramaximal concentrations of the agonist may be required to overcome the counterbalances of the homeostatic system. Finally, of perhaps greatest importance to this issue is the fact that intracerebral and intravenous perfusion of compounds may represent markedly different situations in terms of access to receptors. Circulating hormones have relatively easy access to receptors on the vasculature or in glandular tissue. Intraventricular, cortical, or local tissue perfusion of hormones or neurotransmitters may necessitate that these substances reach receptors at some unknown distance from the site of the perfusion, that they escape a potentially rich concentration of degradative enzymes, and most likely that they enter a synaptic cleft, which can be a very "protected" space. The quantity of substance that actually reaches the receptor is without doubt a fraction of that in the perfusate.

Because it is often necessary to deliver large doses of exogenous substances to tissue in order to elicit a physiological effect, appropriate pharmacological controls to guard against nonspecific effects are of paramount importance. A dose–response curve should be constructed, keeping in mind that responses from neural tissue often go from minimal to maximal within one to three log units. The use of specific analogs and antagonists can do much to convince the skeptic of the physiological relevance of the data.

## 2.4. Collection of Endogenous Substances

Neurochemical analysis of putative neurotransmitters generally has utilized postmortem or in vitro analysis of neural tissue to demonstrate synthesis, storage, release, and reuptake. Perfusion techniques, on the other hand, allow in vivo ongoing evaluation of neurochemical events in the central nervous system of conscious (or anesthetized) animals as expressed through diffusion of endogenous substances to an accessible site (e.g., ventricular system or push-pull perfusion site). Perfusion techniques have proven valuable in helping to fulfill the criterion of "appropriate release" for neurotransmitter candidates. However, in the evaluation of data on release of endogenous substances from brain tissue, several theoretical and methodological considerations must be taken into account.

The first problem to be addressed is the type of methodology to be employed, whether it be local tissue perfusion or cortical or ventricular perfusion. Local tissue perfusion has certain advantages related to proximity to the actual site of release, thereby leading to more discrete and potentially higher levels of neurotransmitter. Furthermore, there may be greater temporal fidelity in relation to the stimulus responsible for the release. A ventricular perfusion may "smooth out" the troughs and peaks of transmitter release (e.g., Levine and Ramirez, 1980) because of its collection from a greater area and the possible involvement of a number of different neural systems utilizing the same transmitter. It is also possible that the ventricular route may only collect from substances released near the ventricles, although there is evidence that substances microinjected up to 1–2 mm from the ventricular wall can diffuse relatively quickly to the ventricular space (Cooper and Veale, 1973). Somewhat unsettling is the fact that studies have reported contrasting results of neurotransmitter release into push-pull perfusates and into ventricular perfusates (Cooper et al., 1979; Kasting et al., 1984); it is tempting to speculate that the data collected from push-pull perfusates most proximal to the release site will give the best indication of the dynamics of transmitter release.

Verification that substances that appear in a perfusate of neural tissue do in fact arise from the area under perfusion is particularly important. As indicated earlier, there is some evidence that substances in the circulation may diffuse in detectable quantities into push-pull perfusates. In order to verify that the release is indeed "local," each site from which a release is elicited should be perfused with a high potassium solution to elicit local release; omission of calcium from this perfusate should reduce the stimulated release. Another way of determining that the release is indeed local is to measure release of a preloaded labeled transmitter or precursor. This technique involves microinjection of a labeled compound into the site to be perfused, and then perfusion of the site and measurement of the washout of the label as the synthesized or stored neurotransmitter (e.g., Martin and Myers, 1975; Nieoullon et al., 1977; Riddell and Szerb, 1971). Although this technique does insure "local release" it can raise interpretative problems as to the relative release of recently synthesized vs older pools of neurotransmitters (Besson et al., 1969; Arnold et al., 1977); it also raises the problem that the label progressively disappears throughout the experiment.

Another problem often encountered in perfusion studies aimed at collecting endogenous substances is that, given available methodology, levels of the substance of interest may be too low to detect. This may arise because of dilutional factors, which can be minimized by slowing the perfusion rate; in recent experiments involving brain dialysis, perfusion rates as low as 1–2 µL/min have been utilized successfully to measure endogenous substances (Zetterstrom et al., 1983). Other studies have addressed this problem by including various blockers of reuptake mechanisms (Hery et al., 1982) or of degradative enzymes (e.g., Bayon et al., 1981; Elghozi et al., 1981).

A final consideration with respect to collection of endogenous substances relates to the assay system utilized and the decision as to what to measure. Recent technological advances have
virtually eliminated the bioassay as a means of measurement, and
radioimmunoassays, radioreceptor assays, thin layer chromatography, gas chromatography–mass spectrometry, and high pressure liquid chromatography (HPLC) now appear to be the methods of choice. These techniques vary in their sensitivity and
specificity, and the best perfusion experiments will be of limited
value in the absence of a reliable assay. A decision must also be
made as to what is an appropriate substance to measure—the actual released neurotransmitter or one of its metabolic products.
With respect to the catecholamines, it is often possible to detect
substantially higher levels of their metabolites than of the native
amine (Loullis et al., 1980). Although it is likely that increased
quantities of a neurotransmitter will also be accompanied by increased levels of its metabolites, many drugs may act to alter the
metabolic processing of a released neurotransmitter. For example,
amphetamine may increase dopamine levels whereas the metabolites 3,4-dihydroxyphenylacetic acid and homovanillic acid are actually decreased (Chen et al., 1984). A study by Palfreyman et al.
(1983) indicated that brain γ-aminobutyric acid (GABA) concentration was better reflected by cerebrospinal fluid levels of GABA
than of its conjugate, homocarnosine. Therefore, it appears preferable to measure the actual endogenous active compound rather
than one of its metabolites when attempting to correlate release
with a physiological function (Van der Gugten and Slangen, 1977)
or pharmacological stimulus (Philips et al., 1982; Loullis et al.,
1980; Chen et al., 1984).

## 3. Application of Techniques

### 3.1. Ventricular Perfusion

Perfusion of the ventricular system of dogs was first reported by
Leuson (1950) and the technique was subsequently popularized
in the 1950s and '60s by Feldberg and Pappenheimer and colleagues (Feldberg and Fleishauer, 1960; Pappenheimer et al.,
1962; Battacharva and Feldberg, 1958). Ventricular perfusion
methods have proven useful for examination of the formation
and clearance of cerebrospinal fluid, indications for the potential
of certain substances as neurotransmitters involved in various
physiological functions, alteration of the brain ionic environment,

and for the collection of endogenous substances from brain tissue. The utility of ventricular perfusions lies in their relative ease of use and in their ability to perfuse widespread areas of the CNS; the latter feature could also be considered their major drawback, since little anatomical resolution can be provided.

Myers (1972) extensively reviewed the techniques for ventricular perfusion, and the methodology has changed little since then. Essentially the technique consists of implanting an inflow and outflow cannula in the ventricular system and perfusing artificial CSF through the ventricular system at a controlled rate (usually 20–60 µL/min). As demonstrated by Carmichael et al. (1964), it is possible, by judicious placement of cannulae into appropriate areas, to selectively perfuse a specific part of the ventricular system. For example, one cannula can be placed at the anterior-dorsal part of the third ventricle (near the foramen of Munro) and another near the point at which the third ventricle flows into the aqueduct; perfusion then can be limited to the third ventricle. Alternately, an inflow can be placed at the third ventricle and an outflow in the cisterna magna to perfuse the length of the diencephalon and brain stem.

Ventricular perfusions have been carried out in almost all common laboratory animals, and the techniques, with minor modifications, are essentially similar. Basically, an inflow cannula consisting of a thin-walled stainless steel tubing (usually 23–30 gage diameter) is positioned at an appropriate sterotaxic coordinate. For acute experiments, the cannula can be mounted on a stereotaxic manipulator and introduced directly into the ventricular system. In chronic preparations, a guide tube sufficiently large to accomodate the inflow cannula is usually implanted so as to rest immediately above the ventricular system and is occluded, when not in use, with a stainless steel stylet. Commercial models of such cannulae are available, but it is easy and inexpensive to construct a homemade cannula system. Implantation of the outflow cannula in most areas of the ventricular system presents no major difficulties; however, positioning of a cisternal or fourth ventricular cannula can often cause difficulties due to the angle of the base of most skulls. Some investigators have overcome this problem by implanting a fourth ventricular or cisternal cannula at a caudal angle so that the cerebellum is punctured by the cannula (e.g., Pappenheimer et al., 1962). Others have found it useful to build up a small bridge and support of stainless steel wire and acrylic and position the cannula external to the skull so as to puncture the atlantooccipital membrane (e.g., Moir and Dow, 1970; Feldberg et al., 1970). Recently, ventriculocisternal perfusion has been reported in unanesthetized fetal lambs using a modification of the latter procedure (Bissonnette et al., 1981).

Following appropriate positioning of the cannulae, perfusion at the inflow tube is initiated. The perfusion pressure is usually provided by a pump, and we have found that the Harvard infusion pumps provide accurate and controlled delivery of solutions, although roller-type pumps have also proved acceptable. It does not appear necessary to utilize a pump for extraction of fluid from the ventricular system. Positioning of the tip of outflow cannula at a position 15–20 cm below the level at which the fluid is collected will facilitate flow. A recent publication (Chang et al., 1984) indicates that a negative collecting pressure of $-15$ to $-20$ cm water is required to insure that neuropeptides do not diffuse from the ventricular system into the peripheral circulation.

### 3.2. Spinal Cord Perfusion

Intrathecal drug administration was introduced by Yaksh and Rudy (1976) as a means of pharmacological stimulation of spinal structures, and they (Yaksh and Tyce, 1980) and others (Morton et al., 1977; Jordan and Webster, 1978) subsequently developed perfusion techniques for collection of endogenous spinal cord substances. The perfusion of a length of spinal cord is particularly advantageous, since afferent or efferent fibers subserving certain physiological functions may enter or exit the cord at many levels throughout its length. Therefore, the perfusion of cord tissue allows one to perfuse simultaneously the area from the cervical through the thoracic and down to lower lumbar levels. Spinal cord perfusion studies have been described for rats (Yaksh and Tyce, 1980; Pittman et al., 1984), rabbits (Morton et al., 1977), and cats (Yaksh and Tyce, 1980; Jordan and Webster, 1978). Many of the interpretative problems associated with ventricular perfusion apply equally well to spinal cord perfusion studies. It is not generally appreciated that relatively large molecules can pass quickly from spinal subarachnoid space to the peripheral circulation, and care must be taken to insure that effects are indeed central in origin.

The technique for perfusion of the in vivo spinal cord is easily mastered (*see* Fig. 2). Basically, the animal's neck is flexed so that the cisterna magna can be approached surgically. An opening in the cisterna magna and puncture of the arachnoid matter (accompanied by a welling of CSF) allows one to insert a polyethylene tubing down the dorsal aspect of the spinal cord within the subarachnoid space. In rats, this is usually a piece of stretched "PE-10" tubing (Clay-Adams); in larger animals such as rabbits, an appropriately larger and stiffer piece of tubing may be em-

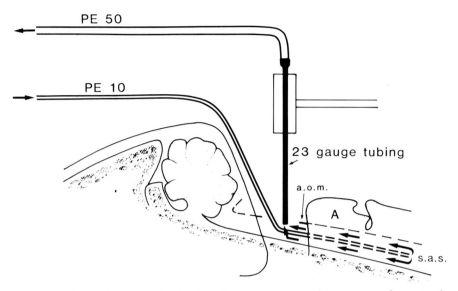

Fig. 2. Diagram illustrating the experimental apparatus for spinal cord perfusion. The rat's head is angled, the atlantooccipital membrane (AOM) is opened, and the PE-10 cannula is introduced into the subarachnoid space (SAS). THE PE-50 "pull" tubing is attached to 23-gage thin-walled metal tubing positioned by micromanipulation at the cisterna magna. Arrows show direction of flow of perfusate. A denotes the atlas bone.

ployed. In the rat, it is necessary to maintain the body rigid with the spinal column stretched to allow the cannula to be advanced using a slight turning motion. This cannula serves as the inflow cannula and can be inserted as far as 7–8 cm in a rat. The outflow cannula is usually mounted on a micromanipulator and placed at the opening of the cisterna magna. We often place a small plug of petroleum jelly in the fourth ventricle to prevent contamination of the spinal perfusate by cerebral spinal fluid from more rostral areas (note: this may increase intracranial pressure!).

Although all studies reported to date have been carried out in anesthetized animals, the technique of maintaining a chronic cannula in the intrathecal space has been described (Yaksh and Rudy, 1976), and it should be quite possible to implant a withdrawal cannula at the level of the cisterna magna to carry out spinal perfusions in unanesthetized animals as well. With the use of a spinal perfusion, it is possible to detect release of presumed neurotransmitters within the spinal cord in response to local depolarization, peripheral afferent input, or stimulation in rostral areas (Pittman et al., 1984). As well, perfusion of appropriate antagonists or agonists can be utilized to modify the physiological responses of animals to peripheral stimuli.

### 3.3. Cup Superfusion

In a manner similar to that employed for ventricular and spinal cord perfusion, it is possible to use a cup superfusion to collect from or stimulate a localized area of tissue on the surface of a structure. Therefore, in 1963, Mitchell introduced the cortical cup for detecting the spontaneous and evoked release of acetylcholine from the cerebral cortex. The cortical cup subsequently has been used in a variety of acute and chronic preparations for the perfusion of cerebral cortex (Celesia and Jasper, 1966; Phillis, 1968), or even for underlying subcortical tissues, such as the caudate nucleus, which can be approached from the ventricular space (e.g., Besson et al., 1971). Basically, the technique consists of implanting or positioning a small cylinder on the pial surface of the cortex or on the ependyma of the ventricular structure so that fluid can be introduced and withdrawn from the cup. This may be carried out either in sequential washes (Mitchell, 1963) or as a continuous flow (Celesia and Jasper, 1966). One important consideration when carrying out cup superfusions of cortical tissue is that it may be necessary to preheat the perfusate because there is no transit through brain tissue to enable the fluid to be warmed. In our experience, cortical electrical activity decreases markedly if the perfusate is not maintained near 37°C (Pittman, Renaud, Blume, and Lamour, unpublished observations).

Further references to methodology employed in cup superfusion can be found in the papers of Mitchell (1963) and Celesia and Jasper (1966), and the review by Myers (1972).

### 3.4. Local Tissue Perfusion

Another perfusion technique has been developed that overcomes many problems associated with the in vivo perfusion analyses described above. Push-pull perfusion allows for an ongoing analysis of the CNS neurochemical activity in discrete brain regions in the awake, behaving animal model. In addition, it provides a route for continuous and accurate administration of drugs into precise brain areas. As the list of putative neurotransmitters grows, the push-pull perfusion technique has been valuable in helping to demonstrate release, coordination of release with physiological function, and identification of site of action for numerous neurotransmitter candidates. As well, a pharmacological profile can be obtained easily by concurrently administering agonists or antagonists and by evaluating behavior or physiological response.

In brief, the principle of the push-pull perfusion technique is to infuse into a small brain region a physiological solution via a "push" cannula and to withdraw through a second "pull" cannula in close proximity to the first cannula. It is assumed that any diffusable substances such as neurotransmitters and metabolites present in the tissue will enter the perfusate; conversely, substances in the perfusate will diffuse into the tissue and down their concentration gradient.

The original concept of tissue perfusion was introduced by Fox and Hilton (1958) and was later utilized by Gaddum (1961) for brain tissue. Delgado and Rubenstein (1964) also experimented with push-pull perfusion techniques, but it was Myers (1970) who was largely responsible for modifying Gaddum's push-pull perfusion procedure into the technique that is presently in use today in many laboratories for collection of brain tissue extracellular samples and for drug administration.

A variety of push-pull perfusion cannulae models has been developed and used successfully in many different laboratories (*see* Myers, 1972). Those in use in our laboratory (Malkinson et al., 1977) are similar to those described by Myers (1970). That for perfusion in larger animals such as cats is illustrated in Fig. 3, and consists of an outer pull cannula of 20 gage (stainless steel) tubing soldered to a Collison cannula. The inflow tube is of 27 gage stainless steel. We usually implant a chronic guide tube (17 gage stainless steel), which allows one to carry out perfusions in unanesthetized animals on successive days and also permits one to perfuse sequentially several different sites by simply lowering the cannula to a greater depth on subsequent experimental days. For rats, a smaller 23 gage outer cannula is used, and the inner, push cannula consists of 30 gage stainless steel tubing (Fig. 4). To reduce the dead space on the pull side of the circuit, the Collison cannula has been omitted and the push and pull cannulae are soldered together. One disadvantage of this structure is that it is somewhat more difficult to clean, since the apparatus cannot be disassembled like that of the larger cannula described by Myers.

While carrying out a push-pull perfusion, it is important not to disturb the cannula following its insertion into the brain tissue and to monitor the push and pull tubings and the animal's behavior throughout the perfusion. A major criticism of the push-pull perfusion technique is that it creates tissue lesions at the site at which the neurochemical analysis is in progress. The primary cause of the lesion is occlusion of the pull cannula by tissue debris. When this occurs, the perfused solution is not recovered, and Myers (1972) has calculated that a 10 µL droplet unrecovered

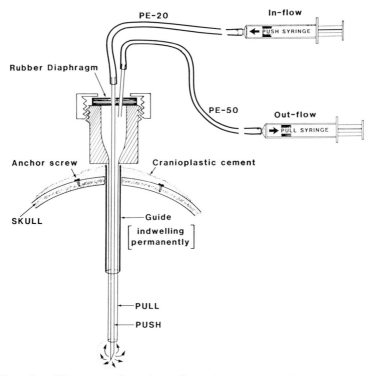

Fig. 3.    Diagram of push-pull perfusion cannula used for larger mammals, such as cats and rabbits. This cannula is similar to that described by Myers (1970) (not drawn to scale).

at the cannula tip displaces an area of over 2 mm of tissue. This can cause a large lesion in less than 1 min if not corrected. Blockage of the pull cannula can be detected when the negative pressure in the cannula increases so that small bubbles appear in the pull tubing. Redgrave (1977) has suggested a modification to allow early detection of blockage by introducing a three-way tap and a manometer at the pull syringe. This three-way tap permits the pull syringe to pull fluid from either the animal or the monitor tube. If the pull tube becomes blocked, the pull syringe pulls from the monitor tube and immediate correction can be made. Nieoullon et al. (1977) also addressed this problem by introducing an "open pulling" system. This is accomplished by placing a small hole in the pull cannula to allow air into the system and prevent local suction of the cannula tip. This modification, however, does not allow early detection of an occlusion and does not prevent the accumulation of the perfusate at the cannula tip should the pull cannula become blocked.

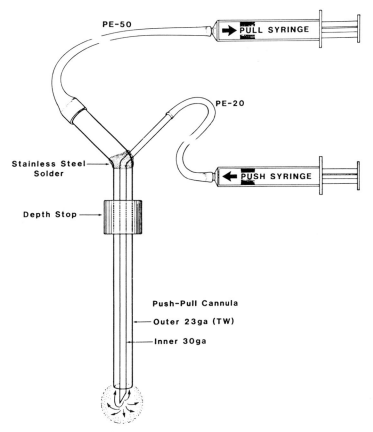

Fig. 4. Push-pull cannula utilized for smaller mammals, such as rats (not drawn to scale).

The rate of the infusion is also an important consideration. Very slow perfusion rates below 25 μL/min in the rat, cat, and monkey increase the risk of cannula blockage. Perfusion rates faster than 100 μL/min result in reduced recovery due to the large dilutional factor. In our laboratory, we successfully carry out perfusions at 25–40 μL/min.

In evaluating push-pull perfusion technology, it is important to address the criticisms relating to tissue damage. Bloom and Giarman (1968) stated that when using push-pull perfusion to show release "the cannula is likely to lead to sufficient tissue damage to make the significance of such types of release somewhat dubious." Izqiuerdo and Izqiuerdo (1971) incorporated that untested assumption into their evaluation of push-pull perfusion and suggested that this technology be discouraged because of the tissue damage. In both cases, no experimental studies appear to have been performed to validate these statements.

Szerb (1967) was the first to examine experimental factors that may influence the recovery of substances through the use of push-pull cannulae. He concluded that the distance of protrusion of the inside needle is directly proportional to the area perfused and the amount of the substance extracted from an agar gel. Yaksh and Yamamura (1974) systematically evaluated numerous factors affecting the performance of the push-pull cannulae in brain tissue. The authors injected metabolically inert $^{14}$C-labeled urea into the caudate nucleus and perfused the area using the concentric design push-pull cannula. Their histological examination showed that, with careful use of the push-pull perfusion cannula, widespread destruction of the tissue does not occur. They also found that maximal uptake of $^{14}$C-labeled urea from the caudate nucleus occurred at a tip extension of 0.75 mm and at a rate of perfusion between 19 μL/min and 80μL/min. This is in general agreement with Myers (1972) who states that, depending on the animal model and experimental conditions, a tip extension of between 0.8 and 1.2 mm and rate of perfusion between 25 and 100 μL/min gives the best recovery of perfusate. In dye perfusion studies, Myers has found that with the use of these recommended parameters, a spherical area of tissue 1.2–1.5 mm diameter can be shown to stain for the dye. Yaksh and Yamamura (1974) also demonstrated a consistent and constant exchange between perfusate and extracellular fluid at the cannula tip if the recommended parameters are maintained and occlusion or partial blockage is prevented. Honchar et al. (1979) showed that insertion of the cannula and interruptions in pump operation disrupt tissue and vascular integrity for only short periods of time, after which changes in perfusate composition are less apparent. As Yaksh and Yamamura (1974) also observed, marked increases in uptake of the tissue label follow insertion of the cannula, and these studies both suggest that longer perfusion periods will minimize the impact of these cannula-insertion artifacts.

A more recent advance based on the same principles as push-pull perfusion is the use of dialysis tubing in brain. Tossman and Ungerstedt (1981) reported the use of a very thin dialysis tubing folded and inserted into the brain of anesthetized rats. They were able to show an increase in the appearance of amino acids into the perfusate when a high potassium solution was perfused. Johnston and Justice (1983) further adapted the dialysis tubing into a cannula assembly very similar to the push-pull cannula. A series of model experiments was performed to characterize the cannula for perfusion of deep brain structures. These authors demonstrated that very slow perfusion rates (less than 1 μL/min) gave 90% recovery of a label, and there was a direct relationship between recovery and perfusion rate. Zetterstrom and Ungerstedt (1984) were able to calibrate their dialysis cannula, in vitro, and then use these values to estimate % recovery and actual extracellular concentrations of brain monoamines. The advantage of using dialysis tubing is that tubing with different cutoffs can be utilized to pick up selectively small molecules and exclude larger enzymes that may destroy the substance to be analyzed. Also, much larger areas can be perfused since loss of perfusate is not a problem; a recent publication reported a transstriatal dialysis in which both striata of a rat were dialyzed (Imperato and DiChiara, 1984). It is quite possible that the dialysis cannula may eventually supplant the classical push-pull perfusion cannula for brain perfusion studies.

## 3.5. In Vitro Perfusion

Whereas the previous discussions have been concerned with the perfusion of a variety of central nervous system areas in vivo, a recent innovation has involved the maintenance of brain tissue under in vitro perfusion conditions. The use of brain slices for neuropharmacological studies has been in use for some years (McIlwain and Rodnight. 1962) and has recently become popular for electrophysiological studies (Dingledine et al., 1980). Problems with brain slices include the disruption of a large amount of tissue and the lack of information as to the connectivity of the cells under study. Therefore, successful attempts recently have been undertaken to maintain entire areas of the brain viable by superfusing the tissue (Otsuka and Konishi, 1974) or by perfusing through the vascular system. Llinas et al. (1981) have described a preparation of the adult guinea pig brainstem-cerebellum that can be maintained by perfusion through the vertebral-basilar arteries and Bourque and Renaud (1983) have described a similar in vitro preparation of the rat hypothalamus, perfused through the carotids. In both preparations, intracellular studies revealed neurons with stable resting potentials, overshooting action potentials, and synaptic potentials. Use of in vitro perfused preparations gives great stability for electrophysiological recording and also permits manipulation of the extracellular environment. It is also possible to pharmacologically activate neurons by adding agents to the perfusate; this raises certain questions as to the integrity of the blood–brain barrier under these conditions.

Perfusion of isolated brain raises certain difficulties not encountered during in vivo studies. The solutions must be oxygenated, since this is the sole source of tissue oxygen. This requires bubbling of the solution; transport of oxygen may be further improved by addition of hydrogen peroxide (Walton and Fulton, 1983). The solution must also be heated and it is possible that addition of protein may be required since plasma oncotic (colloid osmotic) pressure is high relative to that of brain extracellular fluid.

There are numerous compromises that have to be made when undertaking in vitro studies with brain tissue. However, the advantages of this technique for electrophysiological studies and possibly other neuropharmacological studies indicate it has potential as a tool for the neuroscientist.

## 4. Summary

A number of perfusion techniques for neural tissue have been described and attempts have been made to point out some of the technical and interpretative difficulties associated with each approach. In experienced hands and with careful attention to certain details, it appears that the appropriate perfusion experiments can yield timely and important information about brain neurochemistry. In particular, perfusion methods have made it possible to shed some light on the complex relationship between neurotransmitter function and behavioral or physiological processes. There is every reason to believe that perfusion studies will continue to maintain an important position in the repertoire of the modern neuroscientist.

## Acknowledgments

Thanks to C. Von Niessen for typing the manuscript. This work was supported by the Medical Research Council of Canada and the Canadian Heart Foundation.

## References

Arnold E. B., Molinoff P. B., and Rutledge C.O. (1977). The release of endogenous norepinephrine and dopamine from cerebral cortex by amphetamine. *J. Pharmacol. Exp. Ther.* **202**, 544–557.
Bartholini G., Stadler H., Gadea Ciria M., and Lloyd K .G. (1976). The use of the push-pull cannula to estimate the dynamics of acetylcholine and catecholamines within various brain areas. *Neuropharmacology* **15**, 515–519.
Battacharva B. K. and Feldberg W. (1958). Perfusion of cerebral ventricles: effects of drugs on outflow from the cisterna and aqueduct. *Br. J. Pharmacol.* **13**, 156–162.

Bayon A., Shoemaker W. J., Lugo L., Azad R., Ling N., Drucker-Colin R. R., and Bloom F. E. (1981). In vivo release of enkephalin from the globus pallidus. *Neurosci. Lett.* **24**, 65–70.

Besson M. J., Cheramy A., Feltz P., and Glowinski J. (1969). Release of newly synthesized dopamine from dopamine containing terminals in the striatum of the rat. *Proc. Natl. Acad. Sci. USA* **62**, 741–748.

Besson M. J., Cheramy A., Feltz P., and Glowinski J. (1971). Dopamine: spontaneous and drug-induced release from the caudate nucleus in the cat. *Brain Res.* **32**, 407–424.

Bissonnette J. M., Hohimer A. R., and Richardson B. S. (1981). Ventriculocisternal cerebrospinal perfusion in unanesthetized fetal lambs. *J. Appl. Physiol.* **50**, 880–883.

Bloom F. E. and Giarman N. J. (1968). Physiologic and pharmacologic considerations of biogenic amines in the nervous system. *Ann. Rev. Pharmacol.* **8**, 229–258.

Bourque C. W. and Renaud L. P. (1983). A perfused in vitro preparation of hypothalamus for electrophysiological studies on neurosecretory neurons. *J. Neurosci. Methods* **7**, 203–214.

Carmichael. E. A., Feldberg W., and Fleischhauer K. (1964). Methods for perfusing different parts of the cat's cerebral ventricles with drugs. *J. Physiol.* (Lond.) **173**, 354–367.

Celesia G. G. and Jasper H. H. (1966). Acetylcholine released from cerebral cortex in relation to state of activation. *Neurology* **16**, 1053–1063.

Chang T. M., Passaro E. Jr., Debas H., Yamada T., and Oldendorf W.H. (1984). Influence of cisternal pressure on passage of neuropeptides from the cerebrospinal fluid into the peripheral circulation. *Brain Res.* **300**, 172–174.

Chase, T. N. and Kopin, I. J.(1968). Stimulus induced release of substances from the olfactory bulb using the push- pull cannula. *Nature* (Lond.) **217**, 466–467.

Chen J. C., Rhee K. K., Beaudry D.M., and Ramirez, V.D. (1984). In vivo output of dopamine and metabolites from the rat caudate nucleus as estimated with push-pull perfusion on-line with HPLC-EC in unrestrained, conscious rats. *Neuroendocrinology* **38**, 362–370.

Cooper J. F., Hochstein H. D., and Seligmann E. B. Jr. (1972). The limulus test for endotoxin (pyrogen) in radiopharmaceuticals and biologicals. *Bull. Parenter. Drug Assoc.* **26**, 153–162.

Cooper K. E., Kasting, N. W. Lederis, K., and Veale, W. L. (1979). Evidence supporting a role for vasopressin in natural suppression of fever in sheep. *J. Physiol.* (Lond.) **195**, 33–45.

Cooper K. E. and Veale W. L. (1972). Exchange between the blood brain and cerebrospinal fluid of substances which can induce or modify febrile responses. *The Pharmacology of Thermoregulation Symposium,* San Francisco, 277–288.

Davson H. and Welch K. (1971). The Relations of Blood, Brain and Cerebrospinal Fluid, in *Ion Homeostasis of the Brain.* (Siesjo, B. K. and Sorensen, S. C., eds.), pp. 9–21. Academic Press, New York.

Delgado J. M. R. and Rubenstein L. (1964). Intracranial release of neurohumors in unanesthetized monkeys. *Arch. Int. Pharmacodyn. Ther.* **150**, 530–546.

Dingledine R., Dodd J., and Kelly J. S. (1980). The in vitro brain slice as a useful neurophysiological preparation for intracellular recording. *J. Neurosci. Methods* **2**, 323–362.

Elghozi J. L., Le Quan-Bui K. H., Earnhardt J. T., Meyer P. and Devynck M. A. (1981). In vivo dopamine release from the anterior hypothalamus of the rat. *Eur. J. Pharmacol.* **73**, 199–208.

Fox R. H. and Hilton S. M. (1958). Bradykinin formation in human skin as a factor in heart vasodilatation. *J. Physiol.* (Lond.) **142**, 219–232.

Feldberg W. and Fleischhaur K. (1960). Penetration of homo-phenol blue from the perfused cerebral ventricles into the brain tissue. *J. Physiol.* (Lond.) **150, 451**–462.

Feldberg W., Myers R. D., and Veale W. L. (1970). Perfusion from cerebral ventricle to cisterna magna in the unanaesthetized cat. Effect of calcium on body temperature. *J. Physiol.* (Lond.) **207**, 403–416.

Fuxe K. and Ungerstedt U. (1968). Histochemical studies on the distribution of catecholamines and 5-hydroxytryptamine after intraventricular injections. *Histochemie.* **13**, 16–28.

Gaddum J. H. (1961). Push–pull cannulae. *J. Physiol.* (Lond.) **155**, P1–P2.

Heinemann U., Lux H. D., and Gutnick M. J. (1977). Extracellular free calcium and potassium during paroxysmal activity in the cerebral cortex of the cat. *Exp. Brain Res.* **27**, 237–243.

Hery F., Faudon M., and Ternaux J. P. (1982). In vivo release of serotonin in two raphe nuclei (raphe dorsalis and magnus) of the cat. *Brain Res. Bull.* **8**, 123–129.

Honchar M. P., Hartman B. K., and Sharpe L. G. (1979). Evaluation of *in vivo* brain site perfusion with the push-pull cannula. *Am. J. Physiol.* **236**, R48–R56.

Imperato A. and Di Chiara G. (1984). Trans-striatal dialysis coupled to reverse phase high performance liquid chromatography with electrochemical detection: a new method for the study of the in vivo release of endogenous dopamine and metabolites. *Neurosci.* **4**, 966–977.

Izquierdo I. and Izquierdo J. A. (1971). Effects of drugs on deep brain centers. *Ann. Rev. Pharmacol.* **11**, 189–208.

Johnson R. D. and Justice J. B. (1983). Model studies for brain dialysis. *Brain Res. Bull.* **10**, 567–571.

Jordan C. C. and Webster R. A. (1978). The release of acetylcholine in the perfused cat spinal cord in vivo. *Neuropharmacology* **17**, 321–327.

Kasting N. W., Carr D. B., Martin J. B., Blume H., and Bergland R. (1983). Changes in cerebrospinal fluid and plasma vasopressin in the febrile sheep. *Can. J. Physiol. Pharmacol.* **61**, 427–431.

Leusen I. (1950). The influence of calcium, potassium and magnesium ions in cerebrospinal fluid on vasomotor system. *J. Physiol.* (Lond.) **110**, 319–329.

Levine J. E. and Ramirez V. D. (1980). In vivo release of luteinizing hormone-releasing hormone estimated with push-pull cannulae from the mediobasal hypothalami of ovariectomized, steroid-primed rats. *Endocrinology* **107,** 1782–1790.

Llinas R., Yarom Y., and Sugimoni M. (1981). The isolated mammalian brain in vitro: a new technique for the analysis of the electrical activity of neuronal circuit function. *Fed. Proc.* **40**, 2240–2245.

Loullis C. C., Hingtgen J. N., Shea P. A., and Aprison M. H. (1980). In vivo determination of endogenous biogenic amines in rat brain using HPLC and push-pull cannula. *Pharmacol. Biochem. Behav.* **12**, 959–963.

Malkinson T. J., Jackson-Middelkoop L. M., and Veale W. L. (1977). A simple multi-purpose cannula system for access to the brain and/or systemic vascular system of unanesthetized animals. *Brain Res. Bull.* **2**, 57–59.

Martin G. E. and Myers R. D. (1975). Evoked release of [$^{14}$C] norepinephrine from the rat hypothalamus during feeding. *Am. J. Physiol.* **229**, 1547–1555.

McIlwain H. and Rodnight R. (1962). *Practical Neurochemistry*, Churchill-Livingstone, London.

Mitchell J. F. (1963). The spontaneous and evoked release of acetylcholine from the cerebral cortex. *J. Physiol.* (Lond.) **165**, 98–116.

Moir A. T. B. and Dow R. C. (1970). A simple method allowing perfusion of cerebral ventricles of the conscious rabbit. *J. Appl. Physiol.* **28**, 528–529.

Morton I. K. M., Stagg C. J., and Webster R. A. (1977). Perfusion of the central canal and subarachnoid space of the cat and rabbit spinal cord in vivo. *Neuropharmacology* **16**, 1–6.

Muhlethaler M., Sawyer W. H., Manning M. M., and Dreifuss J. J. (1983). Characterization of a uterine-type oxytocin receptor in the rat hippocampus. *Proc. Natl. Acad. Sci. USA* **80**, 6713–6717.

Myers R. D. (1970). An improved push-pull cannula system for perfusing an isolated region of the brain. *Physiol. Behav.* **5**, 243–246.

Myers R. D. (1972). Methods for Perfusing Different Structures of the Brain, in *Methods in Psychobiology* vol. 2, (R. D. Myers, ed.) pp. 169–211, Academic Press, New York.

Nieoullon A., Cheramy A., and Glowinski J. (1977). An adaptation of the push-pull cannula method to study the in vivo release of [$^{3}$H] dopamine synthesized from [$^{3}$H]tyrosine in the cat caudate nucleus: effects of various physical and pharmacological treatments. *J. Neurochem.* **28**, 819–828.

Otsuka M. and Konishi S. (1974). Electrophysiology of mammalian spinal cord in vitro. *Nature* (Lond.) **252**, 733–734.

Palfreyman M. G., Huot S., and Grove J. (1983). Total GABA and homocarnosine in CSF as indices of brain GABA concentrations. *Neurosci. Lett.* **35**, 161–166.

Pappenheimer J. R., Heisey S. R., Jordan E. F., and DeC. Downer J. (1962). Perfusion of the cerebral ventricular system in unanesthetized goats. *Am. J. Physiol.* **203**, 763–774.

Philips S. R., Robson A. M., and Boulton A. A. (1982). Unstimulated and amphetamine-stimulated release of endogenous noradrenaline and dopamine from rat brain in vivo. *J. Neurochem.* **38**, 1106–1110.

Phillis, J. W. (1968). Acetylcholine release from the cerebral cortex: its role in cortical arousal. *Brain Res.* **7**, 378–389.

Pierau F. R. K., Klee M. R., and Klussmann F. W. (1976). Effect of temperature on postsynaptic potentials of cat spinal motoneurones. *Brain Res.* **144**, 21–34.

Pittman Q. J., Hatton J. D., and Bloom F. E. (1980). Morphine and opioid peptides reduce paraventricular neuronal activity: studies on the rat hypothalamic slice preparation. *Proc. Natl. Acad. Sci. USA* **77**, 5527–5531.

Pittman Q. J., Lawrence D., and McLean L. (1982). Central effects of arginine vasopressin on blood pressure in rats. *Endocrinology* **110**, 1058–1060.

Pittman Q. J., Riphagen C. L., and Lederis K. (1984). Release of immunoassayable neurohypophyeal peptides from rat spinal cord, in vivo. *Brain Res.* **300**, 321–326.

Redgrave, Peter (1977). A modified push-pull system for the localised perfusion of brain tissue. *Pharmacol. Biochem. Behav.* **6**, 471–474.

Riddell D. and Szerb J. C. (1971). The release in vivo of dopamine synthesized from labeled precursors in the caudate nucleus of the cat. *J. Neurochem.* **18**, 989–1006.

Robinson, I. C. A. F. (1983). Neurohypophysial Peptides in the Cerebrospinal Fluid, in *The Neurohypophysis: Structure, Function and Control, Progress in Brain Research*, vol. 60, (Cross B. A. and Lenz G., eds.), pp. 129–145, Elsevier, Amsterdam.

Szerb J. C. (1967). Model experiments with Gaddum's push-pull cannulas. *Can. J. Physiol. Pharmacol.* **45**, 613–620.

Tossman U. and Ungerstedt U. (1981). Neuroleptic action on putative amino acid neurotransmitters in the brain studied with a new technique of brain dialysis. *Neurosci. Lett.* Suppl **7**, S479.

Van der Gugten J. and Slangen J. L. (1977). Release of endogenous catecholamines from rat hypothalamus *in vivo* related to feeding and other behaviors. *Pharmacol. Biochem. Behav.* **7**, 211–219.

Walton K. and Fulton B. (1983). Hydrogen peroxide as a source of molecular oxygen for in vitro mammalian CNS preparations. *Brain Res.* **278**, 387–393.

Wood J. H. (1980). Physiology, Pharmacology and Dynamics of Cerebrospinal Fluid, in *Neurobiology of Cerebrospinal Fluid*, (Wood, J.H. ed.) pp. 1–16, Plenum Press, New York.

Yaksh T. L. and Rudy T. A. (1976). Chronic catheterization of the spinal subarachnoid space. *Physiol. Behav.* **17**, 1031–1036.

Yaksh T. L. and Tyce G. M. (1980). Resting and $K^+$ -evoked release of serotonin and norepinephrine in vivo from the rat and cat spinal cord. *Brain Res.* **192**, 133–146.

Yaksh T. L. and Yamamura H. I. (1974). Factors affecting performance of the push-pull cannula in brain. *J. Appl. Physiol.* **37**, 428–434.

Yamamoto C. (1972). Activation of hippocampal neurons by mossy fiber stimulation in thin brain sections in vitro. *Exp. Brain Res.* **14**, 423–435.

Zetterstrom T. and Ungerstedt U. (1984). Effects of apomorphine on the in vivo release of dopamine and its metabolites, studied by brain dialysis. *Eur. J. Pharmacol.* **97**, 29–36.

Zetterstrom T., Sharp T., Marsden C. A., and Ungerstedt U. (1983). In v ivo measurement of dopamine and its metabolites by intracerebral dialysis: changes after *d*-amphetamine. *J. Neurochem.* **41**, 1769–1773.

# Chapter 9

# Brain Electrical Activity in Relation to Behavior

## C. H. Vanderwolf and L.-W. S. Leung

## 1. Introduction

Electrophysiological studies of the brain in mammals have made extensive use of immobilized preparations. Experimental animals are commonly anesthetized or immobilized by curare or by section of the brain stem (cerveau isolé) (Bremer, 1935) or spinal cord at level C1 (encéphale isolé) (Bremer, 1936a,b). Such preparations are preferable to freely moving animals for many types of research since they simplify the technical problems of electrical recording and make it easy to record or control physiological variables, such as blood pressure or blood gas concentrations. Some procedures, such as intracellular recording, are virtually impossible unless the preparation is rigidly immobilized. Furthermore, since animals that have been anesthetized or had the neuraxis transected have a greatly reduced capacity to feel pain, many procedures that could not be used in intact conscious animals can be carried out freely.

However, other types of investigation in neurobiology require the use of freely moving animals. A good case can be made for the view that the chief function of the brain is the regulation of motor activity or behavior (Sperry 1952; Vanderwolf, 1983). Agents such as anesthetics that eliminate nearly all behavior must produce gross disturbances in the overall function of the brain. Consequently, studies carried out in anesthetized animals may provide incomplete or misleading information with respect to normal brain function (see below). The situation is somewhat different in encéphale isolé or curarized preparations in which many aspects of brain function appear to be relatively intact. However, since overt behavior is largely or entirely abolished in these preparations, the possibilities for relating brain activity to behavior are very limited. Furthermore, phenomena observed in curarized preparations may be very different from those observed in freely moving animals. For example, in curarized rats the systemic administration of $d$-amphetamine produces a brief increase followed by a long depression in unit activity in the striatum (Rebec and Groves, 1975). A similar experiment in freely moving rats showed that multiunit activity in the striatum was increased by $d$-amphetamine and remained at high levels as long as the behavioral effects of the drug persisted (Hansen and McKenzie, 1979).

In this paper, we shall discuss the basic concepts and methods used in the study of brain electrical activity as a means of elucidating the neural mechanisms of behavior. Illustrative examples will be drawn from work relating hippocampal and neocortical electrical activity to behavior and to the effects of drugs.


## 2. Behavior

One of the major difficulties confronting a neuroscientist who wishes to add behavioral studies to an ongoing program of neuroanatomical, neuropharmacological, or neurophysiological work is that there seems to be very little general agreement on how behavior should be studied. Current studies in animal behavior emphasize the importance of extensive observation of spontaneous behavior under natural or seminatural conditions (Alcock, 1984; Hinde, 1970). Laboratory studies of behavior frequently involve training animals in various types of apparatus, such as mazes or Skinner boxes (Bureš et al., 1976; Mackintosh, 1974; Munn, 1950). Human behavior is usually assessed by means of various paper and pencil tests (Anastasi, 1982; Cronbach, 1984). On each of these topics there is a vast literature that seems rather opaque to an outsider. How should one begin?

The first problem is to decide how behavior should be described and classified. According to Hinde (1970) there are only two basic ways of describing behavior: (1) Behavior can be described in terms that refer to spatiotemporal patterns of muscular activity. (2) Behavior can be described in terms of its consequences regardless of the pattern of muscular activity involved.

The first approach has been used extensively to describe spinal reflexes and other simple motor patterns. For example, the term "flexion reflex" in a spinal dog refers ultimately to the pattern of contraction of the muscles of the hind limb (Sherrington, 1910; Denny-Brown, 1979). The same tradition is being followed, in principle, when behavior is referred to in such terms as "walking," "rearing on the hind legs," "wet dog shake," or "standing immobile with the head held up and the eyes open." However, the actual patterns of muscle activity responsible for various behaviors have been studied in only a limited number of cases (Basmajian, 1978). When using the second means of describing behavior, one might for example state that "the rat pressed the lever," without specifying whether the lever was pressed with one, the other, or both forepaws, or by gripping it in the teeth. Descriptions of behavior in terms of its consequences are frequently used in situations in which a number of different motor patterns lead to the same result.

Despite the apparent simplicity of the description of behavior, many different schemes for its classification are in use in contemporary neuroscience. Sometimes the criteria for classification are stated explicitly, but in many cases classifications are implicit and taken for granted by their users. For example, many workers in the brain–behavior field accept without question the proposition that behavior can be adequately classified in terms of the different mental processes presumed to cause the behavior. Thus, animals or humans may be described as conscious or unconscious, attentive or inattentive, emotional or calm, motivated or unmotivated, having normal or flat affect, and so forth. It is important to note that terms of this type do not refer to any particular pattern of muscular activity. For example, a man may be described as closely attentive while sitting motionless listening to a lecture or while walking down the street listening to the talk of a friend.

Many other classifications are possible. Behavior is often considered to be divisible into "learned" and "unlearned" categories. Unlearned behavior is then further subdivided into various reflexes and instinctive behaviors, whereas learned behavior is subdivided into classical conditioning, operant conditioning, latent learning, imprinting, and so forth. Another scheme divides behavior into such categories as feeding, courtship, sexual behavior, agonistic behavior, exploratory behavior, and thermoregulatory behavior. These categories comprise both learned and unlearned behavior. A third type of classification that has a long history in neurobiology distinguishes reflexes (including involuntary conditioned reflexes) from voluntary movement (Fearing, 1930).

It is apparent that behavior can be classified in different ways to suit different purposes. From the point of view of neurobiology, it is important to have a classification of behavior that reflects natural divisions of function in the central nervous system. One simple classification that is widely accepted as corresponding to a natural division of function in the brain is the division of behavior into three major states: waking, quiet (slow wave) sleep, and active (rapid eye movement) sleep. An extensive attempt has been made to relate the electrophysiological and neurochemical activity of the brain to these three behavioral states (Jouvet, 1972, 1977). A great deal of related work, spanning several decades, has attempted to interpret hippocampal and neocortical activity in terms of general psychological concepts, such as learning, memory, attention, and motivation. However, more recent work has shown that such concepts do not adequately reflect the relation between brain activity and behavior. It appears that activity in the hippocampus and neocortex, as well as the reticular formation, is strongly related to the spatiotemporal pattern of the concurrent motor activity (Siegel, 1979; Vanderwolf, 1969; Vanderwolf and Robinson, 1981). The fact that this simple conclusion was arrived at only after many years of previous research may be largely attributable to the implicit assumption of most neuroscientists that brain activity should be interpreted in terms of the traditional philosophical and psychological categories of the mind (Vanderwolf, 1983). This assumption has diverted many investigators from studying the relation between brain electrical activity and concurrent spatiotemporal patterns of muscular activity.

## 2.1. Behavioral Methods

The basic method of studying behavior is very simple. It consists of nothing more than patient observation of the behavior of animals or humans in a variety of situations. Simple description has been greatly undervalued in contemporary science (Lorenz, 1973; Dement and Mitler, 1974) and, when beginning a new project, it is usually well worth the time it takes to "look things over" for some time before attempting systematic quantitative experimental work. Premature quantification often turns out to be irrelevant quantification. Hutt and Hutt (1970) and Lehner (1979) provide an introduction to the study of behavior by direct observation. Barnett (1975) discusses the naturally occurring behavior of rats, and a summary of what is known of such behavior in relation to brain function is provided by Whishaw et al. (1983).

In order to correlate spontaneous behavior with electrical events occurring in the brain, it is necessary to make an accurate record of what an animal does. Vanderwolf (1969) used an ink-writing polygraph to record hippocampal and neocortical slow wave activity and a set of manually operated signal markers to code different behaviors on the polygraph chart as they occurred. A variation of this procedure is to record brain activity on a multichannel FM tape recorder together with a running verbal description of the behavior on one of the channels (Leung et al., 1982b). The problem introduced into such records by the variable reaction time of the observer can be partially solved by using one channel of the recording device to record the output of a movement sensor. Correlation of the deflections produced by the movement sensor with the code provided by the signal markers or voice channel usually makes it possible to give accurate onset and offset times for identified behaviors. A simple movement-sensing device that suffices for many purposes consists of a light-weight square platform with a slightly raised edge (to help prevent animals from falling off) that is mounted on foam rubber blocks. A bar magnet attached to the center of the underside of the platform is inserted in a wire coil obtained from an electromechanical relay. Movement of the experimental animal on the platform generates a voltage in the coil. It is also possible to make use of the output of a piezoelectric crystal or phonograph cartridge as a means of monitoring the movements of the platform. The entire device should be isolated from environmental vibration by a heavy vibration-reducing table (Vanderwolf, 1975).

Movement sensors of this type are sensitive enough to record most small movements such as isolated movements of the head or slight changes in posture. The signal appears instantaneously when the platform is tapped and the oscillations are damped out in less than 0.5 s. However, the sensitivity of the sensor depends on multiple factors. It varies in different parts of the platform, is greater for vertical than horizontal movements, decreases with the weight of the animal, and may decrease with time as the foam rubber pads become compressed. A simple alternative system is to include a wire with a free end in the leads to the animal's head. Movement of the head or the leads generates potentials in the wire that can be recorded (Olds et al., 1972). However, when recordings were made using the two systems simultaneously in comparison with direct visual observation of the animal, the platform movement sensor proved to be somewhat the better of the two (Vanderwolf, unpublished data). The potentials generated in the free wire are somewhat unpredictable, occurring occasionally during periods of immobility or after a delay when a movement is begun.

A variety of other platform or cage-mounted, movement-sensing devices has been described (Davis, 1970; Iversen, 1973; Mundl, 1966). Mundl and Malmo (1979) have made use of a small accelerometer mounted on the head of the experimental animal. Chapin et al. (1980) describe a method for recording the contact of the feet with a conductive surface.

It should be emphasized that movement sensors are not a substitute for direct observation, but only a supplementary technique to be used with it. An experimenter should formulate a limited number of nonoverlapping categories of behavior prior to making a record. If there appears to be little correlation between brain electrical activity and the recorded behavior, it may be that the categories of behavior that were selected were inappropriate. In work on the hippocampus (Vanderwolf, 1969; Ranck, 1973a), considerable trial and error was necessary to develop behavioral categories that correlated consistently with the different patterns of slow waves (see below). Consequently, recordings had to be repeated many times in order to try out different ways of categorizing behavior. This particular problem does not arise if one makes videotapes or photographic records of behavior together with the concurrent brain electrical activity. The same record can be analyzed repeatedly in order to examine different aspects of both the behavior and the concurrent brain activity and the relations between them.

Siegel et al. (1979) describe an arrangement in which an animal is filmed directly and in a mirror placed on the side opposite the camera, so that both sides are visible simultaneously while brainstem unit activity is displayed on a counter that increments with each unit discharge. A second counter (both are in the field of view of the camera) displays a time code to permit later correlation with a tape recording of unit and other electrophysiological data.

A videotape recorder has several advantages as compared to film. Disparate images can be readily combined on the same record by using two cameras and a special effects generator. Thus, one camera can be focused on an oscilloscope screen, whereas the other is focused on the moving preparation. The same thing can be accomplished on film by using mirrors, but this is sometimes rather cumbersome. Further, unlike film, once a videotape record has been made, it is immediately available for replay and analysis. Many modern videotape recorders permit slow-motion playback, single-frame advance, and the ability to "freeze" an image for short periods. Some disadvantages are that the temporal resolution of most videotape recorders is only 60 fields/s (two fields are interlaced to give 30 frames per second) and that conventional silicon vidicon tubes display considerable persistence of images. This results in simultaneous multiple images (blurring) of rapidly moving objects on the TV screen. Video cameras equipped with a rotary shutter overcome this problem by a stroboscope-like effect. We have found that a videotape recorder controlled by a specially made tape motion controller is very helpful in analyzing the details of brain–behavior relations. As a means of eliminating all possibility of experimenter bias in such work, it is possible to carry out double-blind analyses of brain activity and behavior (Leung et al., 1982a). Behavior is recorded on a videotape while stimulus-evoked potentials are recorded on the audio channel by means of an FM adapter. The behaviors occurring during successive numbered stimuli are classified by an observer who does not view events on the audio channel. Subsequently, the bioelectrical responses to selected stimulus presentations are averaged by a second observer who does not view the videotapes. Finally, the two analyses are correlated.

With the development of inexpensive but fast video digitizing boards for microcomputers, we believe that in the near future the hardware and software for the analysis of video signals of behaving animals will be readily available to many researchers. For example, a light pen can be used to locate a certain point on the animal shown on a video screen, and with a push of a key, the spatial coordinates of the light pen are stored into computer memory. Subsequently, the computer can plot the sequence of movement of points of the animal. Some simple movements, e.g., the motion of a fish (Wieland, 1983) or the location of the brightest spot on the screen (e.g., a light on an animal's head) may be processed automatically.

### 2.1.1. Specific Behavioral Tests

During spontaneous behavior, many different types of activity occur in an irregular, rapidly changing sequence. Brain activity changes in close relation with the changes in behavior. If stable conditions are required over a long period in order, for example, to average electrical activity or to collect superfusates from the brain (Dudar et al., 1979), it is useful to be able to control behavior so that a given activity will occur continuously with minimal intrusion from other behavior. Whishaw and Vanderwolf (1973) and Rudell et al. (1980) used a motor-driven treadmill to study brain electrical activity during long, continued walking. In contrast, long periods of waking immobility (freezing) can be obtained if rats are placed in an apparatus in which they have previously received 1–2 painful electric shocks to the feet. A shock avoidance procedure in which a rat is trained to jump out of a box provides an abrupt onset of vigorous motor activity following a period of near total immobility. This can be useful in studying movement-related brain activity (Vanderwolf, 1969). Long periods of behavioral immobility are rare in food reinforcement training procedures even when immobility is specifically reinforced (Blough, 1958). This can be a disadvantage since brain events that are correlated with movement may be difficult to detect in an animal that is continually moving about.

Sleep, especially active sleep, occurs in rats in an experimental apparatus only after considerable familiarization. However, both quiet sleep and active sleep occur promptly in rats after their removal from a sleep deprivation apparatus consisting of inverted flower pots in a partially filled aquarium (Morden et al., 1967).

The foregoing discussion has emphasized the study of the correlation of behavior with spontaneously occurring patterns of brain electrical activity. A complementary procedure consists of the modification of brain electrical activity by an experimental manipulation (such as electrical stimulation, surgical removal of selected brain regions, or treatment with a drug) followed by study of: (a) changes in the correlation of brain activity with concurrent behavior, and (b) changes in the adaptiveness of behavior. It is commonly the case that agents that disrupt the function of the cerebral hemispheres in animals do not produce a loss of any specific movement, but rather produce a disorganization of behavior and a lessening of its adaptive value. Decorticated rats, for example, run about very actively but accomplish little of any use to themselves (Lashley, 1929, 1935). There are a large number of behavioral tests that are sensitive to cerebral dysfunction in rats. Learning tests such as various types of mazes as well as such taxon-specific behaviors as nest building, food hoarding, or maternal behavior have been studied extensively. Although an extensive discussion of such tests is beyond the scope of this paper, the following references provide an introduction to current and classical work (Bureš et al., 1976; Munn, 1950; O'Keefe and Nadel, 1978; Morris, 1983; Olton, 1983; Sutherland et al., 1983; Whishaw et al., 1983).

# 3. Electrophysiological Methods

## 3.1. General Recording Methods

### 3.1.1. Units and Slow Waves

In the freely moving animals, two types of extracellularly recorded neural signals are commonly recognized: unit activities
and slow waves. Unit discharges and slow waves have different
generating mechanisms and require different recording and analysis techniques (Table 1). In the following we shall first discuss a
general recording setup and then the specifics of the generation,
recording, and analysis of slow waves and units. It should be emphasized that a lot can be done with the use of a basic setup and
simple analysis or inspection of the records.

### 3.1.2. General Recording Setup

There are many reviews of systems for amplifying and recording biological signals (Fox and Rosenfeld, 1972; Geddes, 1972;
Schoenfeld, 1964; Katz et al., 1964). A recording system may
consist of three main types of equipment: amplifier, signal selection, and storage devices (Fig. 1).

Biological signals are commonly small ($\mu$V or mV range) and
require voltage amplification (about 1 V range) and current amplification before they can be recorded. Commonly, amplification of
small signals is achieved in two stages: a preamplifier and a main
amplifier stage. The preamplifier stage is preferably close to the
signal source, and should have an input impedance of at least 100
times that of the electrode used (Fig. 1). For small, behaving animals, field-effect transistors (FET) or operation amplifiers can be

TABLE 1
Characteristics of Two Types of Extracellular Neural Signals

|  | Units | Slow waves |
|---|---|---|
| Electrode used | Microelectrode (<10 $\mu$m tip) | Micro- and macro-electrode |
| Signal recorded | Action potentials | Mainly postsynaptic potentials |
| Duration | 0.3–2 ms | 10–2000 ms |
| Main frequencies | 0.5–3 kHz | 0.5–100 Hz (EEG) 0.5 Hz–1 kHz (EP) |
| Analysis technique | Digital (point process) | Analog (continuous signal) |

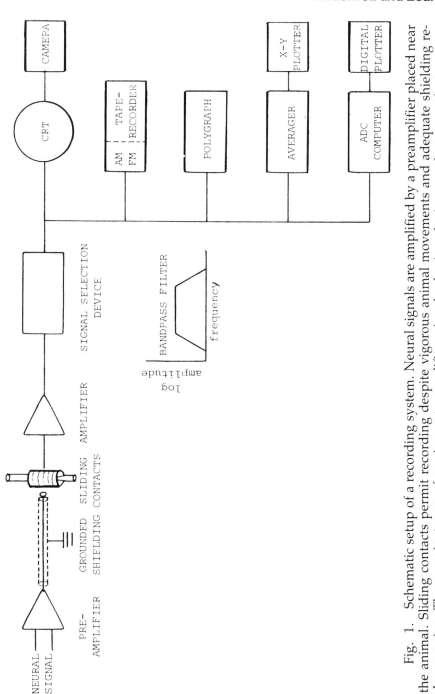

Fig. 1. Schematic setup of a recording system. Neural signals are amplified by a preamplifier placed near the animal. Sliding contacts permit recording despite vigorous animal movements and adequate shielding reduces noise. Three main types of equipment—amplifier, signal-selection devices, and storage devices—are shown. The bandpass filter is an example of a signal-selection device. Storage medium may be film, photograph, paper, magnetic tape, or disk. CRT: cathode ray tube or oscilloscope; ADC: analog to digital converter; AM: amplitude modulation; FM: frequency modulation.

placed on the head of the animal (Fox and Rosenfeld, 1972; Rosetto and Vandercar, 1972). The preamplifier acts mainly as an impedance-matching device (cathode or source follower) so that the low impedance output is not susceptible to various interferences. Signals may be directly coupled for dc recording or capacitance-coupled for ac recording. Differential is preferred over single-ended recording in order to remove noise common to both active and referential leads, e.g., line frequency noise.

Signal selection devices are designed to maximize wanted signals and screen out unwanted noise. A frequency filter is a type of signal selection device. A filter is described as low-pass if it passes dc (0 Hz) and low-frequency signals, high-pass if it passes only high-frequency signals, and band-pass if it passes only signals of a selected bandwidth (Fig. 1, inset). The limits of a filter are indicated by 3 dB (decibel) points, or the frequencies (called corner frequencies) at which the signals are attenuated by 70.7% of the maximal plateau response. The roll-off of a filter is the attenuation of signals beyond the corner frequency expressed as the logarithmic amplitude attenuation (20 dB = 10 times) per decade (10 times) or per octave (2 times) change of frequency. A filter consisting of a single resistance–capacitance combination has a 20 dB/decade or 6 dB/octave roll-off.

The average noise from an electrode depends on the square root of the product of electrode resistance and frequency bandwidth (Geddes, 1972; Schoenfeld, 1964). Therefore, unneeded frequency bands should be filtered at source. For example, units are commonly filtered between 0.3–5 kHz, EEG from 0.5–100 Hz, and evoked potentials 0.5 Hz–1 kHz (Table 1). Some amplifiers provide line-frequency (50 or 60 Hz notch) filters that can filter effectively the interference caused by the ac power supply. However, the line-frequency filter is seldom restricted to 50 or 60 Hz only, and signals of 30–80 Hz can be attenuated and strongly distorted in phase. Therefore, it is inadvisable to use this filter if signals of 30–80 Hz are of importance. Besides, there are often other ways to reduce 50/60 Hz interference by adequate shielding and by connecting high conductance ground leads to a single common point to prevent ground loops (Wolbarsht, 1964).

Storage devices include oscilloscopes with camera or storage capability, tape recorders, polygraphs, and a variety of digital storage devices. The medium of storage may be film, photograph, paper, or magnetic tape or disk. Perhaps the cheapest storage medium on a cost-per-time recorded basis is the magnetic tape. Selected signals can then be photographed or written on a polygraph by playing back the tape. The frequency limits of various recording devices should be noted: An oscilloscope or a cathode ray tube (CRT) can usually follow all frequencies of a biological signal. A mechanically moving pen of a polygraph cannot follow frequencies higher than about 75 Hz. On a tape recorder, frequencies above 50 Hz may be recorded by an amplitude-modulated (AM) audio channel, whereas frequencies of 0 (dc)–50 Hz require the use of an AM–FM (frequency-modulated) adaptor that is an integral part of an instrumentation tape recorder. In AM recording, the amplitude of the signal is coded directly by the strength of magnetization. In FM recording, the amplitude of the signal remains constant, and the frequency deviates from a center, carrier frequency in a manner proportional to the original amplitude of the input signal. The frequency response of a tape-recorded signal at a particular tape speed and the distortions caused by wow and flutter (Katz et al., 1964) should be tested by inputing pure sine waves of different frequencies and inspecting and analyzing the tape response (e.g., using spectral analysis discussed below).

An analog-to-digital converter (ADC) converts a continuous signal to a series of discrete, digital values. Since a digital value can only take discrete values at discrete times (e.g., an 8-bit ADC can take $2^8$, or 256, discrete values), a continuous analog signal is approximated by a staircase of digital values. Voltage digitization creates a resolution approximation, i.e., voltages smaller than a certain value (1/256 or 0.39% of the maximum value in an 8-bit ADC) cannot be resolved. The digitization in time is commonly referred to as sampling. The sampling or digitizing rate must be at least twice the signal frequency to be studied (Souček, 1972). All frequencies higher than half the sampling frequency (Nyquist frequency), if present, must be strongly attenuated by filters to prevent aliasing or Faltung (folding back). The latter is a phenomenon in which signals higher or lower than the Nyquist frequency will be equally represented in the sampled signal (Souček, 1972).

Special devices may be used for recording from freely moving animals. A flexible, low-noise cable (Cooley and Vanderwolf, 1978) and tight connectors are necessary. For active animals and long-term recording, a commutator can help to maintain electrical contact despite rotations of an animal. We prefer a slide-wire, slip-ring commutator (Micco, 1977) to the mercury-filled type. The former appears to have lower noise, is more compact, and contains no toxic mercury.

Telemetry, or the transmission of signals via radio waves, should become more popular in small animals as small, light multichannel telemetry devices become commercially available (Eichenbaum, et al., 1977). The recording of brain activity without restraining cables will allow the study of a larger variety of behaviors, especially in natural settings.

## 3.2. Spontaneous Slow Waves and Evoked Potentials

### 3.2.1. Mechanisms of Generation

Spontaneous slow waves, the electroencephalogram (EEG), or the electrocorticogram (ECG) is commonly recorded by macroelectrodes, or electrodes that are larger than the size and extent of a neuron. Evoked potentials (EP) are responses to sensory or electrical stimuli, usually recorded by macroelectrodes. Much of the frequency of EEG is below 100 Hz. Slow waves recorded extracellularly represent a sum total of potentials generated by various currents that traverse both intracellular and extracellular media (Lorente de Nó, 1947; Freeman, 1975). In regular, layered structures (e.g., cortex), extracellular currents sum together to give high-amplitude slow waves. Slow waves are mainly generated by postsynaptic potentials that, on account of their duration and dipole field characteristics (Elul, 1972; Freeman, 1975; Purpura, 1959; Creutzfeldt, 1974; Humphrey, 1968), are more likely to sum together than unit potentials. The latter, however, may contribute during paroxysmal activities or during a highly synchronous activation. Very slow (0–1 Hz) potentials can be generated by glial cells (Somjen, 1973).

### 3.2.2. Recording Methods

Tungsten, platinum, and steel have proved to be suitable for chronic indwelling electrodes (Cooper, 1971; Delgado, 1964). Wires of 60–250 μm diameter, coated with Teflon or several coats of baked-on varnish, can be cut and used. To provide for a large recording (or stimulation) surface and low impedance, insulation at the cut end is sometimes scraped away. It is advisable to check the solder connections of a wire and look for breaks in the insulation immediately before implantation.

Various methods of chronic implantation of electrodes have been described (Cooley and Vanderwolf, 1978; Delgado, 1964; John, 1973; Olds, 1973; Skinner, 1971). The essential parts for such an implantation are electrodes, small screws, miniature connecting pins with or without a miniature plug, dental cement, and solvent. The skull should be scraped clean of connective tissues. Jeweler's screws are inserted into the skull as anchors and can be used as ground or reference electrodes. Small holes are drilled for insertion of an electrode, or multiple electrodes. The electrodes or wires may be crimped or soldered onto pins (or sockets) that may or may not be inserted into a miniature plug.

The advantage of a plug is to allow quick connection during recording, but longer wires may be required to connect all electrodes at different parts of the brain to a single plug. All screws, wires, pins, and plug must be securely embedded in dental cement.

Electrodes can be placed by means of stereotaxic coordinates. Stereotaxic atlases for various animals have been referred to in DeValois and Pease (1974) and Pellegrino and Cushman (1973). Paxinos and Watson's (1982) atlas for the rat brain is a recent addition. Electrophysiological criteria may be used as a supplement to placement by stereotaxic techniques. For macroelectrodes, the profile of evoked potentials may allow an accurate ($\pm 100$ $\mu$m) placement near cortical cell layers during surgery (Freeman, 1963; Leung, 1980).

Rats usually recover rapidly and without incident after 2–4 h surgery under sodium pentobarbital. By the next day, they will walk and eat normally. However, a minimum of 7 d should be allowed for complete recovery. In our experience, during the first 7–10 d following electrode implantations, electrophysiological (especially evoked) responses tend to be more labile, after which stable recordings can be obtained for weeks or months.

During recording, the animal is connected with mating connectors to the electrode pins (or sockets) on its head. The preamplifier in the recording setup (Fig. 1) may not be necessary for low-impedance ($<20$ k$\Omega$) macroelectrodes. However, it is still a good idea to keep cables between the animal and the first stage of amplification as short as possible. Cooper (1971) and Geddes (1972) have described procedures for measuring impedance of electrodes. It should be noted that metal electrodes have very high impedance at low ($<10$ Hz) frequencies (Geddes, 1972).

It is important that a good signal be obtained from the brain structures of interest. If electrodes are implanted in the neocortex or hippocampus it should be possible to obtain clear artifact-free signals with an amplitude up to 1–3 mV. Tiny signals, of the order of 50 $\mu$V, suggest that the tissue has been damaged during surgery or that the electrodes are incorrectly placed.

Differential monopolar recordings, referred to an indifferent electrode, are more easily interpreted than bipolar recordings in which both electrodes are placed in active tissue. A steel screw placed in the skull over the cerebellum provides a good reference since the cerebellum generates very little low-frequency activity. Reference electrodes placed in the frontal or nasal bones tend to pick up neocortical slow waves or the rhythmical potentials of the olfactory bulb. Bipolar recordings, however, possess certain advantages. Artifacts are fewer than in monopolar recordings and the activity recorded is more likely to be restricted to the region of the electrode tips. For example, monopolar recordings from the parietal neocortex in rats or rabbits contain a mixture of waveforms generated in the neocortex with others generated in the hippocampus. Surface-to-depth bipolar electrodes (one tip on the pial surface and one inserted to a depth of 1–1.5 mm) make it possible to reject hippocampal waveforms as common-mode signals, resulting in an uncontaminated record of neocortical activity. Since the waves recorded at the pial surface are phase-reversed with respect to those in the depths of the neocortex (a dipole field), a surface-to-depth recording will sum the amplitudes of the two, yielding a large signal. Similarly, surface-to-depth electrodes in the hippocampus (one tip near the alveus and the other inserted to the hippocampal fissure) yield a record of hippocampal activity that is relatively free of neocortical waveforms. Some workers, ignoring such precautions, have confused hippocampal and neocortical activity (Vanderwolf and Ossenkopp, 1982).

Intracranial electrical stimulation should be delivered via optical or radiofrequency isolation units. Stimulating wires can run beside recording wires without separate shielding. However, the stimulating current return electrode should be different from the recording ground. Single electrical pulses to the brain do not usually cause behavioral changes. For more details, readers should consult reviews on stimulation (Ervin and Kenney, 1971; Ranck, 1975).

### 3.2.3. Analysis of EEG and EPs

Spontaneous EEG can be recorded on a polygraph. Visual inspection and measurement of polygraph records can be used for dominant EEG rhythms, e.g., alpha and theta rhythms. For a decomposition of EEG into component frequencies, discrete Fourier spectral analysis is commonly used (Jenkins and Watts, 1968; Niedermeyer and Lopes da Silva, 1982; Matoušek, 1973). The latter method decomposes an analog signal into sine and cosine waves. Sine and cosine waves of the same frequency are represented by an amplitude (or power that is the square of the amplitude) and a phase (with respect to some reference time). In practice, one or more EEG signals are digitized at twice the maximal frequency (Nyquist frequency) of interest, with higher frequencies attenuated by antialiasing filters. A segment of EEG is commonly multiplied by a function that tapers off the contribution of both ends of the segment. Various segments of EEG may be averaged for a single spectrum. An autopower spectrum is the power of a single signal as a function of frequency. A cross-power spectrum measures the shared power of two separate signals as a function of frequency. The cross-power spectrum can be represented by a cross-phase and a coherence spectrum. The cross-phase spectrum measures the phase shift and the coherence spectrum estimates the linear relation (similar to a correlation coefficient) between the two signals, as functions of frequency. For gradual changes of spectral estimates, e.g., following a drug and during behavioral changes, a time lapse-compressed spectral display (Matoušek, 1973) can be used.

Averaging is used to improve the signal-to-noise ratio of EPs (Glaser and Ruchkin, 1976). Averaging or summing is triggered by a pulse that usually precedes the actual stimulus. Averaging usually assumes that a signal is a time-locked response to a stimulus, and the background activity is random or white, Gaussian noise. Then the evoked signal-to-noise ratio will increase as $N^{1/2}$ where $N$ = number of stimuli.

Specific-purpose electronic averagers or general-purpose laboratory computers perform averaging by digitizing at discrete time points separated by an interbin (sampling) interval, and then summing the newly acquired values with the existing values at each bin. Obviously, the resolution and sampling problems (above) apply to digital averaging devices.

More elaborate statistical analyses of EPs are described by John (1973), Glaser and Ruchkin (1976), and Freeman (1975).

### 3.3. Unit Activities

#### 3.3.1. Mechanism of Generation

A propagating action potential recorded in an extracellular medium near to an axon is a positive–negative–positive triphasic wave, corresponding to source–sink–source of a traveling active sink (Lorente do Nó, 1947). Extracellular recordings near neuronal cell bodies are commonly positive–negative waves of peak amplitude of 30–2000 μV and duration of 0.3–1 ms (Towe, 1973; Phillips, 1973). Axonal action potentials are usually shorter in duration (< 0.8 ms) (Phillips, 1973), whereas the relatively rare dendritic spike may exceed 4 ms (Llinas and Nicholson, 1971).

#### 3.3.2. Recording Methods

Microelectrodes made of tungsten, steel, or platinum–iridium are commonly used for recording in freely moving animals. The fabrication of these electrodes and their main characteristics have been reviewed previously (Geddes, 1972; DeValois and Pease, 1973; Snodderly, 1973). Unit activity is best recorded differentially between two microelectrodes that are fairly close together in the brain (Sasaki et al., 1983). This permits the rejection of jaw muscle action potentials and other unwanted signals. Brief monopolar records from each of these electrodes serve to identify the one that is carrying unit potentials.

Two methods are commonly used for recording of units in freely moving animals. The first method is to implant many microwires in a particular brain location, perhaps targeted for cell layers (as in the hippocampus) and to hope that some microwires may be close to units when the animal recovers (Olds, 1973; Eichenbaum et al., 1977). The implantation of microwires is generally similar to macroelectrodes as described above. The second method is to implant a receptacle or well over a hole in the skull during surgery. During recording, the dura can be punctured, using a brief period of ether anesthesia, and a microelectrode lowered by means of a microdrive. Various types of microdrives for small animals have been described including simple, rotating types (Ranck, 1973b) and nonrotating types of varying complexity (Bland et al., 1980; Deadwyler et al., 1979; Harper, 1973). Kubie (1984) described a system for driving a bundle of microwires.

Large neurons are believed to generate large potentials that are better selected than smaller neurons (Towe, 1973). It must be noted that some neurons, like the bipolar and horizontal cells in the retina and the granule cell in the olfactory bulb, do not generate an action potential (Schmitt et al., 1976).

A signal selection device for units combines the use of a high-pass (>0.5 kHz) filter, a window discriminator, and an audio monitor (DeValois and Pease, 1973). The voltage window should be adjustable to select a unit of a particular spike height, perhaps a single unit, or multiple units of particular spike heights. A single unit should have a constant waveform and a reasonable (~0.5 ms) refractory period. However, multiple spikes from the same neuron may have a decreasing spike height (Ranck, 1973a), and a refractory period may be difficult to demonstrate in an infrequently firing cell. In some regions of the brain in which neurons are densely packed, e.g., granule cells in the dentate gyrus or the cerebellum, single units may be difficult to isolate.

Classification of a unit may make use of its duration, firing pattern, and site of recording. Positive identification of long-axon projection neurons is made by antidromic invasion of the cell body by stimulation of projection fibers (Fuller and Schlag, 1976). However, the lack of antidromic firing may mean an unknown projection pathway, or the impossibility of antidromic invasion of a cell because of its extensive branching collaterals or its large soma-to-axon diameter ratio. Interneurons are usually tentatively identified by their fast, repetitive firing after a single stimulus (Eccles, 1969). Physiological characteristics of neurons and their anatomy can better be studied by intracellular recording and staining in anesthetized animals.

### 3.3.3. Analysis of Unit Activities

For analytical purposes, units are commonly regarded as point processes, or as standard pulses occurring at an instant of time. In other words, the detailed waveform of a unit is normally disregarded. Various analyses of unit interval have been developed since 1960 (DeValois and Pease, 1973; Glaser and Ruchkin, 1976; Moore et al., 1966). Various neurons have different interspike interval histograms, perhaps generated by different underlying statistical processes. An autocorrelogram or the autocorrelation function is the time-domain equivalent (Fourier transform) of the autopower spectrum. It shows the probability of the firing of a second spike as a function of time after the first one has fired. In a similar manner, the cross-correlation function measures the probability of firing of a spike of a second neuron as a function of time after a spike of the first neuron has fired.[a] The cross-correlation function is the Fourier transform of the cross-power spectrum.

Various analyses of stimulus-elicited spike discharge have been described (Moore et al., 1966). Instantaneous frequency can be calculated as the inverse of the interspike interval. The total number of spike discharges after a stimulus can be counted. A common technique uses the poststimulus (or peristimulus) time histogram (PSTH). At a time before and after a stimulus, the number of times a unit fires in each time bin (e.g., of 1 ms duration) is recorded. Many sweeps of the same stimuli can be summed and then averaged to obtain a mean probability of firing per bin. If sufficient background firing exists and a prestimulus baseline is used, decreases as well as increases of firing probability can be demonstrated.

## 4. Reticulocortical Electrical Activity, Pharmacology, and Behavior

### 4.1. Introduction

The following sections provide a brief discussion of the electrical activity of the hippocampal formation and neocortex and its study

---

[a]An autocorrelogram is constructed by acquiring a postspike time histogram (see PSTH description) for every spike in the spike train. A poor man's autocorrelogram may be acquired on-line by summing a PSTH triggered from a spike. Spikes within the sweep are obviously not used as triggers and the resulting PSTH is thus not a true autocorrelogram. A cross-correlogram can be approximated on-line by using a similar strategy.

in relation to pharmacology and behavior. Additional details and background are provided in a series of recent books, reviews, and symposia on the hippocampal formation (Ciba Foundation Symposium #58, 1978; Gray, 1982; Isaacson and Pribram, 1975; Lopes da Silva and Arnolds, 1978; O'Keefe and Nadel, 1978; Seifert, 1983). Research on neocortical activity in relation to behavior and pharmacology has focused primarily on the sleep–waking cycle. Reviews of the classical concepts and findings in this field are given by Jouvet (1972, 1977); Moruzzi (1972); and Steriade and Hobson (1976), and a critique has been published by Vanderwolf and Robinson (1981). Siegel (1979) has prepared an excellent critical review of studies of unit activity in the reticular formation in relation to behavior.

## 4.2. Hippocampal Activity

Since the research of Green and Arduini (1954), it has been accepted that hippocampal slow wave activity consists of two basic patterns: (1) rhythmical, nearly sinusoidal waves with a frequency range of about 4–12 Hz, usually referred to as rhythmical slow activity (RSA) or theta waves, and (2) a more irregular, slow waveform referred to as large-amplitude irregular activity (LIA). A third pattern that is seen occasionally is small-amplitude irregular activity (SIA). Finally, all these patterns may be associated with varying fast rhythms with a frequency of about 20–70 Hz (Vanderwolf et al., 1975; Leung et al., 1982b) (Fig. 2).

The LIA pattern can be regarded as the basic resting rhythm of the hippocampus. It is not clear whether LIA arises as a result of spontaneous activity in intrahippocampal circuitry or whether the pattern is imposed by influences from other brain structures.

Green and Arduini (1954) were the first to show that RSA could be elicited in the hippocampus by electrical stimulation of the reticular formation. The effect was abolished by destruction of the septal nuclei, suggesting that activity in a reticulo–septo–hippocampal pathway was responsible for the occurrence of RSA. More recently much evidence has accumulated to show that cholinergic neurons in the medial septal nucleus and diagonal band send fibers to the hippocampal formation. It is probable that this cholinergic input is responsible for the RSA that occurs in anesthetized animals. Low frequency (4–7 Hz) RSA occurs in anesthetized animals, either spontaneously or in response to sensory stimuli, electrical stimulation of the reticular formation, or treatment with eserine. Antimuscarinic drugs such as atropine or scopolamine (or hemicholinium) eliminate all such RSA (Vanderwolf et al., 1978; Vanderwolf and Robinson, 1981).

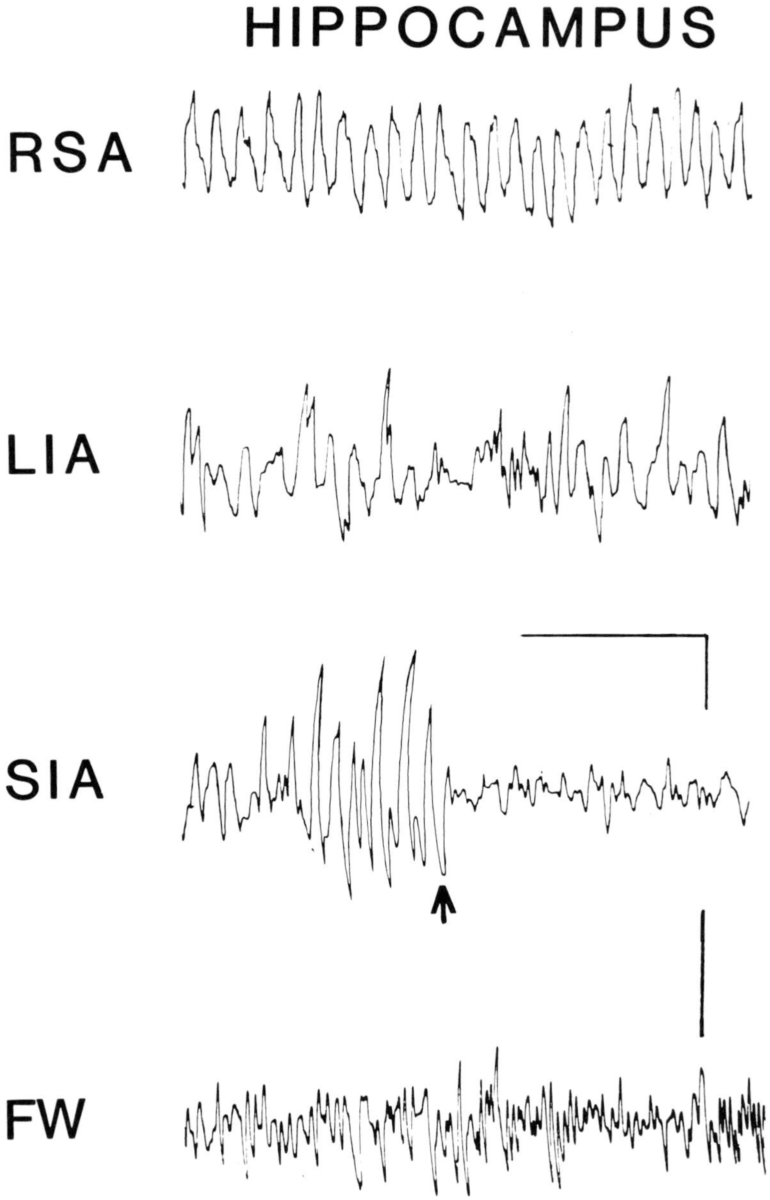

# HIPPOCAMPUS

Fig. 2.    Patterns of slow waves in the rat hippocampal formation. RSA, rhythmical slow activity, during swimming behavior; LIA, large-amplitude irregular activity, during waking immobility; SIA, small-amplitude irregular activity, as the rat was awakened from quiet sleep (a pencil was tapped on the table at the point marked by the arrow); FW, fast waves, during waking immobility. The first three records (RSA, LIA, SIA) were extracted from Whishaw and Vanderwolf (1973) and

Although the RSA in anesthetized animals appears to be dependent on activity in a cholinergic septo-hippocampal pathway, the RSA pattern that appears in freely moving animals is more complex. Even very large systemic doses of atropine or scopolamine or intraventricular injections of hemicholinium in freely moving rats or rabbits do not abolish RSA, although some modifications of the pattern may be produced. This suggests the existence of a noncholinergic input to the hippocampus that generates RSA in freely moving animals. It appears that cholinergic and noncholinergic inputs are usually active concurrently and that the RSA occurring during behavior is the product of their joint action. The pathway responsible for the atropine-resistant input has not been precisely identified, but it appears to originate in the brainstem and to run partly through the cingulum and partly through the white matter subjacent to the neocortex to enter the hippocampus via the entorhinal cortex (Vanderwolf and Leung, 1982, 1983; Buzsáki et al., 1983). Activity in this pathway is suppressed by anesthetics (thus accounting for the cholinergic nature of RSA in anesthetized animals) and by morphine and phencyclidine.

The transition from LIA to RSA is an indication of a very general change in the activity of the entire hippocampal formation. Buzsáki et al. (1981) have shown that transmission through the trisynaptic pathway from the entorhinal cortex to field CA1 of the hippocampus varies widely in correlation with the pattern of spontaneous slow wave activity. Leung et al. (1982b) have shown that the appearance of RSA in freely moving rats is correlated with an increase in fast waves (20–70 Hz), as well as changes in the morphology of the potentials evoked in the hippocampus by stimulation of the Schaffer collaterals or the commissural fibers (Leung, 1980). Pharmacological studies suggest that these changes are due to both cholinergic and noncholinergic inputs (Leung and Vanderwolf, 1980; Leung, 1984b).

It is probable that hippocampal slow wave patterns and evoked potentials are generated largely as a result of postsynaptic potentials occurring in pyramidal cells (Fujita and Sato, 1964). Pyramidal cells are more numerous than interneurons, and their elongated shape and parallel orientation enables them to generate

Fig. 2 (*Cont.*) originally recorded from a fixed bipolar electrode straddling the CA1 pyramidal cell layer. The FW pattern was recorded from an electrode in the hilus of the dentate gyrus in a different rat. Calibration: 1 mV, 1 s. RSA, LIA, and SIA were recorded at the same gain, but gain was increased for FW.

large field potentials. Interneurons, being less numerous and having radially arranged dendrites, would be expected to make a smaller contribution to field potentials. However, the mechanisms that generate LIA and RSA are not fully understood. Recent work indicates that large-amplitude sharp waves (up to 3.5 mV and with a duration of 40–100 ms) that are a component of LIA recorded in the CA1 region, may be postsynaptic potentials produced by sudden volleys of action potentials rising in CA3 and conveyed to CA1 by the Schaffer collaterals (Buzsáki et al., 1983). The atropine-sensitive RSA that appears in anesthetized animals may be produced by rhythmical pacemaker cells in the medial septal nucleus and diagonal band whose axons activate inhibitory interneurons in the hippocampus, thus producing a rhythmical sequence of inhibitory postsynaptic potentials in hippocampal pyramidal cells (Buzsáki et al., 1983; Leung and Yim, 1984). On the other hand, the atropine-resistant input that produces RSA in freely moving animals may generate a rhythmical series of excitatory postsynaptic potentials in dentate granule cells and in the distal apical dendrites of CA1 pyramidal cells as a result of activity in cells in the entorhinal cortex (Buzsáki et al., 1983; Leung, 1984a). As a result, dentate granule cells fire rhythmically at a high rate whenever RSA occurs in a freely moving rat (Buzsáki et al., 1983; Rose, 1983).

## 4.3. Neocortical Activity

The electrical activity recordable from the human scalp (electroencephalogram, EEG) has been classified into a variety of different waveforms (such as alpha, beta, or delta waves) on the basis of frequency and amplitude. The EEG patterns in slow wave sleep have been classified into four stages by Dement and Kleitman or five stages by Loomis (Dement and Mitler, 1974; Hess, 1964).

Although similar classifications have been proposed for use in experimental work in animals (Ursin, 1970) they have not been systematically adopted. In work relating neocortical activity to behavior and drug effects, we have found it useful to classify the patterns seen in the rat into three main groups: (1) low-voltage fast activity (LVFA); (2) large, irregular, slow activity (LISA) with an amplitude two- to eightfold greater than the amplitude of LVFA at the same electrode site; and (3) rhythmical spindle activity of 6–10 Hz and an amplitude approximately equal to the amplitude of LISA at the same electrode site (Fig. 3). Spindle-shaped waveforms with a much smaller amplitude and a frequency of 10–16 Hz are sometimes associated with the LISA pattern.

The LISA pattern is widely regarded as a resting or spontaneous form of activity in the neocortex, comparable to LIA in the hippocampus. It is prominent during quiet sleep, but can also occur in waking animals, especially following treatment with antimuscarinic drugs. In anesthetized animals there may be large slow waves that are similar but not identical to the slow waves that occur in the undrugged state. During anesthesia, LVFA may also occur, spontaneously or as a result of sensory stimulation of the reticular formation or treatment with muscarinic agonists such as eserine. It appears that all LVFA occurring during anesthesia (urethane anesthesia is an exception) can be abolished (resulting in a pattern of continuous LISA) by treatment with antimuscarinic drugs such as atropine or scopolamine. Therefore, analogous to the situation in the hippocampus, it seems that the LVFA occurring in the neocortex of anesthetized animals is dependent on a cholinergic input (Vanderwolf et al., 1978; Vanderwolf and Robinson, 1981). A component of the LVFA occurring during the waking state also appears to be dependent on a cholinergic input. Recent evidence indicates that this cholinergic input originates in the cholinergic cells of the substantia innominata that project to the neocortex (Stewart et al., 1984).

# NEOCORTEX

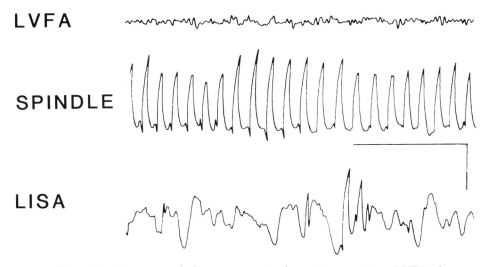

Fig. 3.   Patterns of slow waves in the rat neocortex. LVFA, low voltage fast activity, during waking immobility; spindle, during waking immobility accompanied by tremor of the vibrissae and head; LISA, large irregular slow activity, during waking immobility following atropine $SO_4$ (50 mg/kg, ip). A similar LISA pattern occurs in undrugged rats during quiet sleep. The records were taken from a bipolar surface-to-depth electrode in parietal neocortex. Calibration: 1 mV, 1 s.

As in the case of the hippocampus, the situation is more complex in freely moving animals since, in this state, even very large doses of antimuscarinic drugs fail to abolish all LVFA. This demonstrates that there is an atropine-resistant form of LVFA. It appears that the neocortex, like the hippocampus, receives two activating inputs. In addition to the cholinergic input that may be active during anesthesia as well as in the waking state, there is a noncholinergic activating input that can produce LVFA in the waking state. Recent evidence suggests that this atropine-resistant input may be dependent on ascending serotonergic fibers (Vanderwolf, 1984).

Microphysiological studies of the postulated joint control of LVFA by cholinergic and serotonergic mechanisms have not yet been carried out. A general review of work on cellular mechanisms underlying neocortical slow wave activity can be found in Creutzfeldt (1974).

## 4.4. Reticulocortical Activity in Relation to Behavior

For many years it has been widely assumed that close relations between cortical activity and overt motor activity are to be found only in the motor areas of the neocortex. Thus, research has focused rather narrowly on such topics as the relation between the activity of pyramidal tract neurons and the performance of skilled movements of the forelimb in monkeys (Henneman, 1980). Such work is of great importance. However, skilled movements of the forelimb constitute only a small part of the total behavior of an animal. It is essential to gain an understanding of the organization of brain activity in relation to behavior in a general sense. In this context, it is of considerable interest that electrical activity throughout the reticular formation and the cerebral cortex shows a high degree of correlation with motor activity.

When a rat is allowed to move about spontaneously in a laboratory environment, the pattern of electrical activity in the hippocampus varies from moment to moment in close relation to what the rat does (Fig. 4). An RSA of about 7–9 Hz is prominent during walking and rearing or during struggling if the rat is pushed or handled. The same pattern occurs during swimming, walking in a treadmill, or jumping. An RSA of slightly lower frequency and amplitude occurs if the rat makes extensive lateral or vertical movements of the head or changes posture while standing motionless. An RSA persists during all these behaviors following treatment with atropine (although the RSA accompanying small head movements tends to be abolished by atropine). During behavioral immobility, LIA is generally present, although com-

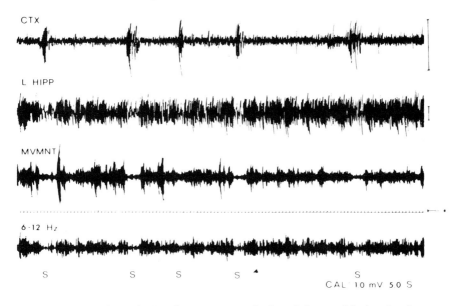

CTX

L HIPP

MVMNT

6-12 Hz

S          S     S        S   ▲                      S

CAL 10 mV 50 S

Fig. 4. Correlation between cortical activity and behavior in a rat following atropine $SO_4$ (50 mg/kg, ip). CTX, neocortex; L.Hipp., left hippocampus; MVMNT, output of magnet-and-coil type of movement sensor; 6–12 Hz, band-pass filtered record of hippocampal activity. Rat moves the head, steps, or walks except during brief periods of immobility (S, still). Note LVFA in neocortex and high levels of 6–12 Hz activity in hippocampus during movement; LISA in neocortex and low levels of 6–12 Hz activity in hippocampus during immobility.

puted spectral analyses may reveal small RSA peaks at about 6 Hz in some animals (Leung et al., 1982b). These small RSA peaks are sensitive to atropine, suggesting their dependence on the cholinergic septo-hippocampal input. A predominant LIA pattern occurs during face-washing, licking, or biting the fur without extensive head movement, chattering the teeth in a threatening manner, licking water, or chewing food. Behaviors of the first group (consistently associated with RSA) are referred to as Type 1 behavior and those of the second group are Type 2 behavior.

Ranck (1973a) divided hippocampal cells into two groups: complex spike cells and theta cells. Complex spike cells sometimes fire in bursts ("complex spikes") of 2–7 individual spikes of progressively declining amplitude and an interspike interval of 1.5–6 ms. At other times these cells may discharge isolated spikes. Complex spike cells are probably hippocampal pyramidal cells, whereas theta cells appear to be interneurons in most cases (Fox and Ranck, 1981). Theta cells never fire in bursts of declining spike amplitude although they often do produce clusters of spikes, of constant amplitude, in phase with the accompanying RSA. Complex spike cells also tend to fire in phase with RSA. The defining characteristic of theta cells, according to Ranck, is that they fire at a high rate if, and only if, a theta or RSA rhythm is present in the slow wave activity of the hippocampus. Theta cells fire arrhythmically at a low rate during LIA. Complex spike cells, on the other hand, tend to fire at a higher rate during LIA than during RSA.

Theta cells in Ammon's horn and granule cells in the dentate gyrus have the same behavioral correlates as RSA since these units increase their firing rates dramatically whenever RSA occurs in a freely moving animal (Ranck, 1973a; Feder and Ranck, 1973; O'Keefe and Nadel, 1978; Sinclair et al., 1982; Rose, 1983; Buzsáki et al., 1983). The same units appear to take part in both atropine-sensitive and atropine-resistant RSA. Complex spike cells often display a different type of correlation with behavior. Most of these cells fire maximally when an animal is in a particular spatial location in the test apparatus (O'Keefe and Nadel, 1978; O'Keefe, 1979). Perhaps such cells play a role in directing locomotion in a spatially extended environment.

Spontaneous neocortical slow wave activity in a normal rat does not reveal the close relation to behavior that is seen in the hippocampus. Similar patterns of LVFA usually occur during both Type 1 and Type 2 behaviors. However, in rats, large amplitude (6–10 Hz) spindles sometimes occur spontaneously during immobility although they apparently *never* occur during Type 1 behavior. Cats frequently display large amplitude slow waves in the neocortex during such behaviors as immobility, drinking milk, or face-washing, but not during head movements or walking (Vanderwolf and Robinson, 1981). These facts suggest that activating inputs to the neocortex are increased during Type 1 behavior as compared to Type 2 behavior. This suggestion is supported by studies showing that whereas atropinic drugs suppress LVFA during Type 2 behavior, replacing it by LISA (Fig. 4), the LVFA accompanying Type 1 behavior is resistant to atropine. An additional atropine-resistant input during Type 1 behavior appears to be dependent on serotonin (Vanderwolf, 1984). Data from a variety of sources suggest that both cholinergic and serotonergic inputs produce LVFA in the neocortex during Type 1 behavior and that cholinergic inputs alone produce LVFA during Type 2 behavior. The cerebral accompaniments of Type 1 behavior do not appear to result primarily from proprioceptive feedback.

How should one interpret these findings? One might suggest that the increased LVFA-producing input during Type 1 behavior indicates that animals are more "aroused" or more "attentive" during Type 1 behavior and more "relaxed" during Type 2 behavior. However, some Type 2 behaviors require considerable muscular exertion. During face-washing a rat has a higher heart rate than during ordinary walking (Vanderwolf and Vanderwart, 1970). Since heart rate correlates highly with $O_2$ utilization, this suggests that the metabolic demands of the musculature are greater during face washing than during walking. Further, a few seconds after having received a painful electric shock to the feet, a rat is likely to stand immobile, chattering its teeth, eyes wide open and slightly protruding, and hair partially erected. Despite these signs of arousal and distress, hippocampal and neocortical activity are much the same as they are during quiet "relaxed" immobility.

Such examples demonstrate that the interpretation of brain events in relation to behavior or psychological processes cannot be done casually. Intuitive concepts derived from everyday experience may be quite misleading when applied uncritically in the brain–behavior field. Valid generalizations can be arrived at only on the basis of a thorough study of many different behaviors in a variety of situations. The history of the study of hippocampal activity in relation to learning, memory, and other cognitive processes is a good example of this. Initially, the experimental procedures necessary for advances in this field seemed rather obvious. Animals were placed in a training box of some sort and baseline records of hippocampal activity were taken. Subsequently, studies were undertaken of habituation, pseudoconditioning, classical or operant conditioning, discrimination, and extinction, according to well recognized procedures. Behavior was usually assessed only in terms of "errors" or time taken to perform the task. The results lead to a variety of proposals. Hippocampal activity was interpreted in terms of arousal, attention, learning, memory, frustration, or motivation (Black, 1975; Izquierdo, 1975). As was noted above, such hypotheses take no account of the actual motor activity displayed. Thus, in a passive avoidance task an animal is said to *learn* to avoid shock by sitting motionless on a small wooden platform placed on a grid floor, whereas in an active avoidance task an animal is said to *learn* to avoid shock by running or jumping away from the location where shock is administered. This terminology suggests, perhaps falsely, that the same process occurs in the two situations. In all studies that have examined the question, hippocampal slow wave activity has been found to be closely related to concurrent motor activity rather than the supposed presence or absence of various inferred psychological processes such as learning and memory (Black, 1975; Vanderwolf and Ossenkopp, 1982). Thus, RSA accompanies walking in a rat regardless of whether the walking is "spontaneous" or occurs as a result of special training. This does not necessarily mean that the hippocampus is not involved in neuroplastic changes, but merely that its electrical activity does not correspond with expectations based on intuitive ideas about learning and memory.

The history of research on the unit activity of the reticular formation has followed a similar course. Early work, in which overt behavior was largely ignored, resulted in a great variety of hypotheses relating the activity of reticular units to sensory input (olfaction, vision, audition, touch, pain), conditioning, habituation, active sleep, arousal, fear, anticipation of reward, and so on. Recent more analytical studies indicate that "discharge in most RF (reticular formation) cells is primarily related to the excitation of specific muscle groups" (Siegel, 1979). As in the case of the hippocampus, the earlier hypotheses can be understood in the light of the fact that a vast number of factors influence motor activity.

If one wishes to test the effect of some factor on brain activity it is important to avoid confusion of direct effects with more indirect effects that are associated with changes in behavior. For example, treatment with a neuroleptic drug reduces the occurrence of RSA, whereas treatment with amphetamine increases it. These phenomena might be interpreted as a direct effect of the drugs on the mechanisms that generate RSA, but if the concurrent behavior is taken into account it becomes evident that they are entirely nonspecific. The RSA accompanying walking is not altered by either neuroleptics or amphetamine. The drugs change the probability of occurrence of specific behaviors and the apparent change in brain activity is simply a reflection of this.

Other drugs have more specific effects, producing clear-cut changes in brain activity even though behavior is held constant across normal and drugged conditions. Thus, the frequency of RSA accompanying walking is reduced by 1–2 Hz by a subanesthetic dose of pentobarbital (Kramis et al., 1975). (Body temperature must also be held constant in this experiment.) Atropine or scopolamine virtually eliminate the LVFA that is usually present during waking immobility (head held up, eyes open). Phencyclidine eliminates atropine-resistant RSA in an animal that walks about very actively (Vanderwolf and Leung, 1983).

Sleep research is another field in which failure to hold behavior constant (or even observe it closely) may have led to errors in interpretation. Cats display a great deal of large-amplitude slow activity in the neocortex when they are immobile. This does not necessarily mean that they are asleep, since similar waves may occur during such behaviors as face-washing or drinking milk (see Vanderwolf and Robinson, 1981). In experiments in which drugs are given and sleep is assessed only by examination of the ECG, it is quite possible that a state of waking immobility may be misinterpreted as sleep. As recommended by Kleitman (1963) a number of years ago, the presence or absence of sleep should be decided primarily on the basis of behavioral and autonomic functions and not on the state of the ECG.

In summary, the investigation of brain activity and behavior does not differ in principle from work in any other scientific field. One should deal with observable phenomena rather than indefinable constructs. Hidden assumptions should be sought out and their validity examined. The phenomena observed should be described accurately and completely and sweeping conclusions should not be drawn on the basis of a limited number of observations. When a number of factors are suspected of having an effect on a phenomenon, each one should be varied alone while all others are held as constant as possible. If a few simple rules of this type were applied consistently, it seems to us that work in the brain–behavior field would progress more rapidly than it has in the past.

## Acknowledgment

Preparation of this paper was supported by grants to the authors from the Natural Sciences and Engineering Research Council.

## References

Alcock J. (1984) *Animal Behavior: An Evolutionary Approach*, 3rd Ed., Sinauer Assoc. Inc., Sunderland, Mass.

Anastasi A. (1982) *Psychological Testing*, 5th Ed., MacMillan Publishing Co. Inc., New York.

Barnett S. A. (1975) *The Rat: A Study in Behavior*, Revised Ed., University of Chicago Press, Chicago.

Basmajian, J. V. (1978) *Muscles Alive: Their Functions Revealed by Electromyography*, 4th Ed., Williams & Wilkins Co., Baltimore.

Black A. H. (1975) Hippocampal Electrical Activity and Behavior, in *The Hippocampus*. vol. 2 *Neurophysiology and Behavior* (Isaacson R. L. and Pribram K. H., eds.) pp. 129–167, Plenum Press, New York.

Bland B. H., Sinclair B. R., Jorgenson R. G., and Keen R. (1980) A direct-drive, nonrotating version of Ranck's microdrive. *Physiol. Behav.* **24,** 395–397.

Blough D. S. (1958) New test of tranquilizers. *Science,* **127,** 586–587.

Bremer F. (1935) Cerveau "isolé" et physiologie du sommeil. *C. R. Soc. Biol.* (Paris) **118,** 1235–1241.

Bremer F. (1936a) Nouvelles recherches sur le mécanisme du sommeil. *C. R. Soc. Biol.* (Paris) **122,** 460–464.

Bremer F. (1936b) Activité électrique du cortex cérébral dan les états de sommeil et de veille chez le chat. *C. R. Soc. Biol.* (Paris) **122,** 464–467.

Bureš J., Burešová O., and Huston J. (1976) *Techniques and Basic Experiments for the Study of Brain and Behavior.* Elsevier Scientific Publishing Co., Amsterdam.

Buzsáki G., Grastyán E., Czopf J., Kellényi L., and Prohaska O. (1981) Changes in neuronal transmission in the rat hippocampus during behavior. *Brain Res.* **225,** 235–247.

Buzsáki G., Leung L. -W. S., and Vanderwolf C. H. (1983) Cellular bases of hippocampal EEG in the behaving rat. *Brain Res. Rev.* **6,** 139–171.

Chapin J. K., Loeb G. E., and Woodward D. J. (1980) A simple technique for determination of footfall patterns of animals during treadmill locomotion. *J. Neurosci. Meth.* **2,** 97–102.

Ciba Foundation Symposium #58 (1978) *Functions of the Septohippocampal System.* Elsevier Scientific Publishing Co., Amsterdam.

Cooley R. K. and Vanderwolf C. H. (1978) Construction of wire leads and electrodes for use in slow wave recording in small animals. *Brain Res. Bull.* **3,** 175–179.

Cooper R. (1971) Recording Changes in Electrical Properties in the Brain: The EEG, in *Methods in Psychobiology,* vol. 1 (Myers R. D., ed.). pp. 155–203, Academic Press, London.

Creutzfeldt O. (1974) The Neuronal Generation of the EEG, in *Handbook of Electroencephalography and Clinical Neurophysiology,* vol. 2(C) (Creutzfeldt O., ed.). pp. 1–157, Elsevier Scientific Publishing Co., Amsterdam.

Cronbach L. J. (1984) *Essentials of Psychological Testing,* 4th Ed. Harper & Row, New York, 157 pp.

Davis S. D. (1970) A continuous activity monitor for small caged animals. *Physiol. Behav.* **5,** 953–954.

Deadwyler S. A., Bielar J., Rose G., West M. and Lynch G. (1979) A microdrive for use with glass or metal microelectrodes in recording from freely moving rats. *Electroenceph. clin. Neurophysiol.* **47,** 752–754.

Delgado J. M. R. (1964) Electrodes for Extracellular Recording and Stimulation, in *Physical Techniques in Biological Research,* vol. V. (Nastuk W. L., ed.). pp. 89–143, Academic Press, New York.

Dement W. C. and Mitler M. M. (1974) An Introduction to Sleep, in *Basic Sleep Mechanisms* (Petre-Quadens O. & Schlag J. D., eds). pp. 271–296, Academic Press: New York.

Denny-Brown D. (1979) *Selected Writings of Sir Charles Sherrington.* Oxford University Press, Oxford, 532 pp.

DeValois R. L. and Pease P. L. (1973) Extracellular Unit Recording, in *Bioelectric Recording Techniques,* Part A. (Thompson R. F. and Patterson M. M., eds.). pp. 95–135, Academic Press, New York.

Dudar J. D., Whishaw I. Q., and Szerb J. C. (1979) Release of acetylcholine from the hippocampus of freely moving rats during sensory stimulation and running. *Neuropharmacol.* **18,** 673–678.

Eccles J. C. (1969) *The Inhibitory Pathways of the Central Nervous System.* C. C. Thomas, Springfield, 135 pp.

Eichenbaum H., Pettijohn D., DeLuca A. M., and Chorover S. L. (1977) Compact miniature microelectrode-telemetry system. *Physiol. Behav.* **18,** 1175–1178.

Elul R. (1972) The genesis of the EEG. *Int. Rev. Neurobiol.* **15,** 227–272.

Ervin F. R. and Kenney G. J. (1971) Electrical Stimulation of the Brain, in *Methods in Psychobiology* (Myers R. D., ed.). pp. 207–246, Academic Press, New York.

Fearing F. (1930) *Reflex Action: A Study in the History of Physiological Psychology.* Williams and Wilkins, Baltimore, 350 pp.

Feder R. and Ranck J. B. Jr. (1973) Studies on single neurons in dorsal hippocampal formation and septum in unrestrained rats. Part II. Hippocampal slow waves and theta cell firing during bar pressing and other behaviors. *Exp. Neurol.* **41,** 532–555.

Fox S. E. and Ranck J. B. Jr. (1981) Electrophysiological characteristics of hippocampal complex-spike cells and theta cells. *Exp. Brain Res.* **41,** 399–410.

Fox S. S. and Rosenfeld J. P. (1972) Recording Evoked Potentials, in *Methods in Psychobiology,* vol. 2 (Myers R. D., ed.). pp. 345–369, Academic Press, New York.

Freeman W. J. (1963) The electrical activity of a primary sensory cortex: Analysis of EEG waves. *Int. Rev. Neurobiol.* **5,** 53–119.

Freeman W. J. (1975) *Mass Action in the Nervous System.* Academic Press, New York, 489 pp.

Fuller J. H. and Schlag J. P (1976) Determination of antidromic excitation by the collision test: Problems of interpretation. *Brain Res.* **112,** 283–298.

Fujita Y. and Sato T. (1964) Intracellular records from hippocampal pyramidal cells in rabbit during theta rhythm activity. *J. Neurophysiol.* **27,** 1011–1025.

Geddes L. A. (1972) *Electrodes and the Measurement of Bioelectric Events.* John Wiley, New York, 364 pp.

Glaser E. M. and Ruchkin D. S. (1976) *Principles of Neurobiological Signal Analysis.* Academic Press, New York, 471 pp.

Gray J. A. (1982) *The Neuropsychology of Anxiety: An Enquiry into the Functions of the Septo-Hippocampal System.* Clarendon Press, Oxford, 548 pp.

Green J. D. and Arduini A. A. (1954) Hippocampal electrical activity in arousal *J. Neurophysiol.* **17,** 533–557.

Hansen E. L. & McKenzie G. M. (1979) Dexamphetamine increases striatal neuronal firing in freely moving rats. *Neuropharmacol.* **18,** 547–552.

Harper R. M. (1973) Relationship of Neuronal Activity to EEG Waves During Sleep and Wakefulness, in *Brain Unit Activity During Behavior* (Phillips M. I., ed.). pp. 130–154, C. C. Thomas, Springfield.

Henneman E. (1980) Motor Functions of the Cerebral Cortex, in *Medical Physiology*, vol. 1 (Mountcastle V. B., ed.). pp. 859–889, C. V. Mosby Co., St. Louis.

Hess R. (1964) The electroencephalogram in sleep. *Electroenceph. Clin. Neurophysiol.* **16**, 44–55.

Hinde R. A. (1970) *Animal Behavior: A Synthesis of Ethology and Comparative Psychology*, 2nd Ed., McGraw-Hill, New York, p. 10.

Humphrey D. R. (1968) Reanalysis of the antidromic cortical response I. Potentials evoked by stimulation of the isolated pyramidal tract. *Electroenceph. Clin. Neurophysiol.* **24**, 116–129.

Hutt S. J. and Hutt C. (1970) *Direct Observation and Measurement of Behavior*. Charles C. Thomas, Springfield, 224 pp.

Isaacson R. L. and Pribram K. H. (1975) *The Hippocampus*, Vols. 1 and 2. Plenum Press, New York.

Iversen J. A. (1973) A flexible system for recording activity of caged animals. *Physiol. Behav.* **10**, 971–972.

Izquierdo I. (1975) The hippocampus and learning. *Prog. Neurobiol.* **5**, 39–75.

Jenkins G. M. and Watts D. G. (1968) *Spectral Analysis and its Application*. Holden Day, San Francisco, 525 pp.

John E. R. (1973) Brain evoked potentials: Acquisition and analysis, in *Bioelectric Recording Techniques*, Part A (Thompson R. F. and Patterson M. M., eds.). pp. 317–356, Academic Press, New York.

Jouvet M. (1972) The role of monoamines and acetylcholine containing neurons in the regulation of the sleep waking cycle. *Ergeb. Physiol.* **64**, 166–307.

Jouvet M. (1977) Neuropharmacology of the Sleep-Waking Cycle, in *Handbook of Psychopharmacology*, vol. 8, *Drugs, Neurotransmitters and Behavior* (Iverson L. L., Iverson S. D., and Snyder S. H., eds.). pp. 233–293, Plenum Press, New York.

Katz G., Webb G., and Sorem A. (1964) Recording and Display, in *Physical Techniques in Biological Research*, vol. V (Nastuk W. L., ed.). pp. 373–447, Academic Press, New York.

Kleitman N. (1963) *Sleep and Wakefulness*. University of Chicago Press, Chicago, 552 pp.

Kramis R., Vanderwolf C. H., and Bland B. H. (1975) Two types of hippocampal rhythmical slow activity in both the rabbit and the rat: Relations to behavior and effects of atropine, diethyl ether, urethane, and pentobarbital. *Exp. Neurol.* **49**, 58–85.

Kubie J. L. (1984) A driveable bundle of microwires for collecting single-unit data from freely moving rats. *Physiol. Behav.* **32**, 115–118.

Lashley K. S. (1929) *Brain Mechanisms and Intelligence*. University of Chicago Press, Chicago, 186 pp.

Lashley K. S. (1935) Studies of cerebral function in learning XI: The behavior of the rat in latch box situations. *Comp. Psychol. Monog. XI* **2**, 1–42.

Lehner P. N. (1979) *Handbook of Ethological Methods.* Garland STPM Press, New York, 403 pp.

Leung L. S. (1980) Behavior-dependent evoked potentials in the hippocampal CA1 region of the rat. I. Correlation with behavior and EEG. *Brain Res.* **198,** 95–117.

Leung L. S. and Vanderwolf C. H. (1980) Behavior-dependent evoked potentials in the hippocampal CA1 region of the rat. II. Effect of eserine, atropine, ether, and pentobarbital. *Brain Res.* **198,** 119–133.

Leung L. S., Harvey G. C., and Vanderwolf C. H. (1982a) Combined video and computer analysis of the relation between the interhemispheric response and behavior. *Behav. Brain Res.* **6,** 195–200.

Leung L.-W. S., Lopes da Silva F. H., and Wadman W. J. (1982b) Spectral characteristics of the hippocampal EEG in the freely moving rat. *Electroenceph. clin. Neurophysiol.* **54,** 203–219.

Leung L. -W. S. (1984a) Model of gradual phase shift of the theta rhythm in the rat. *J. Neurophysiol.* **52,** 1051–1065.

Leung, L. -W. S. (1984b) Pharmacology of theta phase shift in the hippocampal CA1 region of freely moving rats. *Electroenceph. clin. Neurophysiol.* **58,** 457–466.

Leung L. -W. S. and Yim C. Y. C. (1984) Intracellular theta rhythm in hippocampal CA1 cells in the urethane anesthetized rats. *Proc. Canad. Fed. Biol. Soc.* **27,** 40.

Llinas R. and Nicholson C. (1971) Electrophysiological properties of dendrites and somata in alligator Purkinje cells. *J. Neurophysiol.* **34,** 532–551.

Lopes da Silva F. H. and Arnolds D. E. A. T. (1978) Physiology of the hippocampus and related structures. *Ann. Rev. Physiol.* **40,** 185–216.

Lorente de Nó R. (1947) *A Study of Nerve Physiology,* vol. 132(2). Rockefeller Inst. Med. Res., New York.

Lorenz K. Z. (1973) The fashionable fallacy of dispensing with description. *Naturwiss.* **60,** 1–9.

Mackintosh N. J. (1974) *The Psychology of Animal Learning.* Academic Press, London, 730 pp.

Matoušek M. (1973) Frequency and Correlation Analysism, in *Handbook of Electroencephalography and Clinical Neurophysiology,* vol. 5, Part A (Matousek M., ed.). Elsevier Scientific Publishing Co., Amsterdam, 137 pp.

Micco D. J. (1977) Lightweight, multicontact, slip-ring commutator for recording and stimulation with small animals. *Brain Res. Bull.* **2,** 499–502.

Moore G. P., Perkel D. H., and Segundo J. P (1966) Statistical analysis and functional interpretation of neuronal spike data. *Ann. Rev. Physiol.* **28,** 493–522.

Morden B., Mitchell G., and Dement W. (1967) Selective REM sleep deprivation and compensation phenomena in the rat. *Brain Res.* **5,** 339–349.

Morris R. G. M. (1983) An Attempt to Dissociate "Spatial-Mapping" and "Working-Memory" Theories of Hippocampal Function, in *Neuro-*

*biology of the Hippocampus* (Seifert W., ed.). pp. 405–432, Academic Press, London.

Moruzzi G. (1972) The sleep-waking cycle. *Ergeb. Physiol.* **64**, 1–165.

Mundl W. J. (1966) Activity of small animals measured with accelerometer. *Med. Biol. Engng.* **4**, 209–212.

Mundl W. J. and Malmo H. P. (1979) An accelerometer for recording head movement of laboratory animals. *Physiol. Behav.* **23**, 391–393.

Munn N. L. (1950) *Handbook of Psychological Research on the Rat.* Houghton Mifflin Co., Boston, 598 pp.

Niedermeyer E. and Lopes da Silva F. H. (1982) *Electroencephalography.* Urban-Schwarzenberg, Baltimore, 762 pp.

O'Keefe J. and Nadel L. (1978) *The Hippocampus as a Cognitive Map.* Clarendon Press, Oxford, 570 pp.

O'Keefe J. (1979) A review of the hippocampal place cells. *Prog. Neurobiol.* **13**, 419–439.

Olds J., Disterhoft J. F., Segal M., Kornblith C. L., and Hirsch R. (1972) Learning centres of rat brain mapped by measuring latencies of conditioned unit responses. *J. Neurophysiol.* **35**, 202–219.

Olds J. (1973) Multiple Unit Recordings From Behaving Rats, in *Bioelectric Recording Techniques,* Part A. (Thompson R. F. and Patterson M. M., eds.). pp. 165–198, Academic Press, New York.

Olton D. S. (1983) Memory Functions and the Hippocampus, in *Neurobiology of the Hippocampus* (Seifert W., ed.). pp. 335–373, Academic Press, London.

Paxinos G. and Watson C. (1982) *The Rat Brain in Stereotaxic Coordinates.* Academic Press, Sidney, 12 pp., 70 plates and figures.

Pelligrino L. J. and Cushman A. J. (1973) Use of Stereotaxic Technique, in *Methods in Psychobiology,* vol. 1. (Myers R. D., ed.). pp. 67–91, Academic Press, New York.

Phillips M. I. (1973) Unit Activity Recordings in Freely Moving Animals: Some Principles and Theory, in *Brain Unit Activity During Behavior* (Phillips M. I., ed.). pp. 5–40, C. C. Thomas, Springfield.

Purpura D. P. (1959) Nature of electrocortical potentials and synaptic organizations in cerebral and cerebellar cortex. *Int. Rev. Neurobiol.* **1**, 47–163.

Ranck J. B. Jr. (1973a) Studies on single neurons in dorsal hippocampal formation and septum in unrestrained rats. Part 1. Behavioral correlates and firing repertoires. *Exp. Neurol.* **41**, 461–531.

Ranck J. B. Jr. (1973b) A Moveable Microelectrode for Recording From Single Neurons in Unrestrained Rats, in *Brain Unit Activity During Behavior* (Phillips M. E., ed.). pp. 76–79, University of Iowa Press, Iowa City.

Ranck J. B. Jr. (1975) Which elements are excited in electrical stimulation of mammalian CNS: A review. *Brain Res.* **98**, 417–440.

Rebec G. V. and Groves P. M. (1975) Differential effects of the optical isomers of amphetamine on neuronal activity in the reticular formation and caudate nucleus. *Brain Res.* **83**, 301–318.

Rose G. (1983) Physiological and Behavioral Characteristics of Dentate

Granule Cells, in *Neurobiology of the Hippocampus* (Seifert W., ed.). pp. 449–472, Academic Press, London.

Rosetto M. A. and Vandercar D. H. (1972) Lightweight FET circuit for differential or single-ended recording in free-moving animals. *Physiol. Behav.* **9**, 105–106.

Rudell A. P, Fox S. E., and Ranck J. B. Jr. (1980) Hippocampal excitability phase-locked to the theta rhythm in walking rats. *Exp. Neurol.* **68**, 87–96.

Sasaki K., Ono T., Nishino H., Fukuda M., and Muramoto K. -I. (1983) A method for long-term artifact-free recording of single unit activity in freely moving, eating, and drinking animals. *J. Neurosci. Meth.* **7**, 43–47.

Schmitt F. O., Dev P., and Smith B. H. (1976) Electrotonic processing of information by brain cells. *Science* **193**, 114–120.

Schoenfeld R. L. (1964) Bioelectric Amplifiers, in *Physical Techniques in Biological Research*, vol. V (Nastuk W. L., ed.). pp. 277–352, Academic Press, New York.

Seifert W. (ed.) (1983) *Neurobiology of the Hippocampus.* Academic Press, London, 632 pp.

Sherrington C. S. (1910) Flexion-reflex of the limb, crossed extension-reflex, and reflex stepping and standing. *J. Physiol.* **40**, 28–121.

Siegel J. M. (1979) Behavioral functions of the reticular formation. *Brain Res. Rev.* **1**, 69–105.

Siegel J. M., Breedlove S. M., and McGinty D. J. (1979) Photographic analysis of relation between unit activity and movement. *J. Neurosci. Meth.* **1**, 159–164.

Sinclair B. R., Seto M. G., and Bland B. H. (1982) θ-Cells in CA1 and dentate layers of hippocampal formation: Relations to slow-wave activity and motor behavior in the freely moving rabbit. *J. Neurophysiol.* **48**, 1214–1225.

Skinner J. E. (1971) *Neuroscience: A Laboratory Manual.* Saunders Co., Philadelphia, 244 pp.

Snodderly D. M. Jr. (1973) Extracellular Single Unit Recording, in *Bioelectric Recording Techniques,* Part A. (Thompson R. F. and Patterson M. M., eds.). pp. 137–163, Academic Press, New York.

Somjen G. G. (1973) Electrogenesis of sustained potentials. *Prog. Neurobiol.* **1**, 199–237.

Souček B. (1972) *Minicomputers in Data Processing and Simulation.* John Wiley: New York, 467 pp.

Sperry R. W. (1952) Neurology and the mind-brain problem. *Am. Sci.* **40**, 291–312.

Steriade M. and Hobson J. A. (1976) Neuronal activity during the sleep-waking cycle. *Prog. Neurobiol.* **6**, 155–376.

Stewart D. J., MacFabe D. F., and Vanderwolf C. H. (1984) Cholinergic activation of the electrocorticogram: Role of the substantia innominata and effects of atropine and quinuclidinyl benzilate. *Brain Res.* **322**, 219–232.

Sutherland R. J., Whishaw I. Q., and Kolb B. (1983) A behavioral analysis of spatial localization following electrolytic, kainate- or co-lchicine-induced damage to the hippocampal formation in the rat. *Behav. Brain Res.* **7,** 133–153.

Towe A. L. (1973) Sampling Single Neuron Activity, in *Bioelectric Recording Techniques*, Part A. (Thompson R. F. and Patterson M. M., eds.). pp. 79–93, Academic Press, New York.

Ursin R. (1970) Sleep stage relations within the sleep cycles of the cat. *Brain Res.* **20,** 91–97.

Vanderwolf C. H. (1969) Hippocampal electrical activity and voluntary movement in the rat. *Electroenceph. clin. Neurophysiol.* **26,** 407–418.

Vanderwolf C. H. and Vanderwart M. L. (1970) Relations of heart rate to motor activity and arousal in the rat. *Canad. J. Psychol.* **24,** 434–441.

Vanderwolf C. H. (1975) Neocortical and hippocampal activation in relation to behavior: Effects of atropine, eserine, phenothiazines, and amphetamine. *J. comp. physiol. Psychol.* **88,** 300–323.

Vanderwolf C. H., Kramis R., Gillespie L. A., and Bland B. H. (1975) Hippocampal Rhythmical Slow Activity and Neocortical Low Voltage Fast Activity: Relations to Behavior, in *The Hippocampus: Neurophysiology and Behavior*, vol. 2 (Isaacson R. L. and Pribram K. H., eds.). pp. 101–128, Plenum Press, New York.

Vanderwolf C. H., Kramis R., and Robinson T. E. (1978) Hippocampal Electrical Activity During Waking Behaviour and Sleep: Analyses Using Centrally Acting Drugs, in *Functions of the Septo-Hippocampal System*. Ciba Foundation Symposium #58. Elsevier Scientific Publishing Co., Excerpta Medica, Amsterdam, 199–226.

Vanderwolf C. H. and Robinson T. E. (1981) Reticulo-cortical activity and behavior: A critique of the arousal theory and a new synthesis. *Behav. Brain Sci.* **4,** 459–514.

Vanderwolf C. H. and Leung L. -W. S. (1982) Effects of entorhinal, cingulate, and neocortical lesions on atropine resistant hippocampal RSA. *Neurosci. Lett.* (Suppl. 10,) S501.

Vanderwolf C. H. and Ossenkopp K. P. (1982) Are There Patterns of Brain Slow Wave Activity Which are Specifically Related to Learning and Memory?, in *Neuronal Plasticity and Memory Formation* (Ajmone Marsan C. and Matthies H., eds.). pp. 25–35, Raven Press, New York.

Vanderwolf C. H. (1983) The Influence of Psychological Concepts on Brain-Behavior Research, in *Behavioral Approaches to Brain Research* (Robinson T. E., ed.). pp. 3–13, Oxford University Press, New York.

Vanderwolf C. H. and Leung L. -W. S. (1983) Hippocampal Rhythmical Slow Activity: A Brief History and the Effects of Entorhinal Lesions and Phencyclidine, in *Neurobiology of the Hippocampus* (Seifert W., ed.). pp. 275–302, Academic Press, London.

Vanderwolf C. H. (1984) Aminergic Control of the Electrocorticogram: A Progress Report, in *Neurobiology of the Trace Amines* (Boulton A. A., Baker G. B., Dewhurst W., and Sandler M., eds.). pp. 163–183, Humana Press, Clifton, NJ.

Whishaw I. Q. and Vanderwolf C. H. (1973) Hippocampal EEG and behavior: Changes in amplitude and frequency of RSA (theta rhythm) associated with spontaneous and learned movement patterns in rats and cats. *Behav. Biol.* **8,** 461–484.

Whishaw I. Q., Kolb B., and Sutherland R. J. (1983) The Analysis of Behavior in the Laboratory Rat, in *Behavioral Approaches to Brain Research* (Robinson T. E., ed.). pp. 141–211, Oxford University Press, New York.

Wieland C. (1983) The microneye. *Byte* **8,** 316–320.

Wolbarsht M. L. (1964) Interference and its Elimination, in *Physical Techniques in Biological Research,* vol. V. (Nastuk W. L., ed.). pp. 353–372, Academic Press, New York.

# Chapter 10

# Neurotransmitter-Selective Brain Lesions

TIMOTHY SCHALLERT AND RICHARD E. WILCOX

## 1. Historical Background

### 1.1. The Flourens Era

As neuroanatomical discoveries in the brain are made, there follows an intense curiosity about function. Frequently, the two initial questions asked are: What happens if the tissue is removed, and what happens if it is activated? Perhaps the first to use an experimental ablation method in animals to study brain function was DuVerney in 1697 (according to Walker, 1957). However, it was not until Flourens published his influential book in 1824, and its revision in 1842, that the method became accepted widely enough to displace cranioscopy popularized by the phrenologists (Luciani, 1915). Flourens presented the ablation method in great detail and his descriptions of the behavioral effects of the brain damage were far more complete than those of many of today's investigators. Although many of his conclusions regarding localization of function were soon vigorously challenged, particularly by clinicians, he stimulated a great deal of interest in brain–behavior relationships (Kolb and Whishaw, 1980; Young, 1970). Fritsch and Hitzig (1870) and Ferrier (1873) combined the ablation method with electrical stimulation techniques to study the motor functions of the cerebral cortex. When selective cortical areas were removed surgically, specific motor pareses were obtained that matched the topographic maps defined by the stimulation. Their approach quickly opened a new chapter in experimental brain research.

## 1.2. Knife Cuts and Suction Techniques

Mechanical lesions (knife cuts) were first tried by Veyssiere and by Nothnagel in 1874 (Luciani, 1915; Carpenter and Whittier, 1952). In 1876 and 1877, Nothnagel ran a series of experiments in which he passed a fine wire blade through a cannula lowered unilaterally into the head of the caudate nucleus in rabbits. By rotating the curved or angled wire, he was able to cut the neural projections to and from this region. The sensorimotor disturbances resulting from this procedure were in many ways comparable to those observed after ablation of the sensorimotor zone of the cerebral cortex. Nothnagel also described a gradual recovery of function that he attributed to the compensatory action of the homonymous nucleus on the contralateral side of the brain. Ten years later, Baginski and Lehmann (1886) found similar behavioral effects upon aspirating the caudate via a thin glass tube lowered through a small hole in the skull (Luciani, 1915).

## 1.3. Electrolytic Lesions

According to Roussey (1907), Golsinger was the first to make electrolytic lesions in animals in 1895, though this work was not published. In 1898, Sellier and Verger destroyed discrete areas ("about the size of a pea") in the caudate and anterior segment of the internal capsule in dogs by passing current through double-needle electrodes insulated except for the cross section at the tips. Their experiments confirmed the results of Nothnagel and supported Luciani's view (advanced in 1885) that the basal ganglia and the sensorimotor cortex have highly related functions (Luciani, 1915). In 1908, Horsley and Clark developed the stereotaxic method and combined it with electrolytic lesions to improve the localization, precision, and reliability of brain damage in subcortical structures. At about this time, Trendelenburg (1910) began developing the use of local, reversible brain cooling in behaving animals (Brooks, 1983). However, electrolytic lesions became the most widely used method of producing brain damage.

## 1.4. Early Chemical Lesions

Almost from the start investigators were not satisfied with the degree of specificity provided by the ablation and stimulation techniques. Baglioni and Magnini (1909) searched for potential chemical tools that might depress or stimulate cerebral function more selectively. They applied to the surface of cortex varying

doses of substances that were thought potentially to have selective properties based on research in the spinal cord and peripheral
nervous system. According to Luciani (1915), those substances included carbolic acid, which "picks out the motor cells of the ventral horn of the spinal cord; strychnine, which affects the cells of
the dorsal horn; acetic, citric, and glycerine acids; urea, sodium
chloride; sodium sulphate; glucose; picrotoxin; and curare." They
reported that some of the "specific poisons exerted electively on
the cortical ganglion cells," whereas the nerve fibers of the corona
radiata were spared. However, there were technical problems
with chemical lesions that were known even before the experiments of Baglioni and Magnini. "Chemonecrosis" was a term introduced by Beaunis in 1868 and used extensively by Fournie in
1873 and by Nothnagel in 1876 (Carpenter and Whittier, 1952;
Luciani, 1915). Nothnagel injected zinc chloride, chromic acid,
silver nitrate, mercuric chloride, and aluminum hydroxide
through cannulae inserted into subcortical structures such as the
caudate and pallidum. Upon tinting zinc chloride with aniline
blue to estimate diffusion, it became clear that the method was
unsatisfactory. Even small amounts of the chemical spread un-
controllably, resulting in irregular-shaped lesions of unpredictable size.

Some investigators were concerned that even well-localized
electrolytic lesions would affect tissue distant from the intended
brain site because of diaschisis (shock) and/or vascular damage.
To prevent this, Edwards and Bagg (1923) implanted slow-acting
radon seeds into the striatum of dogs. They reported that lesions
even as large as 4 mm in diameter could be produced gradually
without vascular damage and, presumably, with reduced
diaschisis.

## 1.5. Anatomically Specific Chemical Lesions

The first report of a structure-sensitive experimental neurotoxin
may have been by Mella in 1924. Mella found that multiple
intraperitoneal injections of manganese chloride in rhesus monkeys caused severe, and nearly symmetrical, damage to the pallidum and striatum, with relatively little or no damage to other
brain structures. Alpers and Lewy (1940) later found that these
same brain regions were selectively vulnerable in animals exposed to carbon disulfide gas for prolonged periods, a finding
confirmed by Richter in 1945.

In 1949, Brecher and Waxler reported that the neurotoxin
goldthioglucose injected intraperitoneally in mice was taken up
with high selectivity by cells located in the general area of the

ventromedial hypothalamus. The behavioral effects of the damage (hyperphagia and obesity) were comparable to those observed by many investigators after electrolytic lesions (Marshall and Mayer, 1954). Although it was soon realized that goldthioglucose significantly damaged some brain stem regions as well, this technique had a large impact because it dramatically supported Mayer's glucostatic hypothesis that the hypothalamus contained glucoreceptors that initiate feeding whenever the concentration of sugar in the blood falls below a minimum level.

## *1.6. Transmitter-Selective Lesions*

In 1960, a very influential paper by Grossman created the first obvious need for neurotransmitter-selective neurotoxins. He introduced chemical agents directly into the lateral hypothalamus through a cannula. These agents were neurotransmitters or drugs that stimulated postsynaptic receptors. Acetylcholine (mixed with eserine to retard its enzymatic destruction) or the cholinergic agent carbachol caused sated rats to start drinking large amounts of water within 5–10 min, and even to press a bar repeatedly to obtain water. The most important aspect of the study was that the behavioral response seemed to be neurotransmitter-selective. Noradrenaline (NA) placed in exactly the same brain area, through the same cannula, caused rats to start eating large amounts of food and to press a different bar that delivered food. The administration of various control agents ruled out osmotic pressure, pH, vasoconstriction, or vasodilation as factors in these behavioral effects. From then on researchers using lesion techniques to study brain function became much more concerned about the existence of functionally distinct systems that were spatially overlapping and chemically independent. Work since then of course has complicated the status of the neurochemical mechanism of feeding and drinking, as Grossman himself knew it would. Still, in the context of the present discussion, one of the impacts of this study was to prepare investigators to welcome the advent of transmitter-selective neurotoxins enthusiastically and to incorporate them into their work.

Meanwhile, neuroanatomists were mapping the brain in terms of its neurochemical organization. Electrolytic ablation techniques and newly developed chemical assays were being used to describe serotonergic pathways (Heller et al., 1962; Harvey et al., 1963), catecholaminergic pathways (Heller and Harvey, 1963; Dahlstrom and Fuxe, 1964), and cholinergic pathways (Shute and Lewis, 1967). In 1967, the first and most popular of the transmitter-specific neurotoxins, 6-hydroxydopamine (6-OHDA), was introduced. This agent is accumulated by, and destroys, catecholaminergic neurons relatively selectively (Thoenen and Tranzer, 1968; Tranzer and Thoenen, 1967; Ungerstedt, 1968). Because it was known by then that the primary etiology of Parkinson's disease involved the degeneration of nigrostriatal dopamine-containing cells, the 6-OHDA-treated rat was viewed as a potentially excellent animal model of this neurological disorder.

Ungerstedt (1968) was the first to inject 6-OHDA stereotaxically into discrete brain sites. In 1971 he showed that 6-OHDA-induced destruction of primarily catecholaminergic projections passing through the lateral hypothalamic area yielded aphagia, adipsia, and akinesia (Ungerstedt, 1971a). This report drew a great amount of attention because these behavioral signs had been well established as correlates of nonspecific damage to this complex region of the brain (Anand and Brobeck, 1951; Gladfelter and Brobeck, 1962; Teitelbaum and Epstein, 1962). Noradrenaline uptake inhibitors administered during surgery allowed even greater selectivity. These uptake inhibitors help prevent 6-OHDA from accumulating in noradrenergic neurons that are thereby spared relative to dopamine (DA) neurons (Breese and Traylor, 1971; Jonsson and Sachs, 1975). Using this procedure, Stricker and Zigmond (1976) confirmed the work and conclusions of Ungerstedt that the effects of electrolytic lesions were caused largely by the destruction of dopaminergic neurons.

In 1973, Ahlskog and Hoebel infused 6-OHDA into the ventral central tegmental tract that is comprised of adrenergic projections to the medial hypothalamus. They obtained overeating and obesity comparable to that reported after electrolytic lesions of the medial hypothalamus. Thus, 6-OHDA appeared to offer vastly improved selectivity over conventional lesion techniques.

## 1.7. Axon-Sparing Lesions

While 6-OHDA was being widely exploited (and, by some, overly criticized for not being catecholamine-specific: Poirer et al., 1972; Butcher et al., 1974), a complementary neurotoxin was being developed. Olney et al. (1971, 1974) injected kainic acid peripherally into neonatal and adult mice. This neuroexcitatory analog of glutamate destroyed neuronal dendrites and perikarya of certain brain structures (primarily in the limbic system), but spared axons of passage. Coyle and Schwarcz (1976) and McGeer and McGeer (1976) took advantage of this selectivity and injected kainic acid directly into the striatum in an attempt to develop an animal model of Huntington's disease. As in Huntington's disease, GABA-ergic, cholinergic, and other neurons whose cell bodies are intrinsic to the striatum were profoundly reduced by kainic acid, whereas dopaminergic terminals were spared. Soon Grossman et al. (1978) used kainic acid to isolate the effects of damage to cell bodies and dendrites of neurons in the lateral hypothalamus. They noted the appearance of aphagia and adipsia without akinesia or other obvious neurological impairments. This report thus dissociated a DA-unrelated ingestive impairment from one linked to the sensorimotor dysfunctions caused by DA depletion. Balagura et al. (1969) and Schallert and Whishaw (1978) promoted similar dissociations.

Kainic acid is not selective in all brain regions (McGeer et al., 1978; Mason and Fibiger, 1979), and since the work of Schwarcz and his colleagues (1979), ibotenic acid is gradually replacing kainic acid as the neurotoxin of choice for ablating discrete populations of cell bodies and dendrites in the brain (Jonsson, 1980; Kohler and Schwarcz, 1983). Ibotenic acid is a glutamic acid analog that, unlike kainic acid (Leach et al., 1980), does not induce seizures (and therefore may not require coadministration of anticonvulsant drugs) and does not cause neuronal damage remote from the intended injection site (Fuxe et al., 1983). Even massive lesions apparently do not damage axons of passage (Guldin and Markowitsch, 1981). Although there have been other potentially useful axon-sparing compounds described in the literature, including cobalt (Malpeli and Burch, 1982) and quinolinic acid (Fuxe et al., 1983), ibotenic acid may be the most versatile.

## 1.8. Current Status

Quite a number of other transmitter-selective neurotoxins have now been developed. Among the most promising are 5,7-dihydroxytryptamine, (5,7-DHT), which destroys serotonergic neurons (Baumgarten and Lachenmayer, 1972), N-2-chloroethyl-N-ethylbromobenzylamine HCl (DSP-4), which can be applied peripherally or locally to destroy noradrenergic terminals (Jaim-Etcheverry and Zieher, 1980; Hallman et al., 1984; Dooley et al., 1984; Ross, 1976), 1-methyl-4-phenyl-1,2,3,6-tetrahydropyridine (MPTP) or its metabolite $MPP^+$ which may preferentially deplete dopaminergic neurons in the pars compacta of the substantia nigra in primates and mice when applied peripherally (Davis et al., 1979; Langston et al., 1983, 1984a, b; Burns et al., 1983; Heikkila et al., 1984; Steranka et al., 1983), and the choline mustard analog AF64A (ethylcholine aziridinium ion) that according to preliminary reports appears to destroy cholinergic neurons somewhat selectively (Fisher and Hanin, 1980; Fisher et al., 1980; Mantione et al., 1981; Sandberg et al., 1984a, b).

In the next section, methods for producing transmitter-selective brain lesions in four different transmitter systems via neurotoxins will be described. The major emphasis will be on 6-OHDA, which is the most extensively studied of these neurotoxins and remains the best for selectively damaging DA or NA neurons. We also will discuss 5,7-DHT, a good neurotoxin for serotonin, and AF64A, a potentially useful neurotoxin for acetylcholine (ACh). Each neurotoxin will be characterized in the following four ways: (1) general uses; (2) procedures for its application; (3) the degree of specificity potentially achievable and suggested mechanisms of action; and (4) a review of neurochemical methods for evaluating the success of the lesion technique. We will not discuss DSP-4 or MPTP, since they have not been shown to be generally more selective or useful than intracerebrally applied 6-OHDA [MPTP has received much attention because it can be applied peripherally, it appears to have species- and strain-specific properties, and its Parkinson effects have been observed in people (Burns et al., 1984; Chiueh et al., 1983, 1984; Hallman et al., 1985; Heikkila et al., 1984)]. We will also not discuss nonspecific lesions or axon-sparing lesions any further; however, these methods have been reviewed thoroughly elsewhere (Fuxe et al., 1984; Singh, 1975; Spiegel, 1982; Finger and Stein, 1982).

# 2. 6-OHDA (DA or NA Neurotoxin)

## 2.1. General Uses

As noted above, 6-OHDA is a neurotoxin that can be used to destroy dopaminergic and/or noradrenergic neurons relatively selectively. By knowing the anatomy of DA and NA projections and their sensitivity to 6-OHDA, one can infuse 6-OHDA into different brain areas to obtain regional depletions of either catecholamine (Robbins and Everitt, 1982). In addition, by using uptake inhibitors during surgery, reduced accumulation of 6-OHDA in, and therefore greater protection of, NA or DA neurons that are not of interest can be achieved. For example, a common procedure used to destroy nigrostriatal DA neurons is to infuse 6-OHDA into the ventral tegmental area medial to the rostral substantia nigra. The mesotelencephalic dopaminergic axons are concentrated at this point, making them vulnerable to small volumes and low concentrations of 6-OHDA (Fink and Smith, 1979; Marshall et al., 1980). Because NA axons lie in close proximity, desmethylimipramine (DMI), an NA uptake inhibitor, typically is administered prior to 6-OHDA infusion. This procedure not only greatly limits the destruction of NA cells, it also appears to enhance the destruction of DA cells (Simansky and Harvey, 1981). The DA neurons projecting to forebrain areas other than the caudate will also be destroyed by 6-OHDA infused into the ventral tegmental area. If caudate depletion alone is desired, 6-OHDA must be infused locally. Because the caudate is large, it is necessary to infuse 6-OHDA into several subregions to achieve depletion of the entire structure. A subregion, of course, can be depleted selectively as well.

To destroy noradrenergic neurons and spare dopaminergic neurons, 6-OHDA can be infused into the locus ceruleus (Harik, 1984). This area contains the cell bodies for most noradrenergic neurons in the brain and spinal cord. No dopaminergic neurons are located here. This procedure will deplete NA in many areas, including cerebellum and spinal cord. The lesion focus must be varied to restrict the regions of depletion. If, for example, primarily forebrain depletion is desired, 6-OHDA can be infused more rostrally into the dorsal noradrenergic bundle (Sahakian et al., 1983). Dopamine uptake inhibitors such as benztropine may be administered prior to 6-OHDA infusion to help protect dopaminergic neurons. Since benztropine and many other transmitter uptake inhibitors have anticholinergic properties, one should be extremely cautious about using atropine to reduce parasympathetic functions because of a possible additive effect of the uptake inhibitor with the atropine.

The catecholaminergic pathways are highly collateralized and their organization is complex. The pattern of transmitter loss will depend not only on the placement of 6-OHDA and the type of drug pretreatment, but on many other factors, such as the age of the animal. Moreover, the action of 6-OHDA is not precisely predictable from lab to lab. It is essential, therefore, to carry out manipulation checks in the form of regional evaluation of the morphological and neurochemical integrity of the brain in every experiment. The review by Robbins and Everitt (1982) should be consulted as an excellent guide to many of the anatomical and behavioral considerations relevant to the use of 6-OHDA.

## 2.2. Procedures for Application

Effective procedures for local infusion of 6-OHDA into brain tissue are described in this section. These procedures are representative of those used successfully by many investigators over the last decade to deplete DA or NA in various brain areas. This discussion should provide a useful starting point: Results can then be modified depending on the results of neurochemical and histological analyses. .

## 2.2.1. Presurgical Treatment

The most commonly used pretreatment drug for protecting NA neurons is desmethylimipramine HCl (DMI), a noradrenergic uptake inhibitor, whereas for protecting DA neurons, benztropine or amfonelic acid, DA uptake inhibitors, are commonly employed. Zimelidine, a serotonin uptake inhibitor, may also be used, but generally is not necessary unless 6-OHDA is introduced by the intraventricular route (Reader and Gauthier, 1984). The doses of DMI used by various investigators have ranged, with increasing effectiveness, from 10 to 25 mg/kg ip; of benztropine, 0.1–30 mg/kg ip; and amfonelic acid, 5 mg/kg ip. If zimelidine is used, the dose should be 20 mg/kg ip. These doses apply to young adult rats. The maximal degree of protection may be limited in aged animals (Marshall et al., 1983). When using DMI, it should be suspended in an emulsifier, such as Tween-80, or warmed slightly and administered 30 min prior to 6-OHDA. Slightly less anesthesia should be used. It is important to note that there is a greatly increased risk of postoperative death. A primary reason for this is a gradual and uncontrolled hypothermia that is exaggerated by 6-OHDA (particularly if 6-OHDA is infused into the hypothalamus). Therefore, body temperature should be continuously monitored until the animal is fully recovered from the anesthesia. An infrared heat lamp or other warming device should be used to maintain the core temperature at least above 35°C. DMI also causes pathological gastrointestinal transit, including intestinal hypertrophy and malabsorption of nutrients (Saller and Stricker, 1978). This problem can be controlled by infusing the DMI intracerebrally through the infusion cannula (12 μg in 0.8 μL) 15 min prior to 6-OHDA (Hernandez and Hoebel, 1982). Other uptake inhibitors, including benztropine (3 μg in 0.8 μL) are effectively administered this way. The value of a local application of an uptake inhibitor will be magnified after bilateral lesions of DA or NA neurons involved in ingestive behavior, such as those in mesostriatal or mesohypothalamic pathways (Stricker, 1983; Schallert et al., 1978; Marshall et al., 1974; Ungerstedt, 1971a).

Because 6-OHDA can be inactivated by monoamine oxidase (MAO), the effects of 6-OHDA can be potentiated by pretreatment with an MAO inhibitor, usually pargyline, 40–50 mg/kg ip (Breese and Traylor, 1971). The MAO inhibitor should be administered no more than 30 min prior to 6-OHDA. Otherwise, intraneuronal buildup of catecholamines will counteract the effects of 6-OHDA by affecting the rate of enzymatic oxidation of the toxin (Sachs et al., 1975).

The choice of anesthesia may be a factor. The duration, sites, and mechanisms of action may differ enough among anesthetics to influence the action and potency of 6-OHDA. Indeed, Bosland et al. (1981) have shown that 6-OHDA injected into the subarachnoid space of the spinal cord is much more effective when ether is used rather than pentobarbital (NA depletion = 5 vs 33% of control).

### 2.2.2. Infusion Procedures

The concentration and volume of the 6-OHDA solution and the type of vehicle in which the 6-OHDA is dissolved can all make a difference in the degree of DA or NA depletion relative to the amount of nonspecific damage obtained. For example, Willis et al. (1976) systematically varied the concentration and volume of 6-OHDA solution infused into the ventral tegmental area. Based on neurochemical and histological evaluations of their method and of methods commonly used in the recent literature, it is possible to estimate (but by no means predict with precision) the parameters necessary to effectively deplete catecholamines without yielding excessive nonspecific damage (Marshall et al., 1983; Schallert et al., 1982, 1983; Carey, 1982; Reader and Briere, 1983; Heikkila et al., 1981; Altar et al., 1984; Jeste and Smith, 1980).

For local application, a solution should be prepared such that 4–8 µg of 6-OHDA (expressed as the base of either 6-OHDA HBr or 6-OHDA HCl, the latter being more expensive to purchase) dissolved in 2–3 µL of ice-cold saline or artificial cerebral spinal fluid can be delivered. To expel dissolved oxygen, the solution can be bubbled for 1 min with nitrogen (Simansky and Harvey, 1981). To prevent oxidation, most investigators also add ascorbic acid to the vehicle, usually at a concentration of 0.1% w/v (1 mg/mL). Because ascorbic acid, by itself, can cause considerable nonspecific neural damage (Waddington and Crow, 1979; Wolfarth et al., 1977) a few investigators prefer not to use it. The 6-OHDA should be kept cold and in a dark bottle. It must be delivered within about 20 min after it is prepared. If ascorbic acid is not used, delivery should begin in less than about 2 min. The cannula should have a beveled tip and its opening should face a consistent direction for every animal (e.g., caudally). The cannula should be slowly lowered to the desired coordinates and promptly raised 0.1–0.2 mm to provide a small well above the target tissue into which the solution can be delivered. The solution is commonly injected at a rate of about 0.5–1 µL/min (the effects of varying the infusion rate have not been examined). To minimize the variability of injected volumes, the movement of an air bubble through the polyethylene tubing can be measured (Swanson et al., 1972). The cannula should be left in place for a constant amount of time, and for at least 5 min to allow for diffusion of the solution, after which it should be raised very slowly to prevent the solution from being drawn up the cannula track.

When unilateral lesions are produced, it is worthwhile to make sham operations in the opposite hemisphere of experimental animals, in addition to sham operations in control animals. This procedure reduces the effects of asymmetrical nonspecific damage, which can be considerable, especially that occurring as a result of cannula tract penetration through the cortex. It also is important to know that there are asymmetries in catecholaminergic function that can differentially influence the pattern of depletion and behavioral consequences, depending on which hemisphere is selected (Glick, 1976; Jerussi and Glick, 1975; Robinson and Coyle, 1979). For example, Robinson and Stitt (1981) found that 6 μg of 6-OHDA injected into the left frontal neocortex caused a significant bilateral depletion of cortical NA concentrations, whereas the same dose of 6-OHDA injected into the right frontal cortex caused only an ipsilateral depletion of NA. Recently Kubos et al. (1984) repeated this effect using 10–20 μg of the neurotoxin DSP-4. Finally, some asymmetries may even depend on the sex of the animal (Robinson et al., 1983).

The methods of intraventricular or intracisternal infusion are often used because they are believed to cause less nonspecific damage and a more uniform and widespread depletion (Stricker and Zigmond, 1976; Ondrusek et al., 1981; but *see* Reader and Gauthier, 1984). For many experiments these methods would not be as useful because the catecholamine depletion occurs throughout the brain and spinal cord. Typically, 10–20 μL of a 20 μg/μL solution of 6-OHDA are infused into each lateral ventricle (Whishaw et al., 1978; Herman et al., 1983). There is a gradient of depletion descending from medial to lateral in the brain, and anterior to posterior in the spinal cord. Asymmetrical depletions are possible and are common even if both ventricles receive the neurotoxin (Schallert et al., 1979). It is also possible to deplete only the spinal cord by infusing about 5 μl of a solution of 10–20 μg/μL of 6-OHDA via 30-gage polyethylene tubing inserted through a 23-gage guide cannula directed at a posterior angle into the cisterna magna (Schallert et al., 1981; Bosland et al., 1981) using techniques modified from Yaksh and Rudy (1976).

The infusion procedures described above apply to young adult animals. The age of the animal may be a factor. Marshall et al. (1983) reported that 6-OHDA is more toxic in aged animals (27–28 mo) than in young adult animals, at least at doses lower than 4 μg. There is a large volume of literature covering the unique effects of 6-OHDA in neonatal and prenatal animals, in which 6-OHDA readily crosses the blood–brain barrier to deplete NA permanently (Jonsson et al., 1976; Lidov and Molliver, 1982; Schmidt and Bhatnagar, 1979; Jonsson and Hallman, 1982; Sutherland et al., 1982a; Martin-Iverson et al., 1983). Because of the dynamic character of the developing brain, particularly its capacity for reorganization, interpretations of the effects of neurotoxins will be quite different from those for adult-operated animals. For example, recent work by Stachowiak et al. (1984) suggests that new serotonergic terminals in the caudate proliferate to take the place of the missing dopaminergic cells in animals treated neonatally with bilateral intraventicular 6-OHDA (100–200 μg in 10 μL) 30 min after administration of DMI (25 mg/kg ip) and pargyline (40 mg/kg). Depletion of DA in adult animals does not affect serotonin levels (Stricker, 1983).

## 2.3. Specificity and Mechanisms

Local applications of 8 μg of 6-OHDA will probably destroy 95–99% of the DA in about 30% of young adult rats, and 85–95% of the DA in about 40% of those animals. Slightly less 6-OHDA may be used for NA depletions of comparable magnitude. Note that these parameters apply to the highly sensitive DA-containing neurons in the nigrostriatal projections (Marshall, 1984) and to neurons in the dorsal or ventral noradrenergic bundles (Sahakian et al., 1983). Other systems, such as the tuberoinfundibular DA pathway, appear to be resistant to 6-OHDA (Jonsson, 1980). The median eminence appears vulnerable to intravenous administration (Smith et al., 1982). Adjustments may be necessary depending on the lesion placements, even within a given system. Subareas proximal to, or within, a terminal field may be less sensitive to 6-OHDA because the axon density is relatively sparse (Jeste and Smith, 1980; Marshall et al., 1980) despite the fact that the sensitivity of a given neuron appears to be greatest at the terminal and least at the cell body (Jonsson, 1980). Multiple small injections of larger cumulative doses via the same cannula may be required for consistently maximal depletions accompanied by minimal nonspecific damage. (The nonspecific damage in this case would be measured as the volume of necrotic tissue observable during the first week after surgery; the apparent size of the area of nonspecific damage recedes over time.) However, although multiple injections can yield a high degree of depletion (up to 99%), the behavioral sequelae may be far less severe than for large single injections possibly because the small multiple injections do not cause catecholaminergic accumulation (Willis and Smith, 1982), nerve impulse blocking (Altar et al., 1984), or other 6-OHDA-induced events unlinked to catecholamine depletion *per se*, or because small multiple lesions permit the occurrence of compensatory neuronal processes associated with slow-developing lesions (Zigmond et al., 1984).

The mechanisms of the degenerative action of 6-OHDA have been reviewed recently elsewhere (Jonsson, 1980; Jonsson and Sachs, 1975; Altar et al., 1984; Willis and Smith, 1984). It is difficult to summarize these reviews briefly because there is a large amount of information and many conflicting hypotheses that deserve extensive discussion. To destroy catecholaminergic neurons, at least 50–100 m$M$ of 6-OHDA must accumulate in the extragranular cytoplasm (Jonsson, 1976). It is not necessary for 6-OHDA to be taken up into the storage vesicles (Jonsson et al., 1972). According to Jonsson (1980), the DA or NA can be depleted in less than an hour after 6-OHDA infusion if the dose is high enough. Altar et al. (1984) report that DA synthesis is increased within 10 min because of 6-OHDA-induced neuronal impulse block. They suggest that DA levels increase in the terminals to such an extent that there is overflow from the storage granules. The overflow is not released. Catabolism rapidly occurs in the intraneuronal space, resulting in the formation of excess 3,4-dihydroxyphenylacetic acid (DOPAC) that in turn undergoes catabolism into homovanillic acid (HVA) outside the neuron. They concluded that the initial behavioral deficits, that begin at 10 min postinfusion and are maximal by 50 min, are due to the loss of DA release at the terminals, whereas the subsequent (prerecovery) behavioral effects are caused by the degeneration of DA terminals that occurs within 24–48 h. Others have found that it takes at least 72 h for the effects of 6-OHDA to be maximal (Barth et al., 1983; Lidbrink and Jonsson, 1974).

At present there is little agreement on how 6-OHDA destroys catecholaminergic neurons selectively, what neuronal processes are damaged, what behavioral effects are associated with catecholamine loss *per se* vs secondary effects of 6-OHDA, or the temporal aspects of these events. As reviewed by Jonsson (1980), 6-OHDA is readily oxidized following infusion and leads to the formation of several potentially cytotoxic products, including hydrogen peroxide (Heikkila and Cohen, 1971), quinones (Saner and Thoenen, 1971), 5,6-dihydroxyindole (Blank et al., 1972), the superoxide and hydroxy radicals (Heikkila and Cohen, 1973; Cohen and Heikkila, 1974), and singlet oxygen (Heikkila and Cabbat, 1977). Soon after infusion, one or more of these products or 6-OHDA itself causes chronic depolarization associated with increased permeability to $Ca^{2+}$; the $Ca^{2+}$ accumulation is believed to be the mechanism underlying the rapid loss of neuron impulse capacity noted above (Haeusler, 1971; Furness et al., 1970; Jonsson, 1980). Eventually there is a total destruction of uptake mechanisms, synthesizing enzymes, and in 1–4 h, there are ultrastructural signs of degeneration in a percentage of neurons proportional to the dose of 6-OHDA administered (Jonsson, 1980).

Willis and Smith (1982, 1984) have focused on the role of catecholamine buildup as a potential source of the behavioral effects of 6-OHDA. It is well known that DA (or NA) accumulates within the degenerating axons in the first 24–48 h after 6-OHDA infusion (Jonsson and Sachs, 1973; Ungerstedt, 1968). The axons may remain swollen for weeks, during which time the transmitter is proposed to be released into brain structures all along the pathways involved. According to Willis and Smith (1984), the gradual reduction of axon distention and associated amine release may contribute to apparent recovery of function. This event is not specific to 6-OHDA. Amine accumulation occurs also following destruction of aminergic neurons by electrolytic lesions, knife cuts, or other lesion techniques (Ungerstedt, 1971b).

## 2.4. Evaluation of 6-OHDA Neurotoxicity

An excellent review of the methods necessary for a comprehensive evaluation of the primary and secondary anatomical sequelae common to most forms of brain damage has been written by Schoenfeld and Hamilton (1977). Analyses of nonspecific damage at the injection site have been reviewed adequately by Agid et al. (1973b), Sotelo et al. (1973), and Willis et al. (1976). These methods will not be discussed here. Verification of the neurochemical effects of neurotransmitter-selective lesions such as those produced by 6-OHDA requires two minimal criteria, namely extent of depletion and regional specificity.

### 2.4.1. Catecholamine Levels

The extent of the catecholamine depletion induced by 6-OHDA is best established by evaluating directly the catecholamine content of the terminal beds that represent the targets of the neurotoxin administration, e.g., the corpus striatum. The definitive procedure for the determination involves high pressure liquid chromatography (HPLC) with electrochemical detection (EC). In essence, NA, DA, and an internal standard such as 3,4-dihydroxybenzylamine injected into the system are retained on an analytical column with relatively unique retention times. The oxidation of these catecholamine compounds at the electrode of the system is directly proportional to the amounts of the catecholamine present. Thus, by comparing the chromatographic peak area (done electronically by an integrator) for catecholamine standards of known concentration with the areas obtained from samples of brain homogenates, the amount of each catecholamine in a given brain region may be determined without derivitization or radioligand techniques. The sensitivity of the procedure is high, in the 500 fg range for NA and in the 800 fg range for DA. Since DA exists in microgram amounts in striatum, and NA also exists in high levels in areas such as cerebral cortex, cerebellum, and hypothalamus, it is not difficult, using HPLC–EC, to determine accurately as much as a 99% depletion of DA or NA. Definitive discussions of the basic procedure are provided in the chapter by Adams and Marsden (1982) and the text by Kissinger and Heineman (1984).

### 2.4.2. Regional Assays

Lesion specificity is a broad term that has been used in reference to both the anatomical and neurochemical accuracy of the procedure. Because the effects of the 6-OHDA neurotoxin are well established, most investigators (and manuscript reviewers) have found it acceptable to determine the integrity of DA vs NA neurons without examining other neurotransmitter systems. Thus, following 6-OHDA injection, specificity usually is evaluated anatomically by determining changes in catecholamine content in nearby brain areas. For example, damage to the nigrostriatal DA bundle resulting from 6-OHDA injections in the region of the zona compacta of the nigra may also cause destruction of DA projections to frontal cortex or nucleus accumbens. Determination of DA and NA may be carried out as above in each region or subregion of interest. Assays can be done in frozen tissue using very small amounts of sample from each brain region. Also, when samples are stored at −80°C, catecholamines are stable for months. With the use of an internal standard (a catecholamine compound not endogenous to the brain added in known amount to the sample) samples can be processed later.

### 2.4.3. Catecholamine Function

The procedures described above constitute the minimum verification for extent and specificity of 6-OHDA lesions. Yet such assays provide little power in terms of establishing functionally significant neurochemical changes following brain damage. For this purpose it becomes important to recall the steps involved in synaptic transmission that represent the sites for which dynamic changes in neural activity may occur to alter behavior. The disruption of catecholamine synthesis that occurs following 6-OHDA injection may be established by assaying the activity of the rate-limiting enzyme for both DA and NA synthesis, tyrosine hydroxylase. This enzyme catalyzes the conversion of tyrosine to L-dopa, and its activity may be calculated in two ways using the HPLC–EC procedures described above. In vivo activity following brain damage may be determined by inhibiting tyrosine hydroxylase with administration of α-methyl-*para*-tyrosine (a competitive inhibitor of the enzyme) and evaluating the disappearance of either DA or NA over time or by inhibiting the activity of the next enzyme in the synthetic scheme (dopa decarboxylase) with an agent such as NSD-1015 (3,4-dihydroxybenzylhydrazine) and measuring how much L-dopa is produced. L-Dopa is also a catecholic compound and so can be readily detected by the same procedures outlined above. In vitro the activity of the enzyme may also be assessed as a function of the concentration of its substrate (tyrosine) or cofactor (tetrahydropterin) and the affinity of the substrate for the enzyme ($K_m$) and the velocity of the reaction ($V_{max}$) determined. Changes in synthesis of catecholamines in the remaining nerve terminals occur during the course of behavioral recovery and may be documented quantitatively using this procedure. General discussions of background can be found in Cooper et al. (1982a, b) and in Segal (1976). Changes in the use of DA or NA (metabolism, utilization, or turnover) may similarly be documented by measuring levels of the transmitters vs metabolites. General discussions of these issues may be found in Sharman (1981). Specific metabolites of DA (3,4-dihydroxyphenylacetic acid, DOPAC and homovanillic acid, HVA) and of NA (3-methoxy-4-hydroxyphenylethyleneglycol, MHPG) that may be assayed in a nonmicrowaved brain may exist in both free and bound forms in rats. Hunt and Dalton (1983) and Westerink (1984) provide concise discussions of the relevant electrochemical methods for analysis of these metabolites. In general, bound forms of the metabolite are hydrolyzed and then either filtered or run through Sephadex G10 columns for sample cleanup prior to injection onto the HPLC apparatus (to eliminate proteins that clog the system), as above. Since substances are not extracted, recoveries of metabolites are usually nearly 100%, allowing for analysis of extremely small amounts of brain tissue.

Both tyrosine hydroxylase activity and DA levels have been used as indices of lesion severity, and these appear to be highly correlated (Hefti et al., 1980). However, consideration must be given to the postoperative timing of at least some assays and assay conditions (e.g., substrate and cofactor concentrations). For example, lesions that destroy 60% or more of nigrostriatal neurons will lead gradually to a severity-dependent *increase* in DA synthesis, turnover, and release in surviving neurons (Agid et al., 1973a; Hefti et al., 1980).

A somewhat more novel electrochemical detection procedure for assessing the functional capacity of catecholamine neurons following neurotoxic lesions measures the "fast-phase" release of the endogenous transmitter. This in vitro procedure utilizes a potassium depolarizing stimulus to induce transmitter release in synaptosomes. Because the periods of exposure to the depolarizing stimulus are short (1–60 s, rather than many minutes as in older procedures) and because the endogenous transmitter is utilized (rather than an exogenously applied tritiated form of the catecholamine), the procedure is extremely sensitive (Wilcox et al., 1983; Woodward et al., in preparation) and has been demonstrated to be tightly coupled to the uptake of calcium in the same synaptosomal preparation (Leslie et al., 1984). Details of the procedure are provided in the above references. The released catecholamine (DA or NA that can be similarly assayed) is analyzed by EC detection exactly as above to provide an extremely useful index of the ability of the remaining neurons to alter the synaptic concentrations of transmitter in response to normal action potentials.

A final common way in which depletion of catecholamines may be estimated involves an indirect measure using the changes in the uptake of [$^3$H]-DA or [$^3$H]-NA into synaptosomes. Whereas uptake of transmitter may occur into its own neurons as well as into other neurons, changes in catecholamine uptake into synaptosomes following lesions correlates highly with a number of other measures of lesion severity (Spirduso et al., 1984). Recently a procedure has been developed for measuring the "fast-phase" uptake of [$^3$H]-DA into striatal synaptosomes (Woodward et al., submitted). The procedure is similar to more standard assays except that the incubation period allowed for transmitter uptake is extremely short (1–60 s, rather than several minutes). The procedure is more sensitive than standard procedures for detecting a variety of changes in striatal DA function because of less "noise" in the system, and promises to be a useful addition to the menu of procedures for studying functional changes after brain damage. In essence, the procedure involves exposing brain tissue to a [$^3$H]-DA or -NA solution in the absence of calcium that would induce transmitter release and obscure the uptake measurements. The uptake process is a saturable, high-affinity, sodium-dependent process mediated by a carrier system and constitutes an important means of preserving normal synaptic concentrations of the catecholamines (Cooper et al., 1982b). Actual uptake is determined as that amount of radioactivity (determined by liquid scintillation counting) remaining on a washed filter in the presence vs that occurring in the absence of sodium. Thus, it is not necessary to use an expensive HPLC system to estimate depletion of transmitter; one merely needs to have access to a liquid scintillation counter.

### 2.4.4. Postsynaptic Assays

Besides techniques for estimating actual loss of catecholamine-containing terminals following brain damage, there is a standard method for estimating the downstream effect of the lesion. This method is the very popular receptor binding technique that has been used and abused by so many individuals in the neurosciences during the last several years. The basic rationale for applying receptor binding methods to the study of lesioned preparations is that a decreased synaptic content of catecholamine should result in a compensatory increase in the number of receptors for the transmitter postsynaptic to the site of the damage (denervation supersensitivity as a means of maintaining homeostasis; Creese and Snyder, 1979; Seeman, 1980; Creese et al., 1982, 1983; Schallert et al., 1983). The procedure is straightforward. Tissue is incubated in the presence of various concentrations of an antagonist drug such as $[^3H]$-spiperone to determine *total* binding of the drug. Parallel sets of tubes are incubated with the same amount of tissue and $[^3H]$-spiperone, but in the presence of an appropriate excess of a competing dopaminergic drug that is not radioactive. Since all the receptor sites should be occupied by the nonradioactive dopaminergic drug, the only radioactivity that should be observed in these "nonspecific" binding tubes should be to nonsaturable sites not associated with receptors (binding sites on other protein, on glass, and so on). The difference in radioactivity between that occurring in the "total" and "nonspecific" binding tubes is defined as "specific" binding to the DA receptor (Bennett, 1978; Enna, 1983; Yamamura et al., 1985). As above, radioactivity is determined by liquid scintillation counting. The equilibrium dissociation constant for binding (the reciprocal of the affinity of binding or the estimate of the tendency of the ligand to combine with the receptor) and the maximal density of binding sites (respectively, $K_d$ and $B_{max}$) are usually determined from such studies. Following DA-depleting lesions of striatum, increases in binding occur as reflected by a greater number of binding sites with no change in the overall affinity (Schallert et al., 1980b;

Spirduso et al., 1984). "Grind and bind" methods are easy to apply (although not easy to obtain really consistent data with), but often difficult to interpret (Creese, 1981). Basic discussions of the methods are found in articles by Burt (1980) and Enna (1980), and in the volume by Yamamura et al. (1978). Key issues in binding are discussed with regard to the DA receptor by Creese et al. (1983) and Hamblin et al. (1984). Crucial to obtaining meaningful data are the appropriate definition of specific binding (by using the correct amount of the appropriate competing drug in the "nonspecific" binding tubes) and ensuring that unwanted sites are "masked" by adding a different drug to all tubes. Thus, in the rat striatum, [$^3$H]-spiperone should label a single DA $D_2$ receptor population with multiple agonist affinity states (the condition desired by the experimenter) when 100 n$M$ $d$-butaclamol is used to define the "blank" (nonspecific binding), and a 200-fold excess of a serotonergic antagonist, ketanserin, is added to all tubes as a mask for the serotonin receptors otherwise labeled by [$^3$H]-spiperone (Hamblin et al., 1984).

Analysis of the effects of NA depletion on noradrenergic receptor mechanisms is often accomplished by measuring the binding of [$^3$H]-dihydroalprenolol (DHA), a beta-noradrenergic antagonist. Propranolol, a nonspecific beta receptor blocker, is often used to define specific binding of [$^3$H]-DHA in a lesioned area, such as cerebral cortex. Discussions of beta receptor mechanisms may be found in the volume by Williams and Lefkowitz (1978), the review by Minneman (1981), and the reviews by Lefkowitz et al. (1983, 1984). A useful discussion of general issues in receptor theory may be found in the introductory papers by Richelson (1984) and Ruffolo (1982) and articles for more advanced readers in papers by Weiland and Molinoff (1981) and Molinoff et al., (1981).

### 2.4.5. Behavioral and Psychopharmacological Assays

There are many behavioral tests that can be used to estimate the neurotoxic effects of transmitter-selective brain damage with a good degree of accuracy. Indeed, the behavioral indices are often discovered before, and predict the result of, more direct assays of neurochemical changes. A distinct advantage is that they can be used repeatedly over a long postoperative period. For unilateral lesions, these tests include the circling response to DA agonists (Ungerstedt and Arbuthnott, 1970), the von Frey hair sensorimotor test (Marshall et al., 1974; Ljungberg and Ungerstedt, 1976a), and the posture-independent tactile extinction tests (Schallert et al., 1982; Schallert and Whishaw, 1984). For bilateral lesions, they include ingestive behavior, active avoidance, movement initiation, and stereotyped responses to anticholinergic drugs and DA agonists (Ungerstedt, 1971a; Ljungberg and Ungerstedt, 1976b; Schallert et al., 1978, 1979, 1980b; Stricker and Zigmond, 1976; Spirduso et al., 1985).

### 2.4.6. In Summary

The extent of catecholamine depletion caused by 6-OHDA administration may be simply assessed biochemically by means of several electrochemical detection procedures that directly measure the losses of transmitter or the synthetic, metabolic, or release capacity of the neuron. Functional changes in catecholamine neurons after lesion may be measured as synthesis, turnover, release, uptake, receptor binding, or behavioral responses before and after certain drugs. The chromatographic procedures require more "tooling up" and day-to-day maintenance than do the radioligand procedures, as well as the purchase of expensive equipment. On the other hand, the radioligand procedures are relatively tedious, measure only indirectly the extent of the lesion, and require continual outlays for expensive supplies. On theoretical grounds, no single procedure is preferred over any other as a "best" index of the state of the catecholamine neuron after neurotoxin administration. In fact a current wise trend in the literature is to use at least two independent ways of verifying each lesion. Only in this way can the true functional significance of such damage to the nervous system be assessed in relation to behavior. Finally, although 6-OHDA effects on catecholaminergic neurons are usually permanent, this does not mean that all neurochemical assays will yield comparable results at all postoperative periods. It may not always be appropriate to match short-term effects of 6-OHDA with the results of assays of tissue from animals sacrificed long after 6-OHDA treatment.

# 3. 5,7-DHT (Serotonin Neurotoxin)

## 3.1. General Uses

5,7-Dihydroxytryptamine creatinine sulfate (5,7-DHT) has been used to destroy serotonergic neurons relatively selectively (Baumgarten and Lachenmayer, 1972). It must be used in combination with DMI to prevent damage to NA neurons (Bjorklund et al., 1975). 5,7-DHT has generally become the serotonergic neurotoxin of choice over 5,6-DHT and the halogenated amphetamines because it is more potent at serotonergic neurons and causes less damage to nonmonoaminergic cells (Jonsson, 1980; Daly et al.,

1974; Baumgarten et al., 1977; Jacoby and Lytle, 1978). Like 6-OHDA, it may be infused intraventricularly or locally in the medial forebrain bundle (ascending serotonergic projections) or midbrain raphe nuclei (origin of serotonergic cell bodies), or into more rostral branches of serotonin-containing pathways (Clewans and Azmitia, 1984).

## 3.2. Procedures for Application

### 3.2.1. Presurgical Treatment

DMI or another NA uptake inhibitor should be given 30–45 min prior to infusion to protect NA neurons, although 5,7-DHT is not as toxic to NA neurons as is 6-OHDA. The DA uptake inhibitors are not normally necessary. Dosage and side effects of DMI are similar to those described above in the section on 6-OHDA. The MAO inhibitors are seldom used as pretreatments because, unlike the case with 6-OHDA, there is little evidence that 5,7-DHT neurotoxicity is enhanced by MAO inhibition. However, MAO inhibitors may be useful because, like DMI, they protect NA neurons from the neurotoxic action of 5,7-DHT (Baumgarten et al., 1975; Breese and Cooper, 1975). The reason for this is not clear.

### 3.2.2. Infusion Procedures

The procedures for preparing and administering 5,7-DHT are generally similar to those described for 6-OHDA above. Most investigators have infused 3–6 μg of 5,7-DHT into the medial forebrain bundle (Simansky and Harvey, 1981) or the dorsal or median raphe (Robinson, 1983), or 200 μg in 20 μL into each lateral ventricle (deMontigny et al., 1980), although some have used the intracisternal route (Breese and Cooper, 1975). Clewans and Azmitia (1984) recently conducted a careful time-course analysis of the bilateral neurochemical effects of 0, 1, 2, 3, 4, and 5 μg of 5,7-DHT delivered unilaterally to the cingulum bundle.

The age of the animal must be considered, though there is less information about this and other potential confounding procedural factors (Breese and Mueller, 1978; Martin-Iverson et al., 1983). As noted for catecholaminergic systems, if unilateral lesions are desired, consideration must be given to intrinsic lateralization of serotonergic function (Rosen et al., 1984).

## 3.3. Specificity and Mechanisms

The above described procedure has been shown to deplete 75–90% of telencephalic serotonin in all surviving animals. As noted for 6-OHDA above, the timing of the assay is important. Oberlander et al. (1981) found that the neurotoxic effects of 5,7-DHT continue to increase at least up to 12 d after surgery, as measured by in vitro serotonin uptake in the forebrain. This finding was generally confirmed by Clewans and Azmitia (1984), who estimated serotonin function by measuring tryptophan hydroxylase activity in the hippocampus.

According to Lorens et al. (1976), intracerebral infusion of the 5,7-DHT solution causes more nonspecific damage than injections of vehicle alone. However, Simansky and Harvey (1981) found that equivolumetric vehicle and 5,7-DHT solution produce comparable amounts of gliosis. Neither study controlled for time-dependent changes. In either case, it is clear that 5,7-DHT leaves far less nonspecific damage than that produced by electrolytic lesions large enough to cause equivalent depletion of serotonin. Robinson (1983) provides a good summary of the regional serotonin levels after dorsal vs median raphe infusions.

For a survey of the advantages of using 5,7-DHT and its specificity, see the articles by Jacoby and Lytle (1978) and Nobin and Bjorklund (1978). 5,7-DHT apparently is concentrated in serotonergic neurons by a high-affinity uptake process, but does not react with oxygen as 6-OHDA does (Jonsson, 1980). There is less known about the mechanisms of action of 5,7-DHT than about those of 6-OHDA (Baumgarten et al., 1975; Allis and Cohen, 1977; Creveling and Rotman, 1978).

### 3.4. Evaluation of 5,7-DHT Neurotoxicity

Serotonin (5-hydroxytryptamine; 5-HT) is an indolealkylamine. Like the catecholamines, serotonin is electrochemically active and so may be readily detected by HPLC–EC techniques, often using essentially the same chromatographic conditions that are used for the catecholamines (see above). Thus, for example, if one wished to determine the losses of catecholamines and serotonin in striatum following local application of 6-OHDA or 5,7-DHT, the levels of DA, NA, and 5-HT could be determined directly by electrochemical detection procedures in the same sample if so desired. In practice, the retention times for NA and 5-HT are different enough so that either two portions of the same brain tissue sample should be injected into two instruments (or one instrument operating at different times with slightly different mobile phases), or a gradient system (in which mobile phase composition can be adjusted during a sample run) should be utilized. Serotonin does not adsorb onto alumina (the procedure used for cleaning up and concentrating a sample of catecholamines prior to injection onto the HPLC). Thus, the typical procedure for serotonin is to deproteinate the sample by acid precipitation and filtration followed by injection onto the HPLC. General discussions of the procedure may be found in the previously mentioned chapter by Adams and Marsden (1982) and in the chapter by Davis et al. (1981). Also extremely useful in this regard is the publication *Current Separations* (free from Bioanalytical Systems Inc., West Lafayette, IN, that provides detailed "how to" information on electrochemical detection techniques. Estimates of serotonin synthesis and turnover may similarly be accomplished by electrochemical detection procedures following inhibition of tryptophan hydroxylase (by NSD-1015; see above discussion under 6-OHDA) or assay of 5-hydroxyindoleacetic acid (5-HIAA) levels, respectively. A good discussion of serotonin turnover may be found in Curzon (1981). Other relevant discussions are given in Diggory et al. (1981). Should the serotonin need to be concentrated prior to injection onto the HPLC (and this is probably not necessary given the current high sensitivity of HPLC–EC) a procedure by Snyder et al. (1965) gives greater than 85% recovery.

The uptake of $[^3H]$-5-HT can be carried out by a procedure similar to that described above for DA (Barbaccia et al., 1983). Similarly, changes in serotoninergic $S_2$ receptor binding may be assessed using the same ligand, $[^3H]$-spiperone, that is used for DA binding (List and Seeman, 1981). In this instance, however, one masks the higher affinity DA $D_2$ sites with a 200-fold excess of a DA antagonist such as domperidone (Creese et al., 1983; Hamblin et al., 1984) while using ketanserin in the nonspecific binding tubes as the blank. This indicates how very important it is to define specific binding in an appropriate way, since most radioligands label several sites in brain.

# 4. AF64A (Acetylcholine Neurotoxin)

## 4.1. General Uses

One of the newest candidates to the group of transmitter-selective neurotoxins is the aziridinium ion of ethylcholine, AF64A, that is taken up preferentially by the high-affinity choline transport mechanism and incapacitates cholinergic neurons and apparently destroys them (Rylett and Colhoun, 1980; Fisher et al., 1980; Hanin et al., 1983). Indirect assays of cholinergic function indicate that infusion of low doses of this compound into brain areas rich in cholinergic neurons (e.g., striatum, neocortex, substantia innominata/medial globus pallidus, septal–hippocampal complex) significantly depletes ACh, disrupts the sodium-dependent, high-affinity choline transport mechanisms specifically linked to ACh neurons, and decreases levels of choline acetyltransferase (CAT, the ACh synthesizing enzyme) without greatly affecting postsynaptic ACh receptors or the nonspecific low-affinity choline transport mechanism (Mantione et al., 1981; Sandberg et al., 1984a, b). It has not been established definitely whether the reduction in markers for ACh is caused entirely by ACh cell death. Markers of the function of other neurotransmitters, including DA, NA, or GABA may show relatively little change (Sandberg et al., 1984a). The potency of AF64A for ACh neurons is nearly comparable to that of 6-OHDA for DA or NA neurons, though it appears to be somewhat less specific (Myles et al., 1984). That is, the ratio of nonACh/ACh damage is greater with AF64A than the ratio of nonDA/DA damage with 6-OHDA, at least as typically administered. However, AF64A presently is the only neurotoxin that is relatively selective for ACh neurons. Because ACh loss is known to be involved in Alzheimer's disease (Davies and Maloney, 1976) and a host of behavioral processes, this neurotoxin currently is receiving much attention.

## 4.2. Procedures for Application

### 4.2.1. Presurgical Treatment

Atropine sulfate (20 mg/kg ip) increases the toxicity of AF64A because it enhances choline uptake (Fisher et al., 1981). At this time there are no data to suggest that selectivity can be improved by drug pretreatments. Future research may reveal that uptake drugs can be used to protect nonACh neurons damaged by high doses of AF64A.

### 4.2.2. Infusion Procedures

AF64A should be prepared as described by Clement and Colhoun (1975) and Sandberg et al. (1984a,b), or purchased from Research Biochemicals, Inc., Wayland, MA. A solution of 5–7 n$M$ of acetoxy-ethylcholine mustard HCl may be reacted with 1$N$ NaOH at pH 11.5 for 30 min. AF64A will be formed by lowering the pH to 7.4 with 0.1$N$ HCl and buffering with solid $NaHCO_3$. Within 4 h of AF64A formation, 0.5–1 μL of AF64A or vehicle should be infused over a 1 min period. To achieve a decrease in high affinity choline transport without large areas of nonspecific necrosis, at least two (perhaps three) infusions, separated by 24 h, may improve specificity (total dose = 10–14 nmol). It is important, of course, to conduct a dose–response analysis as part of the preliminary studies.

## 4.3. Specificity and Mechanisms

The above procedure can be expected to yield a greater than 50% reduction in biochemical markers for ACh, with less than a 15%

decrease in markers for GABA and catecholamines. The neurotoxic action of AF64A is gradual. It takes about 7–8 d for AF64A to cause maximal decrease in ACh function (Sandberg et al., 1984a), although by 12 h postsurgery a significant reduction in some markers can be observed. The reduction in markers for ACh lasts at least 3 mo when AF64A is delivered into the striatum (Sandberg et al., 1984a). However, this reduction may be site-specific, since ACh markers in the cortex following ibotenic acid-induced destruction of cells in the nucleus basalis revert back to control levels by 3 mo (Wenk and Olton, 1984). Note further that during old age, ACh markers in similarly treated animals decline once again (Brown et al., 1985; *see* also Schallert et al., 1983, for a potentially related effect in the dopamine system).

Compared to ibotenic acid placed into an area of dense ACh cell bodies, AF64A is far greater in potency (Fuxe et al., 1984). Because it is greater in its specificity for ACh neurons, AF64A (or a related substance; Myles et al., 1984) may become the neurotoxin of choice for cholinergic systems.

As noted above, there is considerable nonspecific necrosis associated with AF64A, particularly with single injections of doses above 1 nmol (Levy et al., 1984; Asante et al., 1983; Myles et al., 1984). More disturbing, however, is the recent report of acetylcholinesterase (the ACh catabolic enzyme that is not localized to ACh neurons) activity in the area of the injection site (Sandberg et al., 1984a). These investigators noted that there was no reduction in the density of intensely stained (presumably cholinergic) cells, even at the border of the necrotic tissue. These authors suggest that the neurons may be dysfunctional and note that conclusions about the integrity of ACh neurons will require quantitative immunocytochemical analysis (Mesulam et al., 1983) and biochemical measures such as those described below.

The neurotoxic action of AF64A is reduced by coincubation with choline; thus, it depends on its uptake via the high affinity choline transport process (Sandberg et al., 1982). However, the precise mechanisms whereby AF64A destroys ACh neurons are not known.

## 4.4. Evaluation of AF64A Neurotoxicity

### 4.4.1. Presynaptic ACh Markers

Cytotoxicity for ACh neurons is typically verified using assays for choline acetyltransferase (CAT), high affinity choline uptake (the rate limiting step in ACh synthesis), or binding of ACh to muscarinic receptors [using the antagonist ($^3$H)-quinuclidinyl benzilate, QNB].

CAT activity is typically assessed through radioligand techniques with a separation step so that the counted radioactivity can be attributed to ACh synthesized by the enzyme (Coyle, 1983; Cooper and Schmidt, 1980). An excellent discussion of theoretical and methodological issues pertinent to the study of ACh is provided in the chapter by Ansell (1981). A procedure for determination of choline and ACh levels by electrochemical detection has recently been described and may represent the most appropriate way for determining ACh depletion directly after AF64A lesions as well as for assessing neurotoxin-induced alterations in ACh synthesis. The choline and ACh are separated by the analytical column, mixing the effluent of the column with acetylcholinesterase and choline oxidase to yield oxidizable products, and monitoring the products electrochemically (Potter et al., 1983). Additional comments pertaining to this procedure may be found in HPLC-EC applications note 66 in *Current Separations.* Although the HPLC procedure is newer than those described above for catechol- and indoleamines, the method appears to be quite reliable. Since the detection method depends on the oxidation of choline- and ACh-derived products, it represents the same practical advantages possessed by the naturally electroactive catechol and indole compounds. Generally, animals need to be sacrificed by microwave irradiation to prevent postmortem changes in ACh.

The high-affinity, choline-uptake procedure is based on the technique of Simon et al. (1976) and is quite robust. It is quite similar to the procedure described above for DA, NA, and 5-HT (Goldman and Erickson, 1982, 1983). One caveat is that it appears to be somewhat more crucial to remove the brain and get the appropriate regions into ice-cold buffer quickly than with other uptake procedures. Speed is always important, however, when working with fresh tissue assays since synaptosomes lose their ability to carry out normal functions at a rate of about 15% per h even when kept on ice.

### 4.4.2. Postsynaptic ACh and non-ACh Markers

The binding of $(^3H)$-QNB to brain regional muscarinic receptors is a delightfully useful assay (Schallert et al., 1980b; Luthin and Wolfe, 1984; Hruska et al., 1984). The drug has such high affinity for muscarinic receptors that one may find it difficult to carry out saturability studies in many brain regions without pooling of tissue (Wastek and Yamamura, 1981; Ehlert et al., 1983). In contrast, such a procedure only recently has been applicable consistently within the DA systems of rat brain because of technical modifications (Huff and Molinoff, 1982). Since in a saturability study, one can measure both the $K_d$ and $B_{max}$ of binding in homologous contralateral brain regions for each animal, the amount of information about putative receptor changes is increased enormously. The assays also should include measures that assess the integrity of cell bodies within the region of microinjection and, in addition, procedures that verify the integrity of terminals of non-ACh cells (that may, for example, show reversible changes in sensitivity; Bannon et al., 1980). Using the striatum as an example, it would be appropriate to assess the effects of lesions not only on cholinergic interneurons, but on GABA projection neurons, and on DA and serotonin terminals (Coyle, 1978; 1983b). Levels of each transmitter, synthesis, and uptake (where appropriate) should be monitored as described above. Receptor binding may be measured in striatum (e.g., DA binding decreases after kainic acid and increases after 6-OHDA; Schallert et al., 1980b, 1983; Spirduso et al., 1984). Binding may also be measured downstream, within the terminal beds of projection neurons (for example, GABA binding in the substantia nigra zona reticulata).

GABA synthesis occurs with the decarboxylation of glutamic acid by the enzyme GAD (glutamic acid decarboxylase). As with ACh levels and CAT activity, GABA levels and GAD activity may be measured by a number of procedures (Coyle and Schwarcz, 1976; Coyle et al. 1983a,b; Fonnum, 1981). However, the availability of an electrochemical detection procedure using a precolumn derivitization to yield an oxidizable product makes the use of HPLC techniques for such determinations of GABA a reasonable alternative to current standard procedures (Holdiness, 1983; Van der Hyden et al., 1979). Determination of GABA levels and in vitro synthesis may be carried out in the same brain region in the same animal. The procedure is also applicable to the determination of levels of other amino acids. The major caution is that the currently available derivitizing agent, $o$-phthaldehyde, produces a GABA derivative with a relatively short lifetime. Thus it is necessary to react one sample at a time and to inject the derivative onto the HPLC at a constant time soon after reaction. In practice, this is not much of a problem since only one sample at a time may be injected onto the instrument in any case. References for the GABA detection procedure may be found in Caudill et al. (1982); specific discussions of the precolumn derivitization device, in the articles by Hodgin et al. (1983) and Venema et al. (1983); and a good presentation of GABA measurement in general, in the chapter by Fonnum (1981).

Binding to the GABA receptors is often carried out using ($^3$H)-muscimol, a GABA agonist, since this ligand–receptor complex is sufficiently stable to permit termination of the binding reaction in the usual way (by dilution in ice-cold buffer and filtration under vacuum). The affinity of [$^3$H]-muscimol for GABA receptors is lower than that of spiperone and QNB for their receptors and about on a par with that of DHA for its binding sites. This presents no problem except that one may have to settle for measuring binding at two concentrations of ligand, rather than at several, to avoid pooling tissue (Yamamura et al., 1985).

### 4.4.3. Behavioral Assays Related to ACh Systems

As with 6-OHDA, there are reliable behavioral effects of cholinergic antagonists that may be useful indices of AF64A-induced presynaptic cholinergic dysfunction. These include effects on memory, especially passive avoidance deficits (Sandberg et al., 1984b), spatial mapping deficits (Sutherland et al., 1982b), and stereotyped behavior (Schallert et al., 1980a). Perhaps the most promising characteristic of unilateral or bilateral cholinergic dysfunction comes out of the work of Vanderwolf and his colleagues who have demonstrated that certain types of behaviors (including immobility, face washing, chewing, licking) are related to atropine-sensitive electrical activity in the hippocampus and neocortex (*see* Vanderwolf and Leung, this volume, for review; Schallert et al., 1980a).

## 5. Conclusion

Investigators have increasingly abandoned the exclusive use of nonspecific ablation techniques to study brain function. The techniques for application of the more selective chemical neurotoxins have been improved rapidly and are now employed widely in anatomical, biochemical, physiological, pharmacological, and behavioral analyses of the central nervous system. We have avoided listing endlessly the general caveats traditionally raised in discussions of brain lesions, such as not leaping to conclusions about the functions of a given system after describing the effects of its ablation. The fact is that thoughtful research using lesion techniques has made, and will continue to make, fundamental contributions to brain science.

One theme of this chapter has been the necessity to use multiple independent means to determine the consequences of selective brain damage. A key aspect of this discussion has been the way that behavioral and biochemical measures of nervous system integrity reinforce one another. Yet beyond even the obvious value of combining tests of the ability of the animal to deal with its environment in adaptive ways with measures of synaptic activity, there is another reason for combining behavioral and biochemical measures of brain damage. In our experience and in the current literature (Yamamoto and Freed, 1984; Morgan et al.,

1983; Wenk et al., 1984), it has become obvious that the nervous system is extraordinarily plastic in its responsiveness to the changing environment. We must realize that not only do lesions change behavior and brain chemistry, but that the behavior occurring as a consequence of the brain damage itself alters brain chemistry even further. The reciprocal nature of brain chemistry–behavior interactions is too frequently ignored and should provide an exciting challenge to those of us who study the loss and recovery of function after brain lesions.

# References

Adams R. N. and Marsden C. A. (1982) Electrochemical Detection Methods for Monoamine Measurements in Vitro and In Vivo, in *Handbook of Psychopharmacology*, vol. 15 (Iversen L. L., Iversen S. D., and Snyder S. H., eds.), pp. 1–74, Plenum Press, New York.

Agid Y., Javoy F., and Glowinski J. (1973a) Hyperactivity of remaining dopaminergic neurons after partial destruction of the nigrostriatal dopaminergic system in the rat. *Nature New Biol.* **245,** 150–151.

Agid Y., Javoy F., Glowinski J., Bouvet D., and Sotelo C. (1973b) Injection of 6-hydroxydopamine in the substantia nigra of the rat. II. Diffusion and specificity. *Brain Res.* **58,** 291–301.

Ahlskog J. E. and Hoebel B. G. (1973) Overeating and obesity from damage to a noradrenergic system in the brain. *Science* **182,** 166–169.

Allis B. and Cohen G. (1977) The neurotoxicity of 5,7-dihydroxytryptamine in the mouse striatum: protection by 1-phenyl-3-(2-thioazolyl)-2-thiourea and by ethanol. *Eur. J. Pharmacol.* **43,** 269–272.

Alpers B. J. and Lewy F. H. (1940) Changes in the nervous system following carbon disulfide poisoning in animals and in man. *Arch. Neurol. Psychiat.* **44,** 725–739.

Alta, C. A., O'Neil S., and Marshall J. F. (1984) Sensorimotor impairment and elevated levels of dopamine metabolites in the neostriatum occur rapidly after intranigral injection of 6-hydroxydopamine or gamma-hydroxybutyrate in awake rats. *Neuropharmacology* **23,** 309–318.

Anand B. K. and Brobeck J. R. (1951) Hypothalamic control of food intake in rats and cats. *Yale J. Biol. Med.* **24,** 123–140.

Ansell G. B. (1981) The Turnover of Acetylcholine, in *Central Transmitter Turnover* (C. J. Pycock and P. V. Taberner, eds.). pp. 81–104, University Park Press, Baltimore.

Asante J. W., Cross A. J., Deakin J. F. W., Johnson J. A., and Slater H. R. (1983) Evaluation of ethylcholine mustard aziridinium ion (ECMA) as a specific neurotoxin of brain cholinergic neurones. *Brit. J. Pharmacol.* **80,** 573P.

Baglioni S. and Magnini M. (1909) Archivio di Fisologia, in Luciani L. *Human Physiology* (Holmes G. M., ed.), 1915, McMillan, London.

Balagura S., Wilcox R. H., and Coscina D. V. (1969) The effect of diencephalic lesions on food intake and motor activity. *Physiol. Behav.* **4**, 629–633.

Bannon M. J., Bunney E. B., Zigun J. R., Skirboll L. R., and Roth R. H. (1980) Presynaptic dopamine receptors: insensitivity to kainic acid and the development of supersensitivity following chronic haloperidol. *Naunyn-Schmiedeberg's Arch. Pharmacol.* **312**, 161–165.

Barbaccia M. L., Gandolfi O., Chuang D. M., and Costa E. (1983) Modulation of neuronal serotonin uptake by a putative endogenous ligand of imipramine recognition sites. *Proc. Natl. Acad. Sci. USA* **80**, 5134–5138.

Barth T., Lindner M. D., and Schallert T. (1983) Sensorimotor asymmetries and tactile extinction in unilateral frontal cortex damaged and striatal dopamine-depleted rats. *Soc. Neurosci. Abst.* **9**, 482.

Baumgarten H. G., Bjorklund A., Nobin A., Rosengren E., and Schlossberger H. G. (1975) Neurotoxicity of hydroxylated tryptamines: structure-activity relationships. *Acta. Physiol. Scand.* Suppl. **429**, 7–27.

Baumgarten H. G., Lachenmayer L., and Bjorklund A. (1977) in *Methods in Psychobiology*, vol. 3 (Myers, R. D., ed.). pp. 47–98, Academic Press, New York.

Baumgarten, H. G. and Lachenmayer L. (1972) 5,7-Dihydroxytryptamine: Improvement in chemical lesioning of indoleamine neurons in mammalian brain. *Z. Zellforsch. Mikrosk* **135**, 399–414.

Bennett G. W., Marsden C. A., Sharp T., and Stolz J. F. (1981) Concomitant Determination of Endogenous Release of Dopamine, Noradrenaline, 5-Hydroxytryptamine, and Thyrotropin-Releasing Hormone (TRH) From Rat Brain Slices and Synaptosomes, in *Central Transmitter Turnover.* (Pycock C. J. and Taberner P. V., eds.). pp. 183–190, University Park Press, Baltimore.

Bennett J. P. (1978) Methods in Binding Studies, in *Neurotransmitter Receptor Binding.* (Yamamura H. I., Enna S. J., and Kuhar M. J., eds.). pp. 57–90, Raven Press, New York.

Bjorklund A., Baumgarten H. G., and Rensch A. (1975) 5,7-Dihydroxytryptamine: improvement of its selectivity for serotonin in the CNS by pretreatment with desipramine. *J. Neurochem.* **24**, 833–835.

Blank C. L., Kissinger P. T., and Adams R. N. (1972) 5,6-Dihydroxyindole formation from oxidized 6-hydroxydopamine. *Eur. J. Pharmacol.* **19**, 391–394.

Bosland M.C., Versteeg D. H. G., van Put J., and Jong W. de (1981) Effect of depletion of spinal noradrenaline by 6-hydroxydopamine on the development of renal hypertension in rats. *Clin. Exper. Pharmacol. Physiol.* **8**, 67–77.

Braestrup C. (1977) Changes in drug-induced stereotyped behavior after 6-OHDA lesions in noradrenaline neurons. *Psycopharmacology* **51**, 199–204.

Brecher G. and Waxler S. H. (1949) Obesity in mice due to single injections of goldthioglucose. *Proc. Soc. Exptl. Biol. Med.* **70**, 498–501.

Breese G. R. and Cooper B. R. (1975) Behavioral and biochemical interactions of 5,7-dihydroxytryptamine with various drugs when administered intracisternally to adult and developing rats. *Brain Res.* **98**, 517–527.

Breese G. R. and Mueller R. A. (1978) Alterations in the neuro-cytotoxicity of 5,7-dihydroxytryptamine by pharmacologic agents in adult and developing rats. *Ann. NY Acad. Sci.* **305**, 160–174.

Breese G. R. and Traylor T. D. (1971) Depletion of brain noradrenaline and dopamine by 6-hydroxydopamine. *Brit. J. Pharmacol.* **42**, 88–99.

Brooks V. B. (1983) Study of brain function by local, reversible cooling. *Rev. Physiol. Biochem. Pharmacol.* **95**, 1–109.

Brown M., Hohmann C. F., Lowenstein P. R., Meck W., Wenk G. L., and Coyle J. T. (1985) Age-related reappearance of partial forebrain cholinergic deficits. *Soc. Neurosci. Abst.* **11**.

Burns R. S., Markey S. P., Phillips J. M., and Chiueh C. C. (1984) The neurotoxicity of 1-methyl-4-phenyl-1,2,3,6-tetrahydropyridine in the monkey and man. *Can. J. Neurol. Sci.* **11**, 166–168.

Burns R. S., Chiueh C. C., Markey S., Ebert M. M., Jacobowitz D., and Kopin I. J. (1983) A primate model of Parkinson's disease: selective destruction of substantia nigra, pars compacta dopaminergic neurons by *N*-methyl-4-phenyl-1,2,3,6-tetrahydropyridine. *Proc. Natl. Acad. Sci. USA,* **80**, 4546–4550.

Burt D. R. (1980) Basic Receptor Methods II. Problems of Interpretation in Binding Studies, in *Receptor Binding Techniques. Society for Neuroscience 1980 Short Course Syllabus.* pp. 53–69, Society for Neuroscience, Bethesda.

Butcher L. L., Eastgate, S. M., and Hodge G. K. (1974) Evidence that punctate intracerebral administration of 6-hydroxydopamine fails to produce selective neuronal degeneration. *Naunyn-Schmiedeberg's Arch. Pharmacol.* **285**, 31–70.

Carey R. J. (1982) Unilateral 6-hydroxydopamine lesions of dopamine neurons produce bilateral self-stimulation deficits. *Behav. Brain Res.* **6**, 101–114.

Carpenter M. B. and Whittier J. R. (1952) Study of methods for producing experimental lesions of the central nervous system with special reference to stereotaxic technique. *J. Comp. Neurol.* **97**, 73–131.

Caudill W., Papach L. A. and Wightman R. M. L. (1982) Measurement of brain GABA with LCEC. *Current Separations* **4**, 59–61.

Chiueh C. C., Markey S. P., Burns R. S., Johannessen J., Jacobowitz D. M., and Kopin I. J. (1983) *N*-methyl-4-phenyl-1,2,3,6-tetrahydropyridine, a parkinsonian syndrome-causing agent in man and monkey, produces different effects in guinea pig and rat. *Pharmacologist* **25**, 131.

Chiueh C. C., Markey S. P., Burns R. S., Johannessen J. N., Pert A., and Kopin I. J. (1984) Neurochemical and behavioral effects of systemic and intranigral administration of *N*-methyl-4-phenyl-1,2,3,6-tetrahyd ropyridine in the rat. *Eur. J. Pharmacol.* **100**, 189–194.

Clement J. G. and Colhoun E. G. (1975) Presynaptic effect of the aziridinium ion of acetylcholine mustard (methyl-2-acetoxyethyl-2-

choloroethylamine) on the phrenic nerve rat diaphragm preparation. *Can. J. Physiol. Pharmacol.* **53,** 264–272.

Clewans C. S. and Azmitia E. (1984) Tryptophan hydroxylase in hippocampus and midbrain following unilateral injection of 5,7-dihydroxytryptamine. *Brain Res.* **307,** 125–133.

Cohen G. and Heikkila R. E. (1974) The generation of hydrogen peroxide, superoxide radical, and hydroxy radical by 6-hydroxydopamine, dialuric acid, and related cytotoxic agents. *J. Biol. Chem.* **249,** 2447–2452.

Cooper D. O. and Schmidt D. E. (1980) The use of choline acetyltransferase as a cholinergic marker in the determination of high-affinity choline uptake. *J. Neurochem.* **34,** 1553–1556.

Cooper J. R., Bloom F. E., and Roth R. H. (1982a) Catecholamines I: General Aspects, *The Biochemical Basis of Neuropharmacology,* 4th edn. (Cooper J. R., Bloom F. E., and Roth R. H., eds.). pp. 109–172, Oxford University Press, New York.

Cooper J. R., Bloom F. E., and Roth R. H. (1982b) Catecholamines II: CNS Aspects, in *The Biochemical Basis of Neuropharmacology,* 4th edn. (Cooper J. R., Bloom F. E., and Roth R. H., eds.). pp. 173–222, Oxford University Press, New York.

Cooper J. R., Bloom F. E., and Roth R. H. (1982c) Serotonin (5-Hydroxytryptamine), in *The Biochemical Basis of Neuropharmacology,* 4th edn. (Cooper J. R., Bloom F. E., and Roth R. H., eds.). pp. 223–248, Oxford University Press, New York.

Coyle J. T. (1983a) Neurotoxic action of kainic acid (review). *J. Neurochem.* **41,** 1–11.

Coyle J. T. (1979). An animal model for Huntington's disease. *Biol. Psychiat.* **14,** 251–276.

Coyle J. T., Ferkany J. W., and Zaczek R. (1983b) Kainic acid: insights from a neurotoxin into the pathophysiology of Huntington's disease. *Neurobehav. Toxicol. Teratol.* **5,** 617-624.

Coyle J. T. and Schwarcz R. (1976) Lesion of striatal neurones with kainic acid provides a model for Huntington's chorea. *Nature* (London) **263,** 244–246.

Creese I., Sibley D. R., Hamblin M. W., and Leff S. E. (1983) The classification of dopamine receptors: relationship to radioligand binding. *Ann. Rev. Neurosci.* **6,** 43–71.

Creese I. (1981) Dopamine Receptors, in *Neurotransmitter Receptors, Part 2. Biogenic Amines.* (Yamamura H. I. and Enna S. J., eds.). pp. 129–184, Chapman and Hall, London.

Creese I., Hamblin M. W., Leff S. E., and Sibley D. R. (1982) CNS Dopamine Receptors, in *Handbook of Psychopharmacology,* vol. 17, (Iversen L. L., Iversen S. D., and Snyder S. H., eds.). pp. 81–138. Plenum Press, New York.

Creese I. and Snyder S. H. (1979) Nigrostriatal lesions enhance striatal [$^3$H]apomorphine and [$^3$H]spiroperidol binding. *Eur. J. Pharmacol.* **56,** 277–281.

Creveling C. R. and Rotman A. (1978) Mechanism of action of dihydroxytryptamines. *Ann. NY Acad. Sci.* **305,** 57–73.

Curzon, G. (1981) The Turnover of 5-Hydroxytryptamine, in *Central Transmitter Turnover.* (Pycock C. J. and Taberner P. V. eds.). pp. 20–58, University Park Press, New York.

Dahlstrom A. and Fuxe K. (1964) Evidence for the existence of monoamine containing neurons in the central nervous system. I. Demonstration of monoamines in the cell bodies of brain stem neurons. *Acta Physiol. Scand.* Suppl. 232, **62,** 1–55.

Daly J., Fuxe K., and Jonsson G. (1974) 5,7-Dihydroxytryptamine as a tool for the morphological and functional analysis of central 5-hydroxytryptamine neurons. *Res. Commun. Chem. Pathol. Pharmacol.* **1,** 175–187.

Davies P. and Maloney A. J. F. (1976) Selective loss of central cholinergic neurons in Alzheimer's disease. *Lancet* **ii,** 1403.

Davis G. C., Koch D. D., Kissinger P. T., Brunlett C. S., and Shoup R. E. (1981) Determination of Tyrosine and Tryptophan Metabolites in Body Fluids Using Electrochemical Detection, in *Liquid Chromatography in Clinical Analysis.* (Kabra P. M. and Morton L. J., eds.). pp. 253–306, Humana Press, Clifton, New Jersey.

Davis, G. C., Williams, A. C., Markey S. P., Ebert M. H., Caine E. D., Reichert C. M., and Kopin I. J. (1979) Chronic parkinsonism secondary to intravenous injection of meperidine analogues. *Psychiat. Res.* **1,** 249–254.

de Montigny C. D., Wang R. Y., Reader T. A., and Aghajanian G. K. (1980) Monoaminergic denervation of the rat hippocampus: Microiontophoretic studies on pre- and postsynaptic supersensitivity to norepinephrine and serotonin. *Brain Res.* **200,** 363–376.

Diggory G. L., Dickison S. E., Wood M. D., and Wyllie M. G. (1981) Changes in Central 5-Hydroxytryptamine Turnover Induced by Acute and Chronic Inhibition of the Reuptake Process, in *Central Transmitter Turnover.* (Pycock C. J. and Taberner P. V., eds.). pp. 149–154, University Park Press, Baltimore.

Dooley D. J., Hunziker G., and Hausler A. (1984) Corticosterone secretion in the rat after DSP-4 treatment. *Neurosci. Lett.* **46,** 271–274.

Edwards D. J. and Bagg H. J. (1923) Lesions of the corpus striatum by radium emanation and the accompanying structural and functional changes. *Am. J. Physiol.* **65,** 162–173.

Ehlert F. J., Roeske W. R., and Yamamura H. I. (1983) The Nature of Muscarinic Receptor Binding, in *Handbook of Psychopharmacology,* vol. 17, (Iversen L. L., Iversen S. D., and Snyder S. H., eds.). pp. 241–284, Plenum Press, New York.

Enna S. J. (1983) Radioreceptor Assays for Neurotransmitters and Drugs, in *Handbook of Psychopharmacology,* vol. 15, (Iversen L. L., Iversen S. D., and Snyder S. H., eds.). pp. 75–94, Plenum Press, New York.

Enna S. J. (1980) Basic Receptor Methods I., in *Receptor Binding Techniques. Society for Neuroscience 1980 Short Course Syllabus.* pp. 33–52, Society for Neuroscience, Bethesda.

Ferrier D. (1873) Experimental researches in cerebral physiology and pathology. *West Riding Lunatic Asylum Medical Reports* **3,** 30–96.

Fillion G. (1983) 5-Hydroxytryptamine Receptors in Brain, in *Handbook of Psychopharmacology,* vol. 17, (Iversen, L. L., Iversen S. D., and Snyder S. H., eds.). pp. 139–166, Plenum Press, New York.

Finger S. and Stein D. G. (1982) *Brain Damage and Recovery: Research and Clinical Perspectives.* Academic Press, New York.

Fink J. S. and Smith G. P. (1979) Decreased locomotor and investigatory exploration after denervation of catecholaminergic terminal fields in the forebrain of rats. *J. Comp. Physiol. Psychol.* **93,** 34–65.

Fisher A. and Hanin I. (1980) Choline analogs as potential tools in developing selective animal models of central cholinergic hypofunction. *Life Sci.* **27,** 1615–1634.

Fisher A., Mantione C. R., Abraham D. J., and Hanin I. (1980) Ethylcholine mustard aziridinium (AF64A): A potential irreversible cholinergic neurotoxin in vivo. *Fed. Proc.* **34,** 411.

Fisher A., Mantione C. R., Bech H., and Hanin I. (1981) Atropine potentiates AF64A-induced pharmacological effects in mice in vivo. *Fed. Proc.* **40,** 269.

Fonnum F. (1981) The Turnover of Transmitter Amino Acids, With Special Reference to GABA, in *Central Transmitter Turnover.* (Pycock C. J. and Taberner P. V., eds.). pp. 105–124, University Park Press, Baltimore.

Fuller R. W. (1984) Serotonin Receptors, in *Monographs in Neural Sciences,* vol. 10 (Cohen M. M., ed.). pp. 158–181.

Fritsch G. and Hitzig E. (1870) On the Electrical Excitability of the Cerebrum, in *The Cerebral Cortex* (von Bonin C., ed.). p. 1960, Charles C. Thomas, Springfield, IL.

Furness J. B., Campbell G. R., Gillard S. M., Malmfors T., Cobb J. L. S., and Burnstock G. (1970) Cellular studies of sympathetic denervation produced by 6-hydroxydopamine in the vas deferens. *J. Pharmacol. Exp. Ther.* **174,** 111–123.

Fuxe K., Roberts P. and Schwarcz R. (1984) *Excitotoxins.* MacMillan Press, London.

Gladfelter W. E. and Brobeck J. R. (1962) Decreased spontaneous locomotor activity in the rat induced by hypothalamic lesions. *Am. J. Physiol.* **203,** 811–817.

Glick S. D. (1976) Behavioral Effects of Amphetamine in Brain-Damaged Animals: Problems in the Search for Sites of Action, in *Cocaine and Other Stimulants* (Ellinwood, E., ed.). pp. 77–96. Plenum Press, New York.

Goldman M. E. and Erickson C. K. (1983) Effects of acute and chronic administration of antidepressant drugs on the central cholinergic nervous system. *Neuropharmacology* **22,** 1215–1222.

Goldman M. E. and Erickson C. K. (1982) Atropine-amitriptyline interactions in the rat central cholinergic nervous system. *Brain Res.* **248,** 188–191.

Grossman S. P. (1960) Eating or drinking elicited by direct adrenergic or cholinergic stimulation of hypothalamus. *Science* **132,** 301–302.

Grossman S. P., Dacey D., Halaris A. E., Collier T., and Routtenberg A.

(1978) Aphagia and adipsia after preferential destruction of nerve cell bodies in hypothalamus. *Science* **202**, 537–539.

Guldin W. O. and Markowitsch H. J. (1981) No detectable remote lesions following massive intrastriatal injections of ibotenic acid. *Brain Res.* **225**, 446–451.

Haeusler G. (1971) Early pre- and postjunctional effects of 6-hydroxydopamine. *J. Pharmacol. Exp. Ther.* **178**, 49–62.

Haigler H. J. (1981) Serotonergic Receptors in the Central Nervous System, in *Neurotransmitter Receptors*. Part 2. *Biogenic Amines*. (Yamamura H. I. and Enna S. J., eds.). pp. 1–70, Chapman and Hall, London.

Hallman H., Lange J., Olson L., Strömberg I., and Jonsson G. (1985) Neurochemical and histochemical characterization of neurotoxic effects of 1-methyl-4-phenyl-1,2,3,6-tetrahydropyridine on brain catecholamine neurones in the mouse. *J. Neurochem.* **44**, 117–127.

Hallman H., Sundström E., and Jonsson G. (1984) Effect of the noradrenaline neurotoxin DSP4 on monoamine neurons and their transmitter turnover in rat CNS. *J. Neural Transm.* **60**, 89–102.

Hamblin M. W., Leff S. E., and Creese I. (1984) Interactions of agonists with D-2 dopamine receptors: evidence for a single receptor population existing in multiple agonist affinity states in rat striatal membranes. *Biochem. Pharmacol.* **33**, 877–887.

Hanin I., DeGroat W. C., Mantione C. R., Coyle J. T., and Fisher A. (1983) Chemically-induced cholinotoxicity in vivo: studies utilizing ethylcholine aziridinium ion (AF64A). Banbury Report **15**, Biological Aspects of Alzheimer's Disease, 243–253.

Harik S. I. (1984) Locus ceruleus lesion by local 6-hydroxydopamine infusion causes marked and specific destruction of noradrenergic neurons, long-term depletion of norepinephrine and the enzymes that synthesize it, and enhanced dopaminergic mechanisms in the ipsilateral cerebral cortex. *J. Neurosci.* **4**, 699–707.

Harvey J. A., Heller A., and Moore R. Y. (1963) The effects of unilateral and bilateral medial forebrain bundle lesions on brain serotonin. *J. Pharmacol. Exper.* **140**, 103–110.

Hefti F., Melamed E., and Wurtman R. J. (1980) Partial lesions of the dopaminergic nigrostriatal system in rat brain: Biochemical characterization. *Brain Res.* **195**, 123–137.

Heikkila R. E., Cabbat F. S., Manzino L., and Duvoisin R. C. (1984) Effects of 1-methyl-4-phenyl-1,2,5,6-tetrahydropyridine on neostriatal dopamine in mice. *Neuropharmacology* **23**, 711–713.

Heikkila R. E., Shapiro B. S. and Duvoisin R. C. (1981) The relationship between loss of dopamine nerve terminals, striatal [$^3$H]spiroperidol binding and rotational behavior in unilaterally 6-hydroxydopamine-lesioned rats. *Brain Res.* **211**, 285–292.

Heikkila R. E. and Cabbat F. S. (1977) Chemiluminescence from 6-hydroxydopamine. Involvement of hydrogen peroxide, the superoxide and the hydroxyl radical; a potential role for singlet oxygen. *Res. Commun. Chem. Pathol. Pharmacol.* **17**, 649–662.

Heikkila R. E. and Cohen G. (1973) 6-Hydroxydopamine: evidence for superoxide radical as an oxidative intermediate. *Science* **181**, 456–457.

Heikkila R. E. and Cohen G. (1971) Inhibition of biogenic amine uptake by hydrogen peroxide: A mechanism for toxic effects of 6-hydroxydopamine. *Science* **172**, 1257–1258.

Heller A. and Harvey J. A. (1963) Effect of CNS lesions on brain norepinephrine. *Pharmacologist* **5**, 261.

Heller A., Harvey J. A., and Moore R. Y. (1962) A demonstration of a fall in brain serotonin following central nervous system lesions in the rat. *Biochem. Pharmacol.* **11**, 859–866.

Herman B. H., Berger S., and Holtzman S. G. (1983) Comparison of electrical resistance, bubble withdrawal, and stereotaxic method for cannulation of cerebral ventricles. *J. Pharmacol. Meth.* **10**, 143–155.

Hernandez L. and Hoebel B. G. (1982) Overeating after midbrain 6-hydroxydopamine: Prevention by central injection of selective catecholamine reuptake blockers. *Brain Res.* **245**, 333–343.

Hodgin J. C., Howard P. Y., Ball D. M., Cleoet C., and De Jager L. (1983) An automated device for *in situ* precolumn derivitization and injection of amino acids for HPLC analysis. *J. Chromatog. Sci.* **21**, 503–507.

Holdiness M. R. (1983) Chromatographic analysis of glutamic acid decarboxylase in biological samples (review). *J. Chromatography* **277**, 1–24.

Horsley V. and Clarke R. H. (1908) The structure and function of the cerebellum examined by a new method. *Brain* **31**, 45–124.

Hruska R. E., Ludmer L. M., Pert A., and Bunney W. E. Jr. (1984) Effects of lithium on [$^3$H](−)-quinuclidinyl benzilate binding to rat brain muscarinic cholinergic receptors. *J. Neurosci. Res.* **11**, 171–177.

Huff R. M. and Molinoff P. B. (1982) Quantitative determination of dopamine receptor subtypes not linked to activation of adenylate cyclase in rat striatum. *Proc. Natl. Acad. Sci.USA* **79**, 7561–7565.

Hunt W. A. and Dalton T. K. (1983) An automated method for the determination of biogenic amines and their metabolites by high-performance liquid chromatography. *Anal. Biochem.,* **135**, 269–274.

Jacoby J. H. and Lytle L. D. (Eds.) (1978) *Ann. NY Acad. Sci.* **305**.

Jaim-Etcheverry G. and Zieher L. M. (1980) DSP-4: A novel compound with neurotoxic effects on noradrenergic neurons of adult and developing rats. *Brain Res.* **188**, 513–523.

Jerussi T. P. and Glick S. D. (1975) Apomorphine-induced rotation in normal rats and interaction with unilateral caudate lesions. *Psychopharmacology* **40**, 329–334.

Jeste D. V. and Smith G. P. (1980) Unilateral mesolimbicocortical dopamine denervation decreases locomotion in the open field and after amphetamine. *Pharmacol. Biochem. Behav.* **12**, 453–457.

Jonsson G. and Hallman H. (1982) Substance P counteracts neurotoxin damage on norepinephrine neurons in rat brain during ontogeny. *Science* **215**, 75–77.

Jonsson G. (1980) Chemical neurotoxins as denervation tools in neurobiology. *Ann. Rev. Neurosci.* **3**, 169–187.

Jonsson G. (1976) Studies on the mechanisms of 6-hydroxydopamine cytotoxicity. *Med. Biol.* **54**, 406–420.

Jonsson G., Pycock C., Fuxe K., and Sachs Ch. (1976) Changes in the development of central noradrenaline neurons following neonatal administration of 6-hydroxydopamine. *J. Neurochem.* **22**, 419.

Jonsson G. and Sachs Ch. (1975) On the Mode of Action of 6-Hydroxydopamine, in *6-Hydroxydopamine as a Denervation Tool in Catecholamine Research*, (Jonsson G., Malmfors T., and Sachs Ch. eds.). pp. 41–50, Elsevier North Holland, Amsterdam.

Jonsson G. and Sachs Ch. (1973) Pharmacological modifications at the 6-hydroxy-dopa-induced degeneration of central noradrenaline neurons. *Biochem. Pharmacol.* **22**, 1709–1716.

Jonsson G., Malmfors T., and Sachs, C. (1972) Effects of drugs on the 6-hydroxydopamine induced degeneration of adrenergic nerves. *Res. Commun. Chem. Pathol. Pharmacol.* **3**, 543–556.

Kissinger P. T. and Heineman W. R. (Eds.) (1984) *Laboratory Techniques in Electroanalytical Chemistry*. Bioanalytical Systems Press, West Lafayette, IN.

Kohler C and Schwarcz R. (1983) Comparison of ibotenate and kainate neurotoxicity in rat brain: a histological study. *Neuroscience* **8**, 819–835.

Kolb B. and Whishaw I. Q. (1980) *Fundamentals of Human Neuropsychology*. Freeman, NY.

Kubos K. L., Moran T. H., Saad K. M., and Robinson R. G. (1984) Asymmetrical locomotor response to unilateral cortical injections of DSP-4. *Pharmacol. Biochem. Behav.* **21**, 163–168.

Langston J. W., Ballard P., Tetrud J. W., and Irwin I. (1983) Chronic parkinsonism in humans due to a product of meperidine-analog synthesis. *Science* **219**, 979–980.

Langston J. W., Forno L. S., Robert C. S., and Irwin I. (1984a) Selective nigral toxicity after systemic administration of 1-methyl-4-phenyl-1,2,5,6-tetrahydropyrine (MPTP) in the squirrel monkey. *Brain Res.* **292**, 390–394.

Langston J. W., Irwin I., Langston E. B., and Forno L. S. (1984b) 1-Methyl-4-phenylpyridinium ion ($MPP^2+$): Identification of a metabolite of MPTP, a toxin selective to the substantia nigra. *Neurosci. Lett.* **48**, 87–92.

Leach L., Whishaw I. Q., and Kolb B. (1980) Effects of kainic acid lesions in the lateral hypothalamus on behavior and hippocampal and neocortical electroencephalographic (EEG) activity in the rat. *Behav. Brain Res.* **1**, 411–431.

Lefkowitz R. J., Caron M. G., and Stiles G. L. (1984) Mechanisms of membrane-receptor regulation. Biochemical, physiological, and clinical insights derived from studies of the adrenergic receptors. *New Engl. J. Med.* **310**, 1570–1579.

Lefkowitz R. J., Stadel J. M., and Caron M. G. (1983) Adenylate cyclase-coupled beta-adrenergic receptors: structure and mechanisms of activation and desensitization. *Ann. Rev. Biochem.* **52,** 159–186.

Leslie S. W., Woodward J. J., and Wilcox R. E. (1985) Correlation of rates of calcium entry and endogenous dopamine release in mouse striatal synaptosomes. *Brain Res.* **325,** 99–105.

Levy A., Kant G. J., Meyerhoff J. L., and Jarrard L. E. (1984) Noncholinergic neurotoxic effects of AF64A in the substantia nigra. *Brain Res.* **305,** 169–172.

Leysen J. E., Geerts R., Gommeren W., Verwimpp M., and Van Gompel P. (1982) Regional distribution of serotonin-2 receptor binding sites in the brain and effects of neuronal lesions. *Arch. Int. Phamacodyn. Ther.* **256,** 301–305.

Lidbrink P. and Jonsson G. (1974) Noradrenaline nerve terminals in the cerebral cortex: Effects on noradrenaline uptake and storage following axonal lesion with 6-hydroxydopamine. *J. Neurochem.* **22** 617–626.

Lidov H. G. W. and Molliver M. E. (1982) The structure of cerebral cortex in the rat following prenatal administration of 6-hydroxydopamine. *Devel. Brain Res.* **3,** 81–108

List S. J. and Seeman P. (1981) Resolution of dopamine and serotonin receptor components of [$^3$H]spiperone binding to rat brain regions. *Proc. Natl. Acad. Sci. USA* **78,** 2620–2624.

Ljundberg T. and Ungerstedt U. (1976a) Sensory inattention produced by 6-hydroxydopamine-induced degeneration of ascending dopamine neurons in the brain. *Exper. Neurol.* **53,** 585–600.

Ljundberg T. and Ungerstedt U. (1976b) Reinstatement of eating by dopamine agonists in aphagic dopamine-denervated rats. *Physiol. Behav.* **16,** 277–283.

Lorens S. A., Guldberg H. C., Hole K., Köhler C., and Srebro B. (1976) Activity, avoidance, learning and regional 5-hydroxytryptamine following intra-brainstem 5,7-dihydroxytryptamine and electrolytic midbrain raphe lesions in the rat. *Brain Res.* **108,** 97–113.

Luciani L. (1915) *Human Physiology* Raven Press, New York.

Luthin G. R. and Wolfe B. B. (1984) Comparison of [$^3$H]pirenzepine and [$^3$H]quinuclidinylbenzilate binding to muscarinic cholinergic receptors in rat brain. *J. Pharmacol. Exp. Ther.* **228,** 648–655.

Malpeli J. G. and Burch B. D. (1982) Cobalt destroys neurons without destroying fibers of passage in the lateral geniculate nucleus of the cat. *Neurosci. Lett.* **32,** 29–34.

Mantione C. R., Fisher A., and Hanin I. (1981) The AF64A-treated mouse: possible model for central cholinergic hypofunction. *Science* **213,** 579–580.

Marshall J. F. (1984) Brain function: Neural adaptations and recovery from injury. *Ann. Rev. Psychol.* **35,** 277–308.

Marshall J. F., Drew M. C., and Neve K. A. (1983) Recovery of function after mesotelencephalic dopaminergic injury in senescence. *Brain Res.* **254,** 249–260.

Marshall J. F., Berrios N., and Sawyer S. (1980) Neostriatal dopamine and sensory inattention. *J. Comp. Physiol. Psychol.* **94,** 833–846.

Marshall J. F., Richardson J. S., and Teitelbaum P. (1974) Nigrostriatal bundle damage and the lateral hypothalamic syndrome. *J. Comp. Physiol. Psychol.* **87,** 808–830.

Marshall N. B. and Mayer J. (1954) Energy balance in goldthioglucose obesity. *Amer. J. Physiol.* **178,** 271–274.

Martin-Iverson M. T., Leclere J. F. and Fibiger H. C. (1983) Cholinergic-dopaminergic interactions and the mechanisms of action of antidepressants. *Eur. J. Pharmacol.* **94,** 193–201.

Mason S. T. and Fibiger H. C. (1979) On the specificity of kainic acid. *Science* **204,** 1339–1341.

McGeer E. G. and McGeer P. L. (1976) Duplication of biochemical changes of Huntington's chorea by intrastriatal injection of glutamic and kainic acids. *Nature* (London), **263,** 517–519.

McGeer P. L., McGeer E. G., and Hattori T. (1978), in *Kainic Acid as a Tool in Neurobiology.* (McGeer E. G., Olney J. W., and McGeer P. L., eds.). pp. 123–138, Raven Press, New York.

Mella H. (1924) The experimental production of basal ganglion symptomatology in Macacus rhesus. *Arch. Neurol. Psychiat.* **11,** 405–417.

Mesulam M. M., Mufson E. J., Wainer B. H., and Levey A. I. (1983) Central cholinergic pathways in the rat: An overview based on alternative nomenclatures (Chs. 1–6). *Neuroscience* **4,** 1185–1201.

Minneman K. P. (1981) Adrenergic Receptor Molecules, in *Neurotransmitter Receptors. Part 2. Biogenic Amines.* (Yamamura H. I. and Enna S. J., eds.). pp. 1–70, Chapman and Hall, London.

Molinoff P. B., Wolfe B. B., and Weiland G. A. (1981) Quantitative analysis of drug-receptor interactions. II. Determination of the properties of receptor subtypes. *Life Sci.* **29,** 427–443.

Morgan S., Huston J. P., and Pritzel M. (1983) Effects of reducing sensory-motor feedback on the appearance of crossed nigrothalamic projections and recovery from turning induced by unilateral substantia nigra lesions. *Brain Res. Bull.* **11,** 721–727.

Myles L. A., Steingart M., Rylett R. J., and Colhoun E. H. (1984) Effect of injection of choline mustard into medial septal area of rat brain on biochemical and behavioral parameters. *Soc. Neurosci. Abst.* **10,** 1069.

Nelson D. L. (1982) Central serotonergic receptors: evidence for heterogeneity and characterization by ligand-binding. *Neurosci. Biobehav. Rev.* **6,** 499–502.

Nobin A. and Bjorklund A. (1978) Degenerative effects of various neurotoxic indoleamines on central monoamine neurons. *Ann. NY Acad. Sci.* **305,** 305–327.

Oberlander C., Hunt P. F., Dumont C., and Boissier J. R. (1981) Dopamine independent rotational response to unilateral intranigral injection of serotonin. *Life Sci.* **28,** 2595–2601.

Olney J. W., Ho O. L., and Rhea V. (1971) Cytotoxic effects of acidic and sulphur containing amino acids on the infant mouse central nervous system. *Exp. Brain Res.* **14,** 61–76.

Olney J. W., Rhee V., and Ho O. L. (1974) Kainic acid: A powerful neurotoxic analogue of glutamate. *Brain Res.* **77,** 507–512.

Ondrusek M. G., Kilts C. D., Frye G. D., Mailman R. B., Mueller R. A., and Breese G. R. (1981) Behavioral and biochemical studies of the scopolamine-induced reversal of neuroleptic activity. *Psychopharmacology* **73,** 17–22.

Poirier L. P., Langelier P., Roberge A., Boucher R., and Kitskis A. (1972) Nonspecific histopathological changes induced by the intracerebral injection of 6-hydroxydopamine (6-OH-DA). *J. Neurol. Sci.* **16,** 401–416.

Potter P. E., Meek J. L., and Neff N. H. (1983) Acetylcholine and choline in neuronal tissue measured by HPLC with electrochemical detection. *J. Neurochem.* **41,** 188–194.

Reader T. A. and Briere R. (1983) Long-term unilateral noradrenergic denervation: monoamine content and [$^3$H]prazosin binding sites in rat neocortex. *Brain Res. Bull.* **11,** 687–692.

Reader T. A. and Gauthier P. (1984) Catecholamines and serotonin in the rat central nervous system after 6-OHDA, 5-7-DHT, and pCPA. *J. Neural. Transm.* **59,** 207–227.

Richelson E. (1984) Studying neurotransmitter receptors: binding and biological assays. *Monogr. Neural. Sci.* **10,** 4–19.

Richter R. (1945) Degeneration of the basal ganglia from carbon disulfide poisoning in monkeys. *J. Neuropath. Exp. Neurol.* **4,** 324–353.

Robbins T. W. and Everitt B. J. (1982) Functional studies of the central catecholamines. *Int. Rev. Neurobiol.* **23,** 303–365.

Robinson R. G. and Coyle J. T. (1979) Lateralization of catecholaminergic and behavioral responses to cerebral infarction in the rat. *Life Sci.* **24,** 943–950.

Robinson R. G. and Stitt T. G. (1981) Intracortical 6-hydroxydopamine induces an asymmetrical behavioral response in the rat. *Brain Res.* **213,** 387–395.

Robinson S. E. (1983) Effect of specific serotonergic lesions on cholinergic neurons in the hippocampus, cortex, and striatum. *Life Sci.* **32,** 345–353.

Robinson T. E., Becker J. B., and Camp D. M. (1983) Sex Differences in Behavioral and Brain Asymmetries, in *Hemisyndromes* (Myslobodsky, M. S., ed.), pp. 91–128, Academic Press, New York.

Rosen G. D., Finklestein S., Stoll A. L., Yutzey D. A., and Denenberg V. H. (1984) Neurochemical asymmetries in the albino rat's cortex, striatum, and nucleus accumbens. *Life Sci.* **34,** 1143–1148.

Ross S. B. (1976) Long-term effects of N-2-chloroethyl-N-ethyl-2-bromobenzylamine hydrochloride on noradrenergic neurons in the rat brain and heart. *Br. J. Pharmacol.* **58,** 521–527.

Roussy G. (1907) *La Couche Optique.* Paris, G. Steinheil.

Ruffolo R. R. Jr. (1982) Important concepts of receptor theory (review). *J. Auton. Pharmacol.* **2,** 272–295.

Rylett B. J. and Colhoun E. H. (1980) Kinetic data on the inhibition of high-affinity choline transport into rat forebrain synaptosomes by choline-like compounds and nitrogen mustard analogs. *J. Neurochem.* **34,** 713–719.

Sachs Ch., Jonsson G., Heikkila R., and Cohen G. (1975) Control of the neurotoxicity of 6-hydroxydopamine by intraneuronal noradrenaline in rat iris. *Acta Physiol. Scand.* **93,** 345–351.

Sahakian B. J., Winn P., Robbins T. W., Deelay R. J., Everitt B. J., Dunn L. T., Wallace M., and James W. P. T. (1983) Changes in body weight and food-related behavior induced by destruction of the ventral or dorsal noradrenergic bundle in the rat. *Neuroscience* **10,** 1405–1420.

Saller C. F. and Stricker E. M. (1978) Gastrointestinal motility and body weight gain in rats after brain serotonin depletion by 5,7-hydroxytryptamine. *Neuropharmacology* **17,** 499–506.

Sandberg K., Hanin I., Fisher A., and Coyle J. T. (1984a) Selective cholinergic neurotoxin: AF64A's effects in rat striatum. *Brain Res.* **293,** 49–55.

Sandberg K., Sanberg P. R., Hanin I., Fisher A., and Coyle J. T. (1984b) Cholinergic lesion of the striatum impairs acquisition and retention of a passive avoidance response. *Behav. Neurosci.* **98,** 162–165.

Sandberg K., Schnaar R. L., Hanin I., and Coyle J. T. (1982) Effects of AF64A on neuroblastoma x glioma hybrid cell line NG-108-15: a neurotoxin selective for cholinergic cells. *Soc. Neurosci. Abstr.* **8,** 616.

Saner A. and Thoenen H. (1971) Model experiments on the molecular mechanism of action of 6-hydroxydopamine. *Mol. Pharmacol.* **7,** 147–157.

Schallert T. (1983) Sensorimotor impairment and recovery of function in brain-damaged rats. Reappearance of symptoms during old age. *Behav. Neurosci.* **97,** 159–164.

Schallert T. J., Upchurch M., Wilcox R. E., and Vaughn D. M. (1983) Posture-independent sensorimotor analysis of inter-hemispheric receptor asymmetries in neostriatum. *Pharmacol. Biochem. Behav.* **18,** 753–759.

Schallert T., Upchurch M., Lobaugh N., Farrar S. B., Spirduso W. W., Gilliam P., Vaughn D., and Wilcox R. E. (1982) Tactile extinction: Distinguishing between sensorimotor and motor asymmetries in rats with unilateral nigrostriatal damage. *Pharmacol. Biochem. Behav.* **16,** 455–462.

Schallert T., Farrar S., Lobaugh N., Wilcox R. E., and Vaughn D. (1981) Anticholinergic-induced excessive forward locomotion in rats treated with 6-OHDA in nucleus accumbens, nigrostriatal pathway, spinal subarachnoid space, or cerebral ventricles. *Soc. Neurosci. Abst.* **7,** 565.

Schallert T. and Whishaw I. Q. (1984) Bilateral cutaneous stimulation of the somatosensory system in hemi-decorticate rats. *Behav. Neurosci.* **98,** 518–540.

Schallert T., DeRyck M., and Teitelbaum P. (1980a) Atropine stereotypy as a behavioral trap: a movement subsystem and electroencephalographic analysis. *J. Comp. Physiol. Psychol.* **94,** 1–24.

Schallert T., Overstreet D. H., and Yamamura H. I. (1980b) Muscarinic receptor binding and behavioral effects of atropine following chronic catecholamine depletion or acetycholinesterase inhibition in rats. *Pharmacol. Biochem. Behav.* **13,** 187–192.

Schallert T., De Ryck M., Whishaw I. Q., Ramirez V. D., and Teitelbaum P. (1979) Excessive bracing reactions and their control in an animal analog of parkinsonism. *Exp. Neurol.* **64,** 33–43.

Schallert T. and Whishaw I. Q. (1978) Two types of aphagia and two types of sensorimotor impairment after lateral hypothalamic lesions: Observations in normal weight, dieted, and fattened rats. *J. Comp. Physiol. Psychol.* **92,** 720–741.

Schallert T., Whishaw I. Q., Ramirez V. D., and Teitelbaum P. (1978) Compulsive, abnormal walking caused by anticholinergics in akinetic 6-hydroxydopamine-treated rats. *Science* **199,** 1461–1463.

Schmidt R. H. and Bhatnagar R. K. (1979) Assessment of the effects of neonatal subcutaneous 6-hydroxydopamine on noradrenergic and dopaminergic innervation of the cerebral cortex. *Brain Res.* **166,** 309–313.

Schoenfeld T. A. and Hamilton L. W. (1977) Secondary brain changes following lesions: A new paradigm for lesion experimentation. *Physiol. Behav.* **18,** 951–967.

Schwarcz R., Köhler C., Fuxe K., Hökfelt T., and Goldstein M. (1979) On the Mechanism of Selective Neuronal Degeneration in the Rat Brain: Studies With Ibotenic Acid, in *Advances in Neurology,* vol. 23 (Chase T. H., Wexler N. S., and Barbeau A., eds.), pp. 655–668, Raven, New York.

Seeman P. (1980) Brain dopamine receptors. *Pharmacol. Rev.* **32,** 229–313.

Segal I. H. (1976) *Biochemical Calculations.* (2nd ed.) pp. 208–323, Wiley, New York.

Sellier J. and Verger H. (1898) Recherches experimentales sur la physiologie de la couche optique. *Arch. de Physiol. Norm. Path.* (S.5) **10,** 706–713.

Sharman D. F. (1981) The Turnover of Catecholamines, in *Central Transmitter Turnover.* (Pycock C. J. and Taberner P. V., eds.) pp. 20–58, University Park Press, Baltimore.

Shute C. C. D. and Lewis P. R. (1967) The ascending cholinergic reticular system: Neocortical, olfactory, and subcortical projections. *Brain* **40,** 497–520.

Simansky K. J. and Harvey J. A. (1981) Altered sensitivity to footshock after selective serotonin depletion: Comparison of electrolytic lesions and neurotoxin injections in the medial forebrain bundle of the rat. *J. Comp. Physiol. Psychol.* **95,** 341–350.

Simon J. R., Atweh S., and Kuhar M. J. (1976) Sodium-dependent high-affinity choline uptake: a regulatory step in the synthesis of acetylcholine. *J. Neurochem.* **26,** 909–922.

Singh D. (1975) Experimental Ablation, in *Physiological Techniques in Behavioral Research.* (Singh D. and Avery D. D., eds.) pp. 44–67, Wadsworth, Belmont, CA.

Smith G. C., Courtney P. G., Wreford, N. G. M., and Walker M. McD. (1982) Further studies on the effects of intravenously administered 6-hydroxydopamine on the median eminence of the rat. *Brain Res.* **234,** 101–110.

Snyder S. H., Axelrod J., and Zweig M. (1965) A sensitive and specific fluorescence assay for tissue serotonin. *Biochem. Pharmacol.* **14,** 831–835.

Sotelo C., Javoy F., Agid Y., and Glowinski J. (1973) Injection of 6-hydroxydopamine in the substantia nigra of the rat. I. Morphological study. *Brain Res.* **58,** 269–290.

Spiegel, E. A. (1982) *Guided Brain Operations,* Karger; Basel, pp. 246.

Spirduso W. W., Gilliam P. E., Schallert, T., Upchurch M., Vaughn D. M., and Wilcox R. E. (1984) Reactive capacity: a sensitive behavioral marker of movement initiation and nigrostriatal dopamine function. *Brain Res.* in press.

Stachowiak M. K., Bruno J. P., Snyder A. M., Stricker E. M., and Zigmond M. J. (1984) Apparent sprouting of striatal serotonergic terminals after dopamine-depleting brain lesions in neonatal rats. *Brain Res.* **291,** 164–167.

Steranka L. R., Polite L. N., Perry K. W., and Fuller R. W. (1983) Dopamine depletion in rat brain by MPTP (1-methyl-4-phenyl-1,2,3,6-tetrahydropyridine). *Res. Commun. Subst. Abuse* **4,** 315–323.

Stricker E. M. (1983) Brain Neurochemistry and the Control of Food Intake, in *Handbook of Behavioral Neurobiology* Satinoff E. and Teitelbaum P., eds.). pp. 329–366, Plenum Press, New York.

Stricker E. M. and Zigmond M. J. (1976) Recovery of Function After Damage to Central Catecholamine-Containing Neurons: A Neurochemical Model for the Lateral Hypothalamic Syndrome, in *Progress in Psychobiology and Physiological Psychology,* vol. 6 (Sprague J. M. and Epstein A. N., eds.). Academic Press, New York.

Sutherland R. J., Kolb B., Becker J. B. and Whishaw I. Q. (1982a) Neonatal 6-hydroxydopamine administration eliminates sparing of function after neonatal frontal cortex damage. *Neurosci. Lett.* **31,** 125–130.

Sutherland R. J., Whishaw I. Q., and Regehr J. C. (1982b) Cholinergic receptor blockade impairs spatial localization using distal cues in the rat. *J. Comp. Physiol. Psychol.* **96,** 563–573.

Swanson L. W., Perez V. J., and Sharpe L. G. (1972) Accurate and reliable intracerebral delivery of minute volumes of drug solutions. *J. Appl. Physiol.* **33,** 247–251.

Teitelbaum P. and Epstein A. N. (1962) The lateral hypothalamic syndrome: Recovery of feeding and drinking after lateral hypothalamic lesions. *Psychol. Rev.* **69,** 74–90.

Thoenen H. and Tranzer J. P. (1968) Chemical sympathectomy by selective destruction of adrenergic nerve endings with 6-hydroxydopamine. *Naunyn-Schmiedeberg's Arch. Pharmakol. Exp. Pathol.* **5,** 261, 271–288.

Tranzer J. P. and Thoenen H. (1967) Ultramorphologische Veranderungen der sympatischen Nervenendigungen der Katzenach Vorbehandling mit 5-and 6-Hydroxy-Dopamin. *Naunyn-Schmiedeberg's Arch. Exp. Pathol. Pharmakol.* **257,** 343.

Ungerstedt U. (1968) 6-Hydroxydopamine-induced degeneration of central monoamine neurons. *Eur. J. Pharmacol.* **5,** 107–110.

Ungerstedt U. (1971a) Adipsia and aphagia after 6-hydroxydopamine-induced degeneration of the nigro-striatal dopamine system. *Acta Physiol. Scand.* Suppl. **367**, 95–122.

Ungerstedt U. (1971b) Use of Intracerebral Injections of 6-Hydroxydopamine as a Tool for Morphological and Functional Studies on Central Catecholamine Neurons, in *6-Hydroxydopamine and Catecholaminergic Neurons* (Malmfors T. and Thoenen H., eds.). pp. 317–332, North Holland Publ. Amsterdam.

Ungerstedt U. and Arbuthnott G. W. (1970) Quantitative recording of rotational behavior in rats after 6-hydroxydopamine lesions of the nigrostriatal dopamine system. *Brain Res.* **24**, 485–493.

Van der Hyden J. A. M., Venema K., and Korf J. (1979) In vivo release of endogenous GABA from rat substantia nigra measured by a novel method. *J. Neurochem.* **32**, 469–476.

Venema K., Leever W., Bakker J. D., Haayer G., and Korf J. (1983) Automated precolumn derivitization device to determine neurotransmitter and other amino acids by reverse-phase high-performance liquid chromatography. *J. Chromatog.* **260**, 371–376.

Waddington J. L. and Crow T. J. (1979) Drug-induced rotational behavior following unilateral intracerebral injection of saline-ascorbate solution: neurotoxicity of ascorbic acid and monoamine-independent circling. *Brain Res.* **161**, 371–376.

Waddington J. L. and Cross A. J. (1980) Characterization of denervation supersensitivity in the striatonigral GABA pathway of the kainic acid-lesioned rat and in Huntington's disease. *Brain Res. Bull. 5,* Suppl. **2**, 825–828.

Walker A. E. (1957) Stimulation and ablation: Their role in the history of cerebral physiology. *J. Neurophysiol. 20*, 435–449.

Wastek, G. J. and Yamamura H. I. (1981) Acetylcholine Receptors, in *Neurotransmitter Receptors. Part 2. Biogenic Amines.* (Yamamura H. I. and Enna S. J., eds.). pp. 1–70, Chapman and Hall, London.

Weiland G. A. and Molinoff P. B. (1981) Quantitative analysis of drug-receptor interactions: I. Determination of kinetic and equilibrium properties. *Life Sci.* **29**, 313–330.

Wenk G., Hepler D., and Olton D. S. (1984) Behavior alters the uptake of [$^3$H] choline into acetylcholinergic neurons of the nucleus basalis magnocellularis and medial septal area. *Behav. Brain Res.* **13**, 129–135.

Wenk G. L. and Olton D. S. (1984) Recovery of neocortical choline acetyltransferase activity following ibotenic acid injection into the nucleus basalis of Meynert in rats. *Brain Res.* **293**, 184–186.

Westerink B. H. C. (1984) Determination of normetanephrine, 3,4-dihydroxyphenylethyleneglycol (free and total), and 3-methoxy-4-hydroxyphenylethyleneglycol (free and total) in rat brain by high-performance liquid chromatography with electrochemical detection and effects of drugs on regional concentrations. *J. Neurochem,* **42**, 934–942.

Whishaw I. Q., Robinson T. E., Schallert T., De Ryck M., and Ramirez V. D. (1978) Electrical activity of the hippocampus and neocortex in rats depleted of brain dopamine and norepinephrine: Relations to behavior and effects of atropine. *Exp. Neurol.* **62**, 748–767.

Wilcox R. E., Woodward J. J., Vaughn D. M., and Riffee W. H. (1983) Selective down-regulation of dopaminergic autoreceptors by dopamine agonist pretreatments. Assessment by endogenous fast-phase release of striatal dopamine. Comparison with behavioral, receptor binding, and tyrosine hydroxylase studies. *Soc. Neurosci. Abstr.* 1110.

Williams L. T. and Lefkowitz R. J. (1978) *Receptor Binding Studies in Adrenergic Pharmacology.* Raven Press, New York.

Willis G. L. and Smith G. C. (1984) A role for amine accumulation in the syndrome of ingestive deficits following lateral hypothalamic lesions. *Appetite* **5**, 239–262.

Willis G. L. and Smith G. C. (1982) The behavioral effects of intrahypothalamic multistage versus single injections of 6-hydroxydopamine. *Brain Res.* **245**, 345–352.

Willis G. L., Singer G., and Evans B. K. (1976) Intracranial injections of 6-OHDA. Comparison of catecholamine depleting effects of different volumes and concentrations. *Pharmacol. Biochem. Behav.* **5**, 207–213.

Wolfarth S., Coelle E.-F., Osborne N. N., and Sontag K.-H. (1977) Evidence for a neurotoxic effect of ascorbic acid after an intranigral injection in the cat. *Neurosci. Lett. 185,* 183–186.

Woodward J. J., Wilcox R. E., Riffee W. H., and Leslie S. W. Single dopamine agonist pretreatments and striatal functions. Relationship between fast-phase endogenous dopamine release in vitro, dopamine synthesis, and metabolism in vivo and behavior. *J. Neurochem.,* in preparation.

Woodward J. J., Wilcox R. E., Leslie S. W., and Riffee W. H. Rapid uptake of dopamine by mouse striatal synaptosomes. *J. Neurochem.,* submitted.

Yaksh T. L. and Rudy T. A. (1976) Chronic catheterization of the spinal subarachnoid space. *Physiol. Behav.* **17**, 1031–1036.

Yamamoto B. K. and Freed C. R. (1984) Asymmetric dopamine and serotonin metabolism in nigrostriatal and limbic structures of the trained circling rat. *Brain Res.* **297**, 115–119.

Yamamura H. I., Enna S. J. and Kuhar M. J. (1985) *Neurotransmitter Receptor Binding* (2nd ed.). Raven Press, New York.

Yamamura H. I., Enna S. J., and Kuhar M. J. (eds.) (1978) *Neurotransmitter Receptor Binding.* Raven Press, New York.

Young R. M. (1970) *Mind, Brain, and Adaptation in the Nineteenth Century. Cerebral Localization and its Biological Context from Gall to Ferrier.* Oxford, Clarendon Press.

Zigmond M. J., Acheson A. L., Stachowiak M. K., and Stricker E. M. (1984) Neurochemical compensation after nigrostriatal bundle injury in an animal model of preclinical parkinsonism. *Arch. Neurol.* **41**, 856–861.

# Chapter 11

# Methods to Determine Blood–Brain Barrier Permeability and Transport

## Quentin R. Smith

## 1. Introduction

The blood–brain barrier is a system of tissue sites that restrict and regulate the movement of hydrophilic solutes between the blood and the central nervous system. The barrier between blood and brain extracellular fluid is located at the brain capillary endothelium, whereas the barrier between blood and cerebrospinal fluid is located at the choroid plexus epithelium and arachnoid membrane (Rapoport, 1976). Morphological evidence indicates that the barrier at each site is formed by a single layer of cells that are connected by rings of tight junctions called zonulae occludens (Reese and Karnovsky, 1967; Brightman and Reese, 1969). Figure 1 illustrates the structure of the barrier at the cerebrovascular endothelium. Because intercellular diffusion is restricted by the tight junctions, solutes move across the blood–brain barrier predominantly by the transcellular route, and thus the barrier displays many permeability and transport properties of a continuous cell membrane. For example, the barrier is permselective for lipid-soluble, as compared to water-soluble, compounds (Crone, 1965; Rapoport et al., 1979a). In addition to the permselectivity, there are specific saturable transport systems at the barrier for rapid exchange of essential nutrients (Pardridge and Oldendorf, 1977; Gjedde, 1983) and for regulated transport of inorganic ions (Bradbury, 1979; Smith and Rapoport, 1984). Thus, the blood–brain bar-

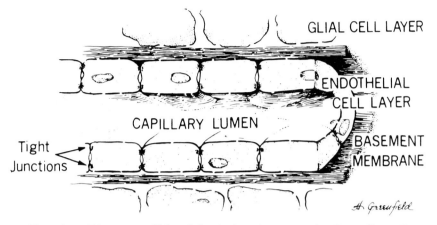

Fig. 1.    Diagram of blood–brain barrier at cerebral capillary (from Rapoport, 1976).

barrier is not only a permeability barrier, but also a regulatory interface that governs the composition of brain extracellular fluid and cerebrospinal fluid.

The rate of entry of a solute into the brain depends on the transport properties of each of the barrier sites because diffusion is not restricted between brain tissue and cerebrospinal fluid (Brightman and Reese, 1969; Patlak and Fenstermacher, 1975). However, the brain capillary surface area, 100–240 $cm^2$/g brain (Crone, 1963; Bradbury, 1979), comprises over 94% of the total surface area of the blood–brain barrier. For comparison, the surface area of the choroid plexus epithelium is approximately 1 $cm^2$/g brain (Welch and Sadler, 1966; Smith et al., 1981a), and the area of the arachnoid membrane is 5 $cm^2$/g brain (Bradbury, 1979). Because of the large surface area of the cerebral capillaries, the cerebrovascular endothelium is the major site for solute entry into the brain. Exceptions to this rule are the inorganic ions $Na^+$, $Cl^-$, $Ca^{2+}$, and $Mg^{2+}$ (Smith et al., 1981a, 1983), the plasma proteins (Rapoport, 1983), and the vitamin ascorbic acid (Spector and Eells, 1984), all of which have a significant choroid plexus component of uptake into the central nervous system.

In the past two decades, considerable effort has been exerted to characterize the permeability and transport properties of the cerebrovasculature. During this time, many methods have been developed to measure transport at the cerebral capillaries (Bradbury, 1979; Fenstermacher et al., 1981; Takasato et al., 1984). Currently, most cerebrovascular transport studies use one of five techniques that measure tracer influx into the brain—the intravenous-administration technique, the *in situ* brain-perfusion technique, the indicator-dilution technique, the brain-uptake index technique, and the single injection-external registration technique. However, efflux from brain to blood can be measured with the brain-washout technique (Bradbury et al., 1975) or the concentration-profile technique (Patlak and Fenstermacher, 1975). In addition, isolated brain capillaries can be used to examine transport into the endothelial cells and to isolate specific capillary transport systems (Goldstein et al., 1984).

In this review, each of the five basic methods for measuring unidirectional transport into the brain will be examined and evaluated. The review will not cover methods for measuring transport at the choroid plexus or at the arachnoid membrane—these are covered in other reviews (Welch, 1975; Wright, 1978; Fenstermacher et al., 1981). For more detailed information on the structure and function of the blood–brain barrier, the reader is referred to Rapoport (1976) and Bradbury (1979).

## 2. Two Methodological Approaches

There are two approaches to measure influx across the cerebrovascular endothelium. In one, a solute is injected as a bolus into the carotid artery of an animal and then brain uptake or extraction is determined from a single pass of the bolus through the brain capillaries. This approach was first applied by Crone (1963) with the indicator-dilution technique, and is the basis of the brain-uptake index and single injection-external registration techniques. It has been extremely popular because of the ease and simplicity of these techniques. However, this approach is not sufficiently sensitive to determine accurate permeability coefficients for poorly penetrating compounds, such as sucrose or mannitol. Furthermore, the permeability coefficients of rapidly penetrating compounds are questionable since cerebral blood flow and brain uptake are not measured simultaneously.

The limitations of these methods have stimulated interest in a second approach, in which uptake time is not limited to that of a single pass (5–15 s). With the intravenous-administration technique, a solute is injected or infused iv, and its brain concentration is determined at any time from 10 s to several hours thereafter, depending on the cerebrovascular permeability coefficient of the solute. Similarly, brain capillaries can be perfused for up to 5 min with the *in situ* brain-perfusion technique. Because uptake time can be prolonged, these techniques are 100 times more sensitive than the single-pass methods. Furthermore, cerebral blood flow and brain uptake can be measured simultaneously with either the intravenous-administration or the brain-perfusion technique, allowing for accurate determination of permeability coefficients for rapidly penetrating compounds. Although the intravenous-administration and brain-perfusion techniques may require more time and effort than single-pass techniques, their additional accuracy, range, and flexibility justify the additional work.

## 3. Continuous Uptake

### 3.1. Intravenous Administration

With this technique a solute is given iv and the arterial plasma concentration is measured until a specific time, when the brain concentration is determined. In animal experiments, the animal is killed and the brain is removed and analyzed for solute content. The intravenous-administration technique can be used in man if brain concentration is determined with positron emission tomography (Raichle, 1979; Kessler et al., 1984). The solute may be administered iv as a single bolus injection, by continuous infusion, or by programmed infusion to maintain a constant plasma concentration (Patlak and Pettigrew, 1976). Three versions of the method are used that differ primarily in the number of time points examined.

### 3.1.1. Single Time Point Analysis

Brain uptake of a diffusible radiotracer after intravenous administration is given as follows by a two-compartment model that incorporates cerebral blood flow (Rapoport et al., 1979a),

$$\frac{dC_{br}}{dt} = k(C_{pl} - C_{br}/V_{br}) \tag{1}$$

where $C_{br}$ = parenchymal or extravascular brain concentration of tracer (dpm/g), $C_{pl}$ = arterial plasma concentration of tracer (dpm/mL), $V_{br}$ = cerebral distribution volume of tracer (mL/g), and $t$ = time (s). In Eq. (1), $k$ is a transfer coefficient defined as:

$$k = F(1 - e^{-PA/F}) \tag{2}$$

where $F$ = regional cerebral blood or plasma flow (mL/s/g), $P$ = cerebrovascular permeability (cm/s), and $A$ = capillary surface area (cm$^2$/g).

With the single time point method, uptake time is limited so that only a small quantity of tracer accumulates in the brain ($C_{br}/V_{br} \approx 0$), and therefore back diffusion from brain can be assumed to be negligible. Furthermore, uptake time is limited to minimize diffusional exchange between brain and cerebrospinal fluid (Smith and Rapoport, 1984). Then, influx across cerebral capillaries is unidirectional and is given as:

$$\frac{dC_{br}}{dt} = kC_{pl} \tag{3}$$

The cerebrovascular permeability–area product, $PA$, can be obtained by integrating Eq. (3) over the uptake time, $T$, and rearranging to give:

$$PA = -F \ln \left[ 1 - \frac{C_{br}(T)}{F \int_0^T C_{pl} \, dt} \right] \tag{4}$$

Cerebrovascular permeability can be calculated by dividing $PA$ by the capillary surface area, $A$.

Figure 2 illustrates a single time point experiment in which a rat was killed 20 min after a bolus injection of $^{14}$C-mannitol into the femoral vein (Ohno et al., 1978). Blood samples were collected at various times after injection from a cannula in the femoral artery. $\int C_{pl} dt$ was obtained by using a nonlinear least-squares procedure to fit the $C_{pl}$ data to the equation

$$C_{pl}(t) = \sum_{i=1}^{n} B_i e^{-b_i t} \tag{5}$$

where $B_i$ (dpm/mL) and $b_i$ (s$^{-1}$) are constants, and then computer integrating Eq. (5) from $t = 0$ to 20 min (Knott, 1979). Alternatively, $\int C_{pl} dt$ can be determined by continuous arterial blood withdrawal at a constant rate with a pump to obtain an average concentration, $\bar{C}_{pl}$, where $\int C_{pl} dt = \bar{C}_{pl} T$ (Gjedde et al., 1980).

Brain tracer concentration must be corrected for residual intravascular tracer in order to obtain parenchymal brain concentration, $C_{br}$. One procedure for doing this is to wash out intravascular tracer from brain–blood vessels at the end of the experiment by perfusion of the brain for 10–30 s with tracer-free fluid (Bachelard et al., 1973). However, this procedure is limited to the uptake of poorly penetrating solutes for which perfusion time is negligible (<5%) compared to uptake time. A second procedure that can be used for all solutes is to calculate $C_{br}$ with the following equation, where $C_{tot}$ = total measured brain content (intravascular plus extravascular) of tracer (dpm/g), $C_{bl}$ = arterial blood concentration of tracer(dpm/mL), and $V_{bl}$ = regional blood or vascular volume of brain (mL/g),

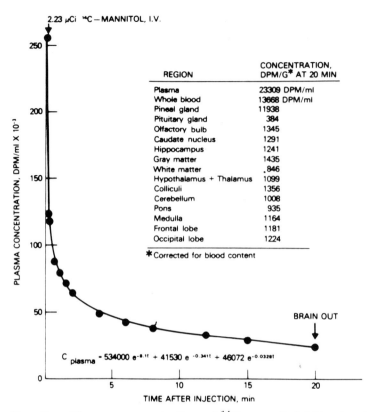

2.23 µCi ¹⁴C — MANNITOL, I.V.

| REGION | CONCENTRATION, DPM/G* AT 20 MIN |
|---|---|
| Plasma | 23309 DPM/ml |
| Whole blood | 13668 DPM/ml |
| Pineal gland | 11938 |
| Pituitary gland | 384 |
| Olfactory bulb | 1345 |
| Caudate nucleus | 1291 |
| Hippocampus | 1241 |
| Gray matter | 1435 |
| White matter | .846 |
| Hypothalamus + Thalamus | 1099 |
| Colliculi | 1356 |
| Cerebellum | 1008 |
| Pons | 935 |
| Medulla | 1164 |
| Frontal lobe | 1181 |
| Occipital lobe | 1224 |

*Corrected for blood content

BRAIN OUT

$$C_{plasma} = 534000\ e^{-8.1t} + 41530\ e^{-0.341t} + 46072\ e^{-0.0328t}$$

Fig. 2.    Plasma concentration of $^{14}$C-mannitol following intravenous injection and regional brain concentration, $C_{br}$, at 20 min after injection (from Ohno et al., 1978).

$$C_{br}(t) = C_{tot}(t) - V_{bl}C_{bl}(t) \qquad (6)$$

With the single time point method, $V_{bl}$ is measured in the same animal or in a second set of animals with a vascular radiotracer or indicator (Ohno et al., 1978; Rapoport et al., 1979a; Gjedde, 1981; Hawkins et al., 1982; Smith and Rapoport, 1984). Radiolabeled solutes that do not measurably cross the blood–brain barrier during the experiment, such as $^{3}$H-inulin, $^{3}$H-dextran, $^{111}$In-transferrin, or $^{125}$I-albumin, are often used to measure blood volume, equal to the brain concentration of vascular indicator (dpm/g) divided by the blood concentration of vascular indicator (dpm/mL). Regional blood volume, expressed as a percent, ranges from 1 to 4% (Ohno et al., 1978). For test solutes that do not measurably enter erythrocytes, intravascular volume can be expressed as a plasma volume, $V_{pl}$, in which case the product of $V_{pl}$ and $C_{pl}$ is subtracted from $C_{tot}$ to obtain $C_{br}$ (Gjedde et al., 1980; Gjedde, 1981).

One advantage of using an intravascular indicator along with the test tracer in each animal is that vascular volume is determined individually, and thus the $C_{br}$ calculation is not based on an average $V_{bl}$ from a second set of animals. This reduces variation in calculated $C_{br}$, particularly for poorly penetrating solutes that may have a significant intravascular fraction of total brain tracer content (Takasato et al., 1984). However, vascular volume may differ among various solutes (Sisson and Oldendorf, 1971). Therefore, the investigator should verify that the volume of the vascular indicator equals that of the test compound, the latter determined with the multiple time point method (*see below*).

An advantage of a bolus iv injection, as compared to an iv infusion, is that $C_{bl}$ decreases during the experiment, so that the contribution of intravascular tracer to measured brain concentration progressively declines (Ohno et al., 1978).

An assumption of the single time point method is that backflux of tracer is negligible and thus tracer uptake is unidirectional during the measurement of $PA$ [Eq. (3)]. For a two-compartment model, influx is unidirectional when $\int(C_{br}/V_{br})dt \ll \int C_{pl}dt$. There are three ways to verify this assumption. In the first, $k$ is determined at several early uptake times, and the time interval of unidirectional uptake is obtained from the initial period in which $k$ is constant. Significant backflux is indicated by a decrease in $k$ (Rapoport et al., 1980). In the second, the period of unidirectional influx is determined from initial uptake data with graphical analysis (Bradbury, 1979; Gjedde, 1981; Blasberg et al., 1983). In the third, compartmental analysis is used to determine $V_{br}$ (Ohno et al., 1978). When $V_{br}$ is known, the assumption of negligible backflux ($\int(C_{br}/V_{br})dt \ll \int C_{pl}dt$) can be verified.

Finally, $F$ must be known to calculate $PA$ for moderately and rapidly penetrating compounds. As given by Eq. (3), brain uptake is equal to the arterial plasma concentration times a transfer constant, $k$, which is less than $PA$ and approaches $F$ to the extent that tracer is extracted during its passage through the capillary bed (Crone, 1963; Johnson and Wilson, 1966; Rapoport et al., 1979a). Figure 3 illustrates the relation between $k$ and $PA$ for three values of $F$. When $F \geq 5PA$, $k \cong PA$ with less than 10% error. Similarly, when $PA \geq 2.3F$, $k \cong F$ with less than 10% error. The transfer constant depends on both $PA$ and $F$ when $0.2 \leq PA/F \leq 2.3$.

Regional cerebral blood flow can be determined in a separate set of animals with $^{14}$C-iodoantipyrine (Sakurada et al., 1978), $^{14}$C-butanol (Schafer et al., 1976), $^{3}$H-nicotine (Ohno et al., 1979), or labeled microspheres (Marcus et al., 1976). In addition, several methods have been developed for the simultaneous measurement of $F$ and $PA$ (Gjedde et al., 1980; Sage et al., 1981; Betz and Iannotti, 1983).

Thus, the single time point version of the intravenous-administration technique has several advantages. $PA$ can be determined in each animal, which reduces time and expense. The method is sufficiently sensitive to determine permeability coefficients as low as $10^{-8}$ cm/s (i.e., the permeability of sucrose) or as high as $10^{-4}$ cm/s (i.e., the permeability of antipyrine or of caffeine) (Fig. 4). Cerebral blood flow and solute uptake can be meas-

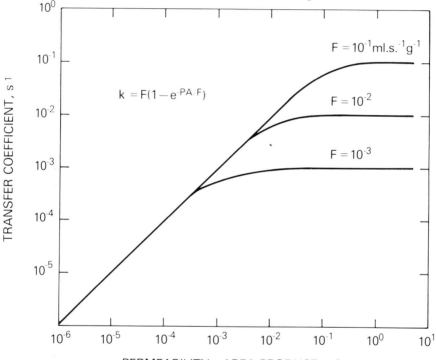

Fig. 3.    Relation between the transfer coefficient and the permeability–area product for three values of cerebral blood flow. The transfer coefficient is defined in Eq. (2), and equals $PA$ (within 10% error) when $PA \leq 0.2F$.

ured independently in order to obtain accurate permeability coefficients for rapidly penetrating compounds. The method can be used on conscious as well as anesthetized animals (Ohno et al., 1978; Rapoport et al., 1980) and can measure $PA$ in specific brain regions, with quantitative autoradiography or tissue dissection (Ohno et al., 1978; Hawkins et al., 1982) or in whole brain (Daniel et al., 1977). In addition, the method can be extended to humans with positron emission tomography (Raichle, 1979; Kessler et al., 1984). A constant arterial concentration of tracer is not required. In fact, no specific time course of arterial concentration is assumed [Eq. (4)]. Last, since *PA* can be determined in individual animals, the method does not require a series of animals with similar regional cerebrovascular permeability or transport properties (Blasberg et al., 1981; Hasegawa et al., 1983).

The limitations of the single time point method include possible error in the determination of $V_{bl}$ and in the assumption of negligible backflux. However, both the vascular correction and the backflux assumption can be verified with either the multiple time point or compartmental analysis methods (Ohno et al., 1978; Rapoport et al., 1980). In addition, the method is subject to errors in the measurement of solute uptake due to radiotracer metabolism by tissues other than the brain, if chromatography of tracer in brain and plasma is not performed (Takasato et al., 1984). Furthermore, regulation of plasma solute concentration to examine the concentration dependence of cerebrovascular transport is not easily performed with this technique (Bachelard et al., 1973; Daniel et al., 1977; Gjedde, 1980).

### 3.1.2. Multiple Time Point Analysis

If initial uptake is measured over two or more time intervals with the intravenous-administration technique, multiple time point analysis can be used to calculate cerebrovascular *PA*. The experimental procedures are the same as those described for the single time point analysis, except that a vascular indicator is not required. After iv administration of the test tracer, $\int C_{pl} dt$ and total brain concentration, $C_{tot}(T)$, are measured for each $T$. When uptake is measured over two time intervals, unidirectional influx is assumed. However, this assumption can be verified with three or more time points.

A transfer coefficient, $k$, is calculated from the $C_{tot}(T)$ and $\int C_{pl} dt$ data with either an iterative procedure (Ohno et al., 1978) or with graphical analysis (Sarna et al., 1977; Gjedde, 1981; Patlak et al., 1983). *PA* is obtained by rearrangement of Eq. (2) (Johnson and Wilson, 1966)

$$PA = -F\ln(1 - k/F) \tag{7}$$

where $F$ is determined in a separate set of animals (Sakurada et al., 1978; Ohno et al., 1979).

Multiple time point analysis requires two or more small experimental animals to calculate *PA*, because each animal is killed in order to measure $C_{tot}$. However, with large animals or humans, *PA* can be determined with a single individual by measuring total brain and plasma concentrations at different times with positron emission tomography.

The iterative procedure determines $k$ with data from two experimental time periods (Ohno et al., 1978; Rapoport et al., 1980). First, $C_{tot}$ at the earlier time point, $T_1$, is used to estimate blood volume, $V'_{bl}$, as $C_{tot}(T_1)/C_{bl}(T_1)$. Then, an initial estimate of parenchymal brain concentration, $C'_{br}$, at the later time, $T_2$, is obtained as:

$$C'_{br}(T_2) = C_{tot}(T_2) - V'_{bl} C_{bl}(T_2) \qquad (8)$$

With $C'_{br}(T_2)$, $k$ can be approximated as:

$$k \cong C'_{br}(T_2) \int_0^{T_2} C_{pl} dt \qquad (9)$$

This value of $k$ is used to obtain a new estimate of $V'_{bl}$ as:

$$V'_{bl} = [(C_{tot}(T_1) - k \int_0^{T_1} C_{pl} \, dt)/C_{bl}(T_1)] \qquad (10)$$

The procedure is repeated until successive estimates of $k$ differ by less than 2% (Ohno et al., 1978). It should be noted that $V_{bl}$ may differ from "true" blood volume because of nonhomogenous distribution of the tracer in blood, differences in hematocrit between brain capillaries and systemic arteries, or binding or accumulation of tracer by the capillary endothelium (Bradbury, 1979; Cremer and Seville, 1983; Blasberg et al., 1983).

A second procedure that can be used with data from two or more time points is graphical analysis (Sarna et al., 1977; Bradbury, 1979; Gjedde, 1981; Patlak et al., 1983). With unidirectional uptake, total brain concentration is given as:

$$C_{tot} (t) = k \int C_{pl} dt + V_{bl} C_{bl}(t) \qquad (11)$$

Division by $C_{bl}(t)$ gives a straight line with slope $k$ and intercept $V_{bl}$

$$\frac{C_{tot}}{C_{bl}} = k \, [\int C_{pl} dt/C_{bl}] + V_{bl} \qquad (12)$$

Alternatively, Eq. (11) can be divided by $C_{pl}(t)$, which gives the intercept as a plasma volume, $V_{pl}$

$$\frac{C_{\text{tot}}}{C_{\text{pl}}} = k[\int C_{\text{pl}} dt / C_{\text{pl}}] + V_{\text{pl}} \tag{13}$$

When $C_{\text{pl}}$ is constant during the experiment, Eq. (13) reduced to:

$$\frac{C_{\text{tot}}}{C_{\text{pl}}} = kt + V_{\text{pl}} \tag{14}$$

The transfer coefficient is calculated with the graphical analysis procedure by fitting the data to Eq. (12), (13), or (14) with linear least-squares regression.

Both the iterative and graphical procedures assume unidirectional uptake. With the iterative procedure, this can be evaluated by measuring $k$ over different time intervals. With the graphical procedure, the period of unidirectional influx is determined from the linear portion of the plot [$C_{\text{tot}}/C_{\text{bl}}$ or $C_{\text{pl}}$] vs [$\int C_{\text{pl}} dt / C_{\text{bl}}$ or $C_{\text{pl}}$], as shown in Fig. 5.

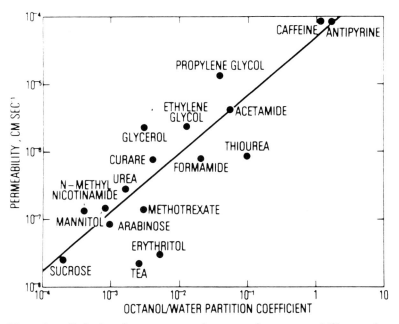

Fig. 4. Relation between cerebrovascular permeability and octanol–water partition coefficient for organic electrolytes and nonelectrolytes. The line is the least-squares fit to the data and is given as: $\log_{10} P = -4.30 \ (\pm 0.34) + 0.866 \ (\pm 0.139) \log_{10}$ partition coefficient (from Rapoport et al., 1979a).

The multiple time point version of the intravenous-administration technique shares many of the advantages of the single time point version. In addition, with the multiple time point method vascular volume is determined directly with the test tracer and the assumption of negligible backflux can be verified. This method is ideally suited for the measurement of cerebrovascular *PA* in humans with positron emission tomography.

Still, the multiple time point method has several limitations. Two or more small animals are required to obtain a *PA* value. The series of animals that are used must have similar values of *PA*, $V_{bl}$, and *F* for the model to be valid. For example, the method cannot be used to determine *PA* in brain tumors or in stroke because of large variability of *PA*, $V_{bl}$, and *F* (Blasberg et al., 1981;

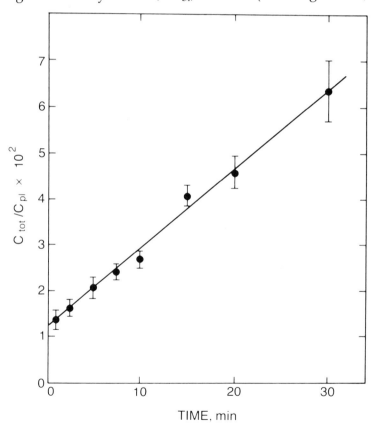

Fig. 5.     Relation of total brain $^{22}$Na concentration divided by plasma $^{22}$Na concentration to infusion time. $^{22}$Na was infused iv into each rat to maintain a constant plasma concentration of tracer ($\int C_{pl}dt/C_{pl} = t$). The blood-to-brain transfer coefficient is obtained from the slope of the line (modified from Sarna et al., 1977).

Hasegawa et al., 1983). Last, this method, like the single time point method, is subject to errors due to tracer metabolism, and provides no simple means of controlling plasma solute concentrations to examine saturable transport.

### 3.1.3. Compartmental Analysis

Compartmental analysis, unlike the single and multiple time point methods, can be used on intravenous-administration data in which backflux is significant. With this method, plasma and brain concentrations of tracer are determined over a wide range of experimental times after administration. Then, data are analyzed with a compartmental model that describes tracer uptake and distribution in brain. The most frequently used model is the simple, linear two-compartment model (plasma and brain) with blood flow (Rapoport et al., 1979a). More complex models have been proposed that describe intracerebral distribution, metabolism, and brain–CSF exchange (Davson and Welch, 1971; Sokoloff et al., 1977; Rapoport et al., 1982).

The solution to the differential equation of a compartmental model, unlike Eq. (4) of the single time point method and Eq. (12) of the multiple time point method, depends on the specific time course of plasma tracer concentration. Because the two-compartment model with flow is commonly used to analyze brain uptake data, the solutions to the model are given below for a constant plasma concentration, for a continuous iv infusion, and for a bolus iv injection (Rapoport et al., 1979a; Rapoport et al., 1982). These solutions are in terms of parenchymal brain concentration, $C_{br}(T)$. Alternatively, they can be extended to include intravascular brain tracer, as $C_{tot}(T) = C_{br}(T) + V_{bl}C_{bl}(T)$.

When the arterial plasma concentration of tracer is constant ($\overline{C}_{pl} = B$), Eq. (1) can be integrated to give:

$$C_{br}(T) = \overline{C}_{pl}\, V_{br}(1 + e^{-kT/V_{br}}) \tag{15}$$

Equation (15) can be rearranged to give $k$ in terms of $C_{br}(T)$, $\overline{C}_{pl}$, $V_{br}$, and $T$ as:

$$k = -\frac{V_{br}}{T} \ln\left[1 - \frac{C_{br}(T)}{\overline{C}_{pl}\, V_{br}}\right] \tag{16}$$

When tracer is infused iv at a constant rate, $C_{pl}$ can be described as (Rapoport et al., 1982):

$$C_{pl}(t) = \sum_{i=1}^{n} B_i(1 - e^{-b_i t}) \tag{17}$$

Substituting Eq. (17) into Eq. (1) and integrating to time $T$ gives:

$$C_{br}(T) = V_{br} \sum_{i=1}^{n} B_i \left[ 1 + \frac{b_i}{(k/V_{br}) - b_i} \left( e^{-kT/V_{br}} - \frac{k}{V_{br} b_i} e^{-b_i T} \right) \right] \quad (18)$$

Following an iv bolus injection, arterial plasma concentration of tracer decreases exponentially as described by Eq. (5) (Ohno et al., 1978; Rapoport et al., 1982). Substituting Eq. (5) into Eq. (1) and integrating gives:

$$C_{br}(T) = \sum_{i=1}^{n} \frac{kB_i}{k/V_{br}} - (e^{-b_i T} - e^{-kT/V_{br}}) \quad (19)$$

In addition to these solutions for a two-compartment model, Rapoport et al. (1982) have derived and applied solutions to a four-compartment model of the central nervous system. This model includes plasma, brain extracellular space, brain intracellular space, and CSF, and incorporates delivery rate as determined by cerebral blood flow.

The recommended method to calculate $k$ with compartmental analysis is nonlinear least-squares regression because it is flexible, unbiased, and simple to perform with a computer (Knott, 1979; Draper and Smith, 1981). First, measured plasma concentrations at various times are fit to the appropriate plasma equation for constant concentration ($\overline{C}_{pl} = B$), continuous infusion [Eq. (17)], or bolus injection [Eq. (5)] to obtain the constants $B_i$ and $b_i$. Then $C_{br}$ data are fit to the model equation [i.e., Eqs. (15), (18), or (19)] with the plasma constants to obtain the best-fit values for the model parameters, $k$ and $V_{br}$ (Ohno et al., 1978; Rapoport et al., 1979b; Ohno et al., 1980; Rapoport et al., 1982). If the variance of the $C_{br}$ data is not constant, weighted nonlinear regression should be used (Draper and Smith, 1981). In addition, the quality of fit of the model and the redundancy of individual parameters can be evaluated statistically (Draper and Smith, 1981). If the investigator does not have access to a computer with a nonlinear regression program, an alternative procedure for data with constant plasma concentration is the graphical analysis method of Solomon (1960) (Riggs, 1963; Levin and Patlak, 1972; Johanson and Woodbury, 1978; Smith et al., 1981b; Smith et al., 1982). Finally, the blood-to-brain transfer coefficient from compartmental analysis can be converted to $PA$ with Eq. (7) if $F$ is measured independently.

Thus, compartmental analysis of brain tracer concentration after iv administration can be used to determine both the cerebrovascular permeability coefficient and the intracerebral distribution of the tracer. Unlike the single time point and multiple time point methods, compartmental analysis does not require unidirectional uptake kinetics. However, initial uptake data from compartmental studies can be analyzed with the single and multiple time poin methods as well, for comparison (Ohno et al., 1978; Rapoport et a , 1979b; Smith et al., 1981a). Intracerebral distribution of tracer ان be examined with multicompartment models that incorporate factors such as brain cell uptake, binding and metabolism, and brain–CSF exchange, in addition to blood–brain exchange.

The limitations of the compartmental analysis method include those of the multiple time point method. In addition, many animals are required to describe the entire time-course of brain tracer concentration. Furthermore, data analysis may require access to a computer with a nonlinear regression program.

## 3.2. *In Situ* Brain Perfusion Method

With the *in situ* brain-perfusion technique, the right cerebral hemisphere of an anesthetized rat is perfused for 5–300 s by retrograde infusion of fluid into the right external carotid artery (Takasato et al., 1984). Cerebrovascular $PA$ is calculated from the brain parenchymal uptake of tracer during perfusion. Because perfusion time can be extended beyond that of a single pass (5–15 s), this method is more sensitive than the indicator-dilution, brain-uptake index, or single injection-external registration techniques. The principal advantage of this method, as compared to the intravenous-administration technique, is that virtually absolute control is permitted over perfusate composition.

A diagram of the perfusion system is shown in Fig. 6. Prior to perfusion, the right external carotid artery of an anesthetized rat is catheterized, and the right pterygopalatine artery is ligated.

The cannula to the right external carotid artery is connected to a syringe containing a test tracer and an impermeant intravascular tracer dissolved in physiological saline, plasma, or blood. One second before perfusion, the right common carotid artery is ligated. Then, perfusion fluid is infused into the external carotid artery at a rate that minimizes (<0.1%) systemic blood flow to the right cerebral hemisphere (Takasato et al., 1984; Takasato et al., 1985). During perfusion with blood, cerebral blood flow and blood volume are comparable to respective values in the conscious rat, whereas perfusion with saline or plasma increases $F$ three- to fourfold because of the low viscosity of these fluids. Perfusion with blood for 300 s or with saline for 60 s does not alter the permeability of the blood–brain barrier (Takasato et al., 1984). The

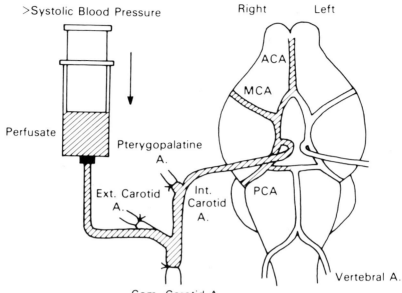

Fig. 6.    Diagram of technique for perfusing the right cerebral hemisphere of a rat. ACA, Anterior cerebral artery; MCA, middle cerebral artery; PCA, posterior cerebral artery (from Takasato et al., 1984).

perfusion is terminated by decapitation of the rat, after which samples from six brain regions and perfusion fluid are analyzed for radiotracer content.

The calculation of cerebrovascular $PA$ is equivalent to that of the intravenous-administration technique, except that perfusion fluid concentration, $C_{pf}$, is inserted for $C_{pl}$ in the equations. If perfusion time is limited to restrict tracer accumulation in the brain and thus minimize back diffusion, then cerebrovascular $PA$ can be calculated with Eq. 4 of the single time point analysis (Takasato et al., 1984). When backflux cannot be ignored ($C_{br}/V_{br} > 0.2\ C_{pf}$), $PA$ can be calculated from a single time point with Eqs. (7) and (16) if $V_{br}$ is known. Similarly, both the multiple time point and compartmental analysis methods can be used with brain-perfusion data when uptake is measured over two or more time intervals.

Regional cerebral perfusion fluid flow can be measured in a separate set of animals with [14]C-iodoantipyrine or [14]C-diazepam (Takasato et al., 1984). Alternatively, both $PA$ and $F$ can be measured simultaneously in short (5–20 s) perfusions by substituting [3]H- or [14]C-diazepam for the intravascular marker.

The *in situ* brain-perfusion technique has several advantages in common with the intravenous-administration technique.

Cerebrovascular permeability coefficients can be measured over a $10^4$-fold range from a minimum of $10^{-8}$ cm/s (Table 1). In addition, $k$ and $F$ can be determined simultaneously in order to obtain accurate permeability coefficients for rapidly penetrating compounds. Lastly, the single time point analysis can be used to obtain a $PA$ value in each experimental animal.

There are a few advantages of the brain-perfusion technique that are not shared by the intravenous-administration technique. First, the perfusion method allows total control of perfusate composition. Specific solute concentrations in the perfusate can be manipulated to examine saturation, competition, and inhibition of carrier-mediated transport (Fig. 7) (Smith et al., 1984). Furthermore, pH, osmolality, ionic content, and protein concentrations can be varied over a greater range than would be tolerated systemically. Second, the brain perfusion technique avoids errors caused by radiotracer metabolism by tissues other than the brain. Last, the three- to fourfold greater $F$ with saline perfusion minimizes flow-related errors in the determination of $PA$ for rapidly penetrating substances, because as $PA/F \to 0$, $k \to PA$ (Takasato et al., 1984).

TABLE 1

Cerebrovascular Permeability Coefficients From the
Intravenous-Administration, Brain-Perfusion, Indicator-Dilution,
and Brain-Uptake Index[a] Techniques

| | Permeability, cm/s $\times$ $10^7$ | | | |
|---|---|---|---|---|
| Solute | Intravenous administration | Brain perfusion | Indicator dilution | Brain uptake index |
| Antipyrine | 700 | 580 | 330 | 290 |
| Acetamide | 44 | 47 | — | 70 |
| Thiourea | 8.0 | 8.2 | 29 | 21 |
| Urea | 3.2 | 2.8 | 44[b] | n.s. |
| Mannitol | 1.1 | 0.8 | 12[b] | n.s. |
| Sucrose | 0.27 | 0.23 | n.s. | n.s. |

[a]Data are from Rapoport et al. (1979a), Takasato et al. (1984), Crone (1965), and Cornford et al. (1982), respectively. Cerebrovascular $P$ was calculated from BUI data by using $C''_{vas}/C''_{ref} = 1.9\%$ (Oldendorf, 1981), $d = 0.75$ (Pardridge and Oldendorf, 1975), $F = 1 \times 10^{-2}$ mL/s/g and $A = 240$ cm$^2$/g (Crone, 1963).

[b]Signifies greater than tenfold difference in $P$ from value with intravenous-administration technique; n.s. signifies not statistically significant. The indicator-dilution and brain-uptake index techniques cannot be used to obtain accurate permeability coefficients when $P \le 10$–$25 \times 10^{-7}$ cm/s.

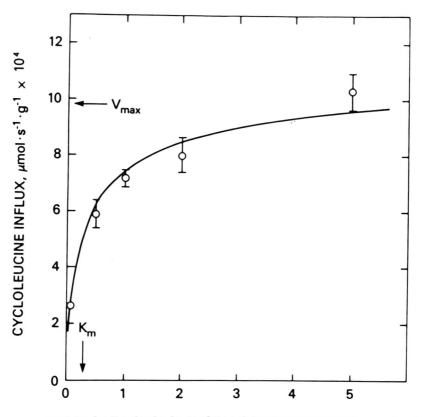

PERFUSATE CYCLOLEUCINE CONCENTRATION, μmol·ml⁻¹

Fig. 7.    Relation of unidirectional cycloleucine influx into the parietal cortex to perfusate cycloleucine concentration. Each point represents a mean ±SEM for three animals. Influx ($J$) is defined as: $J = PA \cdot C_{pf}$. The curve is predicted cycloleucine influx as given by the Michaelis–Menten equation:

$$J = \frac{V_{max} \, C_{pf}}{K_m + C_{pf}}$$

where $V_{max} = 0.93 \times 10^{-3}$ μmol/s/g and $K_m = 0.28$ μmol/mL. Cycloleucine is a nonmetabolized, model large neutral amino acid (from Takasato et al., 1983).

The advantages of the brain-perfusion technique, as compared to the indicator-dilution, brain-uptake index, and single injection-external registration techniques, are that regional cerebral perfusion fluid flow is known and constant during perfusion, and that negligible mixing (<0.1%) of perfusion fluid with blood occurs before the perfusate reaches the brain capillaries.

Whereas several isolated brain perfusion methods have been reported (Woods and Youdim, 1978; Gilboe, 1982), the *in situ* brain-perfusion technique requires less surgery and has a shorter perfusion time than the isolated brain methods.

The limitations of the *in situ* brain-perfusion technique are that, as of now, only one cerebral hemisphere is perfused and the experiments use anesthetized, as opposed to conscious, rats.

# 4. Single Pass Uptake

## 4.1. Indicator Dilution

With the indicator-dilution technique, cerebrovascular permeability is determined from the extraction of tracer during a single pass through the cerebral vasculature. A buffered saline solution containing the test tracer and an impermeant reference tracer is injected as a bolus into the carotid artery. Commonly used reference tracers, like Evans blue-albumin, $^{22}$Na, $^{36}$Cl, and $^{113m}$In-DTPA do not measureably cross the blood–brain barrier in a single pass (Crone, 1965; Yudilevich and DeRose, 1971; Hertz and Paulson, 1980). Immediately following the injection, serial blood samples are collected for 15–30 s from a cannula in the superior sagittal sinus (Fig. 8).

Brain extraction ($E$) is defined as follows, where $C_a$ = arterial concentration and $C_v$ = venous concentration:

$$E = \frac{C_a - C_v}{C_a} \tag{20}$$

The value of $C_a$ is not measured directly with this technique, but is obtained indirectly from the venous concentration of reference tracer. The impermeant reference tracer corrects for dilution of the injectate with blood as the bolus passes through the cerebral vasculature and thus indicates what the concentration of test tracer would have been if no brain uptake occurred. Then Eq. 20 can be expressed as:

$$E = 1 - \frac{C'_{test}}{C'_{ref}} \tag{21}$$

where $C' = (C_v/C_{injectate})$; test = test tracer; and ref = impermeant reference tracer. Cerebrovascular $PA$ is calculated from $E$ as (Crone, 1963):

$$PA = -F \ln (1 - E) \tag{22}$$

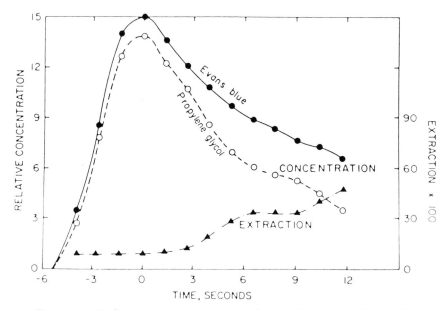

Fig. 8. Relative concentrations of propylene glycol and Evans blue-albumin in samples of sagittal sinus blood after the simultaneous injection of both solutes into the internal carotid artery of the dog. Relative concentration equals $C_v/C_{injectate}$. Zero time is defined as the time of peak concentration in blood. Extraction is given by Eq. (21) (modified from Crone, 1965).

where $F$ equals the flow of the bolus through the brain.

The advantages of the indicator-dilution technique are that several measurements of $E$ can be obtained on the same animal, and that the technique can be used on humans (Lassen et al., 1971; Hertz and Paulson, 1980) as well as large (Crone, 1963) and small animals (Hertz and Bolwig, 1976).

However, the indicator-dilution technique has several limitations. $E$ can be accurately measured only from 0.05 to 0.9, which limits the permeability range of the technique. For example, with $E = 0.05$, $F = 1 \times 10^{-2}$ mL/s/g, and $A = 240$ cm$^2$/g, the lower limit of $P$ for the technique is $25 \times 10^{-7}$ cm/s, which is 100 times the $P$ to sucrose (Table 1). Thus, the method provides permeability coefficients for poorly penetrating solutes, such as urea, mannitol, and sucrose that are either insignificant or spuriously high (Table 1).

An additional problem is that $E$ is not constant at different time points of the venous outflow curve (Fig. 8). $E$ may increase in the rapidly rising part of the outflow curve due to capillary heterogeneity (Hertz and Paulson, 1980; Bass and Robinson, 1982).

Then in the falling part of the outflow curve, $E$ may decrease with time because of back diffusion. Furthermore, $E$ may vary throughout the curve because of separation of test and reference tracers in the blood vessels as a result of intralaminar or Taylor diffusion and of erythrocyte carriage (Crone, 1965; Lassen et al., 1971; Hertz and Paulson, 1980). However, tracer separation can be reduced by using a reference tracer with similar diffusion and erythrocyte-penetration properties to those of the test tracer (Hertz and Paulson, 1980).

Since $E$ is not constant and therefore is ambiguous, Bass and Robinson (1982) have proposed a method to calculate a secure lower bound to cerebrovascular permeability directly from indicator dilution data, by using the following equation:

$$ PA = F\left[\frac{\int_0^T C'_{ref} \ln(C'_{ref}/C'_{test})dt}{\int_0^\infty C'_{ref} \, dt}\right] \tag{23} $$

where $T$ = time when $C'_{ref}$ drops below $C'_{test}$ and where the denominator equals the area under the reference tracer curve corrected, if necessary, for recirculation. Equation (23) gives a lower bound for $PA$ because no explicit correction is made for back diffusion.

Extracerebral contamination of sinus blood can be a significant problem when blood samples are obtained rapidly, and may require the use of a second reference tracer (Hertz and Bolwig, 1976). The rapid intracarotid injection also may transiently elevate cerebral blood flow, which should be known and constant during the measurement period. Furthermore, cerebral blood flow should not be used for $F$ in Eq. (22) or (23), because the flow of the bolus through the brain blood vessels may be greater than blood flow due to the low viscosity of the saline injectate (Takasato et al., 1984). Therefore, since $F$ and $E$ are not determined simultaneously, the method cannot be used to obtain accurate permeability coefficients of rapidly penetrating compounds that are more "flow" than "diffusion" limited. Last, the indicator-dilution technique measures an average cerebrovascular permeability for the brain and cannot be used to examine regional permeability.

## 4.2. Brain-Uptake Index

The brain-uptake index (BUI) technique was introduced by Oldendorf in 1970 as an intracarotid injection–single pass method to measure cerebrovascular transport and permeability. In contrast to the indicator-dilution technique, which measures $E$ from concentration of tracer in venous blood, the BUI method determines $E$ from the brain concentration of tracer after a single pass. Since 1970, the BUI technique has been widely used to study cerebrovascular transport due to the flexibility, speed, and simplicity of the method (Oldendorf, 1977; Pardridge and Oldendorf, 1977; Pardridge, 1983).

A 200-$\mu$L bolus of buffered saline containing a test tracer and a permeant reference tracer is injected rapidly (<1 s) into the carotid artery of an anesthetized rat (Fig. 9). In some studies, an impermeant vascular tracer, such as $^{113m}$In-EDTA, is included in the bolus to correct for residual intravascular test tracer in the brain (Oldendorf and Braun, 1976). After 5–15 s the rat is decapitated, and the brain is removed and analyzed for tracer contents.

Because only a small, variable fraction of injectate goes to the brain, the brain uptake of test tracer is normalized by the use of a permeant reference tracer of known uptake, such as $^{3}$H-H$_2$O, $^{14}$C-butanol, $^{3}$H-tryptamine, or labeled micropheres (Oldendorf, 1981; Clark et al., 1981). Thus, the BUI is defined as:

$$\text{BUI} = 100\left[ \frac{C''_{\text{test}}}{C''_{\text{ref}}} - \frac{C''_{\text{vas}}}{C''_{\text{ref}}} \right] \tag{24}$$

Where $C'' = (C_{\text{tot}}/C_{\text{injectate}})$, ref = permeant reference tracer, and vas = impermeant vascular tracer (Oldendorf, 1981).

The BUI can be converted into an $E$ if the brain extraction of reference tracer is known and if the washout of both test and reference tracers from brain is known for the 5–15-s period before decapitation. Assuming negligible backflux of test tracer, Oldendorf has related BUI to $E$ as:

$$E \simeq d \cdot \text{BUI}/100 \tag{25}$$

where $d$ is a constant that depends on blood flow, reference tracer, decapitation time, and brain region. For $^{3}$H-H$_2$O, $d = 0.84$ at 5 s and $d = 0.43$–0.75 at 15 s in the anesthetized rat (Pardridge and Oldendorf, 1975; Oldendorf and Braun, 1976; Gjedde and Rasmussen, 1980; Pardridge et al., 1982). Cerebrovascular $PA$ can be obtained from the estimated $E$ with Eq. 22 where $F$ equals the flow of the bolus through the brain.

The advantages of the BUI technique are that the experiments are fast and simple to perform, and that absolute control is allowed of injectate composition. Because injection fluid solute concentration can be manipulated easily, this method has been used frequently to examine saturable transport systems at the cerebrovascular endothelium (Pardridge and Oldendorf, 1977).

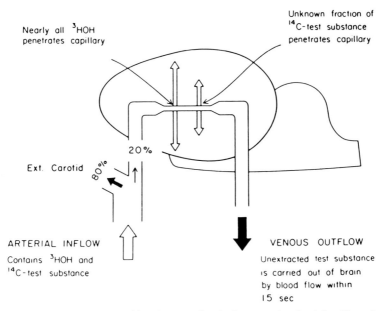

Fig. 9.   Diagram of brain-uptake index method. A buffered solution containing a $^{14}$C-labeled test solute and $^{3}$H$_2$O is injected into the common carotid artery of a rat and the animal is decapitated 15 s later (from Oldendorf, 1981).

The major limitation of the BUI technique is the difficulty in relating the measured BUI to cerebrovascular *PA*. The proportionality constant *d* that relates BUI to cerebrovascular *E*, depends on several variables (blood flow, net reference tracer extraction, test tracer backflux) that may change under different experimental conditions. In addition, the flow of the bolus, which has never been measured, must be known to calculate *PA* (Eq. 22).

Like the indicator-dilution method, cerebral blood flow should not be used for *F* in Eq. (22), because the intracarotid injection may transiently elevate flow (Hardebo and Nilsson, 1979). The fact that a significant fraction of injectate travels down the common carotid into the aorta (Oldendorf, 1981), indicates that the injection elevates carotid pressure above systolic blood pressure. The rise in carotid pressure would increase flow to the brain, which should be constant and known during the experimental procedure. In addition to the injection artifact, *F* should increase transiently because of the low viscosity of the saline injectate (Takasato et al., 1984).

There are additional limitations of the BUI technique. The single pass uptake limits the sensitivity of the method and prevents the method from obtaining accurate permeability coefficients for poorly penetrating compounds (Table 1). The assumption of negligible mixing of the carotid bolus with systemic blood has never been experimentally verified and is probably erroneous. Mixing with blood probably occurs at the injection site and at the Circle of Willis (Fenstermacher et al., 1981). Thus, the fourfold greater $K_m$ value for cerebrovascular leucine transport with the BUI method than with the brain-perfusion method is probably due to the addition of competing amino acids to the carotid bolus through significant mixing with blood (Smith et al., 1984; Takasato et al., 1985). Finally, the BUI technique measures transport in only one cerebral hemisphere, and uses anesthetized as opposed to conscious rats.

## 4.3. Single Injection-External Registration

The single injection-external registration technique of Raichle et al. (1974) is a single pass method that uses external detection to measure brain extraction of a radiotracer. A test solute labeled with a gamma- or positron-emitting radionuclide is injected into the carotid artery of a large animal, such as a monkey, and the time course of total brain radioactivity is measured for 30–60 s with an external NaI detector.

Total brain radioactivity, $C_{tot}$, increases rapidly after injection and reaches a peak within 1–3 s, when all the injected tracer is in either brain blood vessels or brain parenchyma. Thereafter, total brain radioactivity decreases with time. The decline can be resolved into two components; a rapid phase ($t_{1/2} \leq 1$ s) that reflects primarily washout of intravascular brain tracer, and a slow phase ($t_{1/2} \geq 10$ s) that reflects tracer efflux from parenchymal brain tissue into blood (Raichle et al., 1974; Raichle et al., 1976; Tewson et al., 1980).

Brain extraction is calculated as:

$$E = C_{br}(T)/C_{tot}(T) \qquad (26)$$

where $T$ = time of peak total brain radioactivity. $C_{br}(T)$ is obtained by graphically extrapolating the slow component of the total brain radioactivity curve back to time $T$ (Raichle et al., 1974). Cerebrovascular $PA$ is calculated from $E$ with Eq. (22).

One advantage of the single injection-external registration technique is that, unlike the indicator-dilution and BUI methods, a reference tracer is not required. In addition, several measurements of $E$ can be obtained on the same animal.

However, this method is limited to measuring $E$ for moderately to rapidly penetrating compounds ($E \geq 0.05$). The method measures only an average $E$ for whole brain and cannot be used to determine regional cerebrovascular $P$. External detection requires the use of large animals and expensive equipment. Finally, a cyclotron for radionuclide production may be necessary for isotopes with short halflives.

## 5. SUMMARY

The intravenous-administration and *in situ* brain-perfusion techniques are the most versatile and sensitive methods to measure transport into the brain. These techniques are 100 times more sensitive than the single pass methods and can accurately measure $P$ over a $10^4$-fold range from a minimum of $10^{-8}$ cm/s. Cerebral blood flow and brain uptake can be measured independently in order to obtain accurate permeability coefficients for rapidly penetrating compounds. The intravenous administration technique can be used in conscious or anesthetized animals to measure regional $PA$ with either quantitative autoradiography or tissue dissection. In addition, the method can be used on humans with positron emission tomography. The brain-perfusion technique allows total control of perfusate composition to examine saturable transport at the blood–brain barrier.

### Acknowledgments

The author thanks Drs. Stanley I. Rapoport, Peter J. Robinson, and Ananda Weerasuriya for their helpful comments.

### References

Bachelard H. S., Daniel P. M., Love E. R., and Pratt O. E. (1973) The transport of glucose into the brain of the rat in vivo. *Proc. Roy Soc. London B.* **183,** 71–82.

Bass L. and Robinson P. J. (1982) Capillary permeability of heterogeneous organs: A parsimonious interpretation of indicator diffusion data. *Clin. Exp. Pharmacol. Physiol.* **9,** 363–388.

Betz A. L. and Iannoti F. (1983) Simultaneous determination of regional cerebral blood flow and blood–brain glucose transport kinetics in the gerbil. *J. Cereb. Blood Flow Metab.* **3,** 193–199.

Blasberg R. G., Kobayashi T., Patlak C. S., Shinohara M., Miyoaka M., Rice J. M., and Shapiro W. R. (1981) Regional blood flow, capillary permeability, and glucose utilization in two brain tumor models: preliminary observations and pharmacokinetic implications. *Cancer Treat. Rep.* **65,** 3–12.

Blasberg R. G., Fenstermacher J. D., and Patlak C. S. (1983) Transport of α-aminoisobutyric acid across brain capillary and cellular membranes. *J. Cereb. Blood Flow Metabol.* **3**, 8–32.

Bradbury M. W. B. (1979) *The Concept of a Blood–Brain Barrier,* pp. 46–59. John Wiley, Chichester.

Bradbury M. W. B., Patlak C. S., and Oldendorf W. H. (1975) Analysis of brain uptake and loss of radiotracers after intracarotid injection. *Am. J. Physiol.* **229**, 1110–1115.

Brightman M. W. and Reese T. S. (1969) Junctions between intimately apposed cell membranes in the vertebrate brain. *J. Cell Biol.* **40**, 648–677.

Clark H. B., Hartman B. K., Raichle M. E., Preskorn S. H., and Larson K. B. (1981) Measurement of cerebral vascular extraction fractions in the rat using intracarotid injection techniques. *Brain Res.* **208**, 311–323.

Cornford E. M., Braun L. D., Oldendorf W. H., and Hill M. A. (1982) Comparison of lipid-mediated blood–brain barrier penetrability in neonates and adults. *Am. J. Physiol.* **243**, C161–C168.

Cremer J. E. and Seville M. P. (1983) Regional brain–blood flow, blood volume, and haematocrit values in the adult rat. *J. Cereb. Blood Flow Metab.* **3**, 254–256.

Crone C. (1963) The permeability of capillaries in various organs as determined by use of the 'indicator-diffusion' method. *Acta Physiol. Scand.* **58**, 292–305.

Crone C. (1965) The permeability of brain capillaries to nonelectrolytes. *Acta Physiol. Scand.* **64**, 407–417.

Daniel P. M., Pratt O. E., and Wilson P. A. (1977) The transport of L-leucine into the brain of the rat in vivo: saturable and nonsaturable components of influx. *Proc. Roy. Soc. Lond. B.* **196**, 333–346.

Davson H. and Welch K. (1971) The permeation of several materials into the fluids of the rabbit's brain. *J. Physiol.* (Lond.) **218**, 337–351.

Draper N. and Smith H. (1981) *Applied Regression Analysis,* 2nd Edition, pp. 458–517. John Wiley, New York.

Fenstermacher J. D., Blasberg R. G., and Patlak C. S. (1981) Methods for quantifying the transport of drugs across brain barrier systems. *Pharmacol. Ther.* **14**, 217–248.

Gilboe D. D. (1982) Perfusion of the Isolated Brain, in *Handbook of Neurochemistry,* vol. 2 (Lajtha A., ed.). pp. 301–330. Plenum Press, New York.

Gjedde A. (1980) Rapid steady-state analysis of blood–brain glucose transfer in rat. *Acta Physiol. Scand.* **108**, 331–339.

Gjedde A. (1981) High- and low-affinity transport of D-glucose from blood to brain. *J. Neurochem.* **36**, 1463–1471.

Gjedde A. (1983) Modulation of substrate transport to the brain. *Acta Neurol. Scand.* **67**, 3–25.

Gjedde A. and Rasmussen M. (1980) Blood–brain glucose transport in the conscious rat: comparison of the intravenous and intracarotid injection methods. *J. Neurochem.* **35**, 1375–1381.

Gjedde A., Hansen A. J., and Siemkowicz E. (1980) Rapid simultaneous determination of regional blood flow and blood–brain glucose transfer in brain of rat. *Acta Physiol. Scand.* **108**, 321–330.

Goldstein G. W., Betz A. L. and Bowman P. D. (1984) Use of isolated brain capillaries and cultured endothelial cells to study the blood–brain barrier. *Fed. Proc.* **43**, 191–195.

Hardebo J. E. and Nilsson B. (1979) Estimation of cerebral extraction of circulating compounds by the brain uptake index method: influence of circulation time, volume injection, and cerebral blood flow. *Acta Physiol. Scand.* **107**, 153–159.

Hasegawa H., Ushio Y., Hayakawa T., Yamada K. and Mogami H. (1983) Changes of the blood–brain barrier in experimental metastatic brain tumors. *J. Neurosurg.* **59**, 304–310.

Hawkins R. A., Mans A. M., and Biebuyck J. F. (1982) Amino acid supply to individual cerebral structures in awake and anesthetized rats. *Am. J. Physiol.* **242**, E1–E11.

Hertz M. M. and Bolwig T. G. (1976) Blood–brain barrier studies in the rat: an indicator dilution technique with tracer sodium as an internal standard for estimation of extracerebral contamination. *Brain Res.* **107**, 333–343.

Hertz M. M. and Paulson O. B. (1980) Heterogeneity of cerebral capillary flow in man and its consequences for estimation of blood–brain barrier permeability. *J. Clin. Invest.* **65**, 1145–1151.

Johanson C. E. and Woodbury D. M. (1978) Uptake of [$^{14}$C]urea by the in vivo choroid plexus–cerebrospinal fluid–brain system: identification of sites of molecular sieving. *J. Physiol.* (Lond.) **275**, 167–176.

Johnson J. A. and Wilson T. A. (1966) A model for capillary exchange. *Am. J. Physiol.* **210**, 1299–1303.

Kessler R. M., Goble J. C., Bird J. H., Girton M. E., Doppman J. L., Rapoport S. I., and Barranger J. A. (1984) Measurement of blood–brain barrier permeability with positron emission tomography and $^{68}$Ga-EDTA. *J. Cereb. Blood Flow Metab.* **4**, 323–328.

Knott G. D. (1979) M Lab—A mathematical modeling tool. *Comput. Programs Biomed.* **10**, 271–280.

Lassen N. A., Trap-Jensen J., Alexander S. C., Olesen J., and Paulson O. B. (1971) Blood–brain barrier studies in man using the double-indicator method. *Am. J. Physiol.* **220**, 1627–1633.

Levin V. A. and Patlak C. S. (1972) A compartmental analysis of $^{24}$Na kinetics in rat cerebrum, sciatic nerve, and cerebrospinal fluid. *J. Physiol.* **224**, 559–581.

Marcus M. L., Heistad D. D., Ehrhardt J. C., and Abboud F. M. (1976) Total and regional cerebral blood flow measurement with 7-, 10-, 15-, 25-, and 50 µm microspheres. *J. Appl. Physiol.* **40**, 501–507.

Ohno K., Pettigrew K.D., and Rapoport S. I. (1978) Lower limits of cerebrovascular permeability to nonelectrolytes in the conscious rat. *Am. J. Physiol.* **235**, H299–H307.

Ohno K., Pettigrew K. D., and Rapoport S. I. (1979) Local cerebral blood flow in the conscious rat as measured with [14]C-antipyrine, [14]C-iodoantipyrine, and [3]H-nicotine. *Stroke* **10**, 62–67.

Ohno K., Chiueh C. C., Burns E. M., Pettigrew K. D., and Rapoport S. I. (1980) Cerebrovascular integrity in protein-deprived rats. *Brain Res. Bull.* **5**, 251–255.

Oldendorf W. H. (1970) Measurement of brain uptake of radiolabeled substances using a tritiated water internal standard. *Brain Res.* **24**, 372–376.

Oldendorf W. H. (1977) The Blood–Brain Barrier, in *The Occular and Cerebrospinal Fluids* (Bito L. Z., Davson H., and Fenstermacher J. D., eds.). pp. 177–190. Academic Press, New York.

Oldendorf W. H. (1981) Clearance of Radiolabeled Substances by Brain After Arterial Injection Using a Diffusible Internal Standard, in *Research Methods in Neurochemistry* (Marks N. and Rodnight R., eds.). pp. 91–112. Plenum Press, New York.

Oldendorf W. H. and Braun L. D. (1976) [[3]H]Tryptamine and [[3]H]water as diffusible internal standards for measuring brain extraction of radiolabeled substances following carotid injection. *Brain Res.* **113**, 219–224.

Pardridge W. M. (1983) Brain metabolism: a perspective from the blood–brain barrier. *Physiol. Rev.* **63**, 1481–1535.

Pardridge W. M. and Oldendorf W. H. (1975) Kinetics of blood–brain barrier transport of hexoses. *Biochim. Biophys. Acta* **382**, 377–392.

Pardridge W. M. and Oldendorf W. H. (1977) Transport of metabolic substrates through the blood–brain barrier. *J. Neurochem.* **28**, 5–12.

Pardridge W. M., Crane P. D., Mietus L. J., and Oldendorf W. H. (1982) Kinetics of regional blood–brain barrier transport and brain phosphorylation of glucose and 2-deoxyglucose in the barbiturate-anesthetized rat. *J. Neurochem.* **38**, 560–568.

Patlak C. S. and Fenstermacher J. D. (1975) Measurements of dog blood–brain transfer constants by ventriculocisternal pefusion. *Am. J. Physiol.* **229**, 877–884.

Patlak C. S. and Pettigrew K. D. (1976) A method to obtain infusion schedules for prescribed blood concentration time courses. *J. Applied Physiol.* **40**, 458–463.

Patlak C. S., Blasberg R. G., and Fenstermacher J. D. (1983) Graphical evaluation of blood-to-brain transfer constants from multiple-time uptake data. *J. Cereb. Blood Flow Metab.* **3**, 1–7.

Raichle M. E. (1979) Quantitative in vivo autoradiography with positron emission tomography. *Brain Res. Rev.* **1**, 47–68.

Raichle M. E., Eichling J. O., and Grubb R. L. (1974) Brain permeability of water. *Arch. Neurol.* **30**, 319–321.

Raichle M. E., Eichling J. O., Straatmann M. G., Welch M. J., Larson K. B., and Ter-Pogossian M. M. (1976) Blood–brain barrier permeability of [11]C-labeled alcohols and [15]O-labeled water. *Am. J. Physiol.* **230**, 543–552.

Rapoport S. I. (1976) *Blood–Brain Barrier in Physiology and Medicine*, pp. 1-206, Raven Press, New York.

Rapoport S. I. (1983) Passage of Proteins From Blood to Cerebrospinal Fluid. Model for Transfer by Pores and Vesicles, in *Neurobiology of Cerebrospinal Fluid,* vol. 2 (Wood J. H., ed.). pp. 233–245. Plenum Press, New York.

Rapoport S. I., Ohno K., and Pettigrew K. D. (1979a) Drug entry into the brain. *Brain Res.* **172,** 354–359.

Rapoport S. I., Ohno K., and Pettigrew K. D. (1979b) Blood–brain barrier permeability in senescent rats. *J. Geront.* **34,** 162–169.

Rapoport S. I., Fredericks W. R., Ohno K., and Pettigrew K. D. (1980) Quantitative aspects of reversible osmotic opening of the blood–brain barrier. *Am. J. Physiol.* **238,** R421–R431.

Rapoport S. I., Fitzhugh R., Pettigrew K. D., Sundaram U., and Ohno K. (1982) Drug entry into and distribution within brain and cerebrospinal fluid: [$^{14}$C]urea pharmacokinetics. *Am. J. Physiol.* **242,** R339–R348.

Reese T. S. and Karnovsky M. J. (1967) Fine structural localization of a blood–brain barrier to exogenous peroxidase. *J. Cell Biol.* **34,** 207–217.

Riggs D. S. (1963) *The Mathematical Approach to Physiological Problems,* pp. 120–167. M. I. T. Press, Cambridge.

Sage J. I., Van Uitert R. L., and Duffy T. E. (1981) Simultaneous measurement of cerebral blood flow and unidirectional movement of substances across the blood–brain barrier: theory, method, and application to leucine. *J. Neurochem.* **36,** 1731–1738.

Sakurada O., Kennedy C., Jehle J., Brown J. D., Carbin G. L., and Sokoloff L. (1978) Measurement of local cerebral blood flow with iodo-[$^{14}$C]antipyrine. *Am. J. Physiol.* **234,** H59–H66.

Sarna G. S., Bradbury M. W. B., and Cavanagh J. (1977) Permeability of the blood–brain barrier after portocaval anastomosis in the rat. *Brain Res.* **138,** 550–554.

Schafer J. A., Gjedde A., and Plum F. (1976) Regional cerebral blood flow in rat using *n*-[$^{14}$C]butanol. *Neurology* **26,** 394.

Sisson W. B. and Oldendorf W. H. (1971) Brain distribution spaces of mannitol- $^{3}$H, inulin- $^{14}$C, and dextran-$^{14}$C in the rat. *Am. J. Physiol.* **221,** 214–217.

Smith, Q. R. and Rapoport S. I. (1984) Carrier-mediated transport of chloride across the blood–brain barrier. *J. Neurochem.* **42,** 754–763.

Smith Q. R., Johanson C. E., and Woodbury D. M. (1981a) Uptake of $^{36}$Cl and $^{22}$Na by the brain–cerebrospinal fluid system: comparison of the permeability of the blood–brain and blood–cerebrospinal fluid barriers. *J. Neurochem.* **37,** 117–124.

Smith Q. R., Woodbury D. M., and Johanson C. E. (1981b) Uptake of $^{36}$Cl and $^{22}$Na by the choroid plexus–cerebrospinal fluid system: evidence for active chloride transport by the choroidal epithelium. *J. Neurochem.* **37,** 107–116.

Smith Q. R., Woodbury D. M., and Johanson C. E. (1982) Kinetic analysis of [$^{36}$Cl]-, [$^{22}$Na]-, and [$^{3}$H]mannitol uptake into the in vivo choroid plexus–cerebrospinal fluid–brain system: ontogeny of the blood–brain and blood–CSF barriers. *Dev. Brain Res.* **3**, 181–198.

Smith Q. R., Tai C. -Y., and Rapoport S. I. (1983) Brain capillary permeability to inorganic ions. *Soc. Neurosci. Abstr.* **9**, 161

Smith Q. R., Takasato Y., and Rapoport S. I. (1984) Kinetic analysis of L-leucine transport across the blood–brain barrier. *Brain Res.* **311**, 167–170.

Sokoloff L., Reivich M., Kennedy C., Des Rosiers M. H., Patlak C. S., Pettigrew K. D., Sakurada O. and Shinohara M. (1977) The [$^{14}$C]deoxyglucose method of local cerebral glucose utilization: theory, procedure, and normal values in the conscious and anesthetized albino rat. *J. Neurochem.* **28**, 897–916.

Solomon A. K. (1960) Compartmental Methods of Kinetic Analysis, in *Mineral Metabolism,* vol. 1 (Comar C. L. and Bronner F., eds.). pp. 119–167. Academic Press, New York.

Spector R. and Eells J. (1984) Deoxynucleoside and vitamin transport into the central nervous system. *Fed. Proc.* **43**, 196–200.

Takasato Y., Smith Q. R., and Rapoport S. I. (1983) Transport kinetics of large neutral amino acids across the blood–brain barrier. *Soc. Neurosci. Abstr.* **9**, 889.

Takasato Y., Rapoport S. I., and Smith Q. R. (1984) An *in situ* brain perfusion technique to study cerebrovascular transport in the rat. *Am. J. Physiol.* **247**, H484–H493.

Takasato Y., Momma S., and Smith Q. R. (1985) Kinetic analysis of cerebrovascular isoleucine transport from saline and plasma. *J. Neurochem.* (in press).

Tewson T. J., Raichle M. E., and Welch M. J. (1980) Preliminary studies with [$^{18}$F]haloperidol: a radioligand for in vivo studies of the dopamine receptors. *Brain Res.* **192**, 291–295.

Welch K. (1975) The Principles of Physiology of the Cerebrospinal Fluid in Relation to Hydrocephalus Including Normal Pressure Hydrocephalus, in *Advances in Neurology* (Friedlander W. J., ed.). pp. 247–332. North-Holland, Amsterdam.

Welch K. and Sadler K. (1966) Permeability of the choroid plexus of the rabbit to several solutes. *Am. J. Physiol.* **210**, 652–660.

Woods H. F. and Youdim M. B. H. (1978) The isolated perfused rat brain preparation—a critical assessment. *Essays Neurochem. Neuropharmacol.* **3**, 49–69.

Wright E. M. (1978) Transport processes in the formation of the cerebrospinal fluid. *Rev. Physiol. Biochem. Pharmacol.* **83**, 1–34.

Yudilevich D. L. and DeRose N. (1971) Blood–brain transfer of glucose and other molecules measured by rapid indicator dilution. *Am. J. Physiol.* **220**, 841-846.

# Chapter 12

# Axonal Transport
# Methods and Applications

## Scott T. Brady

## 1. Introduction

The processes of axonal transport are in most respects identical to
intracellular transport in other metazoan cells. The shapes and
sizes of neurons do require that intracellular transport be
amplified to an unusual degree in both the amount of material
moved and the distance traveled, but the underlying molecular
mechanisms appear to be shared (Brady, 1984; Grafstein and
Forman, 1980; Heslop, 1975; Lubinska, 1975; Schliwa, 1984;
Weiss, 1982). When considered in the context of studies of the
nervous system, however, it will become apparent that experiments involving axonal transport can provide a unique window
into many aspects of neurobiology. Studies of axonal transport
and related phenomena can produce insights into the cellular and
molecular organization of neurons, the patterns of neuronal connectivity, and the dynamics of the nervous system. Sometimes
these insights will come directly from studying axonal transport
itself and, in other experiments, axonal transport processes are
used as a tool for labeling or experimentally manipulating a group
of neurons.

A number of different approaches, methods, and preparations have been developed to study axonal transport and to utilize
it for exploration of the nervous system. Some of these have been
described in recent reviews and papers (Brady and Lasek, 1982a;

Brady et al., 1985; Snyder and Smith, 1983; Weiss, 1982; Weiss and Gorio, 1982). Axonal transport has become an important tool in the armamentarium of contemporary neurobiology and new applications using the methods of axonal transport continue to be developed. Both qualitative and quantitative analyses are now routine, and manipulation of neuronal functions via axonal transport appears to be within reach. If one knows when and how to use methods based on axonal transport, investigations into the structure and function of the nervous system can be greatly facilitated. The purpose of this review is to provide some general guidelines for detection and measurement of axonal transport and to describe some of the many applications useful for neurobiologists.

## 2. An Overview

The utility of axonal transport as a method derives from fundamental properties of transport, so a brief survey of axonal transport characteristics is helpful. Understanding two key concepts aids in the comprehension of the axonal transport literature. The first of these is the word "transport" itself; in the strictest sense, transport applies only to the actual translocation of cellular materials. However, it is often difficult to delineate the boundaries between translocation, and the packaging of materials for transport or eventual delivery to a specific destination. Whenever possible, the term transport is used here in the strict sense of translocation, but much of the discussion necessarily concerns events associated with commitment of materials to transport and the utilization of those materials. A second key definition is that of "neuronal compartment," which must encompass any structure or region of the neuron that can be distinguished kinetically or metabolically. Well-designed axonal transport experiments are capable of resolving many different cellular and subcellular compartments in the nervous system (Baitinger et al., 1982, 1984; Brady and Lasek, 1982a; Lasek and Brady, 1982a; Lasek and Hoffman, 1976; Lorenz and Willard, 1978; Tytell et al., 1981). As will be seen, some of these compartments are difficult or impossible to identify by other means.

## 3. Six Rate Components and the Structural Hypothesis

When Weiss and Hiscoe (1948) first demonstrated existence of axonal transport in neurons, only the slow flow of material away from the cell body (orthograde or anterograde transport) was described. At the time, this was consistent with the perceived importance of "axonal flow" in the growth and maintenance of axonal volume. Electron microscopy was still in its infancy, so the complexity of subcellular structures inside the axon was not fully recognized. As the organization and the variety of cytological elements in axoplasm were described by electron microscopists (Peters et al., 1976), it was seen that these different structures were not handled equivalently. A more rapidly moving component seemed likely and even essential to explain the speed of neuronal responses to certain stimuli and the rates at which enzyme activities returned after local depletion with irreversible inhibitors (Koenig and Koelle, 1961). These observations could not be explained by bulk flow of axoplasm at a rate of 2 mm/d, which would require months for the response of a cell body to be transported to the distal portions of a long motor neuron (Lasek, 1980). Other properties of the nervous system suggested the possibility of a transport of material from the periphery back toward the neuronal perikaryon (retrograde transport).

During the middle and late 1960s, several investigators demonstrated a faster component by radioisotopic labeling (Grafstein, 1967; Lasek, 1966, 1967, 1968), by histochemistry (Dahlstrom and Haggendal, 1966; Dahlstrom, 1967), and by direct observation of moving organelles (Burdwood, 1965). The existence of a fast retrograde component was also established (Kristensson, 1970a,b; Lasek, 1967; Lubinska et al., 1963, 1964). Later, a combination of biochemical analyses and radioisotopic labeling resolved six discrete rate components of intracellular transport in the axon (Fig. 1) (Hoffman and Lasek, 1975; Lasek and Hoffman, 1976; Willard et al., 1974), each with a characteristic rate, direction, and polypeptide composition. Table 1 summarizes the properties of these six rate components with regard to rate, direction, composition, and structural correlate (*see* also Baitinger et al., 1982, 1984; Black and Lasek, 1980; Brady, 1984; Brady and Lasek, 1982b; Lasek and Brady, 1982b; Tytell et al., 1981).

The different compositions and biochemical properties of the various rate components led to a hypothesis that served to explain the differences between rate components. The *structural hypothesis of axonal transport* states that the relevant units for axonal transport are cytological structures, not the individual molecules that make up those structures (Black and Lasek, 1980; Lasek, 1980; Lasek and Brady, 1982b; Tytell et al., 1981). One important implication is that materials move in the axon only as an integral part of an axonal structure or in long-term association with some axonal structure. Groups of functionally related polypeptides, including proteins important for neuronal structure [the

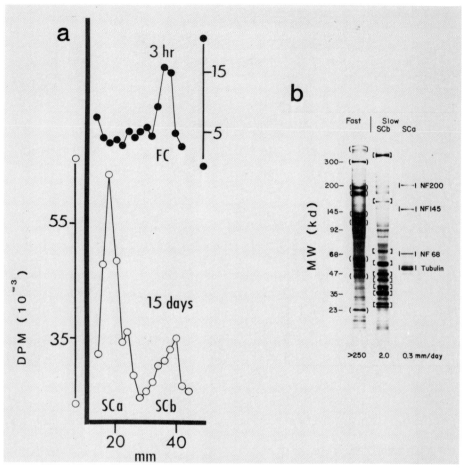

Fig. 1. The major anterograde rate components of axonal transport. When neuron cell bodies are pulse-labeled with radioactive amino acids and the distribution of radioactive polypeptides in the nerve examined at various times after labeling, several waves of radioactivity can be detected moving along the nerve. (a) Distribution of radioactivity in the hypoglossal nerve at 3 h and 15 d after injection of [³H]-amino acids into the medulla of a guinea pig. The only material labeled at 3 h in the nerve is moving as part of the fast component (FC) of axonal transport, whereas at 15 d two peaks of radioactivity can be seen corresponding to Slow Component b (SCb) and a larger peak corresponding to Slow Component a (SCa). In different nerves, the relative amount of radioactivity in these rate components may vary. For example, in the optic nerve of the guinea pig more radioactivity is associated with the SCb peak than the SCa peak for a given injection of precursor. In all nerves, the amount of labeled material associated with the two major slow components is much greater than the amount associated with the fast component. Therefore, the bulk of the protein synthesis in a neuron at any one time is devoted to elements of the cytoskeleton and cytoplasmic enzymes. (b) The polypeptide composition and approximate rates of movement of the three major rate components in the guinea pig optic nerve labeled with

TABLE 1
Rate Components of Axonal Transport[a,b]

| Rate component | Rate mm/day | Examples of protein composition | Cytological structure |
|---|---|---|---|
| *Fast* | 50–400 | Membrane-associated materials | Membranous organelles |
| Orthograde | 200–400 (1–3 μ/s) | Na, K-ATPase, transmitter-associated enzymes, and GAPs | 5-nm Tubulo-vesicular structures and dense-core vesicles |
| Mitochondria | 50–100 | F1 ATPase and a small amount of spectrin | Mitochondria |
| Retrograde | 200 | Lysosomal hydrolases, NGF, and other materials obtained by endocytosis | Prelysosomal structures (multivesicular and multilamellar bodies) |
| *Slow* | 0.2–8 | Cytoskeletal and associated proteins | Cytomatrix |
| SCb | 2–8 | Actin, clathrin, spectrin, NSE, CK, calmodulin, aldolase, and pyruvate kinase | Microfilaments and cytoplasmic matrix |
| SCa | 0.2–1 | Tubulin, neurofilament triplet, tau proteins, and spectrin | Microtubule-neuro-filament network |

[a]Studies on the kinetics of labeled materials and individual proteins in axonal transport permit assignment of specific rate components to specific cytological structures. Shown is a summary of rate, composition, and ultrastructural correlate for each of the rate components.

[b]For general reviews on the composition and organization of fast axonal transport, *see* Baitinger et al., 1982; Brady, 1984; Lasek and Brady, 1982a; Grafstein and Forman, 1980; Heslop, 1975; Wilson and Stone, 1979; Weiss, 1982. For slow axonal transport, *see* Brady and Lasek, 1982b; Grafstein and Forman, 1980. For selected references on transport of specific proteins: Na,K-ATPase (Sweadner, 1983); lysosomal hydrolases (Broadwell et al., 1980); actin (Black and Lasek, 1979; Willard et al., 1979); myosin-like proteins (Willard, 1977); spectrin (fodrin) (Levine and Willard, 1981); neurofilament proteins (Hoffman and Lasek, 1975); tubulin (Karlsson and Sjostrand, 1971; Tashiro and Komiya, 1983); microtubule-associated proteins (tau) (Tytell et al., 1984); clathrin (Garner and Lasek, 1981); growth-associated proteins (GAPs); (Skene and Willard, 1981); calmodulin (Brady et al., 1981; Erickson et al., 1980); nerve-specific enolase (NSE) and creatine kinase (CK) (Brady and Lasek, 1981).

Fig. 1 (*continued*) [³H]-amino acids injected into the vitreous of the eye, separated in SDS gel electrophoresis, and visualized by fluorography. Note that the polypeptide composition for each rate component is distinctive (*see* text and Table 1 for details on the composition) (from Tytell et al., 1981).

microtubule-associated proteins (MAPs) and tubulin] and for metabolism (glycolytic enzymes), can be shown to move coherently in conjunction with larger groups of polypeptides (*see* Figs. 1 and 2) (Brady and Lasek, 1981; Brady et al., 1981; Tytell et al., 1984). This association of material into discrete rate components may be important for integrating physiological functions of the cell, as well as for ease of transport (Brady and Lasek, 1981). Compartmentation thus is inherent in the process of axonal transport, because each rate component represents the sum of interactions between functionally related elements (Lasek and Brady, 1982a).

Now that ultrastructural correlates for the different rate components have been largely identified, determining the rate component for a protein or other compound is equivalent to identifying its axoplasmic compartment (Brady and Lasek, 1982a; Brady et al., 1981). The corresponding physiological context and associated proteins are established automatically and information is gained about the synthesis, metabolism, and location within the axon. Presence of a protein, lipid, or other material in one of the anterograde axonal transport rate components establishes the presence of that material within neurons, which may be difficult to demonstrate for many materials purified from whole brain. Comparisons between different populations of neurons can distinguish between components common to all neurons and those present in only a specific neuronal population (Oblinger et al., 1982; Lasek et al., 1984b). The structural hypothesis provides a conceptual foundation for axonal transport experiments, and the ability to use axonal transport in the identification and labeling of specific neuronal compartments make it a powerful probe of neurons and the nervous system.

## 4. Two Classes of Axonal Transport

Axonal rate components may be grouped into two classes: membranous organelles and elements of the cytoplasmic matrix. The membranous organelles move at rates 1–3 orders of magnitude faster than the cytoskeletal structures and associated proteins of the matrix (Fig. 1), so the traditional split between fast and slow axonal transport remains useful (Lasek and Brady, 1982a). All membrane-associated proteins, glycoproteins, and lipids identified in axonal transport are associated with fast transport (Baitinger et al., 1982; Grafstein and Forman, 1980; Lorenz and Willard, 1978; Tytell et al., 1981; Weiss, 1982). Similarly, cytoskeletal proteins and classically soluble proteins, such as glycolytic enzymes, move in slow transport (Baitinger et al., 1982; Black and Lasek, 1980; Brady and Lasek, 1982b; Lasek et al., 1984a,b; Lasek and Hoffman, 1976). One apparent exception is the small fraction of axonal spectrin (fodrin) that appears to move at a rate comparable to that of mitochondrial proteins (Levine and Willard, 1981). The unusual transport kinetics of spectrin appear to reflect brain spectrin interactions with membranous structures, perhaps comparable to those of erythrocyte spectrin with the plasma membrane (Branton et al., 1981). High-affinity interactions with a membrane would make spectrin a *de facto* membrane-associated protein, and presumably result in the transport of some spectrin with that membranous organelle at fast component rates.

Fast and slow transport can be distinguished even at the level of translation. The cell body can be divided into several compartments with distinct functions: the nucleus (which is the site of transcription); the translational cytoplasm (which includes free and membrane-associated polysomes and the Golgi and associated structures involved in translation), and the expressional cytoplasm (which contains the expressed products of transcription and translation, but contains no protein synthetic machinery) (Lasek and Brady, 1982b). All intrinsic fast transport polypeptides are synthesized on the rough endoplasmic reticulum and pass through the Golgi before commitment to axonal transport (Hammerschlag et al., 1982). A minor exception is presumed to be those few proteins synthesized by mitochondria. The proteins of slow transport are synthesized on the "free" polysomes of the neuron. Both slow and fast transport proteins are subsequently released into the expressional cytoplasmic regions of the neuron, which include the axon. Compartmentation of axonal proteins thus begins in the perikaryon, closely coupled to translation (Lasek and Brady, 1982b).

Many proteins are destined for specific regions of the neuron. For example, some MAPs appear to be preferentially directed to the axon, dendrites, or cell body, depending on the kind of microtubule-associated protein (Bloom et al., 1984; Caceres et al., 1984; Matus et al., 1981). The molecular mechanisms by which this segregation is accomplished have yet to be elucidated. Observations that posttranslational alterations of some proteins or structures are coincident with their transport into the axon, such as glycosylation of fast transport proteins (Hammerschlag, 1983; Hammerschlag et al., 1982) and differences in the subunit composition (Baitinger et al., 1984; Willard and Simon, 1983) or phosphorylation (Sternberger and Sternberger, 1983) of neurofila-

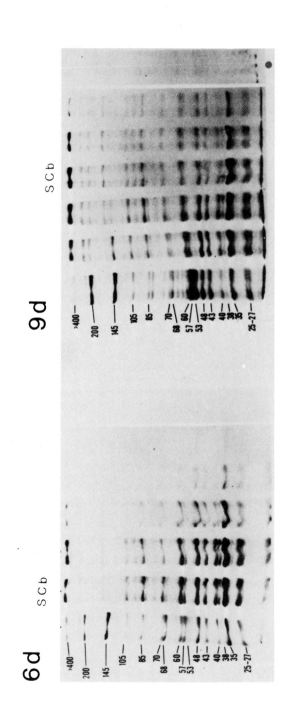

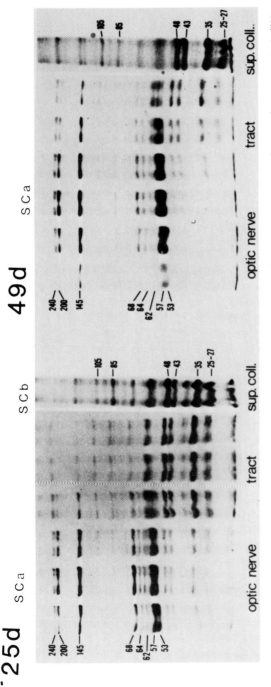

Fig. 2. The distribution of polypeptides associated with the slow components of axonal transport at different times after labeling. Guinea pig optic nerve was labeled by injection of [³H]-amino acids (lysine and proline). At the intervals indicated, nerves were dissected and cut into 3-mm segments. Each segment was processed for electrophoresis and run in sequential wells of a slab gel. Radiolabeled proteins were visualized by fluorography. The coherent movement of groups of polypeptides is well illustrated here and the approximate locations of the SCa and SCb peaks are indicated. Note that at 6 d, only SCb proteins are readily detectable in the nerve and tract. By 9 d, SCa proteins are visible, but only in the first segment of the optic nerve. At 25 d and even 49 d, most SCa proteins remain in the optic nerve, reflecting the much slower rate of transport for this group of proteins. Slow turnover of SCb proteins in the terminal regions (the column labeled 'superior colliculus') is also apparent, demonstrating the value of axonal transport for studies of protein turnover in neurons (from Brady and Lasek, 1982a).

ments, suggests that this could be an important factor in the sorting process, but there may be many factors involved. The large size and discrete, specialized regions of the neuron make it an excellent model system for studying those mechanisms that direct proteins to specific regions of the cell. Neurons are particularly valuable in that membrane- and cytoskeleton-associated proteins may be studied in the same cell by analyzing axonal transport processes.

## 5. Physiological Roles of Fast and Slow Axonal Transport

In much the same way that structures in the axon are assignable to specific rate components of axonal transport, specific physiological functions can be identified with different rate components. For example, many transmitter-related materials (enzymes, peptides, receptors, and so on) are associated with the fast anterograde rate (for examples, *see* Brimijoin et al., 1973, 1980; Brimijoin and Dyck, 1979; Brimijoin and Wiermaa, 1977a,b; Dahlstrom and Heiwall, 1975; Gainer et al., 1977a,b; Gamse et al., 1979; Goldberg et al., 1976; Gulya and Kasa, 1984; Kasa, 1968; Kuhar and Zarbin, 1984; Laduron, 1984; for additional references, *see* Grafstein and Forman, 1980; Weiss, 1982). Integral membrane proteins, such as the Na,K-ATPase, move in fast anterograde transport (Specht, 1983; Specht and Sweadner, 1984), as do acid hydrolases and lysosomal enzymes (Broadwell et al., 1980). Enzymes of oxidative phosphorylation and other mitochondrial-associated pathways are renewed at rates compatible with the movements of mitochondria (Lorenz and Willard, 1978; Partlow et al., 1972).

The cytoskeletal elements moving in slow axonal transport form the basis of neuronal morphologies (Lasek and Hoffman, 1976; Lasek, 1981; Lasek et al., 1983). Changes in axonal diameter that occur during axonal growth can be directly related to changes in the composition of Slow Component a (SCa; microtubule–neurofilament network) (Hoffman et al., 1984). The rate of regeneration correlates with the rate of Slow Component b (SCb; cytoplasmic matrix) (Cancalon, 1979; Wujek and Lasek, 1983) and regenerative potential appears to depend in part on the polypeptide composition in SCb (Lasek et al., 1981; McQuarrie, 1983). Other polypeptides in slow transport include enzymes of intermediary metabolism (Brady and Lasek, 1981) and regulatory proteins like calmodulin (Erickson et al., 1980; Brady et al., 1981). If a particular physiological function of the nervous system is of interest, then our knowledge of the roles played by different rate components can be used as a guide in the design of experiments.

Essentially all polypeptides in slow transport (Lasek and Black, 1977) and a few proteins in fast transport (Bisby, 1977; Tytell et al., 1980) appear to be degraded or released when they reach the synaptic terminal. In some instances, slow component proteins such as the neurofilament subunits may be subjected to partial proteolytic degradation in transit (Nixon et al., 1982). The remaining proteins in fast transport are eventually returned to the cell body via retrograde transport (Bisby, 1977; Bisby and Bulger, 1977; Kristensson, 1978; Kristensson and Olsson, 1971), although some of these may be modified prior to return. Materials from sources outside the neuron can be taken into the axon and also returned to the perikaryon via retrograde transport (Cowan and Cuenod, 1975; Kristensson, 1970b, 1978; LaVail, 1978; LaVail and LaVail, 1972). Those exogenous materials that move in retrograde transport include physiologically important effectors, such as nerve growth factor (Stoeckel et al., 1975a,b; Stoeckel and Thoenen, 1975), clinically important vectors, such as tetanus toxin (Carroll et al., 1979; Schwab and Thoenen, 1978; Stoeckel et al., 1977) or herpesvirus (Cook and Stevens, 1973; Kristensson, 1970a), and artificial labels such as horseradish peroxidase (Hansson, 1973; LaVail and LaVail, 1972, 1974) and lectins (Margolis et al., 1981; Ruda and Coulter, 1979; Schwab et al., 1978). The materials returned to the cell body in retrograde transport can have profound effects on protein synthesis and even the viability of the cell (Grafstein and McQuarrie, 1978; Kristensson and Olsson, 1974). As such, delivery of materials to the cell body by retrograde transport may be a highly specific means for manipulating neuronal populations.

In summary, it has been shown that membrane-associated activities in the axon are dependent on fast transport, whereas changes in the cytoskeleton and many metabolic activities can be related to slow transport (Grafstein and Forman, 1980; Lasek and Brady, 1982a). The properties and functions of each rate component in axonal transport are now sufficiently well understood to permit biologists to use the processes of transport to provide information about the dynamics, organization, and constituents of the axon.

# 6. Methodological Approaches

As methods for the study of cells and cellular components have multiplied, so have the applications of those methods to the study of axonal transport. Most techniques developed for contemporary cellular and molecular biology can be applied to neurobiology if an awareness of the unique spatial organization of neurons and the dynamics of neuronal components are taken into account. For example, neuronal transcription and translation are confined to the cell body, the nucleus, and the associated translational cytoplasm (Lasek and Brady, 1982b), but may be modulated by events that occur as much as a meter away in the synaptic terminals of that neuron. The packaging and return of "signals" to the cell body appear to be major functions of retrograde transport, and some of these signals produce dramatic alterations in transcription and/or translation (Grafstein and McQuarrie, 1978; Hall, 1982; Kristensson and Olsson, 1974). The timing, localization, and nature of this modulatory process are all dependent on retrograde transport.

Several general approaches have proven particularly useful in the design and analysis of axonal transport experiments. Before considering potential applications, some basic tenets for detecting and measuring axonal transport in vivo and in vitro should be surveyed. All studies of axonal transport require some method of detecting the moving elements in a background of other structures and materials moving at the same or different rates. Detection usually involves either an identifying physical and chemical property of the transported materials or specifically labeling the material of interest. Only when the means for detecting moving elements has been chosen does the characterization and quantitation of transport become possible. Therefore, methods for detection of transport will be considered before ways of analyzing the data obtained from axonal transport studies are discussed.

## 7. Direct Visualization of Transported Elements

One approach to the study of axonal transport is to watch individual elements moving in the neuron with the light microscope. In recent years, this has become a particularly valuable way to study the fast components, because our ability to detect the moving elements of fast transport has been greatly extended. The sensitivity and resolution of optical methods for direct measurement of transport have been increased through the use of video technology to enhance light microscopic images (Allen and Allen, 1983; Allen et al., 1981a,b; Inoue, 1981). Several types of light microscopy can be employed, including phase contrast, darkfield, and differential interference contrast (DIC, sometimes referred to as Nomarski) (Allen et al., 1969) light microscopy. Each of these optical techniques take advantage of the fact that membranous organelles differ in refractive index from the surrounding cytoplasm. The resulting light scatter or phase difference can be used to detect structures near or, under some circumstances, below the limit of resolution for the light microscope (Allen et al., 1981b; Miki-Noumura and Kamiya, 1979; Spencer, 1982). Prior to the availability of video methods, the small size of most structures moving in the axon limited analysis to the largest membrane-bound organelles (Breuer et al., 1975; Smith, 1980; Snyder and Smith, 1983). Despite this limitation, optical techniques were used for many years before video to study the movements of large organelles.

The ability to detect objects smaller than the resolution limit of the microscope was not fully appreciated prior to the extensive use of video technology in light microscopy. Both darkfield and DIC microscopy are capable of detecting very small phase objects. It has been shown that darkfield microscopy can detect objects as small as individual microtubules in dilute solutions (Miki-Noumura and Kamiya, 1979), but as an object of interest becomes smaller, less light impinges on it. The resulting image is correspondingly fainter. Diffraction inflates this image to the resolution limit, further reducing the intensity of the image and preventing resolution of two particles closer than the limit of resolution for those optics (i.e., the Rayleigh criterion for resolution). [*See* Allen et al. (1981a,b) or Spencer (1982) for a brief discussion.]

One way to overcome the faintness of very small objects is to increase the intensity of illumination, so much darkfield microscopy utilizes high-intensity light sources (Snyder and Smith, 1983; Spencer, 1982). Unfortunately, high-intensity illumination can have deleterious effects on biological preparations and application of darkfield methods to cells is further limited by the large number of light-scattering objects present in most cells and tissues. This light scatter increases the brightness of the background and interferes with the detection of fine detail. Similar problems result from the light scattered by out-of-focus structures in DIC microscopy. The use of video can largely solve the problems of both light scatter and low-intensity signals.

Allen and his colleagues (1981a,b) recognized that increased background light interferes with detection of fine detail because detection of an object is affected both by the difference in intensity between an image ($I_s$) and its background ($I_b$), and the ratio of that difference to the background intensity [an approximation is $(I_b - I_s)/I_b$]. As a result, detection of small structures can be enhanced either by increasing the intensity of the signal or by reducing the intensity of the background. In DIC, the background can be optically reduced by using the instrument near extinction (Allen et al., 1969; Inoue, 1981; Spencer, 1982), but this limits the magnitude of the signal as well. Allen et al. (1981b) utilized video technology to electronically reduce background intensity without affecting the absolute difference between the image and background intensities. Subsequent processing steps can be used to improve sensitivity still further and, through the use of digital enhancements of the video image (Allen and Allen, 1983), greatly increase the amount of information obtainable from light microscopic images.

Even with the greatly extended capabilities of light microscopic methods, choice of an appropriate preparation remains critical for studies of axonal transport. Most of the structures moving in fast axonal transport are less than 100 nm in diameter (Fahim et al., 1982; Smith, 1980; Tsukita and Ishikawa, 1980), well below the resolution limit, and exist in a highly refractile surround made up of cytoskeletal structures. This is not an ideal situation for obtaining maximum resolution. The greatest success has been achieved with neurites in culture (Breuer et al., 1975; Forman et al., 1977) or with isolated axons (Berlinrood et al., 1972; Kirkpatrick et al., 1972; Smith, 1972; Smith and Koles, 1976). Neurites in culture are generally quite small, with axonal diameters only slightly larger than those of the internal organelles in many cases, so refraction at the edges interferes with detection of internal details. In fact, electron micrographs of neurons in culture (Sasaki-Sherrington et al., 1984) show that neurites display frequent bulges corresponding to the position of internal organelles! Mature vertebrate axons may be much larger, but are surrounded by highly refractile myelin sheaths that interfere with image detail.

At present, the most useful preparations for light microscopic studies of fast axonal transport are large axons from invertebrates, such as those from the walking legs of crabs (Adams, 1982; Adams et al., 1982) or the giant axon of the squid (Allen et al., 1982; Brady et al., 1982, 1985). Each of these axons is exceptionally large (squid giant axons are often >500 μm) (Gilbert, 1974; Lasek, 1984), and invertebrate nerves do not contain myelin. These preparations can also be permeabilized to permit access to the interior of the axon either by using detergents (Forman, 1981; Forman et al., 1984) or high local electrical fields (Adams, 1982; Adams et al., 1982). For the squid axon, axoplasm without a plasma membrane can be readily obtained by extrusion of a cylinder of axoplasm (Bear et al., 1937; Gilbert, 1974; Lasek, 1974, 1984; Brady et al., 1982, 1985). A detailed treatment of the use of extruded axoplasm for studies on the mechanisms of fast axonal transport has recently appeared (Brady et al., 1985).

Fluorescence microscopy, particularly when supplemented by video cameras of very high sensitivity, can also be utilized to follow specific fluorescent structures in axons and other cell types (Inoué, 1981; Keith and Shelanski, 1982; Willingham and Pastan, 1978). Neuronal membranous organelles have been isolated, tagged with fluorescent probes, and shown to move when placed back into the axon (Gilbert and Sloboda, 1984; Schroer et al., 1985). Such approaches permit evaluation of the roles that cytoplasmic membrane surfaces play in the movement of organelles. The fluorescent laser dye Rhodamine 123 has been used as a vital stain for mitochondria in some cell types (Johnson et al., 1980) and other fluorescent tags may prove useful for studies of transport.

Direct visualization of fast transport in these model preparations will probably continue to be especially useful in studies on the mechanisms and microscopic properties of fast axonal transport. When combined with biochemical or pharmacological approaches, direct visualization provides the most sensitive and straightforward assay of membranous organelle transport. A recent review by Snyder and Smith (1983) discusses a variety of methods and techniques for direct visualization of fast axonal transport in some detail.

Models using optical methods for the study of slow axonal transport have received far less attention. The difficulties here are twofold. First, the structures are more difficult to visualize. The largest of the cytoskeletal elements are microtubules, only 25 nm in diameter, although they may be many microns long (Bray and Bunge, 1981; Tsukita and Ishikawa, 1981). Individual microtubules can be visualized under some conditions with darkfield (Miki-Noumura and Kamiya, 1979) or video-enhanced DIC (Allen et al., 1981b) microscopy, but the organization of axonal microtubules into bundles and the number of associated structures such as neurofilaments prevent visualization of individual microtubules in the living axons by these methods. Smaller elements of the cytoplasmic matrix, like microfilaments and neurofilaments, are even more difficult to detect. Second, the rates of movement for slow transport are 2–3 orders of magnitude less than fast transport rates (*see* Table 1). As a result, recording and analyzing the movements require longer observation periods and greater stability of the microscope.

Two instances do exist in which slow axonal transport has been successfully detected using optical methods and cultured neurons. Keith and Shelanski (1982) microinjected fluorescently labeled tubulin into a cultured neurite-bearing cell and found that fluorescence moved into the cell processes consistent with movement at the rate of slow transport. More extensive use of this approach will probably require development of fluorescent probes less subject to bleaching, and more sensitive detection devices.

A second approach to direct visualization of slow axonal transport with the light microscope is not often described as a model for study of slow transport. Extension of neurites in culture is easily detected even in phase contrast microscopy and represents the translocation of neuronal cytoskeletal elements (Harrison, 1910; Lasek, 1982), i.e., slow axonal transport. The end of the cytoskeleton as represented by the growth cone can be visualized at relatively low magnification; its movement may be followed for hours or days without difficulty (Yamada et al., 1971), and it can be physically or pharmacologically manipulated (Bray, 1984). This approach is particularly useful for studies on the role of slow axonal transport in neuronal growth and regeneration because it provides detailed information about the rate and extent of neurite growth. An interesting variant of this model, the isolated neurite, represents a potential in vitro model for studying the mechanisms of slow axonal transport. Under appropriate conditions, neurites that have been severed from their cell bodies will collapse and then reextend a growth cone (Shaw and Bray, 1977; Wessels et al., 1978). The use of isolated neurites separates the processes of cytoskeletal movement from those of protein synthesis and the continued supply of membranous organelles. Reorganization and translocation of the cytoskeleton can then be analyzed directly (George and Lasek, 1983).

## 8. Labeling of Materials in Axonal Transport

As seen in the preceding section, the fundamental physical properties of membranous organelles permit their detection and identification with the light microscope. No comparable physical property permits the study of cytoskeletal structures in transport, and many questions about fast and slow transported materials concern specific polypeptides rather than an entire structure.

Such questions can be answered through the use of specific markers and probes. A wide variety of labeling procedures has been successfully applied to the study of axonal transport. These range from techniques for general labeling of all proteins or lipids with radioactive precursors (metabolic labeling), to highly specific probes that may label only a select class of organelles in a particular population of neurons. Labeling techniques currently in use can be divided into three general categories: endogenous, metabolic, and extrinsic labels.

## 8.1. Endogenous Labels

Changes in the distribution of materials in a nerve can be analyzed to demonstrate movement by axonal transport (Grafstein and Forman, 1980; Lubinska, 1975; Dahlstrom and Heiwall, 1975). Such analyses may measure changes in the absolute amount of material in a segment of axon, but often only relative changes in the distribution of materials can be readily determined. Endogenous markers may be detected in several ways. Histochemical staining procedures may be used to demonstrate reactive materials or enzyme activity in axonal segments (Dahlstrom and Haggendal, 1966; Kasa, 1968; Kasa et al., 1973; Broadwell et al., 1980). When sufficiently sensitive assays of enzyme activity are available, enzyme levels can be measured in consecutive nerve segments (Brimijoin, 1979a; Brimijoin et al., 1973; Couraud and di Giamberardino, 1980; Lubinska et al., 1963; Lubinska, 1964). With this approach, however, it is generally necessary to estimate contributions to enzyme content of a nerve from both axonal and nonneuronal sources, such as glia and fibroblasts (Bisby, 1982; Lubinska, 1975; Snyder and Smith, 1983). Such estimations are far from straightforward, and determinations of "moving" and "stationary" fractions may be subject to a variety of artifactual distortions (Lubinska et al., 1963, 1964; Partlow et al., 1972; Snyder and Smith, 1983). Proteins and other axonal substances without enzymatic activity or histochemical assays can still be assayed. If antibodies suitable for radioimmunoassay or immunohistochemistry are available, then levels may be determined immunochemically (Brimijoin et al., 1980; Gamse et al., 1979).

Other biochemical properties can also be exploited in some cases. For example, catecholamines become fluorescent after fixation and fluorescence may be used to identify and monitor movement of storage granules (Dahlstrom and Haggendal, 1966). Receptor binding activity (Gulya and Kasa, 1984; Kuhar and Zarbin, 1984; Laduron, 1984) and bioassays (Younkin et al., 1978) have also been used to measure the accumulation of neuronal materials at a lesion. In general, any method that can be used to reliably quantitate materials of interest in tissue can be used to detect endogenous markers. However, quantitation of levels must be accompanied by some means for quantitation of movement.

Endogenous materials generally exist at a steady state in the axon, with as much entering each segment as leaving it (Bisby, 1982; Grafstein and Forman, 1980; Lubinska, 1975; Heslop, 1975), so levels are relatively uniform along the axon. Some perturbation of steady-state levels must be made before differential distributions are seen. The most common method for disturbing steady-state levels is a focal lesion, such as a transsection, crush, or ligation, in which accumulations of material can be measured. Another approach is to reduce the amount of some material in an axonal segment. For example, a local region of the axon can be treated with an irreversible inhibitor of an enzyme and the time-course for recovery of activity in that segment measured. Axonal transport of the different isotypes of acetylcholinesterase has been studied using both ligation (Couraud and DiGiamberardino, 1980) and inhibitor approaches (Koenig and Koelle, 1961).

The necessity of perturbing neuronal function and the difficulties in making accurate estimates of neuronal vs nonneuronal constituents based solely on endogenous markers limit the use of endogenous labels for studies of axonal transport. For example, measurements of rate and amount of material moving may differ significantly in the same nerve with different methods of focally blocking transport, such as crush, cut, and so on (Haggendal, 1980; Heiwall et al., 1979). As a result, determinations of rate by measurement of changes in the amount of an endogenous marker at a lesion are particularly difficult and usually unreliable. However, this labeling approach may be the simplest method for demonstrating that a particular substance is transported along the nerve, especially when an antibody or a suitable assay for activity is readily available. In addition, when the molecular identity of material of interest is not well defined (for example, a receptor of uncertain polypeptide composition), measurement of the differential distribution of an endogenous activity may be the only feasible method of analysis.

## 8.2. Metabolic Labels

The geometry and biochemical organization of the neuron are particularly suitable for making use of the neuronal synthetic machinery for labeling axonal transport (Brady and Lasek, 1982; Grafstein and Forman, 1980). Most neuronal anabolic activities are concentrated in the perikaryon of the neuron. This includes the synthesis of proteins and glycoproteins (Wilson and Stone, 1979; Elam, 1979), as well as many carbohydrates (Elam, 1979) and lipids (Currie et al., 1978; Elam, 1979; Gould et al., 1982; Haley et al., 1979; Toews et al., 1979). Introduction of radioisotopically labeled metabolic precursors into or near neuronal perikarya provides perhaps the most flexible and powerful labeling method for the study of axonal transport processes.

Metabolic labels are especially appropriate for studies of neuronal polypeptides, because amino acid precursors are readily available and protein synthesis is associated almost exclusively with the perikaryon (Lasek et al., 1974, 1977; Grafstein and Forman, 1980). Translational cytoplasm (Lasek and Brady, 1982b) includes those regions of the perikaryal cytoplasm containing polysomes (both free and bound), the Golgi apparatus, and associated structures of the endoplasmic reticulum and lysosomal system, all of which are confined to the perikaryon and proximal portions of the dendrites (Steward and Fass, 1983). Despite occasional reports of axonal protein synthesis (Koenig, 1979; Koenig and Adams, 1982), there is no reason to believe that local synthesis in the axon makes a significant contribution to the composition of the axonal regions. There is tRNA (Black and Lasek, 1977; Gunning et al., 1979; Ingoglia et al., 1973; Ingoglia and Tuliszewski, 1976; Lasek et al., 1973) in axons and even some evidence for rRNA or mRNA (Giuditta et al., 1980, 1983; Koenig, 1979) in axoplasm, but the remaining elements of the protein synthesis machinery have not been detected. As a result, the proteins in the axon must be synthesized in the cell body, sorted into appropriate compartments, and transported to the sites of utilization. The size of many neurons and the extent of their axonal regions in particular require that the different steps be separated by truly macroscopic distances (up to a meter or more in the motorneurons of a vertebrate the size of a human). Contributions from local incorporation are often minimal and, when necessary, corrections can be made based on a comparison with the uninjected contralateral side. These features of the nervous system have made the use of metabolic labels a widespread and powerful method for studying axonal transport. Metabolic labeling is particularly useful for measurements of rate and identification of transported proteins.

The choice of precursor for labeling a polypeptide depends both on the properties of the polypeptide of interest and on the metabolism of the precursor in that organism. Obviously, methionine would be a poor choice for labeling a polypeptide that contains little or no methionine, whereas a sugar precursor may be the most specific label for a glycoprotein. The most commonly used [³H]-amino acids are leucine, lysine, and proline (Elam and Agranoff, 1971; Heacock and Agranoff, 1977), primarily for historical reasons (early availability of high-specific-activity forms of these amino acids, early reports using these precursors, and so on). Many others have been used on occasion either singly or in various combinations for more uniform labeling of proteins. Although differences have been reported in efficiency of incorporation using different precursors (for example, *see* Contos and Berkley, 1984; Reparant et al., 1977), it is not always clear whether these variations are peculiar to particular preparations or represent a more fundamental difference in the suitability of various precursors. For most studies, the choice of tritiated amino acid precursor is probably not critical, because the variations in efficiency are usually modest. However, amino acids subject to significant metabolic conversion (such as glutamate) and amino acids present in high concentrations as free amino acid in nervous tissue should generally be avoided. Some proteins have unusual amino acid compositions, so a mix of two or more amino acid precursors may be the safest choice. In recent years, [³⁵S]-methionine has become the precursor of choice for most studies because of the gain in sensitivity obtained by using [³⁵S] as the label (Bonner and Laskey, 1974; Laskey and Mills, 1975). Studies on fast transport of polypeptides must also take into account the observation that some fast transport of free amino acids does occur (Csanyi et al., 1973), presumably inside vesicular structures.

The synthesis of nonprotein constituents of the axon is more variable in location. Addition of carbohydrate or lipid moieties to a polypeptide appears to be primarily associated with the translational cytoplasmic regions (Hammerschlag and Stone, 1982; Grafstein and Forman, 1980), although terminal additions and further processing may occur in the axon (Ambron and Treistman, 1977; Ambron and Schwartz, 1979). The literature concerning axonal transport of glycoproteins and glycosaminoglycans has been reviewed by Elam (1979), but comparatively little is known about axonal lipoproteins. Most of the proteins in fast axonal transport have carbohydrate moieties (Hammerschlag and Stone, 1982; Hammerschag and LaVoie, 1979; Hammerschlag et al., 1982; Wilson and Stone, 1979; Elam, 1979). Little or no glycoprotein or other complex carbohydrate appears to move in the slow components, consistent with evidence that glycoproteins and glycosaminoglycans are membrane-associated proteins.

A variety of precursors for carbohydrate moieties have been used to label glycoproteins in transport, including sulfate (Elam and Peterson, 1976; Elam et al., 1970; Karlsson and Linde, 1977), glucosamine, galactosamine, and mannosamine derivatives (Ambron and Treistman, 1977; Forman and Ledeen, 1972), and fucose (Karlsson and Sjostrand, 1971b; Zatz and Barondes, 1971a). Fucose is the most widely employed as a result of its specificity as a terminal sugar in the carbohydrate moiety of glycoproteins (Quarles and Brady, 1971) (for review, *see* Elam, 1979). In addition, fucosyltransferases appear to be largely confined to the cell body, so relatively little incorporation of fucose occurs in axonal or synaptic regions. Fucose is useful as a specific label for fast transport because it does not label materials in slow axonal transport and is not interconverted to other sugars or otherwise metabolized (Elam, 1979). Precursors for other carbohydrates on glycoproteins, glycosaminoglycans, and gangliosides may be less suitable for labeling materials in axonal transport, because glycosyl transferases for addition of some sugars have been reported in synaptosomes (Barondes, 1968; Zatz and Barondes, 1971b). For example, local incorporation of the amino sugars can be significant (Ambron and Schwartz, 1979).

In some nerves, a significant fraction of the transported glycoproteins may be deposited in the axon (Elam, 1979), presumably left in the plasmalemma or other stationary membrane system of the axon. The amount deposited in the axon ranges from 10 to 20% of the total glycoprotein in a pulse with a large myelinated fiber (Karlsson and Sjostrand, 1971b,c; Karlsson and Linde, 1977) to more than 80% of the total in garfish olfactory nerve (Elam and Peterson, 1976). As a result, carbohydrate precursors as labels of fast axonal transport are particularly useful for studies of the turnover of membrane glycoproteins in axons and terminals (Griffin et al., 1981).

Many lipids and glycolipids are unique to or highly enriched in nervous tissue (Margolis and Margolis, 1979). Some of these are synthesized primarily in the cell body, whereas others are synthesized locally in the axon (Gould et al., 1982, 1983a,b). A variety of precursors for phospholipids (Abe et al., 1973; Gould et al., 1982; Grafstein et al., 1975; Toews et al., 1979), gangliosides and other glycolipids (Forman and Ledeen, 1972), and cholesterol (Blaker et al., 1980) have been used to label lipid components in axonal transport. Unlike the situation with proteins, in which commitment to axonal transport appears to rapidly follow synthesis, the synthesis and commitment of lipid to axonal transport do not appear to be closely coupled. As a result, export of labeled lipid following a pulse of labeled precursor may be prolonged (Grafstein et al., 1975; Toews et al., 1979). All lipids, even lipids released late, move at one of the fast rates, presumably as part of a membrane structure. Inhibiting synthesis of either phospholipid or cholesterol (Longo and Hammerschlag, 1980) results in an inhibition of fast axonal transport for newly synthesized proteins, but does not affect the movement of materials already committed to transport. The characteristic lipid composition of different membrane systems does permit the use of certain lipid labels to follow the movement of specific membranous structures. For example, Morell et al. (1982) have examined the transport of the mitochondrial-specific lipid diphosphatidyl glycerol following application of [$^3$H]-glycerol.

The only other class of metabolic labels that has been successfully used in studies of axonal transport includes precursors of RNA, such as tritiated uridine (Ingoglia et al., 1973, 1975) or orotic acid (Bray and Austin, 1968). In some nerves, and perhaps in all nerves under appropriate circumstances, a significant amount of tRNA can be labeled in the axon with kinetics consistent with slow axonal transport (Gunning et al., 1979; Ingoglia, 1979). The amount of tRNA in axonal transport increases significantly in regenerating axons (Ingoglia, 1979; Ingoglia and Tuliszewski, 1976; Ingoglia et al., 1975) and may be a common characteristic of growing axons. The function of this tRNA is unknown since most studies indicate that no significant protein synthesis occurs in the axon (Lasek et al., 1973, 1974, 1977). It is not generally possible to detect messenger or ribosomal RNA species in the axon (Ingoglia, 1979; Gunning et al., 1979). A few studies have reported small amounts of either mRNA (Giuditta et al., 1983) or rRNA (Koenig, 1979; Giuditta et al., 1980) in certain axons, but these species do not seem to be present in all axons or are present at undetectable levels.

## 8.3. Extrinsic Labels

Contemporary neuroanatomy is largely a product of the development of methods for labeling specific populations of neurons with markers that are subsequently moved by axonal transport to other regions of those neurons. Development of methods for silver impregnation of degenerating axons and terminals (de Olmos et al., 1981) had enormous impact on the field in the 1950s and '60s because they provided reliable protocols for study of connections within the central nervous system. Powerful as these methods are, however, several important limitations inspired a search for other approaches to demonstrating connections in the nervous system. The most obvious limitation was the necessity for destroying the neurons of interest, so that many details of neuronal structure and relationships with other cells were lost or obscured. In addition, specificity was dependent on the selectivity of the method for generating a lesion. For example, in the central nervous system, many neighboring neurons are inevitably affected by physical methods of destruction of a nucleus or axon tract. These neighboring structures may be unrelated to the pathway of interest, leading to ambiguities in the identification of connections (Cowan and Cuenod, 1975; Cowen et al., 1972).

The first critical application of axonal transport to the study of neuronal connectivity was made by Lasek et al. (1968), who labeled fast anterograde transport with metabolic labels ([$^3$H]-amino acids) and identified projections using autoradiography. This idea was soon developed further for both light and electron microscopy by a number of investigators (Cowan and Cuenod, 1975; Cowan et al., 1972). Unfortunately, autoradiography also had limitations, such as the difficulty in attaining uniform labeling of a population of neurons and the dependence on labeling of cell bodies.

Studies on retrograde transport showed that exogenous materials, such as Evans blue-labeled serum albumen (Kristensson, 1970b) and horseradish peroxidase (HRP) (LaVail and LaVail, 1972, 1974; Nauta et al., 1974) can be taken into neurons at the terminals and transported back to the perikaryon. Since the initial demonstration, a large variety of extrinsic labels have been utilized to label both peripheral and central neurons for study at the light and electron microscopic levels (Cowan and Cuenod, 1975; Heimer and Robards, 1981; Jones and Hartman, 1978; Kristensson, 1978; LaVail, 1978). Either fast anterograde or retrograde transport can be labeled depending on where the label is applied (for example, *see* Nauta et al., 1974, 1975; Mesulam and Mufson, 1980; Schwab et al., 1978; Margolis et al., 1981; Margolis and LaVail, 1981), although one direction may be labeled more efficiently than the other by a given labeling agent. Differences in the efficiency of labeling may be especially marked for extrinsic markers that have high-affinity binding sites only at the cell body (preferentially labels anterograde) or the terminals (retrograde).

Extrinsic markers vary considerably in specificity, efficiency of labeling, and most effective method of detection. Specificity need not be very great, if high concentrations of reagent can be used (LaVail, 1978; Cowan and Cuenod, 1975). In both anterograde and retrograde labeling, materials can apparently be taken up in bulk through pinocytosis and committed to transport (Kristensson, 1978; Harper et al., 1980; Nauta et al., 1975; LaVail et al., 1980, 1983). However, it should be noted that not all materials that are taken into the terminal are committed equally well to transport, and there may be differences between different lots of the same reagent or different labeling conditions (LaVail, 1978; Bunt et al., 1976; Hadley and Trachtenberg, 1978). Efficiency of labeling may be closely related to specificity since high-affinity uptake of a marker will greatly increase the efficiency of labeling (LaVail, 1978; Schwab et al., 1978), and it is sometimes useful to prepare a conjugate between a marker that is readily visualized and a protein with high-affinity receptors for rapid internalization (Schwab, 1977; Gonatas et al., 1979). Examples of suitable conjugates range from molecules that are present only in select neuronal populations, such as nerve growth factor (Stoeckel et al., 1975a,b; Stoeckel and Thoenen, 1975) and antibodies to neural antigens (Fillenz et al., 1976; Wenthold et al., 1974), or more general plasma membrane markers on neurons, such as toxins (Stoeckel et al., 1977; Schwab and Thoenen, 1978; Carroll et al., 1979) and lectins (Ruda and Coulter, 1979; Schwab et al., 1978). Even markers as nonspecific as fluorescent latex microspheres (Katz et al., 1984) may successfully label neurons via axonal transport. The methods of detection that have been most widely used include autoradiography (Cowan et al., 1972; Cowan and Cuenod, 1975; Heimer and Robards, 1981; Lasek et al., 1968; Schwab et al., 1978; Margolis et al., 1981), enzymatic reactions that give histologically visible reaction products (Mesulam, 1982; Warr et al., 1981; LaVail, 1978), and fluorescence (Steward, 1981). With the introduction of markers such as colloidal gold (de Mey, 1983a,b) designed for electron microscopy and monoclonal antibodies that identify specific neurons (Zipser et al., 1983), an even wider variety of choices of labels is becoming available.

The most widely used extrinsic label is still horseradish peroxidase (Heimer and Robards, 1981; LaVail, 1978; Mesulam, 1982; Feher, 1984), which may be used either alone or as a conjugate with another protein to increase sensitivity or selectivity of uptake. One advantage of using HRP or HRP-conjugates is the large amount of experience with and understanding of the histochemistry of HRP and its reaction products (Mesulam, 1982; Warr et al., 1981). For histochemical studies, derivatives of benzidine are the preferred chromogens on the basis of sensitivity and ease of use. Many, if not all, of the derivatives of benzidine are potentially carcinogenic, and care should be taken in handling. For light microscopy, tetramethylbenzidine is preferred for sensitivity, simplicity of use, and its relatively weak carcinogenic potential (Mesulam, 1982; Mesulam and Brushart, 1979). Benzidine dihydrochloride is preferred for electron microscopy because of its smaller reaction product (personal communication, J. H. LaVail; Mesulam, 1982; Feher, 1984). Although diffusion of the reaction product limits resolution in electron microscopic use of HRP, the reliability of HRP and the potential to use it in combination with other methods as a double label (Steward, 1981) ensure its continued use. Detailed protocols for use of HRP as a marker for axonal transport have recently been published (Mesulam, 1982; Mesulam and Brushart, 1979; Warr et al., 1981; Steward, 1981).

## 8.4. Labeling Procedures

Endogenous labels are by definition already present in the neuron, requiring only an appropriate method for their detection, but metabolic and extrinsic labels must be introduced into the neuron. Application of label to the appropriate region of the nervous system for entry into the neuron and commitment to axonal transport can be accomplished by several means (Brady and Lasek, 1982a; Grafstein and Forman, 1980; Snyder and Smith, 1983; Schubert and Hollander, 1975). Most often, these procedures require injection of labeled precursor solutions or placement of label in a releasable form in the vicinity of the neuron to be labeled. If anterograde transport is to be labeled (either fast or slow components), label will generally be placed in the vicinity of the neuronal perikarya. Introduction of label into retrograde transport will usually involve the terminal endings or the distal portions of the axon. An exception is the acylating label Bolton-Hunter reagent, which may be applied in the middle of the nerve and label both directions (Fink and Gainer, 1979, 1980). However, Bolton-Hunter reagent does not label neuronal proteins uniformly and labels nonneuronal proteins of the nerve as well (Katz et al., 1982).

For some large neurons, intracellular injection of label is feasible (Ambron and Treistmen, 1977; Goldberg et al., 1976; Isenberg et al., 1980; Koike et al., 1972; Kreutzberg et al., 1973), but intracellular injections are not always practical and it is often desirable to label a population of neurons. Early studies with metabolic labels sometimes involved systemic administration of the marker (Droz and Leblond, 1962; Lubinska, 1964), but this approach was soon abandoned because of inefficiency and high background labeling. Labeling by injection into specific areas of the nervous system or target tissues is almost universally employed. Depending on the site of injection and the nature of the label, labeling with this approach can be highly specific for a class of neurons. Although it is not feasible to describe all possible variants of injection protocols, general guidelines can be provided and some details given for injections to label axonal transport using metabolic precursors in rodent retinal ganglion cells, perhaps the simplest of systems in common use. Unless otherwise stated, the comments on labeling in this section will refer to the use of metabolic labels, though many of these suggestions may be generalized for extrinsic labels.

A wide variety of nerve preparations has been used in studies of axonal transport (for reviews, *see* Brady and Lasek, 1982a; Grafstein and Forman, 1980; Heslop, 1975): vertebrate and invertebrate neurons; peripheral nerves and central nervous system tracts; heterogenous populations and individual identified nerve cells; sensory, motor, autonomic, and neurosecretory neurons have all been utilized. However, although all neurons move materials via axonal transport, some neurons may be more suitable than others for a given experiment. A thoughtful choice of preparation can greatly facilitate understanding of experimental results. The accessibility of the neurons to be labeled, whether cell bodies or terminal fields, is important. Preferably, the target should be readily located and identified, as well as being accessible with a minimum of surgical intervention. The size of the region to be labeled and the ease with which the label can be confined to the region of interest are important for increasing the signal-to-noise ratio. The geometry of the nerve system may affect the ability to make accurate measurements of rate. If a number of different neuronal types are present or several different pathways are possible, analysis may be unduly complicated. Clearly, it is useful to know as much as possible about the anatomy and physiology of the neuronal preparation to be used.

Differences in composition, relative amount of material in transport, velocity, and resolution of rate components have all been reported, revealing variations for homologous nerves in different species (Grafstein and Forman, 1980), for different nerve populations within the same animal (Black and Lasek, 1979a; McQuarrie et al., 1980; Oblinger et al., 1982), and even between two branches of the same set of neurons (Komiya and Kurokawa, 1978; Komiya, 1980; Oblinger and Lasek, 1984; Wujek and Lasek, 1983). It is imperative, therefore, to either use a well-characterized preparation in which the kinetics and other properties of axonal transport have been defined, or determine the relevant parameters according to a consistent set of rules (*see* the segmental analysis section of this paper for some suggested rules). Labeling intervals can then be determined that permit enrichment and, in some cases, isolation of each defined rate component in a given segment of the nerve or region of the neuron (Brady and Lasek, 1982a; Lasek et al., 1983). Identification of specific constituents in a rate component, turnover rates, and associations with other axonal elements is then possible. As an example, in the rat or guinea pig visual system, FC (Fast Component) is in the optic nerve at 2–4 h after labeling; SCb is isolated in the nerve at 4–6 d (the first 1–2 mm must be excluded to avoid contamination from SCa); and SCa is the only material labeled in the optic nerve at 40–50 d.

The volume to be injected should in most cases be as small as is practicable to confine labeling to the cell bodies of interest and reduce blood-borne labeling. This is particularly important when the injection is made into solid tissues, such as a ganglion or a nucleus in the central nervous system, since a large volume can lead to local disruption of tissue organization and labeling of cells other than those of interest. When the neuronal cell bodies are adjacent to a relatively open space, like the vitreous of the eye or a ventricle, somewhat larger volumes may be used. For axonal transport studies, an aqueous buffer should be used as a solvent, although a small percentage of dimethyl sulfoxide as a carrier may be used if the solubility properties of the marker require it. If the injection is a small volume into a reservoir of extracellular fluid, water can be used, but an isotonic physiological saline, suitably buffered, is preferable. Injections may be administered by pressure or iontophoresis of the label into the region of interest.

The rodent (rat, guinea pig, and so on) or lagamorph (rabbit) visual system is the easiest to use and best characterized of the model systems for study of axonal transport processes (Baitinger et al., 1982; Black and Lasek, 1980; Brady and Lasek, 1982a; Karlsson and Sjostrand, 1971c; Lasek and Hoffman, 1976; Willard et al., 1974), although a variety of other visual systems have also been studied extensively (for review, *see* Grafstein and Forman, 1980; Heslop, 1975). Injections are made into the vitreous of the eye following a light anesthesia with no surgical procedures and minimal trauma to the animal. Rodent visual systems have the advantage of well-characterized kinetics, and each of the rate components can be clearly delineated with relatively little ambiguity. In contrast, most other nerve systems that have been studied in these same animals exhibit more complex kinetics with different average rates for homologous rate components, overlap of some rate components, and large differences in the relative amounts of material (Lasek et al., 1983; McQuarrie et al., 1980; Oblinger et al., 1982). Even major rate components may be difficult to identify unambiguously without extensive analysis.

Detailed descriptions of the injection procedure have been published elsewhere (Black and Lasek, 1980; Brady and Lasek, 1982a; Garner and Lasek, 1982), but a brief description follows. In our laboratory, 250–500 µCi of [$^{35}$S]-methionine or a [$^{3}$H]-lysine/ proline mix are lyophilized and resuspended in 5–10 µL of physiological saline (New England Nuclear now recommends that a 50-m$M$ Tricine, pH 7.4, be used to increase stability of methionine, which is subject to oxidation; reducing agents may also be added at low levels) for injection into either the rat or guinea pig eye. A simple, inexpensive device for injection may be constructed from a 50- or 100-µL syringe, PE-20 polyethylene tubing, and a 30-gage needle (taken from a standard disposable, sterile hypodermic needle). To maintain control of volume and rate of flow, a repeating dispenser for the microsyringe capable of delivering 1- or 2-µL aliquots is used. The syringe is filled with distilled water to minimize dead space. One end of the tubing is fitted onto the syringe and the 30-gage needle inserted into the other (stretching the tubing to maintain a tight fit). The tubing is filled almost completely with water, then the desired volume (5–10 µL for a rat eye) of label is drawn into the tubing. The animal is anesthetized with ether or methoxyfluran and the eyelid pulled back to expose the sclera. Angling the needle toward the posterior of the eye in order to avoid the lens, 2–4 mm of the needle is inserted through the sclera into the vitreous (the needle can be seen through the lens if desired). Label is slowly injected, then the needle is withdrawn and the animal is allowed to recover from anesthesia. After an appropriate interval, the animal is sacrificed and the nerve is dissected for processing. Depending on the experimental design, either a segment containing the rate component of interest is taken (Black and Lasek, 1980; Brady and Lasek, 1982a; Willard et al., 1974) or a series of 1–2 mm segments are made of the nerve and analyzed sequentially for kinetic information (Black and Lasek, 1980; Brady and Lasek, 1982a; Garner and Lasek, 1982).

One alternative to injection into tissues that has also been successful is labeling in vitro (Edstrom and Hanson, 1973; Edstrom and Mattsson, 1972; Hanson and Bergqvist, 1982; Theiler and McClure, 1977), which has the disadvantage of being applicable to only a few preparations. Sensory neurons of the dorsal root ganglia from frog (Hammerschlag et al., 1977; Edstrom Mattsson, 1972) and rat (Theiler and McClure, 1977) have been particularly useful in pharmacological studies (Hanson and Edstrom, 1978; Brady et al., 1980). Multicompartment incubation chambers (Brady et al., 1980; Hammerschlag et al., 1975, 1977; Hanson and Bergqvist, 1982) permit isolation of the cell bodies and label from the nerve trunk containing the axons. The nerve itself can pass through multiple chambers, one of which will contain medium and another of which will contain the pharmacological agent being tested. Such an arrangement has several advantages in that it includes an internal control using nerve segments exposed to control media and permits examination of effects of pharmacological agents without the complications of systemic effects from administration to the whole animal. High levels of labeling can be obtained and the effects of a drug on synthesis or commitment to transport can be distinguished from effects on translocation (Hammerschlag et al., 1977; Hammerschlag and Lavoie, 1979). A variant of this approach involves labeling in vivo, then dissecting and incubating the nerve in vitro (Ochs, 1972a,b; Ochs and Ranish, 1969). Unfortunately, only fast transport can be evaluated in vitro and many nerve preparations are unsuitable for in vitro incubations because they cannot be readily dissected and maintained as an intact system.

Primary cultures of neurons have been used on a few occasions for in vitro studies of axonal transport. Earlier studies were somewhat complicated by the lack of clear distinction between dendritic and axonal processes (Estridge and Bunge, 1978). Labeling can also be a problem if nonneuronal cells are present in the culture dish as well. Improvements in primary culture of neurons now permit unambiguous identification of axons and dendrites in the absence of nonneuronal cells (Bartlett and Banker, 1984). Use of specialized incubation chambers, such as those developed by Campenot (1977), have been successfully used to study retrograde transport of nerve growth factor and may prove useful in other studies.

## 9. Accumulation and Segmental Analysis

There are two general approaches for analyzing axonal transport movements of labeled materials: (1) measuring accumulation of label at either a normal destination or a focal block of transport (accumulation), and (2) determining the distribution of label along the axons at various times after labeling (segmental analysis, Fig. 4). Each approach has advantages and disadvantages that are discussed below.

### 9.1. Accumulation

Accumulation may be utilized with any type of label. Measuring accumulation at a block of transport is in effect the only approach suitable for evaluating movements of endogenous labels, because the steady-state flux of endogenous materials present in the intact axon must be disturbed to detect movement. For most studies of neuronal connectivity using extrinsic labels, transport is followed by measuring accumulation in the cell bodies (after labeling of retrograde transport) or in the terminal regions (anterograde transport). A substantial literature exists on the accumulation of metabolically labeled materials at ligatures or other blocks of transport (for reviews, *see* Bisby, 1982; Lubinska, 1975; Snyder and Smith, 1983). This approach has been used somewhat less frequently in recent years, however, because high specific activity precursors and more sensitive methods of detection have become generally available. As a result, more information can generally be obtained by using metabolic labels with segmental anlaysis, but accumulation remains useful in some types of studies.

Changes in the distribution of enzyme or receptor binding activities are readily measured at a focal block of axonal transport. A simple description of the increase over time in label immediately adjacent to a block (on the proximal side for anterograde or distal side for retrograde; *see* Fig. 3) often suffices to demonstrate axonal transport of that activity. Transport may be completely blocked by ligating, crushing, or cutting of the axons (Bisby, 1982; Heiwall et al., 1979). Alternatively, a more or less reversible block can be achieved with pharmacological treatments (Hanson and Edstrom, 1978) and by focal cooling (Brimijoin, 1975). Local application of pressure to the axons will also inhibit transport (Hahnenberger, 1978). The apparent simplicity of measuring accumulation at a block makes it attractive for many investigators, although this simplicity can be deceptive.

In theory, rates can be estimated by measuring accumulations over time and the steady-state amount of label in a unit segment of undisturbed nerve, then calculating the number of unit segments that must be cleared to produce that accumulation. Unfortunately, several complications affect rate estimates by this method. Many endogenous labels are present in noneuronal cells as well as in neurons. The local glia, fibroblasts, and other cell types may react to the injury by locally increasing the synthesis of the marker. Some materials within the axon may be degraded or synthesized at accumulation sites (Bisby, 1982; Dahlstrom and Heiwall, 1975; Fonnum et al., 1973). Materials accumulating on the proximal side may also be turned around and transported away from the accumulation site via retrograde transport (Bisby and Bulger, 1977; Bulger and Bisby, 1978). Finally, not all of the endogenous label along the axon may be moving in axonal transport. A number of studies on the movement of enzyme activities have obtained extremely low estimates of rate until corrections were made for the "mobile" fraction (Fonnum et al., 1973; Lubinska et al., 1964). This last issue is far less of a problem with extrinsic or metabolic labels, although some labeled material may also be deposited in the axon (Cancalon and Beidler, 1975, 1977; Elam and Peterson, 1976; Karlsson and Linde, 1977; Karlsson and Sojostrand, 1971b).

Some of these difficulties can be overcome through the use of a reversible block, the "stop–flow" cold block methods described in detail by Brimijoin (1975, 1979a), or variants employed by others (Fahim et al., 1982; Hanson, 1978; Shield et al., 1977; Tsukita and Ishikawa, 1980). In this method, material is allowed to accumulate for an interval and then released by rewarming the nerve segment, generating a moving wave of material (*see* Fig. 3). Movements of this bolus of marker can be measured directly after rewarming (Brimijoin, 1975, 1979a; Brimijoin and Wiermaa, 1978). However, long-term accumulations cease to be reversible or may only partially recover (Bisby, 1982; Snyder and Smith, 1983) and there are suggestions that transport velocity may be affected by the concentration of transported material (Brimijoin, 1979a).

If only membrane-associated proteins moving in fast anterograde or retrograde transport are of interest, then detailed kinetics are not required. For example, labeling terminals or cell bodies with transport obviously does not require precise knowledge of rates or compositions. A simple series, choosing time-points based on rates in comparable systems, will be sufficient to determine an interval that achieves suitable levels of label at the destinaton. Questions that require knowledge of specific neuronal compartments (i.e., rate components) cannot be adequately answered by measuring total radioactivity at different time-points, nor by considering only the first appearance of labeled material in a given nerve segment.

Accumulations in cell bodies or terminals are most frequently measured in studies of neuronal connections, using either metabolic or extrinsic labels (Cowan and Cuenod, 1975; LaVail, 1978; Steward, 1981). In such studies, quantitation and rates of transport are of little interest as long as transported markers are readily detectable. One exception is when the disposition of labeled material after reaching its destination is of concern. For example, axonal transport can be used to evaluate turnover of proteins in the axons and terminals (Lasek and Black, 1977; Griffin et al., 1981). If a marker moved in retrograde transport has a physiological function of interest, such as nerve growth factor (Campenot, 1977; Stoeckel and Thoenen, 1975), then it may also be useful to

# ACCUMULATION

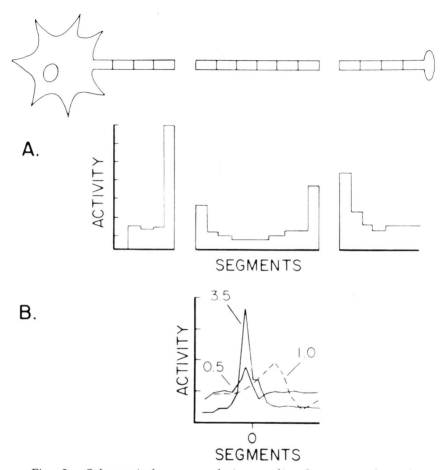

A.

B.

Fig. 3.   Schematic for accumulation studies. In most such studies, a portion of the nerve is isolated from the cell bodies and the terminals in some way (for example, transsection or ligation). Sometimes, transported materials are first labeled with radioactive precursors or by some ligand or marker that remains associated with a rate component in transport (for example, HRP in membranous vesicles). After an incubation to permit accumulation, thus increasing activity above steady-state levels, measurements are made in a series of short segments of the nerve. Activity is expected to be highest at the ends of the piece of nerve (as in A), adjacent to a ligation, or in a region of the nerve that has been locally cooled (as in B) for materials in axonal transport. Depending on the design of the experiment, both anterograde and retrograde transport may be analyzed in the same experiment (as in A). One advantage of focal cooling as a block of transport is the reversibility of such blocks. In B, the nerve segment indicated by O was cooled. Accumulations expected

know the amount of marker delivered by transport. In such cases, many of the considerations for quantitation of accumulation at a block remain relevant. A further complication is that trans-neuronal (Reperant et al., 1977; Specht and Grafstein, 1977) and transsynaptic (Schwab et al., 1979) transfer of labels has been reported, which may result in lower values in quantitative studies.

## 9.2. Segmental Analysis

Labeling of axonal transport with metabolic or extrinsic labels is essentially a pulse/chase procedure, utilizing a relatively brief labeling period, followed by an appropriate interval for the neuron to incorporate, sort, and transport the marker (Fig. 4). The result is a wave of label moving with a rate and direction characteristic for that rate component. The width of the wave is primarily a function of the labeling period, geometry of the labeled region, metabolism of the labeled neurons, and differences in rate transport within the population of labeled neurons. The magnitude of the wave is dependent on the specific activity of the label, efficiency of uptake and incorporation, size of endogenous metabolic pools (for metabolic labels), and amount of label. Despite the large number of variables involved in the generation of a wave, the same population of neurons labeled in comparable animals will give remarkably similar results. Distribution of label at a given time after labeling is highly reproducible. Changes in the distribution of label along the axons can be evaluated by measuring the amount in individual segments of one nerve and comparing it with distribution of label in another nerve after a different chase period. The simplest implementation of segmental analysis requires only that a series of nerves be labeled and examined at different times after labeling. Each nerve is divided into a series of suitably sized sequential segments and the total amount of label in each segment determined (Fig. 4). When it is not feasible to analyze a series of segments, a single defined segment may be analyzed as a "window" for following transport. Use of a single window necessitates the use of more time-points, and re-

Fig. 3 (*continued*) after 0.5 and 3.5 h are indicated, as well as the distribution 1.0 h after release of the block (dashed line). After release, a peak of activity moves along the nerve. However, prolonged cold-blocks cease to be fully reversible. Except for those studies using release after local cooling, accumulation of analyses are not well suited for measurements of rate, being subject to a variety of artifactual distortions that interfere with accurate estimations of rate.

## SEGMENTAL ANALYSIS

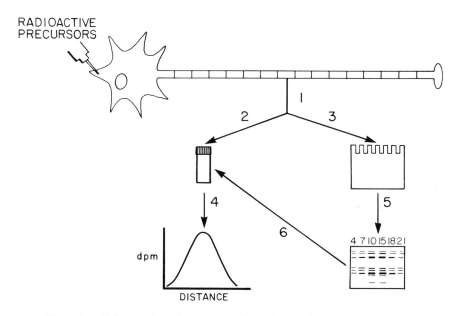

Fig. 4.   Schematic of segmental analysis. In this approach, neuronal proteins are pulse labeled, usually with radioactive precursors.
After an appropriate injection–sacrifice interval for labeling the rate component of interest, the nerve is dissected and cut into a series of segments. For a typical study in our laboratory, each segment is then processed as follows. (1) Each segment is homogenized in an electrophoresis
sample buffer. (2) An aliquot is taken and counted in a liquid scitillation
counter, while (3) another aliquot is loaded into the appropriate well of a
slab gel for electrophoretic analysis. (4) Each segment is handled identically for comparative purposes. The total amount of radioactivity in each
segment may be plotted to give overall distributions of axonally transported proteins at that time-point. (5) After electrophoresis, the gels are
processed for fluorography to permit visualization of the radiolabeled
polypeptides. (6) Such fluorographs may be used directly (as in Fig. 2),
or individual bands may be sliced from the gels, solubilized in 30% hydrogen peroxide, and counted in a liquid scintillation counter. The distribution for polypeptides of a specific molecular weight can then be
plotted. This approach is the most rigorous way of defining rates for a
specific polypeptide, since information can be gained about movements
of the front, the peak, and the trailing edge. This method may also be
combined with other kinds of analysis. For example, if an appropriate
antibody is available, the homogenate from each segment can be
immunoprecipitated and the distribution of only the immunoprecipitated label examined. For further discussion, see text.

sults in loss of resolution in measuring the kinetics of axonal transport. A more powerful approach to segmental analysis includes analysis of each sequential segment by electrophoresis (*see* Figs. 2 and 4) (Black and Lasek, 1980; Brady and Lasek, 1982a; Garner and Lasek, 1982; Hoffman and Lasek, 1975), including both one-dimensional (Laemmli, 1970) and two-dimensional (O'Farrell, 1975; Stone et al., 1978) gel electrophoresis. Combination of electrophoretic and segmental analysis is the only reliable approach for resolution of the slow components of axonal transport in many nerves, because it permits analysis of kinetics at the level of identified polypeptides characteristic for a given rate component (Brady and Lasek, 1982a; Garner and Lasek, 1982). This is particularly valuable for those systems in which the individual rate components of slow transport overlap (Lasek et al., 1983; McQuarrie et al., 1980; Oblinger et al., 1982). Care should be taken in the analysis of proteins labeled metabolically to eliminate contributions from labeled free amino acids, particularly for fast transport. This is not a problem if transported material is analyzed electrophoretically (Stone et al., 1978), but should be considered in cases in which measures are made of total radioactivity in a nerve segment. Precipitation of proteins with 10% trichloroacetic acid or some comparable procedure before measuring total counts will eliminate contributions from free amino acids.

Rigorous determination of the rate at which a given polypeptide or other material moves in axonal transport requires knowledge of the distribution of both the material of interest and appropriate marker proteins in the same nerve. The contemporary definition of a rate component is based on the original observation by Willard et al. (1974) that each of the rate components had a characteristic protein composition. Subsequent studies (Black and Lasek, 1979a; Brady and Lasek, 1981; Brady et al., 1981; Garner and Lasek, 1981; Hoffman and Lasek, 1975; Levine and Willard, 1981; Lorenz and Willard, 1978; Willard, 1977; Willard et al., 1979; Tytell et al., 1984 among others) have demonstrated the unique polypeptide composition of each rate component, i.e., the structural hypothesis (Lasek and Brady, 1982b; Lasek, 1980; Tytell et al., 1981). Proteins have now been identified in each of the rate components and the kinetics of these proteins described in detail for several nerve preparations (Baitinger et al., 1982, 1984; Lasek and Brady, 1982b; Black and Lasek, 1980; McQuarrie et al., 1980; Oblinger, 1984; Oblinger et al., 1982). Assignment of a newly characterized protein to a rate component should either be with reference to these previously identified polypeptides or based on comparably detailed analyses. Kinetic analysis for the proteins of even a single rate component may become quite complex (Garner and Lasek, 1982; Lasek et al., 1983), so the easiest approach is with respect to the marker proteins, whenever possible.

Rate comparisons between studies should be made only when rates are determined by the same method in the same nerve. Many disputes in the literature about the rate at which some material is transported are actually disputes about how to measure rates. A very different result will be obtained if the rate is determined from movements of the front of a wave rather than the peak (Gross and Beidler, 1975; Lasek et al., 1983; Cancalon and Beidler, 1977). A small fraction of the material associated with SCa or SCb may indeed be detected moving at rates two or three times as fast as the main peak. When a complete kinetic analysis is made, however, it often becomes apparent that the faster-moving material is very much part of the same coherently moving wave, rather than a separate wave (Black and Lasek, 1980; Garner and Lasek, 1982; Lasek et al., 1983). This faster-moving material can be viewed as the tail of a gaussian distribution and contributes to widening of the wave, but does not appear to separate into a discrete wave. For some proteins and some preparations, such as fodrin (Levine and Willard, 1981), a discrete wave can be detected. In these cases, each wave has a characteristic spectrum of coherently moving proteins (Garner and Lasek, 1982; Lasek et al., 1983). The situation may be extremely complex for the slow components, for which there is considerable variation in the kinetics of slow transport for different nerves (Oblinger et al., 1982). As detection methods improve in sensitivity and resolution, greater care must be taken in defining the kinetics of movement for elements in axonal transport.

Segmental analysis holds the potential for providing the most information about materials in axonal transport, but is clearly a time-consuming process. For many kinds of experiments in which axonal transport is a tool rather than an end in itself, approaches based on accumulation are quite sufficient. These include studies of neuronal connectivity and studies in which the primary goal is only to label a neuronal structure. For neuronal preparations in which the kinetics of axonal transport are already well defined, the variant of segmental analysis based on a single-window segment may suffice even for identification of the rate component for a polypeptide.

# 10. Uses and Applications of Axonal Transport

The neuroanatomist is already well aware of the power of axonal transport as a tool for probing the nervous system. Before the introduction of axonal transport methods, neuronal pathways had to be laboriously traced using degeneration and silver staining methods (de Olmos et al., 1981). These methods continue to be invaluable to the neuroanatomist, but suffer from the requirement that the cells of interest must be destroyed in order to be studied. A principal advantage of the axonal transport method for labeling neurons is that the neuron remains physiologically intact and essentially undisturbed. Ultrastructural details are retained, as are relationships with other neurons and cells. The availability of fluorescent labels for retrograde transport (Kristensson, 1970b; Katz et al., 1984) permits location of the cell bodies in living preparations so that electrophysiological studies may be performed on the same cells that have been labeled by axonal transport. Labeling with axonal transport can also be quantitative, thereby providing some measure of the number of projections to a given region of the nervous system. Alternatively, neurons fluorescently labeled with retrograde transport can be collected with a fluorescence activated cell sorter and then either cultured or subjected to biochemical analyses. Finally, as the molecular biology and immunochemistry of the nervous system continue to develop (Zipser et al., 1983), even more specific labels of neuronal populations are likely be available—labels that will recognize specific receptors or markers. Eventually, it may be routine to define specific biochemical and physiological properties of a neuron, such as the neurotransmitter(s) used, simply by using several probes of differing specificities and determining which ones will label that neuron in retrograde transport (Cuenod et al., 1982; Fillenz et al., 1976; Streit, 1980). As a result, labeling of neurons by axonal transport becomes a valuable complement to other methods for studying neurons.

Axonal transport represents a straightforward method for establishing whether a particular material is present in neurons or in some other cellular component of nervous tissue. The complex cellular composition of the nervous system makes it difficult to determine in many cases whether a substance obtained from nervous tissue originated in whole or in part from neurons. The structural interrelationships between glia and neurons complicate the issue still further. Demonstration that some material is moved in fast anterograde or slow axonal transport establishes the presence of that material in neurons unequivocally. Taken a step further, comparisons between several different populations of neurons, such as retinal ganglion cells and dorsal root ganglion cells, can determine whether the material of interest is associated with neurons in general or only certain neurons. Axonal transport can answer some of the same questions addressed by immunohistochemistry, but transport studies are not dependent on the availability of suitable antibodies and are not subject to the vagaries of fixation and processing of sections. Information about compartmentation and physiologically important interactions with other polypeptides and structures of the axon is also obtainable from studies of axonal transport. Thus, these two methods serve to complement each other in studies on the distribution of specific materials within the nervous system. As the biochemical complexity of the nervous tissues becomes more and more apparent, such methods will be increasingly important to our understanding of the nervous system.

The structural hypothesis relates the various rate components of axonal transport to specific neuronal subcellular compartments and ultrastructural correlates. Subcellular location and structural associations can therefore be inferred for a polypeptide or other material, once it can be associated with a specific rate component. A physiological context for each rate component has been defined (Brady and Lasek, 1982a; Lasek and Brady, 1982b) (*see* Table 1), so that information about metabolism, associated proteins or structures, and biochemical properties can be obtained from axonal transport studies. In many respects axonal transport processes constitute a subcellular fractionation by the neuron of those materials used in the construction and maintenance of its axonal processes. Almost by definition, any associations defined by axonal transport are physiologically relevant, because the sorting takes place within the neuron with all components at physiological concentrations of protein and small molecular weight constituents (Brady and Lasek, 1982a; Lasek and Brady, 1982a). Interactions between proteins in a given rate component must last for hours (fast component) or even months (SCa) in long axons (Brady et al., 1982; Lasek, 1980), and there can be no adventitious interactions introduced between elements normally kept in separate cellular compartments using this approach.

The complex polypeptide composition of a rate component such as SCb does not invalidate the premise that SCb represents a physiological interaction between these proteins, but only reflects the presence of a hierarchy of interactions within the cell (Brady and Lasek, 1981; Brady et al., 1981). The diversity of such interactions may lead to cytological structures unlike the more familiar membranous and cytoskeletal elements of the axon, but must be assembled according to physical and chemical rules that may be understood through studies of axonal transport (Lasek et al., 1984a). One lesson learned from the study of axonal compartments by axonal transport is that the boundaries between compartments may not always be drawn according to the preconceptions of biologists (for example, the transport of clathrin; *see* Garner and Lasek, 1981).

A fundamental relationship between the growth and regeneration of neuronal processes and axonal transport is well established (Cancalon, 1979; Elam and Cancalon, 1984; Grafstein and McQuarrie, 1978; Lasek et al., 1981; Lasek and Black, 1977; Lasek and Hoffman, 1976; McQuarrie, 1983). Recent studies have begun to identify specific polypeptides associated with axonal growth by labeling with fast axonal transport (Baitinger et al., 1984; Benowitz, 1984; Skene and Willard, 1981). Molecular events associated with the maturation of an axon, such as changes in the composition and amount of neurofilament proteins transported (Hoffman and Lasek, 1980; Hoffman et al., 1984; Willard and Simon, 1983) have been described. Our understanding of the changes that the neuron undergoes during growth and regeneration should continue to be enhanced by studies of axonal transport. Axonal transport processes can also serve as a tool for the study of other aspects of regeneration. For example, labeling of growth cones by fast or slow transport may be used to measure the extent or rate of axonal growth during development and regeneration (Forman and Berenberg, 1978; Wujek and Lasek, 1983). Even extremely fine processes can be detected, and subpopulations within a growing process can be distinguished (Cancalon, 1979; Wujek and Lasek, 1983).

Another use of the axonal transport method is to selectively manipulate specific neuronal populations. Combining extrinsic labeling procedures with specific probes of the nervous system, such as monoclonal antibodies to specific classes of neurons or artificial ligands that bind only specific receptors, has the potential for creating "magic bullets" that find only those targets of interest. Selective labeling has been mentioned, but other uses are also possible. Attachment of toxins to probes for labeling retrograde transport has already been used to create suicide transport agents to selectively kill only those neurons that take up and transport the toxin back to the cell body (Oeltmann and Wiley, 1984). Selective stimulation of specific neuronal populations should be feasible using a similar approach.

Finally, the study of axonal transport has proven to be an invaluable model for investigating intracellular processes and cell motility. The geometry, organization, metabolic activity, and size of neurons are especially well suited for analyses of intracellular transport. Synthesis and eventual disposition of transported materials are spatially and temporally separated from translocation events (Grafstein and Forman, 1980; Lasek, 1982; Lasek and Brady, 1982a). As a result, the movements of both membranous organelles and cytoskeletal structures can be evaluated in the same preparation and an integrated approach can be taken to study these interrelated activities. The molecular mechanisms that underly axonal transport appear to be similar in all cells, so we may learn a great deal about cell biology, as well as neurobiology, from the study of axonal transport.

## 11. Conclusion

Whether employed as a tool or studied for its own sake, the study of axonal transport processes remains a vital area of contemporary biology. An understanding of axonal transport involves the fields of neurobiology, cell biology, and molecular biology. Consequently, methods derived from basic studies of transport can be applied to a wide variety of questions ranging from the cellular organization of the brain to the molecular biology of neurons. This introduction to the uses of axonal transport has described some of the considerations important for rigorous analysis of axonal transport and, it is hoped, will stimulate further development of the field.

## Acknowledgments

The author would like to thank Dr. Raymond Lasek for the many hours of discussion over the years, which have contributed greatly to development of the ideas and procedures described here. The author also wishes to thank Alison K. Hall for her reading of the manuscript and many useful comments. Support during the writing of this manuscript came from a grant to the author by the National Institutes of Health (NS 18361.)

# References

Abe T., Haga T., and Kurokawa M. (1973) Rapid transport of phosphatidylcholine occurring simultaneously with protein transport in the frog sciatic nerve. *Biochem. J.* **136,** 731–740.

Adams R. J. (1982) Organelle movement in axons depends on ATP. *Nature* (Lond.) **297,** 327–329.

Adams R., Baker P., and Bray D. (1982) Particle movement in crustacean axons that have been rendered permeable by exposure to brief intense electric fields. *J. Physiol.* (Lond.) **326,** 7P.

Allen R. D. and Allen N. S. (1983) Video-enhanced microscopy with a computer frame memory. *J. Microsc.* **129,** 3–17.

Allen R. D., David G. B., and Nomarski G. (1969) The Zeiss-Nomarksi differential interference equipment for transmitted light microscopy. *Z. wissen. Mikr. Mikrotech.* **69,** 193–221.

Allen R. D., Travis J. L., Allen N. S., and Yilmaz H. (1981a) Video-enhanced contrast polarization (AVEC-POL) microscopy: A new method applied to the detection of birefringence in the motile reticulopodial network of *Allogromia laticollaris. Cell Motil.* **1,** 275–289.

Allen R. D., Allen N. S., and Travis J. L. (1981b) Video-enhanced contrast, differential interference contrast (AVC-DIC) microscopy: A new method capable of analyzing microtubule-related movement in the reliculopodial network of *Allogromia laticollaris. Cell Motil.* **1,** 291–302.

Allen R. D., Metuzals J., Tasaki I., Brady S. T., and Gilbert S. (1982) Fast axonal transport in squid giant axon. *Science* **218,** 1127–1129.

Ambron R. T. and Schwartz J. H. (1979) Regional Aspects of Neuronal Glycoprotein and Glycolipid Synthesis, in *Complex Carbohydrates of Nervous Tissue.* (Margolis R. and Margolis R., eds), pp. 269–289, Plenum, New York.

Ambron R. T. and Treistman S. N. (1977) Glycoproteins are modified in the axon of R2, the giant neuron of *Aplysia californica,* after intraaxonal injection of [$^3$H]-*N*-acetylgalactosamine. *Brain Res.* **121,** 287–309.

Baitinger C., Cheney R., Clements D., Glicksman M., Hirokawa N., Levine J., Meiri K., Simon C., Skene P., and Willard M. (1984) Axonally transported proteins in axon development, maintenance, and regeneration. *Cold Spring Harbor Symp.* **48,** 791–802.

Baitinger C., Levine J., Lorenz T., Simon C., Skene P., and Willard M. (1982) Characteristics of Axonally Transported Proteins, in *Axonplasmic Transport.* (Weiss D. G., ed.), pp. 110–120, Springer-Verlag, Berlin.

Barondes S. (1968) Incorporation of radioactive glucosamine into macromolecules at nerve endings. *J. Neurochem.* **15,** 699–706.

Bartlett W. and Banker G. (1984) An electron microscopic study of the development of axons and dendrites by hippocampal neurons in culture: Cells which develop without intercellular contacts. *J. Neurosci.* **4,** 1944–1953.

Bear R., Schmitt F., and Young J. Z. (1937) Investigations on the protein constituents of nerve axoplasm. *Proc. Roy. Soc. Lond. (B)* **123,** 520–529.

Benowitz L. (1984) Target-Dependent and Target-Independent Changes in Rapid Axonal Transport During Regeneration of the Goldfish Retinotectal Pathway, in *Axonal Transport in Neuronal Growth and Regeneration* (Elam J. and Cancalon P., eds.), pp. 145–170, Plenum, New York.

Berlinrood M., McGee-Russel S., and Allen R. D. (1972) Pattern of particle movements in nerve fibers *in vitro:* An analysis by photokymography and microscopy. *J. Cell Sci.* **11,** 875–886.

Bisby M. A. (1977) Retrograde axonal transport of endogenous protein: Differences between motor and sensory axons. *J. Neurochem.* **28,** 249–251.

Bisby M. A. (1982) Ligature Techniques, in *Axoplasmic Transport* (Weiss D. G., ed.), pp. 437–441, Springer-Verlag, Berlin.

Bisby M. A. and Bulger V. T. (1977) Reversal of axonal transport at a nerve crush. *J. Neurochem.* **29,** 313–320.

Black M. M. and Lasek R. J. (1977) The presence of transfer RNA in the axoplasm of the squid giant axon. *J. Neurobiol.* **8,** 229–237.

Black M. M. and Lasek R. J. (1979a) Axonal transport of actin: Slow component b is the principal source of actin for the axon. *Brain Res.* **171,** 401–413.

Black M. M. and Lasek R. J. (1979b) A difference between the proteins conveyed in the fast component of axonal transport in guinea pig hypoglossal and vagus motor neurons. *J. Neurobiol.* **9,** 433–443.

Black M. M. and Lasek R. J. (1980) Slow components of axonal transport: Two cytoskeletal networks. *J. Cell Biol.* **86,** 616–623.

Blaker W. D., Toews A. D., and Morell P. (1980) Cholesterol is a component of the rapid phase of axonal transport. *J. Neurobiol.* **11,** 243–250.

Bloom G., Schoenfeld T., and Vallee R. (1984) Widespread distribution of MAP1 (microtubule-associated protein 1) in the nervous system. *J. Cell Biol.* **98,** 320–330.

Bonner W. M. and Laskey R. A. (1974) A film detection method for tritium-labeled proteins and nucleic acids in polyacrylamide gels. *Eur. J. Biochem.* **46,** 83–88.

Brady S. T. (1984) Basic Properties of Fast Axonal Transport and the Role of Fast Transport in Axonal Growth, in *Axonal Transport in Neuronal Growth and Regeneration* (Elam J. and Cancalon P., eds.), pp. 13–29, Plenum, New York.

Brady S. T. and Lasek R. J. (1981) Nerve specific enolase and creatine phosphokinase in axonal transport: Soluble proteins and the axoplasmic matrix. *Cell* **23,** 523–351.

Brady S. T. and Lasek R. J. (1982a) Axonal transport: A cell biological method for studying proteins that associate with the cytoskeleton. *Meth. Cell. Biol.* **25,** 366–398.

Brady S. T. and Lasek R. J. (1982b) The Slow Components of Axonal Transport: Movements, Compositions, and Organization, in *Axoplasmic Transport* (Weiss D. G., ed.), pp. 206–217, Springer-Verlag, Berlin.

Brady S. T., Corthers S., Nosal C., and McClure W. O. (1980) Fast axonal transport in the presence of high $Ca^{2+}$: Evidence that microtubules are not required. *Proc. Natl. Acad. Sci. USA* **77**, 5909–5913.

Brady S. T., Lasek R. J., and Allen R. D. (1982) Fast axonal transport in extruded axoplasm from squid giant axon. *Science* **218**, 1129–1131.

Brady S. T., Lasek R. J., and Allen R. D. (1985) Video microscopy of fast axonal transport in extruded axoplasm: A new model for study of molecular mechansims. *Cell Motil.* **5**, 81–101.

Brady S. T., Tytell M., Heriot K., and Lasek R. J. (1981) Axonal transport of calmodulin: A physiologic approach to identification of long term associations between proteins *J. Cell Biol.* **89**, 607–614.

Branton D., Cohen C., and Tyler J. (1981) Interaction of cytoskeletal proteins on human erythrocyte membrane. *Cell* **24**, 24–32.

Bray D. (1984) Axonal growth in response to experimentally applied mechanical tension. *Dev. Biol.* **102**, 379–389.

Bray J. and Austin L. (1968) Flow of protein and ribonucleic acid in peripheral nerve. *J. Neurochem.* **15**, 731–740.

Bray D. and Bunge M. (1981) Serial analysis of microtubules in cultured rat sensory axons. *J. Neurocytol.* **10**, 589–605.

Breuer A. C., Christian C. M., Henkart M., and Nelson P. G. (1975) Computer analyses of organelle translocation in primary neuronal cultures and continuous cell lines. *J. Cell Biol.* **65**, 562–576.

Brimijoin S. (1975) Stopflow: A new technique for measuring axonal transport and its application to the transport of dopamine-β-hydroxylase. *J. Neurobiol.* **6**, 379–394.

Brimijoin S., (1979a) Axonal transport and subcellular distribution of molecular forms of acetylcholinesterase in rabbit sciatic nerve. *Mol. Pharmacol.* **15**, 641–648.

Brimijoin S., (1979b) On the kinetics and maximal capacity of the system for rapid axonal transport in mammalian neurones. *J. Physiol.* (Lond.) **292**, 325–337.

Brimijoin S. and Dyck P. F. (1979) Axonal transport of dopamine-β-hydroxylase and acetylcholinesterase in human peripheral neuropathy. *Exp. Neurol.* **66**, 467–478.

Brimijoin S. and Wiermaa M. J. (1977a) Rapid axonal transport of tyrosine hydroxylase in rabbit sciatic nerves. *Brain Res.* **121**, 77–96.

Brimijoin S. and Wiermaa M. J. (1977b) Direct comparison of the rapid axonal transport of norepinephrine and dopamine-β-hydroxylase activity. *J. Neurobiol.* **8**, 239–250.

Brimijoin S. and Wiermaa M. J. (1978) Rapid orthograde and retrograde axonal transport of acetylcholinesterase as characgerized by the stop–flow technique. *J. Physiol.* (Lond.) **285**, 129–142.

Brimijoin S., Lundberg J. M., Brodin E., Hokfelt T., and Nilsson G. (1980) Axonal transport of substance P in the vagus and sciatic nerves of the guinea pig. *Brain Res.* **191,** 443–457.

Brimijoin S., Capek P., and Dyck P. J. (1973) Axonal transport of dopamine-β-hydroxylase by human sural nerves in vitro. *Science* **180,** 1295–1297.

Broadwell R., Olver C., and Brightman M. (1980) Neuronal transport of acid hydrolases and peroxidases within the lysosmal system of organelles. Involvement of agranular reticulum-like cisterns. *J. Comp. Neurol.* **190,** 519–532.

Bulger V. T. and Bisby M. A. (1978) Reversal of axonal transport in regenerating nerves. *J. Neurochem.* **331,** 1411–1418.

Bunt A. H. and Haschke R. H. (1978) Features of foreign proteins affecting their retrograde transport in axons of the visual system. *J. Neurocytol.* **7,** 665–678.

Bunt A. H., Haschke R. H., Lund R. D., and Calkins D. F. (1976) Factors affecting retrograde axonal transport of horseradish peroxidase in the visual system. *Brain Res.* **102,** 152–155.

Burdwood W. (1965) Bidirectional particle movement in neurons. *J. Cell Biol.* **27,** 115a.

Caceres A., Binder L., Payne M., Bender P., Rebuhn L., and Steward O. (1984) Differential subcellular localization of tubulin and the microtubule associated protein MAP2 in brain tissue as revealed by immunocytochemistry with monoclonal hybridoma antibodies. *J. Neurosci.* **4,** 394–410.

Campenot R. (1977) Local control of neurite develoment by nerve growth factor. *Proc. Natl. Acad. Sci. USA* **74,** 4516–4519.

Cancalon P. (1979) The Relationship of Slow Axonal Flow to Nerve Elongation and Degeneration, in *Neuronal Growth and Regeneration* (Elam J. and Cancalon P., eds.), pp. 211–242, Plenum, New York.

Cancalon P. and Beidler L. M. (1975) Distribution along the axon and into various subcellular fractions of molecules labeled with [³H]-leucine and rapidly transported in the garfish olfactory nerve. *Brain Res.* **89,** 225–244.

Cancalon P. and Beidler L. M. (1977) Differences in the composition of the polypeptides deposited in the axon and the nerve terminals by fast axonal transport in the garfish olfactory nerve. *Brain Res.* **121,** 215–227.

Carroll P. T., Price D. L., Griffin J. W., and Morris J. R. (1979) Tetanus toxin: Immunocytochemical evidence for retrograde transport. *Neurosci. Lett.* **8,** 335–339.

Cook M. L. and Stevens J. G. (1973) Pathogenesis of herpetic neuritis and ganglionitis in mice: Evidence for intra-axonal transport of infection. *Infect. Immunol.* **7,** 272–288.

Contos N. and Berkley K. J. (1984) Evidence that glial-neuronal transfer of ³H-proline labeled proteins is a component of their transport from the dorsal column nuclei to the interior olive in the cat. *Soc. Neurosci. Abstr.* **10,** 1088.

Courad J. Y. and DiGiamberardino L: (1980) Axonal transport of the molecular forms of acetylcholinesterase in chick sciatic nerve. *J. Neurochem.* **35,** 1053–1066.

Cowan W. M. and Cuenod M., eds. (1975) *The Use of Axonal Transport for Studies of Neuronal Connectivity.* Elsevier, New York.

Cowan W. M., Gottlieb D. I., Hendrickson A. E., Price J. L., and Woolsey T. A. (1972) The autoradiographic demonstration of axonal connections in the central nervous system. *Brain Res.* **37,** 21–51.

Csanyi V., Gervai J., and Lajtha A. (1973) Axoplasmic transport of free amino acids. *Brain Res.* **56,** 271–284.

Cuenod M., Bagnoli P., Beaudet A., Rustioni A., Wiklund L., and Streit P. (1982) Retrograde Migration of Transmitter Related Molecules, in *Axoplasmic Transport in Physiology and Pathology* (Weiss, D. and Gorio A., eds.), pp. 160–166, Springer-Verlag, New York.

Currie J. R., Grafstein B., Whitnall M. H., and Alpert R. (1978) Axonal transport of lipid in goldfish optic axons. *Neurochem. Res.* **3,** 479–492.

Dahlstrom A. (1967) The transport of noradrenaline between two simultaneously performed ligations of the sciatic nerve of rat and cat. *Acta Physiol. Scand.* **69,** 158–166.

Dahlstrom A. and Haggendal J. (1966) Studies on the transport and lifespan of amine storage granules in a peripheral adrenergic neuron system. *Acta Phsiol. Scand.* **67,** 278–288.

Dahlstrom A. and Heiwall P. -O. (1975) Intra-axonal transport of transmitters in mammalian neurons. *J. Neural Transm. Suppl.* **12,** 97–114.

de Olmos J. S., Ebbesson S., and Heimer L. (1981) Silver Methods for the Impregnation of Degenerating Axoplasm, in *Neuroanatomical Tract-tracing Methods* (Heimer L. and Robards M., eds.) pp. 117–170, Plenum, New York.

de Mey J. (1983a) A critical review of light and electron microscopic immunocytochemical techniques used in neurobiology. *J. Neurosci. Meth.* **7,** 1–18.

de Mey J. (1983b) Colloidal Gold Probes in Immunocytochemistry, in *Immunochemistry* (Polak J. and van Noorden S., eds.), pp. 82–112, Wright PSG, Bristol.

Droz B. and Leblond C. P. (1962) Migration of proteins along the axons of the sciatic nerve. *Science* **137,** 1047–1048.

Edstrom A. and Hanson M. (1973) Retrograde axonal transport of proteins in vitro in frog sciatic nerve. *Brain Res.* **61,** 311–321.

Edstrom A. and Mattson H. (1972) Fast axonal transport in vitro in the sciatic system of the frog. *J. Neurochem.* **19,** 205–221.

Elam J. S. (1979) Axonal Transport of Complex Carbohydrates, in *Complex Carbohydrates of Nervous Tissue.* (Margolis R. and Margolis R., eds.) pp. 235–268, Plenum, New York.

Elam J. S. and Agranoff B. (1971) Rapid transport of protein in the optic system of the goldfish. *J. Neurochem.* **18,** 3735–387.

Elam J. S. and Cancalon P., eds. (1984) *Axonal Transport in Neuronal Growth and regeneration.* Plenum, New York.

Elam J. S. and Peterson N. W. (1976) Axonal transport of sulfated glyco-proteins and mucopolysaccharides in the garfish olfactory nerve. *J. Neurochem.* **26**, 845–850.

Elam J. S., Goldberg J., Radin N. S., and Agranoff B. W. (1970) Rapid axonal transport of sulfated mucopolysaccharide proteins. *Science* **170**, 458–460.

Erickson P. F., Seamon K. B., Moore B. W., Lasher R. S., and Miner L. N. (1980) Axonal transport of the $Ca^{2+}$ dependent protein modula-tor of 3′:5′ cyclic AMP phophodiesterase in the rabbit visual system. *J. Neurochem.* **35**, 242–248.

Estridge M. and Bunge R. (1978) Compositional analysis of growing axons from rat sympathetic neurons. *J. Cell Biol.* **79**, 138–155.

Fahim M., Brady S. T., and Lasek R. (1982) Axonal transport of membra-nous organelles in squid giant axons and axoplasm. *J. Cell Biol.* **95**, 330a.

Feher E. (1984) Electron microscopic study of retrograde axonal trans-port of horseradish peroxidase. *Int. Rev. Cytol.* **90**, 1–30.

Fillenz M., Gagnon C., Stoeckel K., and Thoenen H. (1976) Selective uptake and retrograde axonal transport of dopamine-β-hydroxylase antibodies in peripheral adrenergic neurons. *Brain Res.* **114**, 293–303.

Fink D. J. and Gainer H. (1979) The use of a labeled acylating probe for the study of fast axonal transport. *Brain Res.* **177**, 208–213.

Fink D. J. and Gainer H. (1980) Axonal transport of proteins: A new view using in vivo covalent labeling. *J. Cell Biol.* **85**, 175–186.

Fonnum F., Frizell M., and Sjostrand J. (1973) Transport, turnover, and redistribution of choline acetyl transferase and acetylcholinesterase in the vagus and hypoglossal nerves of rabbit. *J. Neurochem.* **21**, 1109–1120.

Forman D. (1981) A permeabilized cell model of saltatory organelle movement. *J. Cell Biol.* **91**, 414a.

Forman D. S. and Berenberg R. A. (1978) Regeneration of motor axons in the rat sciatic nerve studied by labeling with axonally transported radioactive proteins. *Brain Res.* **156**, 213–225.

Forman D. S. and Ledeen R. W. (1972) Axonal transport of gangliosides in the goldfish optic nerve. *Science* **177**, 630–633.

Forman D., Padjen A. L., and Siggins G. (1977) Axonal transport of organelles visualized by light microscopy: Cinemicrographic and computer analysis. *Brain Res.* **136**, 197–213.

Forman D., Brown K., Promersberger M., and Adelman M., (1984) Nucleotide specificity for reactivation of organelle movements of fast axonal transport in permeabilized axons. *Cell Motil.* **4**, 121–128.

Gainer H., Sarne Y., and Brownstein M. J. (1977a) Biosynthesis and axonal transport of rat neurohypophysial proteins and peptides. *J. Cell Biol.* **73**, 366–381.

Gainer H., Sarne Y., and Brownstein M. J. (1977b) Neurophysin biosynthesis: Conversion of a putative precursor during axonal transport. *Science* **195**, 1354–1356.

Gamse R., Lembeck F., and Cuello A. C. (1979) Substance P in the vagus nerve. Immunochemical and immunohistochemical evidence for axoplasmic transport. *Naunyn-Schmiedebergs Arch. Pharmacol.* **306**, 37–44.

Garner J. A. and Lasek R. J. (1981) Clathrin is axonally transported as part of slow component b: The axoplasmic matrix. *J. Cell Biol.* **88**, 172–178.

Garner J. A. and Lasek R. J. (1982) Cohesive axonal transport of the slow component b complex of polypeptides. *J. Neurosci.* **2**, 1824–1835.

George E. B. and Lasek R. J. (1983) Contraction of isolated neural processes: A model for studying cytoskeletal translocation in neurons. *J. Cell Biol.* **97**, 267a.

Gilbert D. (1974) Physiological Uses of the Squid with Special Emphasis on the Use of the Giant Axon, in *A Guide to the Laboratory Use of the Squid Loligo Pealei,* Marine Biological Laboratory Woods Hole, Massachusetts, pp. 45–54.

Gilbert S. and Sloboda R. (1984) Bidirectional transport of fluorescently labeled vesicles introduced into extruded axoplasm of squid *Loligo pealei. J. Cell Biol.* **99**, 445–452.

Giuditta A., Ciysello A., and Lazzarini G(1980) Ribosomal RNA in the axoplasm of the squid giant axon. *J. Neurochem.* **34**, 1757–1760.

Giuditta A., Hunt T., and Scanella L. (1983) Messenger RNA in squid axoplasm. *Biol. Bull.* **165**, 526.

Goldberg D. J., Goldman J. E., and Schwartz J. H. (1976) Alterations in amounts and rates of serotonin transported in an axon of the giant cerebral neurone of *Aplysia californica. J. Physiol.* **259**, 473–490.

Gonatas N. K., Harper C., Mizutani T., and Gonatas J. O. (1979) Superior sensitivity of conjugates of horseradish peroxidase with wheat germ agglutinin for studies of retrograde axonal transport. *J. Histochem. Cytochem.* **27**, 728–734.

Gould R. M., Spivack W., Sinatra R., Lindquist T., and Ingoglia N. (1982) Axonal transport of choline lipids in normal and regenerating rat sciatic nerve. *J. Neurochem.* **39**, 1569–1578.

Gould R. M., Pant H., Gainer H., and Tytell M. (1983a) Phospholipid synthesis in the squid giant axon: Incorporation of lipid precursors. *J. Neurochem.* **40**, 1293–1299.

Gould R. M., Spivack W., Robertson D., and Poznansky M. (1983b) Phospholipid synthesis in the squid giant axon: Enzymes of phosphatidylinositol metabolism. *J. Neurochem.* **40**, 1300–1306.

Grafstein B. (1967) Transport of protein by goldfish optic nerve fibers. *Science* **157**, 196–198.

Grafstein B. and Forman D. (1980) Intracellular transport in neurons. *Physiol. Rev.* **60**, 1167–1283.

Grafstein B. and McQuarrie I. G. (1978) The Role of the Nerve Cell Body in Axonal Regeneration, in *Neuronal Plasticity* (Cotman C., ed.), pp. 155–195, Raven, New York.

Grafstein B., Miller J. A., Ledeen R. W., Haley J., and Specht S. C. (1975) Axonal transport of phospholipid in goldfish optic system. *Exp. Neurol.* **46**, 261–281.

Griffin J. W., Price D. L., Drachman D. B., and Morris J. R. (1981) Incorporation of transported glycoproteins into axolemma during regeneration. *J. Cell Biol.* **88,** 205–214.

Gross G. W. and Beidler L. M. (1975) Fast axonal transport in the C-fibers of goldfish olfactory nerve. *J. Neurobiol.* **4,** 413–428.

Gulya K. and Kasa P. (1984) Transport of muscarinic cholinergic receptors in the sciatic nerve of rat. *Neurochem. Int.* **6,** 123–126.

Gunning P. W., Por S., Langford C., Scheffer J., Austin L., and Jeffrey P. (1979) The direct measurement of the axoplasmic transport of individual RNA species: Transfer but not ribosomal RNA is transported. *J. Neurochem.* **32,** 1737–1743.

Hadley R. T. and Trachtenberg M. C. (1978) Poly-L-ornithine enhances the uptake of horseradish peroxidase. *Brain Res.* **158,** 1–14.

Haggendal J. (1980) Axonal transport of dopamine-β-hydroxylase to rat salivary glands: Studies on enzymatic activity. *J. Neural Transm.* **47,** 163–174.

Hahnenberger R. W. (1978) Effects of pressure on fast axoplasmic flow. An in vitro study in the vagus nerve of rabbits. *Acta Physiol. Scand.* **104,** 299–308.

Haley J. E., Tirri L. J., and Ledeen R. W. (1979) Axonal transport of lipids in the rabbit optic system. *J. Neurochem.* **32,** 727–734.

Hall M. E. (1982) Changes in the synthesis of specific proteins in axotomized dorsal root ganglia. *Exp. Neurol.* **76,** 83–93.

Hammerschlag R. and Stone (1982) Membrane delivery by fast axonal transport. *Trends Neurosci.* **5,** 12–15.

Hammerschlag R. (1983) How do neuronal proteins know where they are going? . . . Speculations on the role of molecular address markers. *Dev. Neurosci.* **6,** 2–17.

Hammerschlag R. and Stone (1982) Membrane delivery by fast the transport system. *Neuroscience.* **4,** 1195–1201.

Hammerschlag R. and Lavoie P. A. (1979) Initiation of fast axonal transport: Involvement of calcium during transfer of proteins from Golgi apparatus to the transport system. *Neuroscience.* **4,** 1195–1201.

Hammerschlag R., Bakhit C., and Chiu A. Y. (1977) Role of calcium in the initiation of fast axonal transport: Effects of divalent cations. *J. Neurobiol.* **8,** 439–451.

Hammerschlag R., Dravid A. R., and Chiu A. Y. (1975) Mechanism of axonal transport: A proposed role for calcium ions. *Science* **188,** 273–275.

Hammerschlag R., Stone G. C., Bolen F., Lindsey J., and Ellisman M. (1982) Evidence that all newly synthesized proteins destined for fast axonal transport pass through the Golgi apparatus. *J. Cell Biol.* **93,** 568–575.

Hanson M. (1978) A new method to study fast axonal transport in vivo. *Brain Res.* **153,** 121–126.

Hanson M. and Bergqvist J. E. (1982) In Vitro Chamber Systems to Study Axonal Transport, in *Axoplasmic Transport.* (Weiss D. G., ed.), pp. 429–436, Springer-Verlag, Berlin.

Hanson M. and Edstrom A. (1978) Mitosis inhibitors and axonal transport. *Int. Rev. Cytol.* (Suppl.) **7**, 373–402.

Hansson H. -A. (1973) Uptake and bidirectional transport of horseradish peroxidase in retinal ganglion cells. *Exp. Eye Res.* **16**, 377–388.

Harper C., Gonatas J. O., Steiber A., and Gonatas N. K. (1980) In vivo uptake of wheat germ agglutinin-horseradish peroxidase conjugates into neruonal GERL and lysosomes. *Brain Res.* **188**, 465–472.

Harrison R. G. (1910) The outgrowth of the nerve fiber as a mode of protoplasmic movement. *J. Exp. Zool.* **9**, 787–848.

Heacock A. and Agranoff B. (1977) Reutilization of precursor following axonal transport of [$^3$H]-proline labeled protein. *Brain Res.* **122**, 243–254.

Heimer L. and Robards M., eds. (1981) *Neuroanatomical Tract-Tracing Methods.* Plenum, New York.

Heiwall P. -O., Dahlstrom A., Larsson P., and Booj S. (1979) The intra-axonal transport of acetylcholine and cholinergic enzymes after various types of axonal trauma. *J. Neurobiol.* **10**, 119–136.

Heslop J. P. (1975) Axonal flow and fast transport in nerves. *Adv. Comp. Physiol. Biochem.* **6**, 75–163.

Hoffman P. N. and Lasek R. J. (1980) Axonal transport of the cytoskeleton in regenerating motor neurons: Constancy and change. *Brain Res.* **202**, 317–333.

Hoffman P. N. and Lasek R. J. (1975) The slow component of axonal transport. Identification of major structural polypeptides of the axon and their generality among mammalian neurons. *J. Cell Biol.* **66**, 351–366.

Hoffman P., Griffin J., and Price D. (1984) Neurofilament Transport in Axonal Regeneration: Implications for the Control of Axonal Caliber, in *Neuronal Growth and Regeneration* (Elam J. and Cancalon P., eds.), pp. 243–260, Plenum, New York.

Ingoglia N. A. (1979) 4S RNA is present in regenerating optic axons of goldfish. *Science* **206**, 73–75.

Ingoglia N. A. and Tuliszewski R. (1976) Transfer RNA may be axonally transported during regeneration of goldfish optic nerves. *Brain Res.* **112**, 371–381.

Ingoglia N. A., Grafstein B., McEwen B., and McQuarrie I. (1973) Axonal transport of radioactivity in the goldfish optic system following intraocular injection of labelled RNA precursors. *J. Neurochem.* **20**, 1605–1615.

Ingoglia N. A., Weis P., and Mycek J. (1975) Axonal transport of RNA during regeneration of the optic nerve of goldfish. *J. Neurobiol.* **6**, 439–563.

Inoue S. (1981) Video image processing greatly enhances contrast, quality, and speed in polarization-based microscopy. *J. Cell Biol.* **89**, 346–356.

Isenberg G., Schubert P., and Kreutzberg G. (1980) Experimental approach to test the role of actin in axonal transport. *Brain Res.* **194**, 588–593.

Johnson L., Walsh M., and Chen L. (1980) Localization of mitochondria in living cells with rhodamine 123. *Proc. Natl. Acad. Sci. USA* **77,** 990–994.

Jones E. G., and Hartman B. K. (1978) Recent advances in neuroanatomical methodology. *Ann. Rev. Neurosci.* **1,** 215–296.

Karlsson J. -O. and Linde A. (1977) Axonal transport of $^{35}$S in retinal ganglion cells of the rabbit. *J. Neurochem.* **28,** 293–297.

Karlsson J. -O. and Sjostrand J. (1971a) Transport of microtubule protein in axons of retinal ganglion cells. *J. Neurochem.* **18,** 975–982.

Karlsson J. -O. and Sjostrand J. (1971b) Rapid intracellular transport of fucose-containing glycoproteins in retinal ganglion cells. *J. Neurochem.* **18,** 2209–2216.

Karlsson J. -O. and Sjostrand J. (1971c) Characterization of the fast and slow components of axonal transport in retinal ganglion cells. *J. Neurobiol.* **2,** 135–143.

Kasa P. (1968) Acetylcholinesterase transport in the central and peripheral nervous tissue: The role of tubules in the enzyme transport. *Nature* (Lond.) **218,** 1265–1267.

Kasa P., Mann S., Karcsu S., Toth L., and Jordan S. (1973) Transport of choline acetyltransferase and acetylcholinesterase in the rat sciatic nerve: A biochemical and electron histochemical study. *J. Neurochem.* **21,** 431–436.

Katz L. C., Burkhalter A., and Dreyer W. J. (1984) Fluorescent latex microspheres as a retrograde neuronal marker for in vivo and in vitro studies of visual cortex. *Nature* (Lond.) **310,** 498–500.

Katz M., Lasek R., Osdoby P., Whittaker J., and Caplan A. (1982) Bolton-Hunter reagent as a vital stain for developing systems. *Dev. Biol.* **90,** 419–429.

Keith C. and Shelanski M. (1982) Direct Visualization of Fluorescently Labeled Microtubules in Living Cells, in *Biological Functions of Microtubules and Related Structures.* (Sakai H., Mohri H., and Borisy G., eds.), pp. 365–376, Academic, New York.

Kirkpatrick J., Bray J., and Palmer S. (1972) Visualization of axoplasmic flow in vitro by Nomarski microscopy. Comparison to rapid flow of radioactive proteins. *Brain Res.* **43,** 1–10.

Koenig E. (1979) Ribosomal RNA in Mauthner axon: Implications for a protein synthesizing machinery in the myelinated axon. *Brain Res.* **174,** 95–107.

Koenig E. and Adams P. (1982) Local protein synthesizing activity in axonal fields regenerating in vitro. *J. Neurochem.* **39,** 386–400.

Koenig E. and Koelle G. B. (1961) Mode of regeneration of acetylcholinesterase in cholinergic neurons following irreversible inactivation. *J. Neurochem.* **8,** 169–188.

Koike H., Eisenstadt M., and Schwartz J. (1972) Axonal transport of newly synthesized acetylcholine in an identified neuron of *Aplysia. Brain Res.* **37,** 152–159.

Komiya Y. (1980) Slowing with age of the rate of slow axonal flow in bifurcating axons of rat dorsal root ganglion cells. *Brain Res.* **183,** 477–480.

Komiya Y., and Kurokawa M. (1978) Asymmetry of protein transport in two branches of bifurcating axons. *Brain Res.* **139,** 354–358.

Kreutzberg G. W., Schubert P., Toth L., and Rieske E. (1973) Intradendritic transport to postsynaptic sites. *Brain Res.* **62,** 399–404.

Kristensson K. (1970a) Morphological studies of the neural spread of herpes simplex virus to the central nervous system. *Acta Neuropathol.* **16,** 54–63.

Kristensson K. (1970b) Transport of fluorescent protein tracer in peripheral nerve. *Acta Neuropathol.* **16,** 293–300.

Kristensson K. (1978) Retrograde transport of macromolecules in axons. *Ann. Rev. Pharmacol. Toxicol.* **18,** 97–110.

Kristensson K. and Olsson Y. (1971) Retrograde axonal transport of protein. *Brain Res.* **29,** 363–365.

Kristensson K. and Olsson Y. (1974) Retrograde transport of horseradish peroxidase in transsected axons. 1. Time relationships between transport and induction of chromatolysis. *Brain Res.* **79,** 101–109.

Kuhar M. and Zarbin M. A. (1984) Axonal transport of muscarinic cholinergic receptors and its implications. *Trends Pharmacol. Sci.* **5,** 53–56.

Laduron P. M. (1984) Axonal transport of receptors: Coexistance with neurotransmitter and recycling. *Biochem. Pharmacol.* **33,** 897–903.

Laemmli U. (1970) Cleavage of structural proteins during the assembly of the head of bacteriophage T4. *Nature* (Lond.) **227,** 680–685.

Lasek R. J. (1966) Axoplasmic streaming in the cat dorsal root ganglion cell and the rat ventral motoneuron. *Anat. Rec.* **154,** 373–374.

Lasek R. J. (1967) Bidirectional transport of radioactively labeled axoplasmic components. *Nature* (Lond.) **216,** 1212–1214.

Lasek R. J. (1968) Axoplasmic transport in cat dorsal root ganglion cells: As studied with [$^3$H]-L-leucine. *Brain Res.* **7,** 360–377.

Lasek R. J. (1974) Biochemistry of the Squid Giant Axon, in *A Guide to the Laboratory Use of the Squid* Loligo pealei Marine Biological Laboratory Woods Hole, Massachusetts.

Lasek R. J. (1980) A dynamic view of neuronal structure. *Trends Neurosci.* **3,** 87–91.

Lasek R. J. (1981) The dynamic ordering of neuronal cytoskeletons. *Neurosci. Res. Prog. Bull.* **19,** 7–32.

Lasek R. J. (1982) Translocation of the neuronal cytoskeleton and axonal locomotion. *Philos. Trans. Roy. Soc. Lond.* (B) **299,** 313–327.

Lasek R. J. (1984) The structure of axoplasm. *Curr. Top. Memb. Trans.* **22,** 39–53.

Lasek R. J. and Black M. M. (1977) How do Axons Grow? Some Clues from the Metabolism of the Proteins in Slow Component of Axonal Transport, in *Mechanisms, Regulation, and Special Functions of Protein Synthesis in the Brain* (Roberts E., ed.), pp. 161–169, Elsevier, Amsterdam.

Lasek R. J. and Brady S. T. (1982) The axon: A prototype for studying expressional cytoplasm. *Cold Spring Harbor Symp. Quant. Biol.* **46,** 113–124.

Lasek R. J. and Brady S. T. (1982b) The Structural Hypothesis of Axonal

Transport: Two Classes of Moving Elements, in *Axoplasmic Transport* (Weiss D. G., ed.) pp. 397–405, Springer-Verlag, Berlin.

Lasek R. J. and Hoffman P. (1976) The Neuronal Cytoskeleton, Axonal Transport, and Axonal Growth, in *Cell Motility* (Goldman R., Pollard T., and Rosenbaum J., eds.) *Cold Spring Harbor Conf. Cell Prolif.* **3,** 1021–1049.

Lasek R. J., Joseph B., and Whitlock D. (1968) Evaluation of a radioautographic neuroanatomical tracing method. *Brain Res.* **8,** 319–336.

Lasek R. J., Dabrowski C., and Nordlander R. (1973) Analysis of axoplasmic RNA from invertebrate giant axons. *Nature New Biol.* **244,** 162–165.

Lasek R. J., Gainer H., and Barker J. (1977) Cell to cell transfer of glial proteins to the squid giant axon. The glial–neuron protein transfer hypothesis. *J. Cell Biol.* **74,** 501–523.

Lasek R. J., Gainer H., and Przybylski, R. (1974) Transfer of newly synthesized proteins from Schwann cells to the squid giant axon. *Proc. Natl. Acad. Sci. USA* **71,** 1188–1192.

Lasek R. J., Garner J., and Brady S. T. (1984a) Axonal transport of the cytoplasmic matrix. *J. Cell Biol.* **99,** 212s–221s.

Lasek R. J., McQuarrie I., and Brady S. T. (1984b) Transport of Cytoskeletal and Soluble Proteins in Neurons, in *Biological Structures and Coupled Flows* (Oplatka A. and Balaban M., eds.), pp. 329–347, Academic, New York.

Lasek R. J., McQuarrie I., and Wujek J. (1981) The Central Nervous System Regeneration Problem: Neuron and Environment, in *Posttraumatic Peripheral Nerve Regeneration: Experimental Basis and Clinical Implications.* (Gorio A., ed.), pp. 59–74, Raven, New York.

Lasek R. J., Oblinger M. M., and Drake P. (1983) Molecular biology of neuronal geometry: Expression of neurofilament genes influences axonal diameter. *Cold Spring Harbor Symp. Quant. Biol.* **48,** 731–744.

Laskey R. and Mills A. (1975) Quantitative film detection of $^3$H and $^{14}$C in polyacrylamide gels by fluorography. *Eur. J. Biochem.* **56,** 335–341.

LaVail J. H. (1978) A Review of the Retrograde Transport Technique, in *Neuroanatomical Research Techniques.* (Robertson R. T., ed.), pp. 355–384. Academic, New York.

LaVail J. H. and LaVail M. M. (1972) Retrograde axonal transport in the central nervous system. *Science* **176,** 1416–1417.

LaVail J. H. and LaVail M. M. (1974) The retrograde intraaxonal transport of horseradish peroxidase in the chick visual system: A light and electron microscopic study. *J. Comp. Neurol.* **157,** 303–358.

LaVail J. H., Rapisardi S., and Sugino I. K. (1980) Evidence against the smooth endoplasmic reticulum as a continuous channel for the retrograde axonal transport of horseradish peroxidase. *Brain Res.* **191,** 3–20.

LaVail J. H., Sugino I. K., and McDonald D. M. (1983) Localization of axonally transported $^{125}$I-wheat germ agglutinin beneath the plasma membrane of chick retinal ganglion cells. *J. Cell Biol.* **96,** 373–381.

Levine J. and Willard M. (1981) Fodrin: Axonally transported polypeptides associated with the internal periphery of many cells. *J. Cell. Biol.* **90,** 631–643.

Longo F. and Hammerschlag R. (1980) Relation of somal lipid synthesis to the fast axonal transport of protein and lipid. *Brain Res.* **193,** 471–485.

Lorenz T. and Willard M. (1978) Subcellular fractionation of intraaxonally transported polypeptides in the rabbit visual system. *Proc. Natl. Acad. Sci. USA* **75,** 505–509.

Lubinska L. (1964) Axoplasmic Streaming in Regenerating and in Normal Nerve Fibres, in *Progress in Brain Research: Mechanisms of Neural Regeneration* (Singer M. and Schade J., eds.), **13,** 1–66, Elsevier, Amsterdam.

Lubinska L. (1975) On axoplasmic flow. *Int. Rev. Neurobiol.* **17,** 241–296.

Lubinska L., Niemierko S., Oderfield B., and Szwarc L. (1964) Behavior of acetylcholinesterase in isolated nerve segments. *J. Neurochem.* **11,** 493–503.

Lubinska L., Niemierko S., Oderfield B., Szwarc L., and Zelena Z. (1963) Bidirectional movements of axoplasm in peripheral nerve fibres. *Acta Biol. Exp.* **23,** 239–247.

Margolis R. and Margolis R. (eds.) (1979) *Complex Carbohydrates of Nervous Tissue.* Plenum, New York.

Margolis T. P and LaVail J. H. (1981) Rate of anterograde axonal transport of [$^{125}$-I]wheat germ agglutinin from retina to optic tectum in the chick. *Brain Res.* **229,** 218–223.

Margolis T. P., Marchand C., Kistler H. B., and LaVail J. H. (1981) Uptake and anterograde axonal transport of wheat germ agglutinin from retina to optic tectum in the chick. *J. Cell Biol.* **89,** 152–156.

Matus A., Bernhardt R., and Hugh-Jones H. (1981) High molecular weight microtubule associated proteins are preferentially associated with dendritic microtubules in brain. *Proc. Natl. Acad. Sci. USA* **78,** 3010–3014.

McQuarrie I. (1983) Role of the Axonal Cytoskeleton in the Regenerating Nervous System, in *Nerve, Organ, and Tissue Regeneration: Research Perspectives.* (Seil F., ed.), pp. 51–88, Academic, New York.

McQuarrie I., Brady S., and Lasek R. (1980) Polypeptide composition and kinetics of SCa and SCb in sciatic nerve motor axons and optic axons of rat. *Soc. Neurosci. Abstr.* **6,** 501.

Mesulam M. (ed.) (1982) *Tracing Neural Connections with Horseradish Peroxidase.* Wiley, New York.

Mesulam M. and Brushart T. (1979) Transganglionic and anterograde transport of horseradish peroxidase across dorsal root ganglia: A tetramethylbenzidine method for tracing central sensory connections of muscles and peripheral nerves. *Neuroscience* **4,** 1107–1117.

Mesulam M. and Mufson E. (1980) The rapid anterograde transport of horseradish peroxidase. *Neuroscience* **5,** 1277–1286.

Miki-Noumura T. and Kamiya R. (1979) Shape of microtubules in solutions. *Exp. Cell Res.* **97**, 451–453.

Morell P., Blaker W., and Goodrum J. (1982) Axonal Transport of a Mitochondria Specific Lipid, in *Axoplasmic Transport* (Weiss D. G., ed.), pp. 175–180, Springer-Verlag, Berlin.

Nauta H. J., Kaiserman-Abramof I., and Lasek R. J. (1975) Electron microscopic observations of horseradish peroxidase transported from the caudoputamen to the substantia nigra in the rat: Possible involvement of the agranular reticulum. *Brain Res.* **85**, 373–384.

Nauta H. J., Pritz M. B., and Lasek R. J. (1974) Afferents to the rat caudoputamen studied with horseradish peroxidase: An evaluation of a retrograde neuroanatomical research method. *Brain Res.* **67**, 219–238.

Nixon R. A., Brown B., and Marotta C. (1982) Posttranslational modification of neurofilament protein during axoplasmic transport: Implications for regional specialization of CNS axons. *J. Cell Biol.* **94**, 150–158.

Oblinger M. (1984) Slow axonal transport in a CNS motor pathway: The protein composition and kinetics of SCa and SCb in hamster corticospinal axons. *Soc. Neurosci. Abs.* **10**, 1087.

Oblinger M. and Lasek R. J. (1984) A conditioning lesion of the peripheral axons of dorsal root ganglion cells accelerates regeneration of only their peripheral axons. *J. Neurosci.* **4**, 1736–1744.

Oblinger M., Brady S., and McQuarrie I. (1982) Comparative compositional analysis of slowly transported axonal proteins in peripheral and central mammalian neurons. *Soc. Neurosci. Abstr.* **8**, 826.

Ochs S. (1972a) Fast transport of materials in mammalian nerve fibers. *Science* **176**, 252–260.

Ochs S. (1972b) Rate of fast axoplasmic transport in mammalian nerve fibers. *J. Physiol.* (Lond.) **227**, 627–645.

Ochs S. and Ranish N. (1969) Characteristics of the fast transport system in mammalian nerve fibers. *J. Neurobiol.* **1**, 247–261.

Oeltmann T. N. and Wiley R. G. (1984) Wheat germ agglutinin-ricin A-chain (WGA-SS-RTA) conjugate: A new semisynthetic suicide transport agent. *Soc. Neurosci. Abstr.* **10**, 352.

O'Farrell P. (1975) High resolution two dimensional electrophoresis of proteins. *J. Biol. Chem.* **250**, 4007–4021.

Partlow L., Ross C., Motwani R., and McDougal D. (1972) Transport of axonal enzymes in surviving segments of frog sciatic nerve. *J. Gen. Physiol.* **60**, 388–405.

Peters A., Palay S., and Webster H. (1976) *The Fine Structure of the Nervous System.* Saunders, Philadelphia.

Quarles R. H. and Brady R. O. (1971) Synthesis of glycoproteins and gangliosides in developing rat brain. *J. Neurochem.* **18**, 1809–1820.

Reperant J., Miceli D., and Raffin J. (1977) Transneuronal transport of tritiated fucose and proline in the avian visual system. *Brain Res.* **121**, 343–347.

Ruda M. A. and Coulter J. D. (1979) Lectins as markers of axoplasmic transport in the nervous system. *J. Histochem. Cytochem.* **28**, 607.

Sasaki-Sherrington S., Jacobs J., and Stevens J. (1984) Intracellular control of axial shape in nonuniform neurites: A serial electron microscopic analysis of organelles and microtubules in AI and AII retinal amacrine neurites. *J. Cell Biol.* **98**, 1279–1290.

Schliwa M. (1984) Mechanisms of intracellular organelle transport. *Cell Musc. Mot.* **5**, 1–82.

Schroer T. A. and Brady S. T. (1984) Fast axonal transport (FAT) of fluorescently labelled elasmobranch synaptic vesicles in isolated axoplasm from squid giant axon. *Biol. Bull.* **167**, 504–505.

Schubert P. and Hollander H. (1975) Methods of Delivery of Tracers to the Central Nervous System, in *The Use of Axonal Transport for Studies of Neuronal Connectivity* (Cowan W. M. and Cuenod M., eds.), pp. 113–126, Elsevier, New York.

Schwab M. E. (1977) Ultrastructural localization of a nerve growth factor–horseradish peroxidase (NGF-HRP) coupling product after retrograde transport in adrenergic neurons. *Brain Res.* **130**, 190–196.

Schwab M. E. and Thoenen H. (1978) Selective binding, uptake, and retrograde transport of tetanus toxin by nerve terminals in the rat iris. *J. Cell Biol.* **77**, 1–13.

Schwab M. E., Javoy-Agid F., and Agid Y. (1978) Labeled wheat germ agglutinin (WGA) as a new, highly sensitive retrograde tracer in the rat brain hippocampal system. *Brain Res.* **152**, 145–150.

Schwab M. E., Suda K., and Thoenen H. (1979) Selective retrograde transsynaptic transfer of a protein, tetanus toxin, subsequent to its retrograde axonal transport. *J. Cell Biol.* **82**, 798–810.

Shaw G. and Bray D. (1977) Movement and extension of isolated growth cones. *Exp. Cell Res.* **104**, 55–62.

Shield L., Griffin J., Drachman D., and Price D. (1977) Retrograde axonal transport: A direct method for measurement of rate. *Neurology* **27**, 393.

Skene P. and Willard M. (1981) Axonally transported proteins associated with growth in rabbit central and peripheral nervous systems. *J. Cell Biol.* **89**, 96–103.

Smith R. S. (1972) Detection of organelles in myelinated nerve fibers by dark field microscopy. *Can. J. Physiol. Pharmacol.* **50**, 467–469.

Smith R. S. (1980) The short term accumulation of axonally transported organelles in the region of localized lesions of single myelinated axons. *J. Neurocytol.* **9**, 39–65.

Smith R. S. and Koles Z. J. (1976) Mean velocity of optically detected intraaxonal particles measured by a cross correlation method. *Can. J. Physiol. Pharmacol.* **54**, 859–869.

Snyder R. and Smith R. S. (1983) Physical methods for the study of the dynamics of axonal transport. *CRC Crit. Rev. Biomed. Eng.* **10**, 89–123.

Specht S. (1983) Axonal transport of Na,K-ATPase in optic nerve of hamster. *Curr. Top. Memb. Res.* **19**, 819–823.

Specht S. and Grafstein B. (1977) Axonal transport and transneuronal transfer in mouse visual system following injection of $^3$H fucose into the eye. *Exp. Neurol.* **41**, 705–722.

Specht S. C. and Sweadner K. J. (1984) Two different NA,K-ATPases in the optic nerve: Cells of origin and axonal transport. *Proc. Nat. Acad. Sci. USA* **81**, 1234–1238.

Spencer M. (1982) *Fundamentals of Light Microscopy.* Cambridge Univ. Press, Cambridge.

Sternberger L. and Sternberger N. (1983) Monoclonal antibodies distinguish phosphorylated and nonphosphorylated forms of neurofilaments *in situ. Proc. Natl. Acad. Sci. USA* **80**, 6126–6130.

Steward O. (1981) Horseradish Peroxidase and Fluorescent Substances and Their Combination With Other Techniques, in *Neuroanatomical Tract–Tracing Methods.* (Heimer L. and Robards M., eds.), pp. 279–310, Plenum, New York.

Steward O. and Fass B. (1983) Polyribosomes associated with dendritic spines in the denervated dentate gyrus: Evidence for local regulation of protein synthesis during reinnervation. *Prog. Brain Res.* **58**, 131–136.

Stoeckel K. and Thoenen H. (1975) Retrograde axonal transport of nerve growth factor: Specificity and biological importance. *Brain Res.* **85**, 337–341.

Stoeckel K., Schwab M., and Thoenen H. (1975a) Specificity of retrograde transport of nerve growth factor (NGF) in sensory neurons: A biochemical and morphological study. *Brain Res.* **89**, 1–14.

Stoeckel K., Schwab M., and Thoenen H. (1975b) Comparison between the retrograde axonal transport of nerve growth factor and tetanus toxin in motor, sensory, and adrenergic neurons. *Brain Res.* **99**, 1–16.

Stoeckel K., Schwab M., and Thoenen H. (1977) Role of gangliosides in the uptake and retrograde axonal transport of cholera and tetanus toxins as compared to nerve growth factor and wheat germ agglutinin. *Brain Res.* **132**, 273–285.

Stone G. C., Wilson D. L., and Hall M. E. (1978) Two-dimensional gel electrophoresis of proteins in rapid axoplasmic transport. *Brain Res.* **144**, 287–302.

Streit P. (1980) Selective retrograde labeling indicating the transmitter of neuronal pathways. *J. Comp. Neurol.* **191**, 429–465.

Tashiro T. and Komiya Y. (1983) Subunit composition specific to axonally transported tubulin. *Neurosci.* **4**, 943–950.

Theiler R. and McCLure W. O. (1977) A comparison of axonally transported proteins in the rat sciatic nerve by in vitro and in vivo techniques. *J. Neurochem.* **28**, 321–330.

Toews A., Goodrum J., and Morell P. (1979) Axonal transport of phospholipids in rat visual system. *J. Neurochem.* **32**, 1165–1173.

Tsukita S. and Ishikawa H. (1980) The movement of membranous organelles in axons: Electron microscopic identification of anterogradely and retrogradely transported organelles. *J. Cell Biol.* **84**, 513–530.

Tsukita S. and Ishikawa H. (1981) The cytoskeleton in myelinated axons: A serial section study. *Biomed. Res.* **2**, 424–437.

Tytell M., Black M., Garner J., and Lasek R. (1981) Axonal transport: Each of the major rate components consist of distinct macromolecular complexes. *Science* **214**, 179–181.

Tytell M., Brady S. T., and Lasek R. (1984) Axonal transport of a subclass of tau proteins: Evidence for the regional differentiation of microtubules in neurons. *Proc. Natl. Acad. Sci. USA* **81**, 1570–1574.

Tytell M., Gulley R., Wenthold R., and Lasek R. (1980) Fast axonal transport in auditory neurons: A rapidly turned over glycoprotein. *Proc. Natl. Acad. Sci. USA* **77**, 3042–3046.

Warr W., deOlmos J., and Heimer L. (1981) Horseradish Peroxidase: The Basic Procedure, in *Neuroanatomical Tracing Methods* (Heimer L. and Robards M., eds.), pp. 207–262, Plenum, New York.

Weiss D. G. (ed.) (1982) *Axoplasmic Transport.* Springer-Verlag, Berlin.

Weiss D. G. and Gorio A. (eds.) (1982) *Axoplasmic Transport in Physiology and Pathology.* Springer Verlag, Berlin.

Weiss P. and Hiscoe H. (1948) Experiments on the mechanism of nerve growth. *J. Exp. Zool.* **107**, 315–395.

Wenthold R., Skaggs K., and Reale R. (1984) Retrograde axonal transport of antibodies to synaptic membrane components. *Brain Res.* **304**, 162–165.

Wessels N., Johnson S., and Nuttall R. (1978) Axon initiation and growth cone regeneration in cultured motor neurons. *Exp. Cell Res.* **117**, 335–345.

Willard M. (1977) The identification of two intraaxonally transported polypeptides resembling myosin in some respects in the rabbit visual system. *J. Cell Biol.* **75**, 1–11.

Willard M. and Simon C. (1983) Modulations in neurofilament axonal transport during development of rabbit retinal ganglion cells. *Cell* **35**, 551–559.

Willard M., Cowan W. M., and Vagelos P. R. (1974) The polypeptide composition of intraaxonal transported proteins: Evidence for four transport velocities. *Proc. Natl. Acad. Sci. USA* **71**, 2183–2187.

Willard M., Wiseman M., Levine J., and Skene P. (1979) Axonal transport of actin in rabbit retinal ganglion cells. *J. Cell Biol.* **81**, 581–591.

Willingham M. and Pastan I. (1978) The visualization of fluorescent proteins in living cells by video intensification microscopy (VIM). *Cell* **13**, 501–507.

Wilson D. L., and Stone G. C. (1979) Axoplasmic transport of proteins. *Ann. Rev. Biophys. Bioeng.* **8**, 27–45.

Wujek J. and Lasek R. J. (1983) Correlation of axonal regeneration and slow component b in two branches of a single axon. *J. Neurosci.* **3**, 243–251.

Yamada K., Spooner B., and Wessels N. (1971) Ultrastructure and function of growth cones and axons of cultured nerve cells. *J. Cell Biol.* **49**, 614–635.

Younkin S., Brett R., Davey B., and Younkin L. (1978) Substances moved by axonal transport and released by nerve stimulation have an innervation-like effect on muscle. *Science* **200**, 1292–1295.

Zatz M. and Barondes S. (1971a) Rapid transport of fucosyl glycoproteins to nerve endings in mouse brain. *J. Neurochem.* **18,** 1125–1133.

Zatz M. and Barondes S. (1971b) Particulate and solubilized fucosyl transferases from mouse brain. *J. Neurochem.* **18,** 1625–1637.

Zipser B., Stewart P. L., Flanagan T., Flaster M., and Macagno E. (1983) Do monoclonal antibodies stain sets of functionally related leech neurons? *Cold Spring Harbor Laboratory Symp. Quat. Biol.* **48,** 551–556.

# Chapter 13

# Neurochemical Studies in Human Postmortem Brain Tissue

## Gavin P. Reynolds

*The proper study of mankind is man.*
Alexander Pope

## 1. Introduction

The use of human postmortem brain tissue in neurochemical and neuropharmacological research has received increasing attention over the past two decades. In fact, there is one work that, more than any other, can be identified as being responsible for the interest in this approach. It was Birkmayer and Hornykiewicz who, having observed a deficit in the content of the (then newly recognized) neurotransmitter dopamine (DA) in brain tissue taken postmortem from patients with Parkinson's disease, set about to counteract this deficit in living patients by treatment with L-dopa. The identification of an abnormally low transmitter concentration and its supplementation by the administration of the appropriate biochemical precursor has revolutionized the treatment of this disease (Ehringer and Hornykiewicz, 1960; Birkmayer and Hornykiewicz, 1961). It has also served to motivate neurochemists to study other neurological and psychiatric diseases using postmortem brain tissue. Despite the large amount of data that this approach has provided, and with it the increased understanding of neurochemical dysfunction in these disorders, such a success story has yet to be repeated.

This defines one major goal for studies involving postmortem human brain tissue: to determine the chemical pathology of diseases of the human brain, the neurological and psychiatric disorders. A further aim is to understand the actions of drugs that act on the brain and relieve, or induce, such disorders. With this and other information from more basic biochemical studies, it is hoped to add to our (admittedly naive) picture of the complex integration of cellular and molecular processes that we call human brain function.

## 2. Human Brain or the Animal "Model"

The small mammal bred for laboratory use has many undeniable advantages over human tissue taken postmortem. Severe neurosurgical procedures, acute or chronic drug treatment, and more subtle manipulations, such as dietary adjustments or behavioral training, can precede neurochemical investigation in animals. Nevertheless, such experiments can be misleading. The importance of results from animal studies is often overemphasized. It is very tempting to draw analogies with human brain function when faced with interesting results from experiments performed on the rat. The danger of such extrapolation is understood by every good scientist and yet, all too often, is overlooked or ignored. There are innumerable examples of metabolic differences between *Homo sapiens* and experimental animals; one very relevant to the study of DA function is the relative amount of this transmitter oxidized to dihydroxyphenylacetic acid, 10–20% in humans, and about 70–80% in small rodents. Similarly the enzymes responsible for conjugation and removal of endogenous and exogenous compounds vary greatly between species (and even between different races in humans); one arbitrary example is the major phenylalanine metabolite phenylacetic acid, removed by conjugation with glutamine in humans but with glycine in the mouse. Similar effects will be responsible for species differences in pharmacokinetics and drug disposition. Even in in-vitro pharmacological studies such differences may well be apparent, because of variation in antagonist affinity for a receptor or inhibition of an enzyme.

### 2.1. Human Neuropharmacology

It is unfortunate, then, that postmortem human brain tissue is still underused in basic pharmacological studies. Seeman (1980), in his comprehensive review of DA receptors, lists data for a wide range of antagonists including the neuroleptic drugs. Very few of these commonly prescribed antipsychotics have been investigated in their in vitro action on human brain. Very simple studies with human tissue often can serve to correct widespread misconceptions as to the pharmacological action or specificity of such drugs (e.g., Reynolds et al., 1982). Not only are the results more relevant to the use of drugs (except in veterinary pharmacology), human tissue is so much more abundant. One brain can provide 10 g of caudate tissue; over 100 rats would typically be required for an equivalent amount of striatum!

In addition to this straightforward in vitro human neuropharmacology, the effects of previously administered drugs can be studied postmortem in patients. Birkmayer and Riederer (1983) quote several such investigations into the biochemical changes induced by L-dopa and/or other drugs in parkinsonian patients. Even single cases can occasionally be instructive, such as the large increase in DA receptor number observed in a hypertensive patient who had received chronic reserpine treatment before death (Reynolds et al., 1981a).

## 3. The Collection of Human Brain Postmortem

Before describing what is involved in the collection of material for studies on human brain tissue, in essence the logistics of "brain banking," I should remind the reader that what follows are procedures carried out in Cambridge, UK, that necessarily represent a compromise among a range of very different factors. The neuroscientist wishing to embark on such work may well find that other considerations, including differences in priorities and facilities, as well as in local regulations and pathology practice, require him or her to follow different procedures.

Once a patient has died and all appropriate consent has been obtained (i.e., permission for autopsy with explicit or implicit consent for removal of tissue for research), preparations can be made to obtain the brain. The body should be moved to refrigerated (4°C) storage, ideally within 4 h of death. Under these conditions the brain still cools slowly, taking some 15 h to cool to below 10°C (Spokes and Koch, 1978). At this point, two opposing factors need some consideration. First, there is the desire to obtain a neurochemical profile that best reflects the premortem state of the brain. This is offset by the fact that most biochemical changes occur in the first few hours after death as the tissue equilibrates with ōts anoxic state and before the temperature has dropped substantially. It is this latter problem, along with the relatively long delay between death and autopsy typically found in a British general hospital, that has prescribed for us an optimum period of 24–72 h between death and brain removal. For certain studies, to be discussed later, this is far too long; the above figures represent a compromise applied to most of the brain tissue taken by this brain bank for subsequent frozen storage, dissection, and measurement of various biochemical species. Investigations in metabolic activity or transmitter uptake and release require fresh tissue taken only a few hours postmortem.

Once the brain has been removed, it is normally cut midsagittally into the two hemispheres. Arbitrarily, one hemisphere is chosen for fixing in 10% formalin, whereas the other is frozen at −20°C with the sagittal surface downwards to preserve the shape as well as possible. This frozen hemisphere will be transferred to storage at −70°C in an airtight bag to await dissection. Clearly, this procedure is inappropriate for many research projects. For example, neurochemical studies of laterality require both hemispheres to be frozen. This is best done by inverting the whole brain without sectioning into a mold formed from a bowl padded with cotton wool and covered with a plastic film. The lack of a formalin-fixed hemisphere available for histological studies prevents the investigator from obtaining a full neuropathological assessment of the brain. Thus, it may be advisable to remove a few small sections of fresh tissue from regions not required for neurochemistry; these sections can be used to give an indication as to the possible presence of confounding factors, such as the neuronal changes of Alzheimer's disease or other microscopic signs of cellular degeneration. Such a procedure is, of course, a compromise. Even when one hemisphere is fixed and carefully studied by an experienced neuropathologist, it is still impossible to ensure that degenerative changes, which inevitably affect neurochemical parameters, are not present in the other, frozen hemisphere.

Most "banked" tissue is stored frozen after dissection, done either from the fresh or (slow) frozen brain. The former approach has the advantage of enabling anatomical landmarks to be more easily identified: obvious color differences in the fresh brain tend to disappear in frozen tissue. However, this requires the dissection to take place within a few hours of the postmortem, an impossibility in cases when a brain bank collects tissue from a large, possibly nationwide, catchment area. For this reason, too, the standard procedure followed requires one hemisphere to be frozen at −20°C for 2–3 d before being stored at −70°C because many pathologists do not have access to a −70°C freezer. Brains taken in other hospitals can thus be collected after a few days using a container of dry ice for transport.

## 4. Dissection

When the frozen brain is to be dissected, it should be placed at −20°C for 12–24 h to warm up; tissue at −70°C is extremely hard. Coronal sections 5 mm thick are then cut using a standard kitchen slicer with an electrically powered rotating blade (Fig. 1). These sections are laid on a refrigerated (approx. −10°C) tray whereupon the individual brain regions and nuclei can be dissected out using a scalpel (Fig. 2). A standard atlas (e.g., Roberts and Hanaway, 1970) can be employed to identify these regions. This dissected tissue is then finely diced and stored at −70°C in screwcap polycarbonate tubes until required for analysis, at which time an aliquot of the tissue may be removed while leaving the remainder frozen. Dicing the tissue before storage is an attempt to provide a fairly homogeneous sample by eliminating the neurochemical gradients present in intact samples of tissue from particular brain regions. Other groups have approached this problem in different ways. Perhaps the best solution, given the

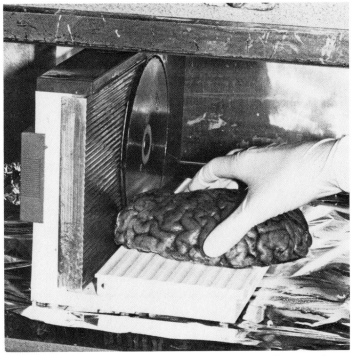

Fig. 1.   Preparation of 5 mm coronal slices from a frozen hemisphere.

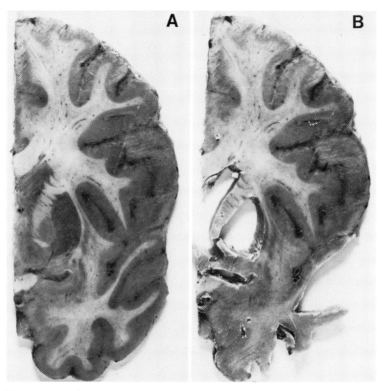

Fig. 2.    A frozen coronal slice from a brain hemisphere at the level of the anterior commissure before (A) and after (B) removal of tissue from regions of the striatum and temporal cortex.

availability of equipment, is that of Winblad et al. (1982), who pulverize the tissue in liquid nitrogen, resulting in a totally homogeneous preparation.

The previous section referred to the impossibility of obtaining histological assessment on the same tissue samples as are frozen for neurochemistry. Two different approaches can be used to minimize this problem. First, tissue can be dissected fresh for neurochemistry, and adjacent sections fixed for histological assessment. This is perhaps more appropriate for tissue from certain brain regions (e.g., cortex, hippocampus) than from other areas in which tissue volume is limited and/or complex neurochemical gradients occur in all three dimensions (e.g., locus ceruleus, hypothalamus, amygdala).

The second approach is to freeze the tissue quickly. This can be done after slicing the brain into coronal sections and freezing the slices between light metal plates surrounded by dry ice. Such fast freezing reportedly (Tourtellotte, unpublished) minimizes the ice crystal artifacts that otherwise limit the use of frozen human brain tissue for histological investigations. Presumably the rate of freezing governs the size of the ice crystals that form within the tissue. On the other hand, a slower rate of freezing has been found to be better at preserving intact, metabolically viable, synaptosomes. This will be discussed later in the chapter.

Blocks of tissue fast-frozen in this way can be cut on a microtome in a cryostat to prepare slices for both histological and neurochemical assessment—thin slices to be stained for microscopy, thin slices for receptor autoradiography, and thicker slices for neurochemical assay using, for example, the grid or punch microdissection techniques. This is clearly a most valuable way of obtaining the maximum amount of information from a tissue sample. It is, however, labor intensive and, thus, less appropriate as a routine "banking" technique. One brain bank, in Los Angeles, does store tissue in frozen coronal sections that may be appropriate for such studies; however, this requires a much larger volume of deep-freeze space.

## 5. Safety

Dissection probably represents one of the more potentially hazardous procedures in the handling of human brain postmortem tissue, so some considerations should be given to the topic of safety. Brain tissue derived from cases of disease that present a substantial danger of infection, such as tuberculosis or hepatitis B, should be treated in the same way as any blood or tissue sample containing such pathogens, by following the precautions normally taken in a clinical laboratory. In addition, special account should be taken of diseases of the brain. Creutzfeldt-Jakob disease is a better known example in which postmorten transmission to an otherwise healthy individual has had particularly tragic results. Similar virus-like vectors have been implicated in other diseases of unknown etiology that are frequently studied using "banked" brain tissue. Parkinson's disease, Alzheimer's disease, and even schizophrenia are notable examples.

The optimist can point at the lack of hard evidence relating to both the transmissibility of such diseases and to morbidity in neuropathologists. Nevertheless, it is better to be safe than sorry, and in this respect it is unfortunate that there are no generally recommended regulations. However, Britain does have advisory guidelines for the handling of unfixed human brain that recognize the potential pathogenicity of all brain tissue. Thus, they recommend that all procedures resulting in tissue disruption (i.e., mechanical slicing, homogenization), and are thus liable to produce an aerosol containing brain tissue, should be carried out in a fume cabinet. This should be of an exhaust-protective, open-fronted design and housed in a room set aside for such purposes. Other manipulations are considered less hazardous and can be done under laboratory conditions at which time more common precautions are observed.

As more research groups are becoming involved in postmortem brain investigations, such considerations are being increasingly ignored. It is clearly expensive (and often prohibitively so) to equip a laboratory to meet these recommendations. It would be unfortunate if, in the future, it were "brain bankers" who were the subjects used to confirm the hypothesis of an infectious vector in one or other neurological disease of unknown etiology!

## 6. Variables Affecting Postmortem Neurochemistry

Table 1 lists some of the factors that are known to have effects on various neurochemical species measured in brain tissue.

TABLE 1
Variables Potentially Affecting Neurochemical
Parameters in Human Brain Tissue

Premortem
  Age
  Sex
  Agonal state
  Cause and manner of death
  Psychiatric and neurological history
  Terminal drug treatment
  Previous long-term drug treatment
  Time of day
  Time of year
  Point in menstrual cycle (younger females)
Postmortem
  Delay before refrigeration
  Delay before brain removal
  Dissection technique
  Storage conditions of frozen tissue

## 6.1. Age

Age and sex of the donor are perhaps the most obvious influences that are easiest to control for when attempting to match experimental groups. In particular, many changes with age have been documented that include reductions in catecholamines and glutamate decarboxylase (GAD) specific to some brain regions (Spokes, 1979). Allen et al. (1983) also have reported specific losses of presynaptic GABA function in the cortex and of cholinergic markers in the striatum. Many of the neurochemical changes associated with aging have emerged from the recent upsurge in interest in dementia of the elderly. Winblad et al. (1982) have summarized the losses in various monoamine-related parameters (although there is an increase in monoamine oxidase activity) that occur with age; many, but not all, of these are more profoundly changed in the dementia of Alzheimer's disease.

Spokes (1979) reports a lack of agreement between different groups regarding several of these observed changes. In this respect it is important to assess the influence of other factors, particularly agonal status and cause of death, which are likely to differ between younger and older patients. Since, for example, bronchopneumonia is more likely to be a cause of death in an older group, a potential influence of this and similar diseases on the measurement of, say, GAD may masquerade as an apparent age effect unless it is specifically controlled for.

## 6.2. Agonal State and Cause of Death

A decrease in GAD does, in fact, occur in brain tissue from patients who die after a protracted illness as opposed to previously healthy subjects who die suddenly (Spokes, 1979; Perry et al., 1982) have undertaken a study of this effect of agonal state on a wide range of biochemical species and find that GAD is the only activity directly associated with neurotransmitter function that shows a significant difference. However, tissue pH drops and several amino acids, notably tryptophan (increased by over 200%), exhibit substantial increases in concentration in the chronically ill group. Whether these changes are in some way related to hypoxia has yet to be determined.

Other causes of death may be thought likely to affect the neurochemical status of the brain; encephalitis, meningitis, and hepatic coma are a few more obvious examples. Certainly the last of these has been found to have profound effects on brain amino acids, particularly tryptophan (Weiser et al., 1978), as well as on the concentrations of 5-HT and its specific binding site (Riederer

et al., 1981). These authors also report losses in brain levels of
DA-stimulated adenylate cyclase in noncomatose cases of liver
cirrhosis.

## 6.3. Neuropsychiatric Disease

Since many studies using human brain tissue are undertaken in
order to test neurochemical hypotheses of neurological and psychiatric conditions, subjects with a history of such disease ideally
should be excluded from any control group or studies with "normal" brain tissue. Not only are potential neurochemical changes
associated with the disease process a factor here; such patients almost invariably receive long-term drug treatment.

## 6.4. Drug Treatment

A wide range of neuroleptics, stimulants, antidepressants, anxiolytics, antiparkinsonian drugs, and other psycho- and neuroactive medications have been shown to have effects on neurochemical parameters in experimental animals. It is not always
clear whether these effects are paralleled by changes in the human brain (for reasons that have been discussed), although there
are indications that the neuroleptic (Reynolds et al., 1981; Mackay
et al., 1980) and antiparkinsonian drugs (Birkmayer and Riederer,
1983), among others, do induce similar neurochemical changes in
human brain tissue.

It should be remembered that drugs active on the cardiovascular system may well also have CNS activity: particularly
notable are the beta blockers. Noradrenaline (NA) or DA may be
given in relatively large quantities shortly before the death of subjects with acute heart failure who might otherwise appear to be
"good" controls. This can lead to massive (i.e., over tenfold) increases in the concentration of these transmitters in certain brain
regions. Administration of opiates, common in chronically ill terminal cases, has been observed to have effects on 5-
hydroxytryptamine (5-HT) systems (Bucht et al., 1981) in addition
to their inevitable influence on the opiate peptides. Cytotoxic
chemotherapy for cancer also is likely to have effects on neuronal
function although, as with most drug treatments, a systematic
study using postmortem brain tissue is lacking.

## 6.5. Cyclic Fluctuations With Time

It is well established that the pineal exhibits a profound circadian
rhythm in the content of melatonin and its synthesizing enzymes,
and human postmortem studies have closely related these con-

centrations to the time of death (Smith et al., 1981). There are indications that other neuronal systems exhibit a dependence on time of day. Perry and Perry (1983) have reported that two enzymic markers of cholinergic transmission, as well as muscarinic receptor binding, exhibit circadian variations. Hypothalamic NA and its metabolites, as well as DA and 5-HT in this region, have also been identified (Carlsson et al., 1980) as showing a dependence on hour of death. These authors have reported seasonal fluctuations in 5-HT and DA in the hypothalamus, with less notable variations in other brain regions. Pineal function also exhibits a circannual rhythmicity (Smith et al., 1981) as do some indicators of DA activity (Karson et al., 1984) although, as the latter authors point out, potential postmortem changes dependent on ambient temperature may well introduce artificial seasonal effects.

The menstrual cycle is likely to evoke neurochemical indications of a central effect; certainly female sex hormones can, in animal experiments, induce changes in several transmitter systems including DA receptors (Hruska and Silbergeld, 1980). Therefore, apart from the menstrual cycle, the sex of the subject would be expected to have effects, and certainly several indicators of 5-HT and DA systems in the brain differ between men and women (Gottfries et al., 1981).

## 6.6. Postmortem Delays

The reluctance of many biochemists to use human tissue as a research tool is based mainly on the delay of many hours (and sometimes days) between death and tissue availability for experimentation. The initial assumption is that in this time so many proteolytic, oxidative, or other changes will have occurred that there is little resemblance to the neurochemical status before death. This view is to some extent understandable when one considers the work done in animal experiments to minimize such changes: For certain studies involving analyses of transmitters some groups consider decapitation and immediate immersal of the brain in liquid nitrogen to be inadequate and recommend killing the animal by high-power microwave heating of the head that inactivates the relevant enzymic processes in much less than a second!

It is, however, wrong to dismiss work on human brain tissue for this reason. Many enzymes and neurotransmitters are surprisingly stable postmortem, particularly when the body is refrigerated shortly after death. How can this stability be assessed? First, a large group of control material with varying postmortem delays can be used to assess variation with time, although this does not indicate what changes occurred during the initial hours after death. Gottfries et al. (1981) have identified changes in some amine transmitters and their metabolites as well as increases in tryptophan and tyrosine with increasing postmortem delay using this method. Second, postmortem and biopsy tissues could be compared, although the very limited availability of the latter, along with even greater limitations in the brain regions that could be studied, prevent this method from being of general use. Third, an animal model of the cooling conditions of the human brain between death and autopsy can be used. Spokes and Koch (1978) first used this method by monitoring brain temperature after death and subsequently constructing a cooling curve for mice killed by cervical dislocation (i.e., without breaking the skin). They observed that while the activities of GAD and choline acetyltransferase in the brains were little changed (GAD stabilizing at 80% after about 24 h), DA concentrations dropped by about 50% and most of this occurred during the first 4 h after death. On the other hand, the neuropeptides exhibit no losses over 72 h (Emson et al., 1981) and a wide range of enzymic or receptor activities are reported (Hardy and Dodd, 1983) to be stable for several hours postmortem.

### 6.7. Controlling for These Factors

The above discussion has mentioned many potential influences on neurochemical measurements, all of which would be difficult, if not impossible, to correct or take fully into account. Hence the importance of choosing, when necessary, appropriate control groups. Matching for age and postmortem delay is usually straightforward, although we often find that from patients dying in a small provincial hospital, the brain is removed faster than would be the norm for our "control" subjects taken from a large general hospital. Many of the other variables, such as sex, time of death, and so on, will often automatically show no significant difference between the groups. However, when the study involves a group of such neurological or psychiatric cases, several factors inevitably will differ. The unknown influence of hospitalization/ institutionalization, and the possible differences in cause of death that may reflect this, are particular examples. A further "institutionalized control" group may permit such influences to be accounted for, albeit only after introducing further confounding factors. Drug treatment is another problem and it is frequently impossible to differentiate between drug effects and the disease process (*see* section 9).

In many sample groups the number of variable parameters will outweigh the number of samples, so it is perhaps advisable to approach such studies with a different mental attitude than if one were planning an equivalent animal experiment. First, one should remember that in general there would be no equivalent animal study for the investigation (parameter $X$ in disease state $Y$ in the human). Second, the variability of the subject matter, which is such a limitation in one sense, may also provide far more information than the "ideal" situation in which the subjects within a group were all the same age, with a similar cause of death and so on. For example, it has been shown that choline acetyltransferase activity in cortical tissue increases with age of death in subjects with Alzheimer's disease but not in controls, an observation that has as much importance for the understanding of this disease as the difference in the enzyme activities between the groups (Rossor et al., 1981a). The fortuitous identification of such a correlation would be impossible in groups of subjects with less individual variability.

## 7. Micromethods

A wide range of neurochemical determinations can be applied to postmortem brain tissue, dissected as described above, using fairly standard biochemical "test tube" techniques. These permit comparisons of, for example, the neurochemistry of different regions of the brain or between the same region in different subject groups. However, there are limitations. The problem of histological assessment of tissue samples has been mentioned above (section 4) along with different freezing and dissection techniques that permit more precise study of the neurochemical anatomy of the human brain. Some of these will be discussed further.

### 7.1. Histochemistry

The various histochemical techniques, in particular immunohistochemistry, are widely used in the neurosciences and are equally applicable to human tissue as to animals except in those cases in which changes postmortem or during tissue preparation remove or inactivate the relevant marker. Histochemistry in general requires fixation of relatively fresh tissue so that cellular integrity is preserved. Peptide transmitters, fairly stable postmortem, are particularly suitable subjects for immunohistochemistry, which has provided some interesting data (e.g., Hunt et al., 1982). Histochemistry, as well as being used to provide detailed anatomy of a transmitter or enzyme, has also been employed to define anatomical nuclei prior to "gross" neurochemical studies, for example identifying the cholinergic cells of the substantia innominata and their changes in Alzheimer's disease (Rossor et al., 1981b).

## 7.2. Receptor Autoradiography

Autoradiography has been used as a visualization technique for immunohistochemical studies by employing radiolabeled antibodies. However, the recent developments in provision of high specific activity radioligands for receptor studies have been followed by the application of autoradiography to the identification of neurotransmitter receptors in sections of brain tissue. A few groups have investigated human brain in this way, using thin sections prepared from frozen tissue (e.g., Palacios et al., 1980). Since receptors are, on the whole, very stable in the postmortem period, the technique as developed in animal tissue can be applied generally to human samples. The same advantages and disadvantages over "test-tube" receptor binding apply; these include the generally lower amounts of nonspecific binding and the greater difficulties in kinetic characterization of the receptor.

## 7.3. Microdissection

The dissection method generally applied to "brain banking" (*see* section 4) can hardly be described as microdissection, involving, as it does, fairly crude dissection of structures from brain sections several mm thick. However, the reader would be reminded that an animal brain of a gram or so in weight would certainly require microscopic techniques to obtain an equivalent accuracy. Nevertheless, even greater anatomical detail can be obtained with classical biochemical assays after using microdissection methods. The use of these methods in the study of animal tissues is described by Palkovits (*see* chapter in this volume), who pioneered punch microdissection. They are in many ways even better suited to subregional mapping of the human brain with its more complex anatomy yet larger volume of most regions and nuclei. The application of the "punch" and "grid" microdissection techniques to frozen sections of human postmortem brain tissue has recently been reviewed (Aquilonius et al., 1983; Kanazawa, 1983). Either small blocks of tissue (e.g., striatum or a cross-section slice of spinal cord) or, with the use of an appropriate large-section cryomicrotome, a whole brain or hemisphere can be used to provide frozen sections of tissue from which punches can be taken or a grid of tissue pieces prepared for biochemical analysis. These techniques have been used to study the heterogeneity of various biochemical species within regions of the brain. One notable study was that of Oke et al. (1978) who, by mapping the distribution of NA within the human thalamus, identified a pattern of lateral asymmetry for this transmitter.

## 8. Studies on Dynamic Neuronal Processes

So far the discussion has been restricted to static parameters in postmortem tissue; the measurement of receptor densities, transmitter concentrations, and enzyme activities. Few realize, however, that a range of functions normally associated with extremely fresh tissue can also be measured in preparations of brain many hours postmortem.

The initial indication that some dynamic cellular processes remain is that synaptosomes prepared conventionally from such tissue exhibit respiratory activity (Hardy et al., 1982). Thus, there is a supporting mechanism for fueling those processes, such as uptake and release of transmitters, which require energy. Uptake of catecholamines, 5-HT, and transmitter amino acids has indeed been observed by several groups, as have active depolarization-sensitive release processes (reviewed by Hardy and Dodd, 1983). These workers also describe an even more surprising phenomenon: the ability to store tissue deep-frozen for long periods without substantial losses of synaptosomal viability. They use a procedure of slow freezing followed by rapid thawing to obtain high yields of functioning synaptosomes (Hardy et al., 1983).

Bowen et al. (1982) have shown that metabolic activity and acetylcholine release occur in cortical tissue prisms taken shortly after death. This group has also been able to preserve such tissue in a viable form by the use of dimethyl sulfoxide to protect against damage from freezing and thawing (Haan and Bowen, 1981).

The use of these procedures to assess neurotransmitter uptake and release is clearly underused. Further work is required, particularly in the application to disease states, before one can assess their value as a research tool. The substantial variation in absolute functional activity ($V_{max}$) between individual samples due, for example, to agonal state and postmortem delay, will probably preclude simple comparisons, although the kinetic parameters ($K_M$) should be more stable and may well permit the identification of functional abnormalities in neuropsychiatric disease.

These preparations can be stored to retain viability almost indefinitely with potential for future use when neuroscience has far more to offer.

## 9. Neuropsychiatric Disease

The major motivation to work with postmortem brain tissue is to contribute to our understanding of the neurological and psychiatric diseases. This has been done with great success in Parkinson's disease. Huntington's chorea, too, has attracted a lot of scientific interest. Yet despite all the research and biochemical data that have emerged over the past decade (Spokes, 1981), no therapeutic breakthrough has been made. Can we identify why this is so?

Several minor transmitter abnormalities have been reported in the parkinsonian brain, but the sole major dysfunction appears to be in the dopaminergic nigro-striatal tract as indicated by a loss of striatal DA of about 90%. Other neurochemical changes are functional differences secondary to this disturbance. On the other hand, in Huntington's chorea a severe atrophy of the striatum occurs along with, to a somewhat lesser extent, a loss of cortical tissue. The substantial neuronal loss in the striatum is paralleled by depleted concentrations of several transmitters including, among others, GABA and substance P. In this case, it would appear that no one transmitter pathway is primarily affected and so no single specific pharmacological strategy will be able to normalize the balance of neurotransmitter function.

Unfortunately similar multiple losses occur in other neurological diseases. In addition to the well-established loss of cholinergic innervation of the cortex in Alzheimer's disease, there are deficits in the NA- and 5-HT-containing mesocortical pathways (Winblad et al., 1982). The degeneration of these latter neuronal systems may be responsible for the lack of therapeutic effect in demented patients of drugs designed to supplement cholinergic function.

The problem with the biochemistry of the psychiatric diseases is quite the opposite: a paucity of confirmed reports of biochemical changes. Nevertheless, few well-planned studies have investigated the affective disorders and one might still hope that postmortem research eventually may yield some identifiable neurochemical dysfunction in depression that will provide a more logical approach to treatment. Schizophrenia, however, has been the subject of more studies. Concordance in one biochemical finding is apparent: the density of DA $D_2$ receptors is increased. It is over the interpretation of this observation that there is much dispute, although the weight of the evidence would appear to indicate that the change in receptor number is a response to antipsychotic medication and not an effect of the disease itself

(Reynolds et al., 1981a; Mackay et al., 1980). This conclusion is reached after observing that a few schizophrenic patients who were not treated with neuroleptic drugs for at least some months before death did not exhibit increased receptor number. However, one must bear in mind the possibility that these patients were treated differently because they either did not respond to neuroleptics or had a milder form of the disease. This is difficult to assess retrospectively; but dividing the drug-treated group into different disease subgroups according to the International Classification did not reveal any difference in DA receptor number (Reynolds et al., 1981b). Finally, one can (cautiously) consider animal experiments, in which chronic neuroleptic drug treatment has been shown to increase the density of these receptors (Clow et al., 1980). This circumstantial evidence is far from conclusive but illustrates the particular difficulties inherent in postmortem research into psychiatric disorders.

## 10. The Future

This chapter has provided a brief review of the neurochemistry of postmortem human brain tissue as an approach, albeit with many limitations and problems, that is essential to a proper understanding of brain function.

An assessment of how research in this field will progress over the next few years is impossible, but there are indications. The growth in molecular biology has made an impact on neurology with the identification of a genetic marker linked to Huntington's chorea (Gusella et al., 1983). The "brain bank" does not provide any special source of DNA, found in all tissues, but it may be the only retrospective source for such studies.

The isolation of active mRNA from human postmortem brain has been demonstrated (Gilbert et al., 1981), and lends itself well to the investigation of hereditary neurological disorders or chromosomal abnormalities such as Down's syndrome; the latter disease has already been studied in this way (Whatley et al., 1984).

Mention has already been made of using small blocks of fast-frozen tissue to integrate classical biochemical assays, histochemistry, histology, and/or autoradiographic methods. The use of the large-section cryomicrotome should permit many of these techniques to be applied to larger sections from whole or half brains (Aquilonius et al., 1983). Neurochemical mapping of the whole brain should now be within reach!

## Acknowledgments

I especially wish to thank Prof. Peter Riederer, in whose laboratory I was introduced to postmortem neurochemistry of the human brain, and Dr. Martin Rossor, who taught me dissection and the logistics of brain banking. I am most grateful to the Schizophrenia Research Fund for many years of financial support, Sue West for typing the manuscript, and Richard Hills for photography.

## References

Allen S. J., Benton J. S., Goodhart M. J., Haan E. A., Sims N. R., Smith C. C. T., Spillane J. A., Bowen D. M., and Davison A. N. (1983) Biochemical evidence of selective nerve cell changes in the normal aging human and rat brain. *J. Neurochem.* **41,** 256–265.

Aquilonius S -M., Eckernäs S -Å., and Gillberg P -G. (1983) Large Section Cryomicrotomy in Human Neuroanatomy and Neurochemistry, in *Brain Microdissection Techniques* (Cuello A.C., ed.) pp. 155-170, IBRO/Wiley, London.

Birkmayer W. and Hornykiewicz O. (1961) Der L-Dioxyphenylanin (L-DOPA) Effekt bei der Parkinson-Akinesie. *Wien Klin. Wochenschr.* **73,** 787.

Birkmayer W. and Riederer P. (1983) *Parkinson's Disease.* Springer, Vienna.

Bowen D. M., Sims N. R., Lee K. A. P., and Marek K. L. (1982) Acetylcholine synthesis and glucose oxidation are preserved in human brain obtained shortly after death. *Neurosci. Lett.* **31,** 195–199.

Bucht G., Adolfsson R., Gottfries C. G., Ross B. E., and Winblad B. (1981) Distribution of 5-hydroxytryptamine and 5-hydroxyindole-acetic acid in human brain in relation to age, drug influence, agonal status and circadian variation. *J. Neural Transm.* **51,** 185–203.

Carlsson A., Svennerholm L., and Winblad B. (1980) Seasonal and circadian monoamine variations in human brain examined postmortem. *Acta Psychiatr. Scand.* **61** Suppl. 280, 75–85.

Clow A., Theodorou A, Jenner P., and Marsden C. D. (1980) Changes in rat striatal dopamine turnover and receptor activity during one year's neuroleptic administration. *Eur. J. Pharmacol.* **63,** 135–144.

Ehringer H. and Hornykiewicz O. (1960) Verteilung von Noradrenalin und Dopamin im Gehirn des Menschen und ihr Verhalten bei Erkrankungen des extrapyramidalen Systems. *Wien Klin. Wochenschr.* **72,** 1236.

Emson P. C., Rossor M., Hunt S. P., Clement-Jones V., Fahrenkrug J., and Rehfeld J. (1981) Neuropeptides in Human Brain, in *Transmitter Biochemistry of Human Brain Tissue* (Riederer P. and Usdin E., eds.), pp. 221–234, Macmillan, London.

Gilbert J. M., Brown B. A., Strocchi P., Bird E. D., and Marotta C.

A.(1981) The preparation of biologically active messenger RNA from human postmortem brain tissue. *J. Neurochem.* **36,** 976–984.

Gottfries C. G., Adolfsson R., and Winblad B. (1981) Analytical Problems in Postmortem Brain Studies, in *Transmitter Biochemistry of Human Brain Tissue* (Riederer P. and Usdin E., eds.) pp. 47-54, Macmillan, London.

Gusella J. F., Wexler N. S., Conneally P. M., Naylor S. L., Anderson M. A., Tanzi R. E., Watkins P. C., Ottina K., Wallace M. R., Sakaguchi A. Y., Young A. B., Shoulson I., Bonilla E., and Martin J. B. (1983) A polymorphic DNA marker genetically linked to Huntington's disease. *Nature* (Lond.) **306,** 234–238.

Haan E. A. and Bowen D. M. (1981) Protection of neocortical tissue prisms from freeze-thaw injury by dimethyl sulphoxide. *J. Neurochem.* **37,** 243–246.

Hardy J. A., Dodd P. R., Oakley A. E., Kidd A. M., Perry R. H., and Edwardson J. A. (1982) Use of postmortem human synaptosomes for studies of metabolism and transmitter amino acid release. *Neurosci. Lett.* **33,** 317–322.

Hardy J. A. and Dodd P. R. (1983) Metabolic and functional studies on postmortem human brain. *Neurochem. Int.* **5,** 253–266.

Hardy J. A., Dodd P. R., Oakley A. E., Perry R. H., Edwardson J. A., and Kidd A. M. (1983) Metabolically active synaptosomes can be prepared from frozen rat and human brain. *J. Neurochem.* **40,** 608–614.

Hruska R. E. and Silbergeld E. K. (1980) Estrogen treatment enhances dopamine receptor sensitivity in the rat striatum. *Eur. J. Pharmacol.* **61,** 397–400.

Hunt S. P., Rossor M. N., Emson P. E., and Clement-Jones V. (1982) Substance P and enkephalins in the spinal cord following limb amputation. *Lancet,* 1023.

Kanazawa I. (1983) Grid Microdissection in Human Brain Areas, in *Brain Microdissection Techniques* (Cuello A.G., ed.) pp. 127–153, IBRO/Wiley, London.

Karson C. N., Berman K. F., Kleinman J., and Karoum F. (1984) Seasonal variation in human central dopamine activity. *Psychiatr. Res.* **11,** 111–117.

Mackay A. V. P., Bird E. D., Spokes E., Rossor, M., Iversen L. L., Creese I., and Snyder S. H. (1980) Dopamine receptors and schizophrenia: drug effect or illness? *Lancet* **ii,** 223–225.

Oke A., Keller R., Mefford I., and Adams R. (1978) Lateralisation of norepinephrine in human thalamus. *Science* **200,** 1411–1413.

Palacios J. M., Young W. S., and Kuhar M. J. (1980) GABA and Benzodiazepine Receptors in Rat and Human Brain: Autoradiographic Localisation by a Novel Technique, in *Enzymes and Neurotransmitters in Mental Disease* (Usdin E., Sourkes T. L., and Youdim M. B. H., eds.) pp. 573–583, Wiley, London.

Perry E. K., Perry R. H., and Tomlinson B. E. (1982) The influence of agonal status on some neurochemical activities of postmortem human brain tissue. *Neurosci. Lett.* **29,** 303–307.

Perry E. K. and Perry R. H. (1983) Human brain neurochemistry—some postmortem problems. *Life Sci.* **33,** 1733–1743.

Reynolds G. P., Riederer P., Jellinger K., and Gabriel E. (1981a) Dopamine receptors and schizophrenia: the neuroleptic drug problem. *Neuropharmacol.* **20,** 1319–1320.

Reynolds G. P., Cowey L., Rossor M. N., and Iversen L. L. (1982) Thioridazine is not specific for limbic dopamine receptors. *Lancet* **ii,** 499–500.

Reynolds, G. P., Riederer, P., Jellinger, K., and Gabriel, E. (1981b) Dopamine Receptors and Schizophrenia: The Influence of Neuroleptic Drug Treatment and Disease Symptoms, in *Biological Psychiatry 1981* (Perris, C., Struwe, G., Jansson, B., eds.) pp. 715–718, Elsevier, Amsterdam.

Riederer P., Kruzik P., Kienzl E., Kleinberger G., Jellinger K., and Wesemann W. (1981) Central Aminergic Function and its Disturbance by Hepatic Disease, in *Transmitter Biochemistry of Human Brain Tissue* (Riederer P. and Usdin E., eds.) pp. 143–182. Macmillan, London.

Roberts M. and Hanaway J. (1970) *The Atlas of the Human Brain in Section.* Lea and Febiger, Philadelphia.

Rossor M. N., Iversen L. L., Johnson A. J., Mountjoy C. Q., and Roth M. (1981a) Cholinergic deficit in frontal cerebral cortex is age dependent. *Lancet* **ii,** 1422.

Rossor M. N., Svendsen C., Hunt S. P., Mountjoy C. Q., Roth M., and Iversen L. L. (1981b) The substantia innominata in Alzheimer's disease: a histochemical and biochemical study of cholinergic marker enzymes. *Neurosci. Lett.* **28,** 217–222.

Seeman P. (1980) Brain dopamine receptors. *Pharmacol. Rev.* **32,** 229–313.

Smith J. A., Mee T. J. X., Padwick D. J., and Spokes E. G. (1981) Human postmortem pineal enzyme activity. *Clin. Endocrinol.* **14,** 75–81.

Spokes E. G. S. and Koch D. J. (1978) Postmortem stability of dopamine, glutamate decarboxylase and choline acetyltransferase in the mouse brain under conditions simulating the handling of human autopsy material. *J. Neurochem.* **31,** 381–383.

Spokes E. G. S. (1979) An analysis of factors influencing measurements of dopamine, noradrenaline, glutamate decarboxylase and choline acetylase in human postmortem brain tissue. *Brain* **102,** 333–346.

Spokes E. G. S. (1981) The neurochemistry of Huntington's chorea. *Trends Neurosci.* **4,** 115–118.

Weiser M., Riederer P., and Kleinberger G. (1978) Human cerebral free amino acids in hepatic coma. *J. Neural Transm.* Suppl. 14, 95–102.

Whatley S. A., Hall C., Davison A. N., and Lim L. (1984) Alterations in the relative amounts of specific mRNA species in the developing human brain in Down's syndrome. *Biochem. J.* **220,** 179–187.

Winblad B., Adolfsson R., Carlsson A., and Gottfries C. G. (1982) Biogenic Amines in Brains of Patients with Alzheimer's Disease, in *Alzheimer's Disease: A Report of Progress* (Corkin S. et al., eds.) pp. 25–33. Raven Press, New York.

# Chapter 14

# Histochemical Mapping of Vertebrate Brains for Study of Evolution

## Harvey B. Sarnat

## 1. Introduction

Radical changes in life have evolved on Earth since the Cambrian period; radical changes in evolutionary theory have developed since the Darwinian period. The concept of gradual physical change in species is yielding to a more credible hypothesis of long epochs of genetic stability interrupted by bursts of multiple mutations, some providing improved survival, others creating physical features that are not necessarily better, but simply different. Of all specialized organ systems, the brain has changed the most in phylogenesis. Texts of comparative anatomy rarely describe the brain with greater perspective than its gross external form. In the present decade, neurochemistry and histochemistry have kindled an interest in comparative neuroanatomy from a wholly new perspective, enabling explanations of cerebral evolution to surface that could not have been previously suspected by examining histologic sections, or even from tracing neural pathways within the brain.

The ventricular system, neural components of the brainstem, and even some forebrain structures are recognized by their anatomy in all vertebrates, despite variations in embryonic neuronal migration among species. The simple, unspecialized brains of lampreys and salamanders undergo minimal neuronal migration during embryonic development, and neurons in the adult remain clustered in the subependymal germinal zone, where they mature *in situ* and develop their synaptic relations. Homologous structures, identified as readily in the most primitive fishes as in hu-

497

mans, provide insight into the *origin* of the brain, but do not denote *evolution* (Fig. 1). Evolved structures are so different among species that their common origin is not obvious.

In acknowledging that all species have evolved from ancestral forms now extinct, the human chauvinism that has created clichés in our language and subconsciously colored our attitude toward other species is gradually being purged from our terms of reference. Though amphibians and reptiles have evolved along different lines of specialization than mammals, we no longer condescendingly refer to these animals as "submammalian vertebrates," but rather more objectively and less emotionally as "nonmammalian vertebrates." Each species remains unique, with a biological and evolutionary history unlike any others.

The relation of the phylogenetic evolution of a species to the embryology of the individual within that species has been a focus of scientific philosophy for more than a century. Haeckel in 1866 and 1875 developed the hypothesis that "ontogeny recapitulates phylogeny" from his observations that the embryos of more advanced species pass through stages in which they resemble the embryos, but not the adult forms, of simpler species. DeBeer (1958) challenged this widely accepted theory, proposing that just the converse was true: new species evolve because of changes in the embryogenesis of existing species. Although phylogeny may not actively determine the embryonic development of an individual, the imprint of earlier evolutionary stages is exhibited too clearly in the embryos of all species to be merely fortuitous. This intimate relation of phylogeny and ontogeny remains one of the more speculative and romantic aspects of the origin of life.

## 2. The Origin of the Nervous System

Only in the phylum Porifera, the sponges, do contractile cells function in the absence of nerve; the muscular tissue of oscular sphincters responds directly to environmental stimuli. Parker (1919) proposed that sponge sphincters, which require several minutes to contract, exemplified the elementary property of protoplasmic conduction. Sea anemones, belonging to the primitive phylum Coelenterata, which also includes the hydra and jellyfishes, are tubular polyps with muscular walls that contract in response to innervation, but also retain the capacity to contract with environmental stimuli in the absence of nerve. The muscular iris of the eyes of some invertebrates and fishes also has direct

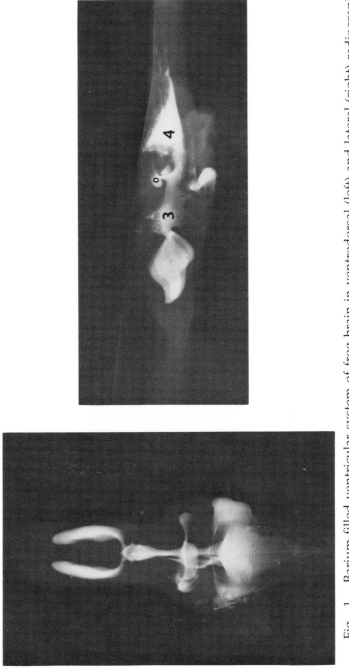

Fig. 1. Barium-filled ventricular system of frog brain in ventrodorsal (left) and lateral (right) radiographic views. 3 = third ventricle; 4 = fourth ventricle; o = optic tectal ventricle; choroid plexus forms roof of third ventricle. The features of the brain that are easily recognized in all vertebrates provide more insight into the origin than into the evolution of the nervous system (courtesy Dr. E. Kier, 1977, reproduced with permission).

photoreceptive properties, contracting on exposure to light even without innervation. In humans, some smooth and cardiac muscles preserve contractile function in the absence of nerve supply.

A neuron is a cell specialized for the reception of stimuli, conduction of excitation, and transmission of the signal to other cells (Lentz, 1968a). Two fundamental properties distinguish the neuron from other cells: electrical polarity of its plasma membrane to allow the propagation of conduction, and the biosynthesis of secretory products. Polarity and conduction are properties shared by other types of cells, particularly muscle cells. Neurosecretion is an early evolved property, exhibited even in the primitive neurons of coelenterates. Clark (1956) and Grundfest (1965) have suggested that nerve cells actually arose from ancestral secretory cells, when the secretory activity became confined to the terminations of their processes to provide a new role in neurotransmission.

The first development in neuronal differentiation was the elaboration of cell processes. In coelenterates, a nerve net consists of primitive nerve cells whose processes are not yet distinguished as dendrites and axons; polarity of direction of conduction either toward or away from the cell body is possible in all processes. Neurons do not aggregate to form a brain or even simple ganglia.

A major advance in the development of the neuron was the specialization of one process as an axon, ensuring undirectional flow of the action potential away from the cell body while receptive data continue to be channeled to the cell by one or more processes. This feature is first found in the phylum Platyhelminthes, flatworms, represented by the planarian. Synaptic junctions developed from further modification of receptive membrane surfaces and from the concentration of neurosecretory products as vesicles at the axonal tip. Electronic "tight junctions," synapses without chemical transmitters, are more abundant in simple species than in more evolved animals, and are reminiscent of an early stage of evolution of the nervous system.

Bullock and Horridge (1965) defined a nervous system as ". . . an organized constellation of cells (neurons) specialized for the repeated conduction of an excited state from receptor sites or from other neurons to effectors or to other neurons." The planarians not only provide the simplest example of animals with mature neuronal forms of nerve cells; they are also the lowest phylum to organize such cells into a cephalic organ fulfilling criteria of a brain (Sarnat and Netsky, 1985). Unlike coelenterates, which exhibit radially symmetric body forms, flatworms are organized with bilateral symmetry and a cephalo–caudal gradient, the body plan of most advanced animals—invertebrate and vertebrate species. The planarian brain is bilobed and symmetric, with a cortex or "rind" of surface neurons and deep commissures crossing the midline to interconnect the two cerebral halves. The planarian has a further importance in phylogeny and relevance to the human nervous system that is little appreciated. Platyhelminthes is the phylum preceding the great divergence in the subsequent evolution of animal life into more complex phyla of invertebrates on one hand and chordates, including vertebrates, on the other.

A closer examination of the neurons of the planarian brain by Golgi silver impregnation and by intracellular electrical recording followed by the microinjection of the fluorescent dye lucifer yellow that fills the processes of the neuron, leads to a surprising conclusion: *the neurons of the planarian display many features more closely resembling those of vertebrates than of higher invertebrates* (Keenan et al., 1981; Koopowitz, 1984.)

The neurons of higher invertebrates are generally unipolar with the soma separated from the processes by a stalk and not participating in the integration of electrical impulses. In vertebrates, by contrast, the soma typically is interposed between dendritic arborizations and the axon, requiring its active participation in conducting electrical impulses to the axon. Vertebrate-like multipolar neurons are a common form of nerve cell in the planarian, and processes resembling dendritic spines are also observed. Of special interest are decussating neurons in the planarian, whose axons cross the midline. Small multipolar glial cells also occur, another feature of the vertebrate brain. Inhibitory as well as exciting effects are developed in the planarian nervous system (Koopowitz et al., 1979). Histochemical and pharmacologic studies have demonstrated that various planarian neurons contain epinephrine, norepinephrine, 5-hydroxytryptamine, perhaps other monoamines, and acetylcholine, the same substances that are known to serve as neurotransmitters in the vertebrate nervous system and in higher invertebrates. In addition, specific neurosecretory cells reminiscent of those specialized neurons of the hypothalamus and neurohypophysis are also detected in the planarian brain (Lender and Klein, 1961; Oosaki and Ishii, 1965; Morita and Best, 1966). Ultrastructural studies of the planarian brain confirm the presence of at least three types of synaptic vesicles originating from Golgi membranes and chemical synaptic junctions in the neuropil (Oosaki and Ishii, 1965; Morita and Best, 1966; Koopowitz and Chien, 1975). The somatic musculature of the planarian is unstriated, but this condition is a primitive stage of development before the differentiation of striated and smooth muscle, rather than being mature smooth muscle (Sarnat, 1984).

In the planarian, we thus recognize the structural, electrophysiological, and neurochemical elements of the primordial vertebrate nervous system. These simple worms are probably little changed from their ancestral forms that were in the line of evolution leading to vertebrates and ultimately to human life.

## 3. When Does a Ganglion Become A Brain?

The neuron is a secretory cell with an electrically polarized, excitable membrane. A ganglion is an aggregate of neurons with similar morphologic, functional, and chemical properties. In primitive animals, such as the planarian, the question arises as to whether the aggregate of nerve cells in the head is a "cephalic ganglion," as it is often called, or whether it constitutes a true brain. The brontosaurus dinosaur is sometimes said to have possessed a "sacral brain" larger than the one in its head. Nineteenth century neuroanatomists such as Huxley spoke of the "cephalic ganglion" of the crayfish, to avoid acknowledging that this animal has a brain, and Robinson terms the celiac ganglion of man the "abdominal brain."

The presence of tubular invagination of the central nervous system and internal, ependymal-lined cavities as a requirement of a brain would render all invertebrates brainless. Moreover, the most advanced fishes, the teleosts, have an everted forebrain unlike most other vertebrates, in which internal ventricles are lacking, but a narrow ependymal cavity lies external to the cerebrum.

To establish a definition of a brain that may be applied to the many divergent lines of evolution of the nervous system in both invertebrates and vertebrates, the following criteria are suggested: (a) the presence of a bilobed structure at the rostral end of an animal, consisting of an aggregate of neurons and nerve fibers and possessing functional sensory and motor connections with the periphery to subserve the whole animal, with these connections not confined to one side or to restricted anatomical segments; (b) the ability of the structure to receive sensory projections from more than one source, such as photoreceptors, chemoreceptors, tactile receptors, proprioceptors, or visceral receptors; (c) the structure is composed of three or more distinct types of neurons, at least one of which has synaptic connections exclusively within the organ and no peripheral processes (i.e., interneurons). By these three criteria, the planarian is the lowest animal to develop a brain, and it is not merely a cephalic ganglion.

Additional features of most brains of even primitive animals, but not essential criteria of a brain, include the presence of a connective tissue enclosure or capsule (e.g., meninges in vertebrates), nonneuronal supportive or glial cells, and subdivisions into two or more anatomically distinct regions characterized by groupings of similar nerve cells. The neurosecretory function is assumed to exist by the inclusion of neurons in the definition of the brain. Systems of neurons with identical or very similar neurosecretory function develop in specialized species. Examples include the dopaminergic, serotoninergic, acetylcholinergic, and GABAergic systems of the brainstem and basal telencephalic nuclei in vertebrates. Specifically excluded as criteria of a brain are behavioral aspects of reflexive, conditioned, learned, and "intelligent" motor responses to stimuli, because these are interpretative and highly subjective aspects of neuron function, difficult to measure in simple species.

A ganglion, by contrast, supplies a restricted anatomical region of the body, may receive sensory impulses from a single source, is composed of only one or two neuronal types, and has few or no interneurons and therefore cannot develop complex intrinsic circuits; monosynaptic relays are characteristic. Isolated spinal cord segments and autonomic ganglia thus are not brains. However, some parts of brains may be equivalent to ganglia.

Although the definition of a brain may seem a moot point in studying highly evolved species, it is fundamental to understanding the origin and evolution of the nervous system as inferred from simple species.

## 4. Brain Size, Brain Weight, and Functional Complexity

The size and weight of the brain are among the most classical measures of comparative neuroanatomy because they are so readily defined by techniques available even to the investigators of the 19th century. Moreover, the size and estimated weight of the brain of extinct species can be calculated from intracranial volumes of fossilized skulls, a small compensation for the lack of preservation of brains that disappeared from the planet before man evolved. Mathematical formulas or indices expressing brain size and encephalization have been developed to assist paleontologists (Jerison, 1973). The small size of the brains of all mammals of the Tertiary period relative to contemporary species was first noted as early as 1874 by Marsh.

Larger brains are associated with increased complexity of function and "intelligence," as a general rule (Fig. 2), but the correlation is flawed by many exceptions. Some insect brains of small size guide much more complex behavior than the larger brains of some molluscs. The elephant and whale have larger brains than humans. Holloway (1979) contends that brain size *per se* is unimportant, except as it predicts the size and complexity of brain substructures. He emphasizes the importance of relative growth and the timing of developmental events in cerebral organization for understanding brain–behavioral relationships.

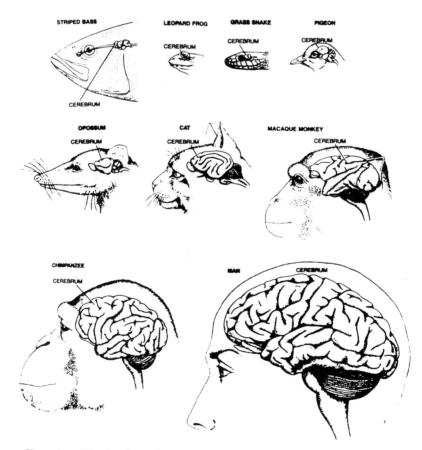

Fig. 2.   The brain enlarges progressively in a series of vertebrates drawn to the same scale. The face decreases in size relative to the brain (from Hubel, 1979, reproduced with permission).

Another aspect of brain growth that may correlate with intellectual capacity is the pattern of increases in brain size during ontogeny. Epstein (1974) found that the human brain grows in spurts and plateaux rather than at a uniform rate, and concluded that mental development is related to the growth spurts.

Larger brains may result from a larger total number of neurons, but the arrangement of those neurons is also important. The reason convolutions develop in the cerebral and cerebellar cortices of higher animals is to efficiently increase the surface area without proportionately increasing the mass of the cortex. Reptiles and birds would not benefit from convolutions of their cerebrum because of a nuclear rather than a columnar and laminar organization of their forebrain, discussed more fully later in this chapter. However, these species are hatched from free eggs; excessively large brain size at birth would not pose the same hazards to mother and child that passage through a mammalian birth canal presents. Human infants with defective gyration of the cerebral cortex (pachygyria), caused by abnormal neuronal migrations in fetal life, generally have smaller or larger brains than normal, although those with macrocephaly are usually mentally retarded and have defective neurologic functions. The complexity of dendritic branching of neurons and the size of the dendritic field as measured by concentric ring analysis radiating from the cell body influences total brain size less than functional capability. Synaptic density and dendritic arborization are found less in abnormally formed brains than in those of normal children or animals, and genetic factors also may account for individual differences in fine structures and performance within a species (Greenough and Juraska, 1979). Ultimately, this becomes an anthropological question because it relates to humans and is actually a question of the evolution of intelligence (Jerison, 1973). The behavioral implications of brain growth and size are addressed in the proceedings of a symposium on this topic (Hahn et al., 1979).

Another reason the brain of one species may be larger than that of another is simply that individual cells are all larger. Whether a large neuron is capable of more diverse functions than a small neuron of the same type in another species is an untested hypothesis. The density of neurons, or compactness of cell packing, also influences net brain size. Cellular density varies greatly in different parts of the brain, however, making this measure secondary to relative development of the various structures of the brain when comparing closely related species. The amount of white matter is another function of neuronal cellularity. Well-myelinated tracts occupy a larger volume than lightly or unmyelinated pathways.

The size of internal cavities or ventricles also has a bearing on the brain size and weight. In most fishes the homolog of the ventricles is external to the brain of an everted telecephalon. Invertebrates lack a ventricular system altogether. Many mammals have more spacious ventricles than do reptiles and birds.

Finally, environmental and nutritional as well as genetic factors must be considered when interpreting brain size of any individual within a species.

## 5. Relation of Evolution to Human Disease

Human diseases resulting from faulty embryonic development are well known. Cerebral malformations are a major cause of chronic neurologic disability in children. Developmental disorders are not generally regarded from a phylogenetic perspective, yet some human diseases are distinguished by an apparent selective involvement of neural structures evolving late or found exclusively in mammals, suggesting a biochemical evolution in parallel with anatomic development. Other pathologic conditions in humans may have a physiologic counterpart as the normal condition in other species of vertebrates (Sarnat and Netsky, 1984).

John Hughlings Jackson in 1884 was the first to clearly articulate a possible relation between evolution and human neurologic disease. Medical aspects of phylogeny were not addressed, however, by the many comparative neuroanatomists of this century whose contributions provided the foundation for further development of this theme. Only recently has the hypothesis of phylogenetic diseases emerged from obscurity to achieve a measure of scientific dignity (Roofe and Matzke, 1968; Sarnat and Netsky, 1981, 1984).

One example of a human degenerative disease selectively affecting recently evolved structures is Krabbe's leukodystrophy, a progressive disease of white matter of the brain and peripheral nerves related to deficiencies of galactocerebroside-beta-galactosidase and psychosine galactosidase, with resulting accumulations of galactolipids in myelin-forming cells followed by death of those cells. The distribution of lesions in the brain and spinal cord suggests an evolutionary pattern because the tracts involved are almost exclusively mammalian structures, while the phylogenetically old and constant pathways found in all classes of vertebrates are spared. Uninvolved structures include the olfactory tracts, fornix, mammillothalamic fasciculus, stria medullaris, and medial longitudinal fasciculus. By contrast, the corticospinal tracts, subcortical white matter of the centrum semiovale, and dorsal columns of the spinal cord are severely affected. Mammals have more galactolipids in cerebral myelin, but less phospholipids, than do nonmammalian vertebrates (Cuzner et al., 1965; Smith, 1967; Ramsey, 1981), even though structural differences between white matter in various species cannot be demonstrated. By contrast with Krabbe's disease, Niemann-Pick disease results from a deficiency of sphingomyelinase and impaired phospholipid metabolism; no selective involvement of new or old structures of the brain is evident. Other examples of selective vulnerabilty of phylogenetically recent structures of the central nervous system in human disease include olivopontocerebellar atrophy and Friedreich's ataxia, the orginal examples cited by Jackson (1884). Leber's optic atrophy may be yet another example. The optic nerve is fundamentally different in mammals than in other vertebrates because more than 90% of its fibers project to the midbrain in nonmammalian species and more than 95% of mammalian optic nerve fibers terminate in the thalamus.

Few examples may be offered of selective vulnerability of phylogenetically old structures of the brain, sparing recent structures, in human disease. The reason for the rarity of this phenomenon may be related to the primordial life-support functions of phylogenetically old brainstem structures, without which embryonic death occurs in early stages of development. One possible example might be Leigh's subacute necrotizing encephalopathy of infancy.

Several developmental malformations of the human brain exhibit a striking similarity to the normal conditions of other species. The Dandy-Walker cyst is a ballooning of the posterior ependymal membrane of the fourth ventricle in humans, often associated with atresia of the foramina of Luschka and Magendie, dysplasia or absence of the posterior cerebellar vermis, and nearly always with obstructive hydrocephalus. Except for the hydrocephalus that causes the clinical symptoms in humans, the condition resembles the normal state in most birds. Birds lack free communication between the fourth ventricle and subarachnoid space, but the large surface area of an expanded pouch of the fourth ventricle allows sufficient transudation of fluid and transport of solutes to substitute for the circulation of cerebrospinal fluid in these animals. Because of expansion of the neocortex, we mammals cannot afford the luxury of a large fluid-filled cavity occupying space within our cranium. Other examples of pathologic conditions in humans reminiscent of normal anatomy in other animals include agenesis of the corpus callosum in marsupials such as the oppossum and kangaroo, congenital muscle fiber-type disproportion in rodents, and the abnormal pyramidal cells with loss of basal dendritic spines in human Pick's disease, which resemble large cortical 'candelabra' neurons normally found in reptiles (Sarnat and Netsky, 1984).

The greater flexibility of nonmammalian vertebrates to utilize anaerobic metabolic pathways as an alternative energy source may make mammals more vulnerable to poisons of oxidative metabolism or genetic deletions of crucial enzymes for mitochondrial metabolic pathways. The comparison of pathologic conditions in man with the normal state of other animals does not imply that similarities of appearance necessarily signify similar mechanisms, but the comparisons are useful in understanding the ways in which nervous systems of other species have developed and function differently from the evolved human brain.

Clues to other human diseases might also be found in studying evolution. Why do invertebrates, even those with long life spans, only rarely develop neoplasms, whereas malignant tumors are common in vertebrates, especially mammals? Indeed, it often is difficult to determine whether invertebrate tumors are neoplasia, hyperplasia, dysplasia, or a secondary reaction to injury. The so-called black tumors of flies and certain other insects are in reality nothing more than granulomas. Melanotic tumors of fruitflies are composed of blood and fat cells (Barigozzi, 1969), but these same insects also develop invasive neuroblastomas of the larval brain (Gateff and Schneiderman, 1969). These rare neoplasms in insects are commonly genetically determined, unlike human cancer. Some species of planarians develop benign tumors spontaneously (Lange, 1966) or upon exposure to chemical carcinogens (Best, 1983). Such organic carcinogens also may be teratogenic in these simple worms, including anomalies such as supernumerary heads and tails (Foster, 1969; Best, 1983). However, some polyaromatic mammalian carcinogens induce malignant tumors in planarians, which are composed of pleomorphic cells, are progressive in size, and ultimately result in death of the animal (Foster, 1963, 1969). This phenomenon has not been observed in nature, outside experimental laboratory conditions. Is the rarity of invertebrate neoplasia in these animals because of differences in the immune system, genetic differences, metabolic differences, or other factors? Comparative neurobiology may provide an insight not previously suspected into many human diseases.

# 6. Evolution of Organization in the Vertebrate Brain

Bilateral symmetry is the fundamental body plan of all vertebrates, by contrast with the radial symmetry of some invertebrates, such as the hydra, jellyfish, sea anemone, and starfish. Among the echinoderms, the larval stage often has bilateral symmetry. The vertebrate brain is a paired structure with only a few unpaired midline structures, such as the pineal, and a few unequal paired structures, such as the left and right vagal nerves, and functional differences between the left and right human cerebral cortex. The asymmetric size of the left and right habenular nuclei of fishes and amphibians may serve as a form of "cerebral dominance" in these species (Sarnat and Netsky, 1981).

The deceptively simple segmental arrangement of the vertebrate central nervous system is in reality only a transitory stage of early embryogenesis, even in the simplest, least specialized species. Functionally similar neurons become grouped into longitudinal columns in the brainstem without reference to the segmental origin before migration of the primitive embryonic neuroblasts commences. For example, the descending trigeminal tract extends from the pons to the cervical cord in mammals, and as far caudally as the lumbar spinal cord in the frog, probably to mediate rapid reflexive jumping responses to tactile stimulation of the large surface of the head in these amphibians. The descending trigeminal nucleus contains somatic sensory neurons contributing not only to the trigeminal nerve, but also to the somatic sensory fibers of the facial, glossopharyngeal, and vagal nerves. The nucleus becomes continuous with the substantia gelatinosa of the spinal cord. Motor columns such as the nucleus ambiguus also contribute to several cranial nerves, and the autonomic columns of the spinal cord are an additional example of the regrouping of neurons into functional columns rather than primordial segments.

In the telencephalon, the nucleus accumbens is incorporated into the caudate nucleus of reptiles, but in mammals this structure is anatomically distinct, though remaining part of the caudate as defined by its fiber connections, cytologic detail, and histochemical profile. This again emphasizes the species-specific variation in embryonic neuronal migrations that make anatomic position within the brain an unreliable criterion for homology of neural structures between different species.

## 6.1. Why One Side of the Brain Controls the Opposite Side of the Body

Control of sensory and motor functions of the opposite side of the body by each half of the symmetric brain is as fundamental to the body plan of all vertebrates as bilateral symmetry itself, yet this unique arrangement is not reproduced in any other body system. However primordial this crossed control may be, the reason for its original evolution and persistence remains speculative. A phylogenetic explanation is proposed (Sarnat and Netsky, 1981), summarized below, and schematically illustrated in Fig. 3.

The origin of decussating tracts of the central nervous system may be found in a contemporary animal little changed from its fossilized ancestors and extinct siblings. This unassuming creature is the personification of simplicity in neural organization; its dorsal notochord and neural groove-like spinal cord disclose its evolutionary kinship to vertebrates, as do its primitive thyroid and many other primitive body organs. This protochordate or prevertebrate is called *amphioxus.* The diminutive brain of amphioxus lacks many of the structures possessed by true vertebrate brains: there are no eyes or visual system, no cranial nerves related to extraocular muscles, no labyrinth or vestibular system, no cerebellum, and no recognizable forebrain; a single midline cerebral ventricle is probably the homolog of the vertebrate third ventricle.

Amphioxus survives by using very few reflexive responses to its environment. It regards any tactile stimulus as potentially harmful and coils away from it. The development of this primitive coiling reflex required the evolution of one of the fundamental neuronal types, reproduced in almost infinite variations in all vertebrate nervous systems: the *decussating interneuron* (the nerve cell that relays an impulse from a sensory neuron on one side of the body to motor neurons on the other side of the body). It is the earliest example of crossed neural control in a chordate (prevertebrate); the decussating neuron of the planarian is not a true interneuron. Amphioxus possesses only a few types of neurons; among them is the cell of Rhode, a decussating interneuron that either ascends or descends in the spinal cord, as well as crossing to the contralateral side. This neuron is also novel in remaining entirely within the central nervous system, projecting neither afferent nor efferent processes to the periphery.

As the visual system evolved, it was similarly necessary to avoid visually perceived objects or changes in light that might be harmful by coiling or swimming away from the threatened side, necessitating another decussating interneuron, in this case the retinal ganglion cell giving rise to the optic nerve fiber. In non-mammalian vertebrates and most unspecialized mammals such as rodents, nearly all optic nerve fibers cross; the ipsilateral projection of fibers from the temporal half of the retina in humans and other advanced primates is the late secondary modification.

With cephalization, long pathways ascended to or descended from progressively more rostral parts of the brain. The decussating interneuron established a fundamental organization of the central nervous system to subserve the coiling reflex. Further evolution built upon the three primordial neuronal types: sensory and motor neurons and decussating interneuron. Once this organization was established, further developments of the brain could not deviate from this arrangement without sacrificing integration with other cerebral systems.

Short interneurons for local circuits that do not possess long or crossing axons developed from specialization of primary sensory and motor neurons that in primitive animals still possess both central and peripheral branches of their main processes (Fig. 3). Loss of the central ramus produced the mature model sensory and motor neuron; loss of the peripheral ramus created the short ipsilateral interneuron that became important for local circuitry in cerebral "nuclei" of clustered neurons specialized for similar function.

Examples of tracts in the human brain that are essentially decussating interneurons include the spinothalamic, tectospinal, and corticospinal tracts.

### 6.2. Divergent Evolution Between Vertebrates Into Nuclear or Laminar Arrangement of Neurons

Functional specialization of neurons in early ancestral vertebrates required that similar nerve cells be grouped for efficiency of afferent and efferent connections and to allow intrinsic synaptic relations to become established. Small interneurons proliferated to mediate coordination and synchrony between functionally similar cells. The synaptic organization of the brain developed not only grouping of similar neurons, but a somatotopic organization within these groups to perform sequential motor activity and for the localization, in addition to the mere perception, of the various sensory stimuli, without which appropriate motor responses would not have been possible to ensure survival.

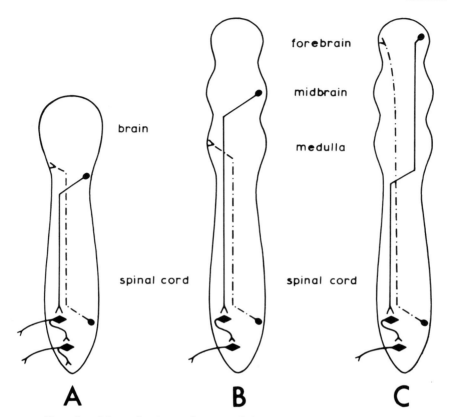

Fig. 3.   Hypothetic evolution of decussated long tracts of the central nervous system. (A) Primitive condition, similar to that in amphioxus; axons of interneurons in rostral part of spinal cord decussate and
descend; those of caudal interneurons decussate and ascend to inner-
vate primitive segmental motor neurons. Only caudal motor neurons
are illustrated. This arrangement provides a mechanism for the defen-
sive coiling reflex. (B) Condition in lower vertebrates: ascending axons
cross and extend to medulla; other fibers arise in the brainstem and descend to the opposite side of the spinal cord. Examples include the
spinovestibular and tectospinal tracts. (C) Condition in advanced vertebrates: further cephalization of crossed long pathways. Examples are the
spinothalamic and corticospinal tracts. The various long tracts are not
homologous, but have similar origin from the primitive decussating
interneuron. Primitive motor neurons (lower part of A) have branched
axons innervating both muscle and adjacent spinal motor neurons. As
these cells differentiate in evolution, loss of the axonal branch to other
neurons results in pure motor neurons; other cells lose the peripheral
axonal branch to give origin to local interneurons that do not have crossing fibers (lower part of panel B and C). These latter cells proliferate to
form spinal interneurons and sensory nuclei in the brain (reproduced
from Sarnat and Netsky, 1981, with permission from Oxford University
Press).

Two types of neuronal organization were developed from embryonic neuronal migrations, each species using both in different parts of the brain. Nuclear organization was the pattern evolved in most of the brainstem, in which similar neurons cluster into compact aggregates, usually still retaining a somototopic arrangement for function, even if this latter feature is difficult to appreciate in histological sections. The second pattern of organization was laminar, an efficient means of arranging neurons in progressive sequence and relating different types of neurons by synaptic connection while preserving a precise somatotopic matrix. This laminar organization first evolved in the cerebellar cortex and optic tectum (superior colliculus of mammals), but eventually was developed most extensively in the mammalian cerebral cortex. An advantage of laminar organization over nuclear organization was that the surface area of the layers could be increased efficiently without a corresponding increase in the mass or volume of tissue by the development of convolutions: cerebellar folia and cerebral gyri.

Another advantage of lamination as a type of cortical organization is that vertical columns extending through all cell layers provide an additional geometric arrangement for serial connections between several neuronal types in close proximity. Each column may then subserve a distinct and limited topographic part of the body or part of a sensory field (e.g., retinal, hence visual, field) with a complete synaptic circuit.

The development of the forebrain is a good example of divergent evolution leading to a similar functional result when avian and mammalian brains are compared. Birds have evolved a nuclear organization of their forebrain, continuing rostrally the arrangement established in the brainstem, whereas mammals developed a laminar cortex.

Birds have a very thin, poorly formed superficial cortical layer of the telencephalon, but very large and complex subcortical aggregates of neurons known as the 'dorsal ventricular ridge.' For many years, this cellular structure was thought to be a highly developed corpus striatum, but we now know from histochemical studies that only the most ventral part of this tissue is homologous with the caudate nucleus and putamen of mammals (Fig. 4). The belief of early comparative neuroanatomists that birds have only a rudimentary cerebral cortex is correct only if we rigidly maintain that cerebral cortex must be a six-layered laminar structure; if we consider the primordial neurons giving origin to cerebral cortex as potentially migrating during ontogenesis into another arrangement to form nuclei instead of laminae, we can then accept the forebrain of the bird as homologous with that of mam-

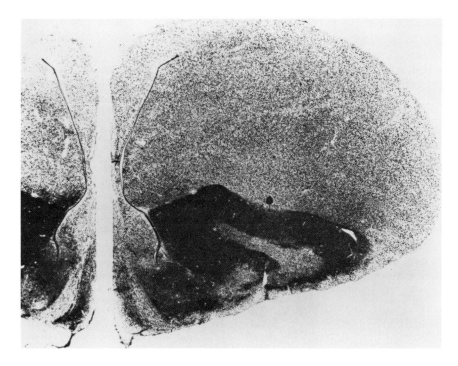

Fig. 4. Coronal section of forebrain of pigeon histochemically incubated for acetylcholinesterase activity and counterstained with cresyl violet for identification of cell groups. The large cellular zone of grey matter, which displaces the lateral ventricle medially into a vertical slit-like structure, does not correspond anatomically to any recognized structure in coronal sections of mammalian brains, and was long believed to be a highly developed corpus striatum. This histochemical stain demonstrates that only its ventral part corresponds to the mammalian basal ganglia and para-olfactory area, whereas the more dorsal area, known as the dorsal ventricular ridge and having little acetylcholinesterase activity, contains neurons homologous with those of the mammalian cerebral cortex organized in a nuclear rather than laminar fashion. This is an example of divergent evolution between the avian and mammalian telecephalon, both beginning with similar primordia (reproduced with permission from Karten and Dubbeldam, 1973).

mals (Karten, 1969; Karten and Dubbeldam, 1973). If all the pyramidal neurons of layer 5 of the mammalian motor cortex were clustered together as one aggregate, adjacent to another aggregate containing the clustered granular cells of layer 3, and synaptic connections established between these aggregates, the fundamental plan of the avian forebrain would be established. The same principle is evident in the visual and auditory systems of the bird, and the total number of synapses in series in these special sensory systems is the same in birds and mammals (Sarnat and Netsky, 1981). The dorsal ventricular ridge of birds is thus composed of many nuclei given names unfamiliar in human neuroanatomy (e.g., accessory hyperstriatum; ectostriatum), but is fundamentally homologous with specific regions and laminae of mammalian neocortex. The functions of the special sensory systems in birds and mammals are remarkably similar despite this divergent evolution, implying a great deal about the plasticity of organization of the nervous system, at least in phylogeny. In mammals, a nuclear organization also is employed in some structures derived from embryonic telecephalon. Examples include the amygdala and caudate nucleus.

Among other classes of vertebrates, the reptilian forebrain is less highly developed, but similar to the avian—not a surprising finding since birds evolved from dinosaur-like reptiles. Amphibians have a less evolved forebrain with both simple nuclear organization and rudimentary lamination. Fishes have an everted forebrain unlike any other vertebrates, and both nuclear and laminar structures are identified. Finally, it is instructive to examine the brains of simple species of vertebrates exhibiting minimal neuronal migration after neural tube formation; in the salamander, neurons that mature in the same embryonic periventricular matrix zone establish neither nuclear nor laminar organization, limiting the complexity of synaptic circuits and of the resultant functional capabilities.

## 6.3. Parallel Duplicate Systems of the Brain

Sensory systems of the brain are duplicated, but these parallel structures and pathways are neither equal in size nor identical in their connections. In every special sensory system, a parallel supplementary system provides a different or modified function of a similar type, providing for greater flexibility in the evolutionary potential of each species and allowing for a broader range of function than would have otherwise been possible. Such duplicate systems occur in the human brain as in other species.

The principal olfactory bulbs are supplemented by a pair of accessory olfactory bulbs. Impulses generated in the olfactory mucosa (or vomeronasal organ, in the case of some reptiles) are relayed to sites in the amygdala different from the terminal targets of the principal olfactory projections.

In the visual system, the large retinotectal and retinogeniculate pathways are accompanied by a parallel visual pathway also originating in the retina but terminating in another part of the thalamus, the nucleus rotundus in reptiles or the pulvinar in mammals. Secondary connections from these diencephalic centers are with visual associative neocortex rather than primary visual cortex in mammals, or with corresponding structures in the dorsal ventricular ridge in reptiles and birds. Some amphibians and reptiles possess even a third parallel visual system arising from the well-formed globe of the median eye, through the pineal body and optic tectum.

The lateral line system of fishes and cochlear auditory system of terrestrial vertebrates might be regarded as a duplicate parallel system in relation to the vestibular system, both originating from the acoustic nerve, but having evolved different specialized functions.

Even the somatic sensory system has at least three or more functional components traveling in parallel and subserving different functions, as with spinothalamic, spinocerebellar, and dorsal column-medial lemniscus systems. Accessory special sensory systems are often associated with their own modified special end-organs for perception.

The somatic motor system exhibits parallel but different pools of motor neurons innervating muscle fibers differing in physiologic and metabolic (histochemical) type in all vertebrates and even in amphioxus. The mammalian system of intrafusal muscle fibers within muscle spindles, innervated by gamma motor neurons, might be regarded as yet another parallel system.

The autonomic nervous system of lower vertebrates has two components that often provide synergistic effects upon visceral end-organs (Sarnat and Netsky, 1981). In higher vertebrates, the sympathetic and parasympathetic systems become antagonistic and specialize in different functions. Even the right and left nerves of the same pair, such as the vagus, may differ in their visceral distribution and function.

## 6.4. Is the Dominant Special Sensory System of Lower Vertebrates Olfactory?

Even in as young a discipline as evolutionary theory, speculation sometimes becomes accepted as fact and misconceptions are difficult to dispel. The myth of olfactory dominance of the brain in less complex creatures than primates began with the observations of J. Herrick, a pioneer comparative neuroanatomist of the early 20th century whose valuable contributions are now classics. Herrick was impressed with the large size of the olfactory bulbs of primitive jawless cyclostome fishes and sharks. In these species the olfactory bulbs indeed constitute nearly half the total volume of the entire forebrain, although they are less complex in the number of neuronal types and synaptic relations than the mammalian olfactory bulb. He also was aware that sharks are attracted by the odor of blood and that more advanced teleostean fishes are very perceptive to changing chemical contents in their ambient water. These dissolved odors in the water are largely detected by gustatory rather than olfactory receptors, as is now known. Herrick developed an hypothesis that olfactory connections dominated the higher cerebral functions of lower vertebrates, and the term "smell brain" became a popular evolutionary designation.

Herrick was wrong about this aspect of evolution. Later studies using methods of fiber tracing not available in Herrick's day have conclusively shown that nerve fibers from the olfactory bulbs in sharks are confined to a small region on the ventrolateral surface of the brain (Heimer, 1969; Ebbesson and Heimer, 1970); the major part of the telencephalon of cartilaginous fishes and all other vertebrates is devoid of olfactory connections. Furthermore, although the olfactory acuity of sharks may be higher than in humans in terms of stimulus threshold, the discrimination of a larger number of odors is clearly a superior function of the mammalian brain, including even humans in whom the olfactory system is actually regressive.

Olfaction was an important evolutionary development because it permitted the tracking of a distant stimulus to its source by intensity or concentration gradient, an essential skill for both predators and those individuals concerned with detecting and avoiding predators. Diffusion of chemical particles in water is slow and influenced by water currents, limiting the precision by which olfaction is a reliable localizing sense for tracking. Also, olfaction is less complex than hearing or vision for localization of stimuli because it does not involve comparing two sides for computing movement.

If olfaction does not then dominate the brains of lower vertebrates as Herrick had believed, which special sensory system is of primordial importance?—the same system that dominates in nearly all higher vertebrates: the visual system (Sarnat and Netsky, 1981).

## 6.5. Functional Adaptation and Cerebral Capabilities Outside Human Experience

Phylogenetically old structures of the central nervous system do not disappear in evolution with new developments. They are supplemented rather than replaced. Functions do sometimes change, however. The cochlear nuclei, decussating fibers of the trapezoid body, lateral lemniscus, inferior colliculi (torus sem-icircularis in amphibians and reptiles), and other structures of the auditory system become acoustic structures only with the evolution of the cochlea in terrestrial vertebrates. In fishes, these structures subserve the lateral line system, a special sensory system that would be useless on land and is unknown to human experience. Once lost, no mammal who has returned to an aquatic life has ever redeveloped this unique sense that allows fishes to navigate free from collision with underwater obstacles, including other fishes, by perceiving pressure waves in the water. The central neural structures of the lateral line system were simply readapted to the new terrestrial auditory system, whereas the vestibular system remained essentially unchanged. This readaptation occurred not only in the evolution of land vertebrates, but is reenacted in the metamorphosis of every tadpole who matures to become a frog. Another example is the pineal, which has a visual function in some reptiles and is an endocrine organ in mammals.

Another specialization in some species, unknown to human experience, is the adaptation of some trigeminal nerve fibers and the gasserian ganglion in certain species of snakes as perceptors of infrared light, associated with the development of unique cutaneous receptors known as pit organs. The infrared band is invisible to most animals except some snakes, and in these species the central infrared pathway ultimately ends in the optic tectum, as well as in the descending and principal trigeminal nuclei; whether such reptiles perceive infrared rays as visual or as somatosensory stimuli is known only to them. In the visible band of light, it is not a mere evolutionary coincidence that the narrow range of wavelength that all animals utilize for vision, 350–750 m$\mu$, is the same energy band used by plants for photosynthesis, and that the ultimate source of energy, the sun, has peak energy emission at 500 m$\mu$ (Wald, 1960).

Many other examples of specializations in other species of vertebrates that exceed those of man could be cited: the exceptional gustatory sense of many fishes, olfactory acuity in bloodhounds and sharks, an extraordinary vestibular control system in birds of flight. Additional capabilities of the central nervous system to which human experience is totally naïve are the adaptation of the cerebellum to emit electrical impulses for orientation or for shocking predators or prey in some electric fishes; the specialization of certain muscles of some electric fishes in which contractile function is renounced for the capability of a voltaic pile. The third or midline pineal eye is functional in some lizards but vestigial in humans.

Two important premises may be concluded from these few examples: (a) that many species have evolved specializations of their nervous system rendering to them unique capabilities outside human experience, but that the human brain has evolved other capabilities beyond those of other species; (b) that the evolution of equivalent structure within the central nervous system cannot be equated with the development of equal functional adaptation in all species.

## 7. The Problem of Homology

Homologous structures are those derived from the same primordium in different species. Their common origin may be evident only in the embryo, since they undergo modifications beyond easy recognition as comparable structures in the adult. Homologous muscles in frog and man, as one example, may be identified by their bony origin and insertion and by their innervation that remain constant throughout phylogeny. Homology is the clue by which evolutionary sequence may be reconstructed from living creatures whose common ancestor is long extinct.

The brain presents special problems in determining homology (Campbell and Hodos, 1970; Sarnat and Netsky, 1981). Not only do species differences in embryonic neuronal migrations and the relative development of surrounding structures make anatomic position of groups of neurons in the adult an unreliable criterion for comparing parts of the brain, but the frequent adaptation of homologous structures to new functions limits the definition of homology to one of structure and excludes function.

The demonstration of homology of a particular neuroanatomic nucleus between two species thus is based on two premises: (a) identical or very similar afferent and efferent fiber connections; and (b) similarity of neurotransmitter substances and the various enzymes mediating their biosynthesis and degradation. Homologous nuclei do not have cholinergic synapses in some animals and monoaminergic synapses in others, even if those species are as divergent as a goldfish and a monkey.

The traditional means of determining homology in the nervous system for almost a century has been the demonstration of the extrinsic fiber connections relating zones of grey matter to other structures. Older techniques of tracing degenerating myelin of axons whose cell bodies are damaged or destroyed in selective lesions of the brain were restricted to heavily myelinated pathways in adult brains. Modern methods of tracing the destination of both myelinated and unmyelinated nerve fibers use silver impregnation techniques that demonstrate degenerating axoplasm, such as those of Nauta and Gygax and of Fink and Heimer (Heimer, 1970). Retrograde axonal transport has also been used recently to great advantage in conjunction with horse-radish peroxidase to identify the origin of nerve terminals (Mesulam, 1982); although a histochemical stain in principle, it is used for fiber tracking rather than as a metabolic marker. The observation of central chromatolysis of motor neurons whose peripheral axons are damaged in lesions has largely been supplanted by these newer methods, but the traditional Golgi impregnations, for which Ramón y Cajal and Golgi were awarded the Nobel Prize in medicine in 1909, retain a vigorous and active role for demonstrating the intrinsic organization of neuroanatomic nuclei and certain structural details, such as dendritic spines, still unmatched by any more recent method. Radioautography using isotope-labeled amino acids is useful for studying neuronal migrations in the embryonic brain, but its value in comparative neuroanatomy is limited mainly to immature stages of development. An additional use of radioautography is to demonstrate the incorporation of tritiated amino acids into certain neurons and not others of the adult brain after intraventricular instillation; the uptake of a particular amino acid suggests, but does not conclusively prove, the synthesis of a related neurotransmitter.

## 8. Application of Histochemistry to the Study of Comparative Neuroanatomy

The metabolic and biochemical mapping of the brain is among the newer methods of studying neuroanatomy, introducing another entire perspective of brain development, both ontogenetic and phylogenetic, not even suspected by most of the pioneer comparative neuroanatomists. The introduction in this decade of positron emission tomography (PET) scanning techniques adds a further dimension invaluable to medicine because it allows the derivation of localized metabolic function in the brain of the living patient (Walker, 1984). The potential application to the study of comparative neuroanatomy is obvious, but the method is too recent to provide even preliminary data.

   Another promising new method for comparative metabolism is the use of isotope-labeled analogs of physiologic substrates, such as desoxyglucose. The incorporated isotope may then be measured in various parts of brain slices at specified periods after administration to the subject, but it requires sacrifice of the experimental animal.

Biochemical analysis of various parts of the brain is an increasingly important aspect of comparative neuroanatomy, but has limitations inherent in the small size of tissue samples available from small brains and the microscopic size of many structures of quite different biochemical character, particularly in the brainstem. Despite these limitations, the contribution of biochemical study to understanding the relative evolution of homologous structures between vertebrates is great, and chemical neuroanatomy is just now developing into a discipline of promising future importance (Emson, 1983).

Histochemistry is not quantitative as with biochemical determinations, and may not necessarily indicate that the presence of a particular enzyme signifies that it is metabolically the most important constituent in a given region of brain. In some areas, more than one neurotransmitter may be present. The advantage of histochemistry is precise microscopic localization of activity, providing a feasible means of identifying metabolic differences between adjacent small structures. Histochemistry has already been helpful in questions of homology, as with the corpus striatum of birds (Fig. 4), but the application to comparative neuroanatomy is only now becoming appreciated. This application is dependent upon the principle of constancy throughout phylogeny of homologous structures of the central nervous system, a premise that histochemistry conversely is helping to verify.

Although some metabolic products and enzymes may be directly demonstrated in tissue sections by histochemical methods and the light microscope, others require the use of UV microscopy to show fluorescent products, or require immunocytochemical methods. An example of the latter is the demonstration in brain of dopamine beta-hydroxylase.

The following are examples of histochemical methods that may be applied to the central nervous system for metabolic mapping of the brains of various vertebrates for the study of evolution.

## 8.1. Mitochondrial Oxidative Enzymes

Specific histochemical methods are available for the demonstration of the enzymes of the tricarboxylic acid (Krebs) cycle, such as succinic dehydrogenase (SDH), malic dehydrogenase, and isocitric dehydrogenase. Other oxidative enzymes that may be demonstrated histochemically include those of the electron transport chain, including nicotinamide adenine dinucleotide (NADH) diaphorase and cytochrome-c-oxidase. Unfixed frozen sections of tissue are required. When applied to sections of brain, these stains demonstrate the areas of highest mitochondrial concentration.

Because mitochondria are present in almost all cells, stains for mitochondrial enzymes are nonspecific in a sense, but their distribution in the nervous system is not uniform; they are most concentrated at nerve terminals. These stains thus are useful for demonstrating the regions of highest synaptic activity. They do not simply show a differential staining of grey and white matter despite the concentration of neurons in grey matter, but rather indicate where the most synapses are found. For example, the molecular layer of the cerebellar cortex is rich in synaptic contacts despite the relatively few neurons within the layer, and oxidative enzymatic activity is quite high (Fig. 5). Oxidative enzymatic acivity is stronger in the outer layers of mammalian neocortex, where more synaptic activity takes place, than in the deeper pyramidal layers, and regional differences are seen in the various parts of the cortex in addition to the laminar gradation (Manocha and Shantha, 1970). In spinal dorsal root ganglia, dehydrogenase activity is more intense in smaller neurons than in large ones (Robain and Jardin, 1972). These histochemical stains are therefore of value in mapping the synaptic density in various parts of the central nervous system. In addition, SDH activity is stronger in small than in large neurons, and may be helpful in distinguishing mixed neuronal types in the spinal cord and certain parts of the brain, such as the thalamus and corpus striatum (Manocha and Shantha, 1970; Campa and Engel, 1970).

## 8.2. Glycogen and Glycolytic Enzymes

Glycogen and the enzymes of glycogenolysis, particularly those of the Embden-Meyerhof pathway, are present in all neurons. Their concentration in motor neurons, however, greatly exceeds that of most other neuronal types. They are prominent in pyramidal cells of the motor cortex, as well as in lower motor neurons of the spinal cord (Fig. 6) and of cranial nerve motor nuclei. Small motor neurons in the ventral horn of the spinal cord are easily distinguished from interneurons (Renshaw cells) by phosphorylase because the latter exhibit little activity, and gamma motor neurons have less phosphorylase activity than alpha motor neurons (Campa and Engel, 1970, 1971). The histochemical reaction for phosphorylase is completely lost from anterior horn neurons rich in phosphorylase within 72 h after proximal or distal axonal sec-

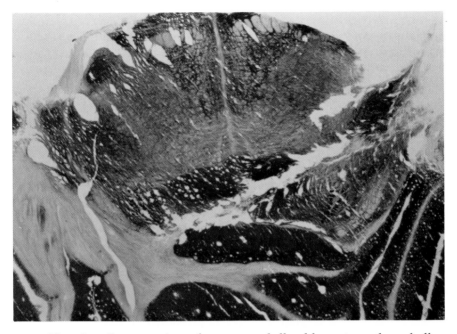

Fig. 5. Cross-section of upper medulla oblongata and cerebellum of mouse, stained histochemically for succinate dehydrogenase activity. This oxidative mitochondrial enzyme demonstrates the distribution of synapses, since axonal terminals are associated with the highest concentration of mitochondria in the nervous system, even though mitochondria are found in all cells. Long tracts and deep cerebellar white matter have only sparse activity, but the molecular zone of the cerebellar cortex is intensely stained because of its numerous synapses, even though histologically it is mainly white matter (×14).

tion in the cat, providing a marking technique for axonal damage (Campa and Engel, 1971). In dorsal root ganglia of the spinal cord, the enzymes of the pentose shunt are more active in small sensory neurons, whereas phosphorylase activity is greater in large nerve cells (Robain and Jardin, 1972).

Phosphorylase is best demonstrated in the nervous system by following its incubation with the periodic acid-Schiff (PAS) reaction for glycogen, a difference from the technique commonly employed in histochemistry of muscle. Phosphorylase also demonstrates neuroepithelial cell processes (Sarnat et al., 1975; see below).

## 8.3. Nucleic Acids

Ribonucleic acid (RNA), the constituent of ribosomes associated with rough endoplasmic reticulum, may be demonstrated in tis-

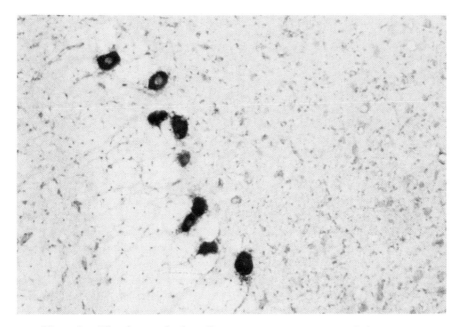

Fig. 6.    The large dark cells are motor neurons of the motor trigeminal nucleus in the pontine tegmentum of a cat. They appear brilliant red against a pale bluish-pink background in the original slide. The
locus ceruleus lies medial to these motor neurons (at the left), but is
poorly defined by this technique because its neurons are not intensely
stained. Compare with figure 8B. Histochemical stain for phosphorylase
activity, counterstained with the periodic acid-Schiff reaction ($\times 272$).

sue sections by basophilia when stained by hemotoxylin or the ca-
pricious methyl green-pyronin stain, but it is most vividly shown
with acridine orange (AO), a compound forming fluorescent complexes with nucleic acids. When excited by UV light, AO–DNA
complexes fluoresce yellow-green, whereas AO–RNA complexes
are a brilliant orange-red. The technique is described by Perl and
Little (1980) and Sarnat (1985).

Acridine orange clearly demonstrates neurons and distinguishes them from the pale green surrounding neuropil and glial
cells in the central nervous system (Sarnat, 1985). Immature nerve
cells and those still in the process of migration are much less intense than the luminous orange of mature neurons. Motor neurons with prominent Nissl granules are most vividly stained. The
technique may be applied to both frozen unfixed sections and to
formalin-fixed, paraffin-embedded tissue.

## 8.4. Histochemical Stains of Neuroepithelial Processes

Nonspecific esterase, unlike the histochemical stain for acetylcholinesterase, is useful not as a metabolic marker, but rather as a selective histologic stain because of its affinity for ependymal and glial processes; this selective staining quality is particularly prominent in lower vertebrates in whom these neuroepithelial cell processes probably fulfill a nutritive function supplanted in mammals by an improved capillary network (Sarnat et al., 1975; Sarnat and Netsky, 1981). The stain is reminiscent of the heavy metal impregnation of Cajal or Golgi (Fig. 7). Phosphorylase and lactic dehydrogenase are other histochemical stains that may be used to demonstrate neuroepithelial processes, apart from their use as metabolic markers within neurons (Sarnat et al., 1975).

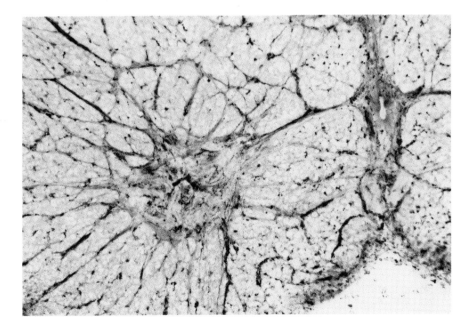

Fig. 7. Nonspecific esterase applied to the spinal cord of a shark pup demonstrates subependymal and glial processes reminiscent of Cajal impregnations. In this case, histochemistry is used not as a metabolic marker, but rather to better demonstrate anatomical features (×13).

## 8.5. Neurotransmitters and Enzymes Mediating
### Their Metabolism

The identification of chemical substances released by terminal axons at synaptic junctions and the enzymes catalyzing their biosynthesis and degradation is the object of most contemporary biochemical investigations of the nervous system, central and peripheral. The current intense interest in neurotransmitters is a result of two major recent discoveries: the first was the development of reliable histochemical methods for demonstrating the transmitters themselves or related metabolic products and enzymes, providing for precise anatomical localization in tissue sections; the second was the discovery that a potentially treatable neurologic disease resulted from deficiency of a demonstrable transmitter. Degeneration of dopaminergic neurons in the substantia nigra that project to the corpus striatum was found to be the pathologic basis of Parkinson's disease, and the disabling symptoms of many afflicted patients could be ameliorated by administering supplementary dopamine precursors.

The original histochemical method for demonstrating monoamines was based on formaldehyde condensation to fluorescent products that could then be visualized under UV light (Falck et al., 1962; Carlsson et al., 1962; Dahlström and Fuxe, 1964; Fuxe, 1965). Modifications and refinements of this method employing glyoxylic acid were later introduced to increase the sensitivity (Björklund et al., 1972; Lindvall and Björklund, 1974; Bloom and Battenberg, 1976; Nygren, 1976; Watson and Barchas, 1977; Maeda et al., 1979). Serotonin (5-hydroxytryptamine) was more difficult to demonstrate by this method than other monoamines, such as norephinephrine and dopamine, so that other techniques were developed to help localize serotonin in the brain. These methods include the autoradiographic demonstration of the uptake of tritiated serotonin after injection into the cerebral ventricles (Descarries et al., 1975; Calas and Ségu, 1976; Chan-Palay, 1977; Léger and Descarries, 1978; Léger et al., 1978; Ségu and Calas, 1978) and immunocytochemical techniques either for the direct localization of serotonin (Steinbusch and Nieuwenhuys, 1983) or using antibodies to the enzymes tryptophan hydroxylase (Pickel et al., 1977) or DOPA decarboxylase (Hökfelt et al., 1973) as markers of serotonin metabolism. Pharmacologic alterations using uptake inhibitors, antagonists, and analogs of monamines are at times useful for verifying the identity of a particular catecholamine, but are generally more applicable to quantitative biochemical than qualitative histochemical reactions. Finally, some less recent, simple histochemical methods, such as those for demonstrating monoamine oxidase (MAO) (Glenner et al., 1957), still retain great value in localizing concentrations of monoaminergic neurons (Fig. 8). Most MAO activity is located in the mitochondria (Tipton and Dawson, 1968; Squires, 1968; Achee and Gabay, 1977) and the strongest MAO activity is localized in the periphery of cell processes, particularly in synaptic regions (Manocha and Shantha, 1970). Histochemical protein markers of monoamine cell bodies also have been described (Panayotacopoulou and Issidorides, 1982).

Unlike the catecholamines, acetylcholine is very difficult to display directly in tissue sections, and a simple histochemical method for its demonstration remains elusive. Fortunately, reliable techniques for exhibiting the enzymes catalyzing the biosynthesis and degradation of acetylcholine (choline acetyltransferase and acetylcholinesterase, respectively) were developed early and continue to be extensively used as markers of cholinergic neurons and synapses in the peripheral autonomic nervous system and at neuromuscular junctions, as well as in the brain and spinal cord (Giacobini, 1959; Shute and Lewis, 1961, 1963; Karnovsky and Roots, 1964; Giacobini et al., 1967; Buckley et al., 1967; Lewis et al., 1967; Baljet and Drukker, 1975). The histochemical technique may also be adapted to electron microscopy for intracellular and axonal localization of acetylcholinesterase (AChE) (Shute and Lewis, 1966). Immunohistochemical localization of choline acetyltransferase also has been successfully demonstrated (Cozzari and Hartman, 1980).

Immunocytochemistry, the production and purification of specific enzymes to peptide substrates and the final addition of a fluorescent tag to the molecule, has provided an alternative solution to the difficult problem of developing specific histochemical reactions to demonstrate simple peptides and other molecules believed to function in vivo as neurotransmitters. This method has proved particularly useful for studying the histologic distribution of certain enzmes for which there is no other specific enough histochemical reaction, such as dopamine beta-hydroxylase (Nagatsu et al., 1979; Grazanna and Molliver, 1980) and glutamic acid decarboxylase (Saito et al., 1974; Barber et al., 1978), and even some transmitters themselves such as gamma-aminobutyric acid (GABA) (Roberts, 1978). Putative neurotransmitter peptides such as substance P, enkephalins, and opioids may also be demonstrated in this way (Pickel et al., 1979, 1980; Watson and Akil, 1980).

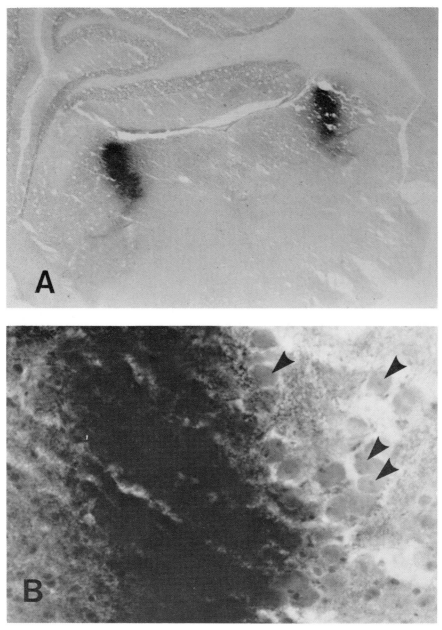

Fig. 8. Cross-section of pons and cerebellum of mouse stained for monoamine oxidase activity. (A) The locus ceruleus, an inconspicuous cell group in histologic sections, is readily identified by its high concentration of this enzyme important in the degradation of catecholamines (×13). (B) The large neurons of the motor trigeminal nucleus (arrowheads) are unstained, whereas the adjacent smaller neurons of the locus ceruleus are so intensely stained that cytologic detail is difficult to see. Compare with Fig. 6 (×268).

Another technique for the localization of substances in neural structures is autoradiography, but the capacity of a particular neuron to incorporate a radiolabeled substance does not necessarily prove that the neuron normally possesses the enzymes for its synthesis. The methodology is reviewed by Marks et al. (1962), Wolfe et al. (1963), and Aghajanian et al. (1966). Tritiated amino acids have been incorporated into neurons after instillation into the cerebral ventricles and into the retina after intraocular injection (Lam et al., 1981).

The interpretation of histochemically localized enzymatic activity in the nervous system is not always as evident as it might appear. For example, AChE staining is at times demonstrated in adrenergic neurons, such as in the locus ceruleus (Lewis and Schon, 1975; Jones and Moore, 1974) and in primary sensory nerve cells (Eränkö and Härkönen, 1964; Jacobowitz and Koelle, 1965; Schlaepfer, 1968). Both MAO and AChE activity may also occur simultaneously in the same neurons (Rodriguez, 1967; Manocha and Shantha, 1970), but MAO may not be as specific for adrenergic nerves as AChE is for cholinergic nerves, a feature first recognized by Koelle and Valk in 1954. Furthermore, enzymes such as MAO may have functions other than those associated with the metabolism of neurotransmitters. Cholinesterases other than AChE, particularly butrylcholinesterase (BChE), also occur in the same tissue, but inhibitors of nonspecific esterases may be employed to verify specific identity of AChE. BChE is present in higher concentrations in satellite cells than in sensory neurons of spinal ganglia, a reciprocal relation to AChE (Robain and Jardin, 1972). AChE activity is higher in the rat than in primates; in general, the smaller the brain the higher the activity of AChE per gram of tissue (Gerebtzoff, 1959; Manocha and Shantha, 1970).

## 8.6. Demonstrating Transmitters for Comparative Neuroanatomy

The lower vertebrates and most invertebrates use electrotonic tight-junction synapses more extensively than do mammals, yet chemical synaptic transmission is also universally found in the animal kingdom among all animals possessing a central nervous system. Furthermore, the putative neurotransmitters identified in even the simplest animals are the same ones demonstrated in the mammalian brain. Free-living flatworms such as the planarian have within their rudimentary nervous systems monoamines (Welsh and Williams, 1970), serotonin (Welsh and Moorhead, 1960), and acetylcholine or at least AChE (Lentz, 1968). Representative species of all higher phyla of invertebrates and all clas-

ses of vertebrates also possess these substances, indicating a very
early evolution of chemically mediated transsynaptic membrane
depolarization. Comparative anatomic study of the distribution of
these substances in widely divergent but highly evolved central
nervous systems, as well as their localization in simple, minimally
evolved nervous systems, has the potential for revealing much
about the phylogenetic history of neurologic development.

Despite the extensive studies of neurotransmitters in mammals and in certain invertebrates, particularly mollusks, as
"model neural systems," the histochemical mapping of the brains
of nonmammalian vertebrates and of most invertebrate species
has been the object of relatively few investigations. Pioneer work
in this approach to the study of evolution was initiated by Hebb
and Ratković (1964), who demonstrated a common pattern in the
distribution of choline acetyltransferase in the brains of various
fishes, amphibians, reptiles, birds, and mammals. Northcutt
(1974, 1978) showed the comparative distributions of AChE and
of MAO (Figs. 9 and 10), and Karten and Dubbeldam (1973) contributed an understanding of the homology of the avian corpus
striatum to that of mammals (Fig. 4) and the true nature of the

Rana
AChE

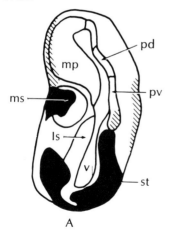
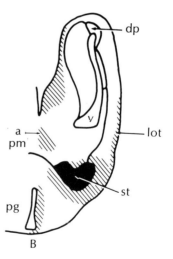

   0.29-0.15

∞-0.30

Fig. 9   (a).

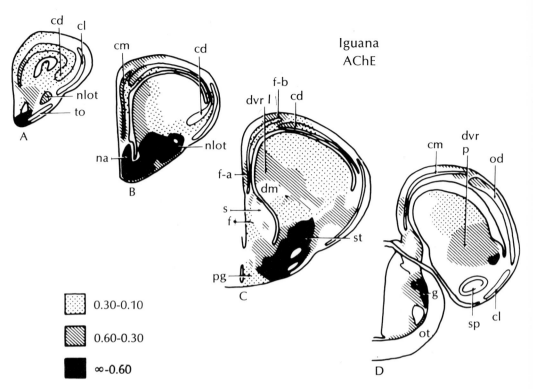

Fig. 9. Drawing of transverse sections of forebrains of frog (Rana)
and lizard (iguana) mapping localization of acteylcholinesterase (AChE)
concentrations. The most intense regions of enzymatic activity are
confined to the inferior pallium, corresponding to the mammalian corpus striatum. The dorsal ventricular ridge has low activity. Numbers are
optic density values of absorbance. Abbreviations: apm, amygdala pars
medialis; cd, dorsal cortex; cl, lateral cortex; cm, medial cortex; dp, dorsal pallium; dvr 1, medial (auditory) zone of anterior dorsal ventricular
ridge; dvr P, posterior part of dorsal ventricular ridge; f, fornix; f-a,b,
areas of moderate density of AChE in dorsal and ventral parts of medial
cortex; g, ventral nucleus of lateral geniculate body; lot, lateral olfactory
tract; ls, lateral septal nucleus; mp, medial pallium; ms, medial septal
nucleus; na, nucleus accumbens; nlot, nucleus of lateral olfactory tract;
on, optic nerve; ot, optic tract; pg, preoptic periventricular gray matter;
s, septal nuclei; sp, nucleus sphericus; st, striatum; to, olfactory tuber-
cle; v, lateral ventricle (reproduced with permission from Northcutt,
1978).

dorsal ventricular ridge as the avian equivalent of the mammalian
neocortex, using histochemistry to supplement other neuroanatomic methods. Parent and his colleagues mapped the distribution of catecholamines in the brains of fishes and amphibians, as
well as of mammals (Parent, 1973, 1975; Parent and Olivier, 1970;

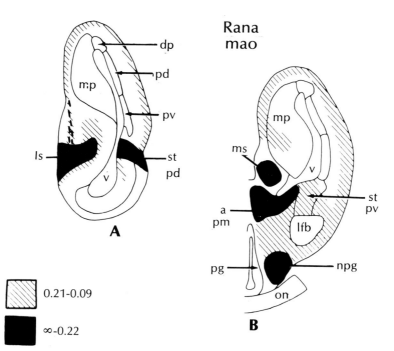

Fig. 10. Map of regional distribution of monoamine oxidase (MAO) in transverse sections of frog brain. Abbreviations: *see* legend to Fig. 9 (reproduced with permission from Northcutt, 1974).

Parent and Poitras, 1978; Parent et al., 1978; Poitras and Parent, 1978). The contribution of the comparative neuroscientists and others both proves the hypothesis that homologous nuclei of the brain use the same primary neurotransmitter in all classes of vertebrates and also provide a sound foundation for histochemistry as a reliable method for the study of neuroanatomic evolution through comparative metabolic anatomy.

At times, histochemical staining discloses an organization not previously suspected. For example, AChE in the molecular layer of the cerebellar cortex in the cat occurs in alternating longitudinal bands of strong and weak activity that may be associated with zones of the inferior olivery nucleus that also show a regular variation in staining (Marani and Voogd, 1977; Marani et al., 1977).

Quantitative biochemical determinations of neurotransmitters in different species pose other kinds of questions than do histochemical stains. Why, for example, would the frog brain need more than three times the amount of serotonin and catecholamines than does the rat brain, and seven times more than the fish brain (Ramsey, 1981)?

Finally, the combination of ontogenetic and phylogenetic studies by comparing embryos and immature specimens of different animals offers yet another facet in evolutionary research even less explored than the comparative histochemistry of adult brains of various species. In rodents, the accumulation of MAO, dopamine, serotonin, and norepinephrine in brainstem nuclei follows a predictable developmental pattern. (Agrawal et al., 1966; Robinson, 1968; Wawrzyniak, 1965). For example, the locus ceruleus and nucleus ambiguus show much stronger MAO activity than other nuclei at birth (Fig. 8) and MAO is also strong in neurons whose axons are becoming myelinated. AChE activity cannot be demonstrated in the layers of the mammalian embryonic cerebral cortex, but AChE-containing fibers are seen in the cortex of adults, suggesting that the cortex develops from noncholinergic primordial cells and that an ingrowth of cholinergic fibers originating in other parts of the brain later appears (Krnjevic and Silver, 1966; Manocha and Shantha, 1970).

## 8.7. Neurosecretion

The ability of neurons to secrete hormones that are carried systemically to remote sites in the body by blood or other body fluids and are endocrinologically active was first conclusively demonstrated in the skate and other fishes by Speidel (1917, 1922), although the theoretic possibility had been considered even earlier. Speidel's careful and objective results and pharmacologic experiments were not accepted or believed by the scientific and medical communities and were actively ridiculed by some of the most prominent authorities of the day, until Ernst and Berta Scharrer (Scharrer, 1934 a,b; Scharrer and Scharrer, 1945) proved that Speidel was correct in his conclusions. The Scharrers and others then extended their studies to demonstrate neurosecretory neurons in all classes of vertebrates and indeed in all phyla of multicellular animals except sponges, which do not possess cells that differentiate as recognizable neurons. Neurosecretion is essential for the metamorphosis of insects and amphibians, just as it is for the onset of puberty in mammals. The interesting history of the discovery of neurosecretion was recently reviewed by the present writer (Sarnat, 1983).

Histochemistry and immunocytochemistry are readily adapted to the demonstration in tissue sections of products of neurosecretion and related metabolic compounds. A large body of literature has now been published in relation to this application of histochemistry, and the results are as promising in the study of comparative metabolism and evolution as are the studies of the histochemical exhibition of neurotransmitters.

## 9. Conclusions

Reconstructing the probable sequence of evolution of the brain depends upon recognizing homologous structures in the brains of modern species, since the fossilized imprints of the brains of extinct ancestral animals provide only crude clues to external form. The nervous system has diverged more widely than any other organ system in the evolution of animal life. Despite anatomical differences, many of the recognized neurotransmitters of the mammalian brain may also be demonstrated in the brains of all vertebrates and in most invertebrates, including even such simple species as the planarian. Enzymes catalyzing the biosynthesis and degradation of these neuronal chemicals also evolved early and are almost universally found in multicellular animals. The histochemical and immunocytochemical exhibition of specific natural compounds and enzymes within the nervous systems of representative species of the various phyla and classes provides one means of demonstrating homology. Both adults and embryos may be studied in this way. Comparative metabolic maps of the brain may thus be constructed, based on the premise that *homologous neural structures that mediate chemical synaptic transmission use the same chemical transmitters and related enzymes throughout phylogeny.*

Histochemistry can also be used to demonstrate the relative concentrations in different types of neurons of metabolic products not directly related to synaptic transmision, such as glycolytic enzymes, nucleic acids, and neurosecretory products. Histochemical methods provide a metabolic marker of the zones of highest mitochondrial, hence synaptic, activity. Although histochemistry is not a quantitative technique and pitfalls in interpretation are possible, these disadvantages are balanced by the precise cytologic localization provided in even the smallest brains and the adaptability of histochemistry to optical, fluorescence, and electron microscopy.

## Acknowledgment

This work has been supported in part by the Alberta Children's Hospital Foundation.

## References

Achee F. M. and Gabay S. (1977) Some aspects of monoamine oxidase activity in brain. *Progr. Neurobiol.* **8,** 325–348.

Aghajanian G. K., Bloom F. E., Lovell R. A., Sheard M. H., and Freeman D. X. (1966) The uptake of 5-hydroxytryptamine-$^3$H from the cerebral ventricles: Autoradiographic localization. *Biochem. Pharmacol.* **15,** 1401–1403.

Agrawal H. C., Glisson S. N., and Himwich W. A. (1966) Changes in monoamines of rat brain during postnatal ontogeny. *Biochem. Biophys. Acta.* **130,** 511–513.

Baljet B. and Drukker J. (1975) An acetylcholinesterase method for *in toto* staining of peripheral nerves. *Stain Technol.* **50,** 31–36.

Barber R. P., Vaughn J. E., Saito K., McLaughlin B. J., and Roberts E. (1978) GABAergic terminals are presynaptic to primary afferent terminals in the substantia gelatinosa of the rat spinal cord. *Brain Res.* **141,** 35–55.

Barigozzi C. (1969) Genetic control of melanotic tumors in *Drosophila. Natl. Cancer Inst. Monograph* **31,** 277–290.

Best J. B. (1983) Transphyletic Animal Similarities and Predictive Toxicology, in *Old and New Questions in Physics, Cosmology, Philosophy and Theoretical Biology* (van der Merwe, A., ed.) pp. 549–591. Plenum Press, New York.

Björklund A., Lindvall O., and Svensson L. -A. (1972) Mechanisms of fluorophore formation in the histochemical glyoxylic method for monoamines. *Histochemie* **32,** 113–131.

Bloom F. E. and Battenberg E. L. F. (1976) A rapid, simple and sensitive method for the demonstration of central catecholamine-containing neurons and axons by glyoxylic acid-induced fluorescence. II. A detailed description of methodology. *J. Histochem. Cytochem.* **24,** 561–571.

Buckley G., Consolo S., and Sjöqvist F. (1967) Cholinacetylase in innervated and denervated sympathetic ganglia and ganglion cells of the cat. *Acta Physiol. Scand.* **71,** 348–356.

Bullock T. H. and Horridge G. A. (1965) *Structure and Function in the Nervous Systems of Invertebrates,* Vol. 2, pp. 809–1719. Freeman, San Francisco.

Calas A. and Ségu L. (1976) Radioautographic localization and identification of monoaminergic neurons in the CNS. *J. Microsc. Biol. Cell* **27,** 249.

Campbell C. B. G. and Hodos W. (1970) The concept of homology and the evolution of the nervous system. *Brain Behav. Evol.* **3,** 353–367.

Campa J. F. and Engel W. K. (1970) Histochemistry of motor neurons and interneurons in the cat lumbar spinal cord. *Neurology* **20,** 559–568.

Campa J. F. and Engel W. K. (1971) Histochemical and functional correlates in anterior horn neurons of the cat spinal cord. *Science* **171,** 198–199.

Carlsson A., Falck B., and Hillarp N.-A. (1962) Cellular localization of brain monoamines. *Acta Physiol. Scand.* **196** (suppl), 1–27.

Chan-Palay, V. (1977) Indolamine neurons and their processes in the normal rat brain and in chronic diet-induced thiamine deficiency demonstrated by uptake of $^3$H-serotonin. *J. Comp. Neurol.* **176,** 467–494.

Clark R. B. (1956) On the origin of neurosecretory cells. *Ann. Sci. Natl. Zool.* **18,** 199–207.

Cozzari C. and Hartman B. K. (1980) Preparation of antibodies specific to choline acetyltransferase from bovine caudate nucleus and immunohistochemical localization of the enzyme. *Proc. Natl. Acad. Sci. USA* **77,** 7453–7457.

Cuzner J. L., Davison A. N., and Gregson N. A. (1965) The chemical composition of vertebrate myelin and microsomes. *J. Neurochem.* **12,** 469–481.

Dahlström A. and Fuxe K. (1964) Evidence for the existence of monoamine-containing neurons in the central nervous system. I. Demonstration of monoamines in the cell bodies of brain stem neurones. *Acta Physiol. Scand.* **232** (suppl), 1–55.

DeBeer G. (1958) *Embryos and Ancestors.* 3rd ed., Oxford University Press, London.

Descarries L., Beaudet A., and Watkins K. C. (1975) Serotonin nerve terminals in adult rat neocortex. *Brain Res.* **100,** 563–588.

Ebbesson S. O. E. and Heimer L. (1970) Projection of the olfactory tract fibers in the nurse shark (*Ginglymostoma cirratum*). *Brain Res.* **17,** 47–55.

Emson P. C. ed. (1983) *Chemical Neuroanatomy.* Raven, New York.

Epstein H. T. (1974) Phrenoblysis: Special brain and mind growth periods. I. Human brain and skull development. II. Human mental development. *Devel. Psychobiol.* **7,** 207–216; 217–224.

Eränkö O. and Härkönen A. (1964) Noradrenaline and acetylcholinesterase in sympathetic ganglia cells of the rat. *Acta Physiol. Scand.* **61,** 299–300.

Falck B., Hillarp N. -Å.., Thieme G., and Torp A. (1962) Fluorescence of catecholamines and related compounds condensed with formaldehyde. *J. Histochem. Cytochem.* **10,** 348–354.

Foster J. (1963) Induction of neoplasia in planarians with carcinogens. *Cancer Res.* **23,** 300–303.

Foster J. (1969) Malformations and lethal growths in planarians treated with carcinogens. *Natl. Cancer Inst. Monograph* **31,** 683–691.

Fuxe K. (1965) Evidence for the existence of monoamine neurons in the central nervous system. IV. Distribution of monoamine nerve terminals in the central nervous system. *Acta Physiol. Scand.* **247** (suppl), 39–85.

Gateff E. and Schneiderman A. (1969) Neoplasms in mutant and cultured wild-type tissues of Drosophila. *Natl. Cancer Inst. Monograph* **31**, 365–397.

Gerebtzoff M. A. (1959) *Cholinesterases.* Pergamon, New York.

Giacobini E. (1959) The distribution and localization of cholinesterase in nerve cells. *Acta Physiol. Scand.* **156** (suppl), 1–45.

Giacobini E., Palmborg B., and Sjöqvist F. (1967) Cholinesterase activity in innervated and denervated sympathetic ganglion cells of the cat. *Acta Physiol. Scand.* **69**, 355–361.

Glenner G. G., Burtner H. J., and Brown G. W., Jr. (1957) The histochemical demonstration of monoamine oxidase activity by tetrazolium salts. *J. Histochem. Cytochem* **5**, 591–600.

Grazanna R. and Molliver M. E. (1980) The locus coeruleus in the rat: An immunocytochemical delineation. *Neuroscience* **5**, 21–41.

Greenough W. T. and Juraska J. M. (1979) Experience-Induced Changes in Brain Fine Structure: Their Behavioural Implications, in *Development and Evolution of Brain Size: Behavioural Implications.* (Hahn M. E., Jensen C., and Dudek B. C., eds.) pp. 295–320, Academic, New York.

Grundfest H. (1965) Evolution of Electrophysiological Properties Among Sensory Receptors Systems, in *Essays on Physiological Evolution,* (Pringle, J. W. S., ed.) pp. 107–138, Pergamon, Oxford.

Haeckel E. (1866) *Generelle Morphologie der Organismen.* Berlin.

Haeckel E. (1875) *Ziele und Wege der heutigen Entwicklungsgeschichte,* Jena.

Hahn M. E., Jensen C., Dudek B. C., eds. (1979) *Development and Evolution of Brain Size. Behavioural Implications.* Academic, New York.

Hebb C. and Ratković D. (1964) Choline Acetylase in the Evolution of the Brain in Vertebrates, in *Comparative Neurochemistry* (Richter D., ed.) MacMillan, New York.

Heimer L. (1970) Selective Silver-Impregnation of Degenerating Axoplasm, in *Contemporary Research Methods in Neuroanatomy.* (Nauta W. J. H. and Ebbesson S. O. E., eds.), pp. 106–131, Springer-Verlag, New York.

Heimer L. (1969) The secondary olfactory connections in mammals, reptiles, and sharks. *Ann. NY Acad. Sci.* **167**, 129–146.

Hökfelt T., Fuxe K., and Goldstein M. (1973) Immunohistochemical localization of aromatic-L-amino acid decarboxylase (DOPA-decarboxylase) in central dopamine and 5-hydroxytryptamine nerve cell bodies of the rat brain. *Brain Res.* **53**, 175–180.

Holloway R. I. (1979) Brain Size, Allometry, and Reorganization: Toward a Synthesis, in *Development and Evolution of Brain Size: Behavioural Implications.* (Hahn M. E., Jensen C., and Dudek B., eds.), pp. 59–88, Academic, New York.

Hubel D. H. (1979) The brain. *Sci. Amer.* **241**, 45–53.

Jackson J. H. (1884) Croonian lectures. *Brit. Med. J.* **1,** 501, 660, 703.

Jacobowitz D. and Koelle G. B. (1965) Histochemical correlations of acetylcholinesterase and catecholamines in postganglionic autonomic nerves of the cat, rabbit, and guinea pig. *J. Pharmacol. Exp. Therap.* **148,** 225–237.

Jerison, H. J. (1973) *Evolution of the Brain and Intelligence.* Academic, New York.

Jones B. E. and Moore R. Y. (1974) Catecholamine-containing neurons of the nucleus locus coeruleus in the cat. *J. Comp. Neurol.* **157,** 43–52.

Karnovsky M. J. and Roots L. (1964) A "direct-coloring" thiocholine method for cholinesterases. *J. Histochem. Cytochem.* **12,** 219–221.

Karten H. J. (1969) The organization of the avian telencephalon and some speculation on the phylogeny of the amniote telecephalon. *Ann. NY Acad. Sci.* **167,** 164–179.

Karten H. J. and Dubbeldam J. L. (1973) The organization and projections of the paleostriatal complex in the pigeon *(Columba livia). J. Comp. Neurol.* **148,** 61–89.

Kennan C. L., Coss R., and Koopowitz H. (1981) Cyroarchitecture of primitive brains: Golgi studies in flatworms. *J. Comp. Neurol.* **195,** 697–716.

Kier E. L. (1977) The Cerebral Ventricles: A Phylogenetic and Ontogenetic Study, in *Radiology of the Brain and Skull,* Vol. 3, Anatomy and Pathology. (Newton T. H. and Potts D. G., eds.), pp. 2787–2914, Mosby, St. Louis.

Koelle G. B. and Valk A. (1954) Physiological implications of the histochemical localization of monamine oxidase. *J. Physiol.* (Lond.) **126,** 434–447.

Koopowitz H. (1984) The evolution of the nervous system in the turbellaria. Fourth International Symposium on the Biology of Turbellaria. Fredericton, New Brunswick, Canada, Aug. 5–10.

Koopowitz H. and Chien P. (1975) Ultrastructure of nerve plexuses in flatworms. II. Sites of synaptic interactions. *Cell Tiss. Rev.* **157,** 207–216.

Koopowitz H., Kennan L,. and Bernardo K. (1979) Primitive nervous systems: Electrophysiology of inhibitory events in flatworm nerve cords. *J. Neurobiol.* **10,** 383–395.

Krnjevic D. and Silver A. (1966) Acetylcholinesterase in the developing forebrain. *J. Anat.* **100,** 63–89.

Lam D. M.-K., Fung S.-C., and Kong, Y.-C. (1981) Postnatal development of dopaminergic neurons in the rabbit retina. *J. Neurosci.* **1,** 1117–1132.

Lange C. S. (1966) Observations on some tumours found in two species of planaria: *Dugesia etrusen* and *D. ilvana. J. Embryol. Exp. Morphol.* **15,** 125–130.

Léger L. and Descarries L. (1978) Serotonin nerve terminals in the locus coeruleus of adult rat: A radioautographic study. *Brain Res.* **146,** 1–13.

Léger L., Mouren-Mathieu A. M., and Descarries L. (1978) Identification radioautographique de neurones monoaminergiques centraux par micro-installation locale de serotonine ou de noradrénaline tritiée chez le chat. *CR Acad. Sci. (D)* (Paris) **286**, 1523–1526.

Lender T. and Klein N. (1961) Mise en évidence de cellules sécrétrices dans le cerveau de la planaire *Polycelis nigra. CR Acad. Sci. Paris.* **253**, 331–333.

Lentz T. L. (1968a) *Primitive Nervous Systems.* Yale Univ. Press; New Haven. 148 pp.

Lentz T. L. (1968b) Histochemical localization of acetylcholinesterase activity in a planarian. *Comp. Biochem. Physiol.* **27**, 715–718.

Lewis P. R. and Schon F. E. G. (1975) The localization of acetylcholinesterase in the locus coeruleus of the normal rat and after 6-hydroxdopamine treatment. *J. Anat.* (Lond.) **120**, 373–385.

Lewis P. R., Shute, C. C. D., and Silver A. (1967) Confirmation from choline acetylase analysis of a massive cholinergic innervation to the rat hippocampus. *J. Physiol.* (Lond.) **191**, 215–224.

Lindvall O. and Björklund A. (1974) The glyoxylic acid fluorescence histochemical method: A detailed account of the methodology for the visualization of central catecholamine neurons. *Histochemistry* **39**, 97–127.

Maeda T., Nagai T., Imamoto K., and Satoh K. (1979) A glyoxylic acid freeze-drying histofluorescence method for central serotonin neuron. *Acta Histochem. Cytochem.* **1**, 572–575.

Manocha S. and Shantha T. R. (1970) *Macaca mulatta. Enzyme Histochemistry of the Nervous System.* Academic, New York.

Marani E. and Voogd J. (1977) An acetylcholinesterase band-pattern in the molecular layer of the cat cerebellum. *J. Anat.* (Lond.) 124, 335–345.

Marani E., Voogd J., and Boekee A. (1977) Acetylcholinesterase staining in subdivisions of the cat's inferior olive. *J. Comp. Neurol.* **174**, 209–226.

Marks B. H., Samorajski T., and Webster E. J. (1962) Radioautographic localization of norepinephrine-$^3$H in the tissues of mice. *J. Pharmacol. Exp. Therap.* **138**, 376–381.

Marsh O. C. (1874) Small size of the brain in Tertiary mammals. *Amer. J. Sci. Arts* **8**, 66–67.

McGinty J. F., Van der Kooy D., and Bloom F. E. (1984) The distribution and morphology of opioid peptide immunoreactive neurons in the cerebral cortex of rats. *J. Neurosci.* **4**, 1104–1117.

Mesulam M. M., ed. (1982) *Tracing Neural Connections with Horseradish Peroxidase.* John Wiley, Chichester, UK.

Morita M. and Best J. B. (1966) Electron microscopic studies of planaria. III. Some observations on the fine structure of planarian nervous tissue. *J. Exp. Zool.* **161**, 391–412.

Nagatsu I., Inagaki S., Kondo Y., Karasawa N., and Nagatsu T. (1979) Immunofluorescent studies on the localization of tyrosine hydroxylase and dopamine-beta-hydroxylase in the mes-, di-, and telencephalon of the rat using unperfused fresh frozen sections. *Acta Histochem. Cytochem.* **12,** 20–37.

Northcutt, R. G. (1974) Some histochemical observations on the telencephalon of the bullfrog *Rana catesbeiana Shaw. J. Comp. Neurol.* **157,** 379–389.

Northcutt R. G. (1978) Forebrain Midbrain Organization in Lizards and Its Phylogenetic Significance, in *Behavior and Neurology of Lizards.* (Greenberg N. and MacLean P. D., eds.) pp. 11–64, National Institutes of Mental Health, Rockville, Maryland.

Nygren L. G. (1976) On the visualization of central dopamine and noradrenaline nerve terminals in cryostat sections. *Med. Biol.* **54,** 278–285.

Oosaki T. and Ishii S. (1965) Observations on the ultrastructure of nerve cells in the brain of the planarian, *Dugesia gonocephala. Ztschr. Zellforsch.* **66,** 782–793.

Panayotacopoulou M. T. and Issidorides M. R. (1982) Histochemical protein markers of monoamine cell bodies in man. *Arch. Neurol.* **39,** 635–639.

Parent A. (1973) Demonstration of catecholaminergic pathway from the midbrain to the strio-amygdaloid complex in the turtle *(Chrysemys picta). J. Anat.* (Lond.) **114,** 379–387.

Parent A. (1975) The monaminergic innervation of the telecephalon of the frog, *Rana pipiens. Brain Res.* **99,** 35–47.

Parent A., Dube L., Braford M. R. Jr., and Northcutt R. G. (1978) The organization of monoamine-containing neurons in the brain of the sunfish *(Lepomis gibbosus)* as revealed by fluorescence microscopy. *J. Comp. Neurol.* **182,** 495–516.

Parent A. and Olivier A. (1970) Comparative histochemical study of the corpus striatum. *J. Hirnforsch* **12,** 73–81.

Parent A. and Poitras D. (1978) The origin and distribution of catecholaminergic axon terminals in the cerebral cortex of the turtle. *(Chrysemys picta) Brain Res.* **78,** 345–358.

Parker G. H. (1919) *The Elementary Nervous System.* Lippincott, Philadelphia.

Perl D. P. and Little B. W. (1980) Acridine orange-nucleic acid fluorescence: Its use in routine diagnostic muscle biopsies. *Arch. Neurol.* **37,** 641–644.

Pickel V. M., Joh T. H., and Reis D. J. (1977) A serotonergic innervation of noradrenergic neurons in nucleus locus coeruleus: Demonstration by immunocytochemical localization of the transmitter specific enzymes tyrosine and trytophan hydroxylase. *Brain Res.* **131,** 197–214.

Pickel V. M., Joh T. H., Reis D. J., Leeman J. E., and Miller R. J. (1979) Electron microscopic localization of substance P and enkephalin in axon terminals related to dendrites of catecholaminergic neurons. *Brain Res.* **160,** 387.

Pickel V. M., Sumai K. K., Beckley S. C., Miller R. J., and Reis D. J. (1980) Immunocytochemical localization of enkephalin in the neostriatum of rat brain: A light and electron microscopic study. *J. Comp. Neurol.* **189,** 721–740.

Poitras D. and Parent A. (1978) Atlas of the distribution of monoamine-containing nerve cell bodies in the brain stem of the cat. *J. Comp. Neurol.* **179,** 699–718.

Ramsey R. B. (1981) Comparative Neurochemistry of the Vertebrates, in *Evolution of the Nervous System.* (2nd ed.) (Sarnat H. B. and Netsky M. G., eds.) pp. 24–38, Oxford University Press, New York.

Robain O. and Jardin L. (1972) Histoenzymologie du ganglion spinal du lapin. *J. Neurol. Sci.* **17,** 419–433.

Roberts E. (1978) Immunocytochemical Visualization of GABA Neurons, in *Psychopharmacology: A Generation of Progress.* (Lipton M. A., DiMascio A., and Killam K. F., eds.) Raven, New York.

Robinson N. (1968) Histochemistry of rat brain stem monoamine oxidase during maturation. *J. Neurochem.* **15,** 1151–1159.

Rodriguez S. (1967) Über die transneuronale Degeneration der optischen Zeutren du Ratte, bensondors der corpus geniculatum laterale. *Anat. Anz.* **120,** 187–197.

Roofe P. G. and Matzke H. A. (1968) Introduction to the Study of Evolution: Its Relation to Neuropathology, in *Pathology of the Nervous System* (Minckler, J., ed.) pp. 14–22. McGraw-Hill, Blakiston Div., NY.

Saito K., Wu J. -Y., Matsuda T., and Roberts E. (1974) Immunochemical comparisons of vertebrate glutamic acid decarboxylase. *Brain Res.* **65,** 277–285.

Sarnat H. B. (1983) The discovery, proof, and reproof of neurosecretion. *Can. J. Neurol. Sci.* **10,** 208–212.

Sarnat H. B. (1984) Muscle histochemistry of the planarian *Dugesia tigrina* (Turbellaria: Tricladida): Implications in the evolution of muscle. *Tr. Am. Microsc. Soc.* **103,** 284–294.

Sarnat H. B. (1985) L'acridine-orange: Un fluorochrome des acides nucléiques pour l'étude de cellules musculaires et nerveuses. *Rev. Neurologique* (Paris), **141,** 120–127.

Sarnat H. B., Campa J. F., and Lloyd J. M. (1975) Inverse prominence of ependyma and capillaries in the spinal cord of vertebrates: A comparative histochemical study. *Am. J. Anat.* **143,** 439–450.

Sarnat H. B. and Netsky M. G. (1981) *Evolution of the Nervous System.* (2nd ed.) Oxford Univ. Press, New York.

Sarnat H. B. and Netsky M. G. (1984) Hypothesis: Phylogenetic diseases of the nervous system. *Can. J. Neurol. Sci.* **11,** 29–33.

Sarnat H. B. and Netsky M. G. (1985) The brain of the planarian as the ancestor of the human brain. *Can. J. Neurol. Sci.* **12,** in press.

Scharrer E. (1934a) Stammt alles Kolloid in Zwischenhirn aus der Hypophyse? *Frankf. Ztschr. Pathol.* **47,** 134–142.

Scharrer E. (1934b) Über die Beteiligung der Zellkerns an sekretorischen Vorgängen in Nervenzellen. *Frankf. Ztschr. Pathol.* **47,** 143–151.

Scharrer E. and Scharrer B. (1945) Neurosecretion. *Physiol. Rev.* **25,** 171–181.

Schlaepfer W. W. (1968) Acetylcholinesterase activity of motor and sensory nerve fibres in the spinal nerve roots of the rat. *Z. Zellforsch. Mikrosk. Anat.* **88,** 441–456.

Ségu L. and Calas A. (1978) The topographical distribution of serotoninergic terminals in the spinal cord of the cat: Quantitative radioautographic studies. *Brain Res.* **153,** 449–464.

Shute C. C. D. and Lewis P. R. (1961) The use of cholinesterase techniques combined with operative procedures to follow nervous pathways in the brain. *Bibliotheca Anat.* **2,** 34–49.

Shute C. C. D. and Lewis P. R. (1963) Cholinesterase-containing systems of the brain of the rat. *Nature* (Lond.) **199,** 1160–1164.

Shute C. C. D. and Lewis P. R. (1966) Electron microscopy of cholinergic terminals and acetylocholinesterase-containing neurons in the hippocampal formation of the rat. *Z. Zellforsch. mikrosk. Anat.* **69,** 334–343.

Smith M. E. (1967) The metabolism of myelin lipids. *Adv. Lipid Res.* **5,** 241–278.

Speidel C. C. (1917) Gland cells of internal secretion in the spinal cord of the skates. Thesis: Princeton University, New Jersey.

Speidel C. C. (1922) Further comparative studies in other fishes of cells that are homologous to the large irregular glandular cells in the spinal cord of skates. *J. Comp. Neurol.* **34,** 303–317.

Squires R. F. (1968) Additional evidence for the existence of several forms of mitochondrial oxidase in the mouse. *Biochem. Pharmacol.* **17,** 1401–1411.

Steinbusch H. W. M. and Nieuwenhuys R. (1983) The Raphé Nuclei of the Rat Brainstem: A Cytoarchitectonic and Immunohistochemical Study in *Chemical Neuroanatomy,* (Emson P. C., ed.) pp. 131–207, Raven, New York.

Tipton K. F. and Dawson A. P. (1968) The distribution of monoamine oxidase and α-glycerophosphate dehydrogenase in pig brain. *Biochem. J.* **108,** 95–101.

Wald G. (1960) The Distribution and Evolution of Visual Systems, in *Comparative Biochemistry.* (Florkin, M. and Mason, H. S., eds.) pp. 311–345. Academic Press, New York.

Walker M. D., ed. (1984) Research issues in positron emission tomography. *Ann. Neurol.* **15,** suppl., 1–204.

Watson S. J. and Akil H. (1980) Alpha-MSH in rat brain: Occurrence within and outside brain beta-endorphin neurons. *Brain Res.* **182,** 217–223.

Watson S. J. and Barchas J. D. (1977) Catecholamine histofluorescence using cryostat sectioning and glyoxylic acid in unperfused frozen brain: A detailed description of the technique. *Histochem. J.* **9,** 183–195.

Wawrzyniak M. (1965) The histochemical activity of some enzymes in the mesencephalon during the ontogenetic development of the rabbit and guinea pig. II. Histochemical development of acetylcholinesterase and monoamine oxidase in the nontectal portion of the midbrain of the guinea pig *Z. Zellforsch.* **72,** 261–305.

Welsh J. H. and Moorhead M. (1960) The quantitative distribution of 5-hydroxytryptamine in the invertebrates, especially in their nervous systems. *J. Neurochem.* **6,** 146–169.

Welsh J. H. and Williams L. D. (1970) Monoamine-containing neurons in planaria. *J. Comp. Neurol.* **138,** 103–116.

Wolfe D. E., Potter L. T., Richardson K. C., and Axelrod J. (1963) Localizing tritiated norepinephrine in sympathetic axons by electron microscopic autoradiography. *Science* **138,** 440–444.

# Index